Inhibition of DNA Repair Enzymes as a Valuable Pharmaceutical Approach

Inhibition of DNA Repair Enzymes as a Valuable Pharmaceutical Approach

Editors

Konstantin Volcho
Olga Lavrik

MDPI • Basel • Beijing • Wuhan • Barcelona • Belgrade • Manchester • Tokyo • Cluj • Tianjin

Editors

Konstantin Volcho
Department of Medicinal Chemistry
Novosibirsk Institute of Organic Chemistry
Novosibirsk
Russia

Olga Lavrik
Laboratory of Bioorganic Chemistry of Enzymes
Institute of Chemical Biology and Fundamental Medicine
Novosibirsk
Russia

Editorial Office
MDPI
St. Alban-Anlage 66
4052 Basel, Switzerland

This is a reprint of articles from the Special Issue published online in the open access journal *International Journal of Molecular Sciences* (ISSN 1422-0067) (available at: www.mdpi.com/journal/ijms/special_issues/DNA_Repair_Enzymes).

For citation purposes, cite each article independently as indicated on the article page online and as indicated below:

LastName, A.A.; LastName, B.B.; LastName, C.C. Article Title. *Journal Name* **Year**, *Volume Number*, Page Range.

ISBN 978-3-0365-7881-1 (Hbk)
ISBN 978-3-0365-7880-4 (PDF)

Contents

About the Editors

Konstantin Volcho

Prof. Konstantin Volcho received his PhD in 1997 from Novosibirsk State University, Russia. Since then, he has been working at the Novosibirsk Institute of Organic Chemistry (Russia). Research of Prof. Volcho is focused on synthetic transformations of natural compounds and their analogues, including catalytic reactions, for the development of new biologically active compounds. The research interest of Prof. Volcho includes the development of novel treatments against nervous system disorders, antivirals, and anticancer agents. Konstantin has authored or co-authored more than 250 peer-reviewed publications. He is an inventor in 50 issued patents He is a Professor of the Russian Academy of Science.

Olga Lavrik

Prof. Olga Lavrik received her PhD in 1972 from M.M. Shemyakin and Y.A. Ovchinnikov Institute of Bioorganic Chemistry, Moscow, Russia. She is the head of the Laboratory of Bioorganic Chemistry of Enzymes of the Institute of Chemical Biology and Fundamental Medicine (Novosibirsk), Professor in the Department of Molecular Biology of Novosibirsk State University, and the head of the Department of Physical–Chemical Biology and Biotechnology, Altai State University, Barnaul, Russia. The current research interest of Dr. O. Lavrik and her laboratory includes the investigation of supramolecular machines of base excision repair (BER) and nucleotide excision repair (NER) and their contributions to human health and longevity. The mechanism of poly(ADP-ribosyl)ation catalyzed with PARP1, PARP2, and PARP3 and an understanding of the role of poly(ADP-ribose) in the regulation of DNA protein and protein–protein interactions in DNA repair is the main research field of interest of Dr. Olga Lavrik. The effective inhibitors of PARP1/PARP2 and tyrosyl-DNA phosphodiesterase 1 were identified with a high potential as anticancer drugs. She is the author or co-author of more than 400 scientific papers and reviews. She is a full member of the Russian Academy of Science.

Preface to "Inhibition of DNA Repair Enzymes as a Valuable Pharmaceutical Approach"

Although DNA repair enzymes play a crucial role in maintaining the integrity of the genome, the hyperactivity of certain enzymes of the DNA repair system can lead to the resistance of tumors to chemo- and radiotherapy, aimed at damaging the DNA of cancer cells. Therefore, the inhibition of DNA repair enzymes could help to overcome this resistance. The reviews and research articles included in this collection describe the molecular mechanisms of action of some important enzymes of the DNA repair system, as well as some new inhibitors of such enzymes and the pharmacological properties of these inhibitors. This reprint clearly demonstrates the importance of the inhibition of DNA repair enzymes to fight various diseases, especially cancer.

Konstantin Volcho and Olga Lavrik

Editors

International Journal of *Molecular Sciences*

Editorial

Inhibition of DNA Repair Enzymes as a Valuable Pharmaceutical Approach

Konstantin P. Volcho [1,*] and Olga I. Lavrik [2,*]

1 N. N. Vorozhtsov Novosibirsk Institute of Organic Chemistry, Siberian Branch of the Russian Academy of Sciences, 9, Akademika Lavrentieva Ave., 630090 Novosibirsk, Russia
2 Institute of Chemical Biology and Fundamental Medicine, Siberian Branch of the Russian Academy of Sciences, 630090 Novosibirsk, Russia
* Correspondence: volcho@nioch.nsc.ru (K.P.V.); lavrik@niboch.nsc.ru (O.I.L.)

Citation: Volcho, K.P.; Lavrik, O.I. Inhibition of DNA Repair Enzymes as a Valuable Pharmaceutical Approach. *Int. J. Mol. Sci.* **2023**, *24*, 7954. https://doi.org/10.3390/ijms24097954

Received: 19 April 2023
Accepted: 24 April 2023
Published: 27 April 2023

The DNA repair system plays a crucial role in maintaining the integrity of the genome. Disturbances in the function of certain DNA repair system enzymes can lead to various diseases and contribute to carcinogenesis. On the other hand, tumor cells are often characterized by the hyperactivity of DNA repair system enzymes, which allows them to resist chemotherapy and radiotherapy for cancer aimed at damaging the DNA of these cells. The ability of cancer cells to recognize DNA damage and initiate DNA repair is one of the key mechanisms for therapeutic resistance of tumors to chemotherapy. Therefore, the targeting of DNA repair enzymes can be used as a strategy to potentiate the cytotoxicity of the currently available DNA damaging agents toward cancer cells. The role of inhibition of DNA repair enzymes in cancer therapy is thoroughly analyzed in the review written by Lodovichi et al. [1].

The understanding of the ways DNA repair enzymes function in the repair processes is the basis for creating any therapeutic agents aimed at inhibiting these enzymes and protein factors. One such protein factor is the fused in sarcoma (FUS), an RNA binding protein which has not yet been used as a pharmacological target. FUS interacts with poly(ADP-ribose) produced by PARP1/2 on DNA damages. This process is pivotal for the formation of nonmembrane compartments to process DNA repair. Sukhanova et al. reviewed the participation of FUS in major DNA repair pathways [2], with an emphasis on the interaction of FUS with poly(ADP-ribose) polymerase 1 (PARP1) during poly(ADP-ribosyl)ation events important for the compartmentalization of DNA strand breaks and DNA repair proteins. Molecular mechanisms of action of human apurinic/apyrimidinic endonuclease 1 (APE1), an important enzyme involved in base excision DNA repair, was studied by Moor et al. [3].

The Special issue highlights the pharmacological effects of PARP inhibitors. PARPs are important for repairing single-strand breaks in DNA. Tumors mutated through alternative pathways of repair (BRCA1, BRCA2 or PALB2) are especially sensitive to PARP inhibitors. It is worth noting that some inhibitors of PARP1, including Olaparib, Rucaparib, Niraparib and Talazoparib, are already clinically utilized as antitumor drugs. The review by Maluchenko et al. considers the mechanisms of PARP1 participation in the development of various pathologies, and summarizes data on the effect of natural polyphenols on PARP-dependent cellular processes [4]. An article written by Nilov et al. [5] describes studies in the mechanism of action of 7-methylguanine, a natural inhibitor of PARP1. Sherstyuk et al. demonstrated that new morpholino-nucleoside adenosine dinucleotides form a novel class of pan-PARP inhibitors, which are active against PARP1, PARP2 and PARP3 [6]. In the research by Ryu et al. [7], PARP1 and PARP2 inhibitor *N*-(3-(hydroxycarbamoyl)phenyl)carboxamide (KJ-28d) have been indicated to act as sensitizers of non-small cell lung cancer cell lines to radio- and chemotherapy. At the same time, the PARP1 inhibitor, Olaparib, did not potentiate a cytotoxic effect of chemotherapeutic agent bleomycin in VERO cells (Perini et al. [8]).

Tyrosyl-DNA phosphodiesterase 1 (TDP1) plays a key role in the removal of DNA damage resulting from the poisoning of topoisomerase 1, with clinically important anticancer drugs irinotecan and topotecan. Novel structural type of TDP1 inhibitors was developed based on natural alkaloid berberine by Gladkova et al. [9]. It was shown that these inhibitors are able to increase the cytotoxicity of topotecan against the HeLa cancer cell line. Another type of TDP1 inhibitors was designed based on 4-arylcoumarin conjugated with monoterpene moieties (Khomenko et al. [10]). The new inhibitors induced a significant increase in the antitumor effect of topotecan on the Krebs-2 ascites tumor model in mice. Dyrkheeva et al. showed that thioethers of natural usnic acid demonstrated good inhibitory activity against three DNA repair enzymes, TDP1, TDP2 and PARP1 [11]. Moreover, the synergy of the leader compound with topotecan on HeLa cells was shown.

The importance of nucleotide excision repair (NER) in the formation of tumor cell resistance to cisplatin is reviewed by Duan et al. [12]. Furthermore, Fidrus et al. demonstrated that inhibitors of NER can decrease UVB-irradiation-induced mutagenesis [13].

AKT is involved in the regulation of DNA damage response and repair. Boichuk et al. found that the inhibition of the AKT signaling pathway sensitizes tumors to anticancer drug, Doxorubicin, by targeting homology-mediated DNA reparation [14].

Finally, in their review, Mechetin et al. demonstrated that the inhibition of DNA glycosylases, the enzymes that initiate the base excision repair (BER) pathway, could be useful not only in the treatment of cancer, but also to cure neurodegenerative diseases, chronic inflammation, as well as bacterial and viral infections [15].

Thus, the articles and reviews published in this Special Issue emphasize the importance of DNA repair enzyme inhibition in the fight against various diseases, with major attention paid to the development of anticancer drugs, as well as the fast progress in this field.

Funding: The research was supported by the Russian Science Foundation (grant 19-13-00040).

Conflicts of Interest: The authors declare no conflict of interest.

References

1. Lodovichi, S.; Cervelli, T.; Pellicioli, A.; Galli, A. Inhibition of DNA Repair in Cancer Therapy: Toward a Multi-Target Approach. *Int. J. Mol. Sci.* **2020**, *21*, 6684. [CrossRef] [PubMed]
2. Sukhanova, M.V.; Singatulina, A.S.; Pastré, D.; Lavrik, O.I. Fused in Sarcoma (FUS) in DNA Repair: Tango with Poly(ADP-ribose) Polymerase 1 and Compartmentalisation of Damaged DNA. *Int. J. Mol. Sci.* **2020**, *21*, 7020. [CrossRef] [PubMed]
3. Moor, N.; Vasil'eva, I.; Lavrik, O. Functional Role of N-Terminal Extension of Human AP Endonuclease 1 In Coordination of Base Excision DNA Repair via Protein–Protein Interactions. *Int. J. Mol. Sci.* **2020**, *21*, 3122. [CrossRef] [PubMed]
4. Maluchenko, N.V.; Feofanov, A.V.; Studitsky, V.M. PARP-1-Associated Pathological Processes: Inhibition by Natural Polyphenols. *Int. J. Mol. Sci.* **2021**, *22*, 11441. [CrossRef] [PubMed]
5. Nilov, D.; Maluchenko, N.; Kurgina, T.; Pushkarev, S.; Lys, A.; Kutuzov, M.; Gerasimova, N.; Feofanov, A.; Švedas, V.; Lavrik, O.; et al. Molecular Mechanisms of PARP-1 Inhibitor 7-Methylguanine. *Int. J. Mol. Sci.* **2020**, *21*, 2159. [CrossRef] [PubMed]
6. Sherstyuk, Y.V.; Ivanisenko, N.V.; Zakharenko, A.L.; Sukhanova, M.V.; Peshkov, R.Y.; Eltsov, I.V.; Kutuzov, M.M.; Kurgina, T.A.; Belousova, E.A.; Ivanisenko, V.A.; et al. Design, Synthesis and Molecular Modeling Study of Conjugates of ADP and Morpholino Nucleosides as A Novel Class of Inhibitors of PARP-1, PARP-2 and PARP-3. *Int. J. Mol. Sci.* **2020**, *21*, 214. [CrossRef] [PubMed]
7. Ryu, H.; Kim, H.J.; Song, J.-Y.; Hwang, S.-G.; Kim, J.-S.; Kim, J.; Bui, T.H.N.; Choi, H.-K.; Ahn, J. A Small Compound KJ-28d Enhances the Sensitivity of Non-Small Cell Lung Cancer to Radio- and Chemotherapy. *Int. J. Mol. Sci.* **2019**, *20*, 6026. [CrossRef] [PubMed]
8. Perini, V.; Schacke, M.; Liddle, P.; Vilchez-Larrea, S.; Keszenman, D.J.; Lafon-Hughes, L. PARP Inhibitor Olaparib Causes No Potentiation of the Bleomycin Effect in VERO Cells, Even in the Presence of Pooled ATM, DNA-PK, and LigIV Inhibitors. *Int. J. Mol. Sci.* **2020**, *21*, 8288. [CrossRef] [PubMed]
9. Gladkova, E.D.; Nechepurenko, I.V.; Bredikhin, R.A.; Chepanova, A.A.; Zakharenko, A.L.; Luzina, O.A.; Ilina, E.S.; Dyrkheeva, N.S.; Mamontova, E.M.; Anarbaev, R.O.; et al. The First Berberine-Based Inhibitors of Tyrosyl-DNA Phosphodiesterase 1 (Tdp1), an Important DNA Repair Enzyme. *Int. J. Mol. Sci.* **2020**, *21*, 7162. [CrossRef] [PubMed]
10. Khomenko, T.M.; Zakharenko, A.L.; Chepanova, A.A.; Ilina, E.S.; Zakharova, O.D.; Kaledin, V.I.; Nikolin, V.P.; Popova, N.A.; Korchagina, D.V.; Reynisson, J.; et al. Promising New Inhibitors of Tyrosyl-DNA Phosphodiesterase I (Tdp 1) Combining 4-Arylcoumarin and Monoterpenoid Moieties as Components of Complex Antitumor Therapy. *Int. J. Mol. Sci.* **2020**, *21*, 126. [CrossRef] [PubMed]

11. Dyrkheeva, N.S.; Filimonov, A.S.; Luzina, O.A.; Orlova, K.A.; Chernyshova, I.A.; Kornienko, T.E.; Malakhova, A.A.; Medvedev, S.P.; Zakharenko, A.L.; Ilina, E.S.; et al. New Hybrid Compounds Combining Fragments of Usnic Acid and Thioether Are Inhibitors of Human Enzymes TDP1, TDP2 and PARP1. *Int. J. Mol. Sci.* **2021**, *22*, 11336. [CrossRef] [PubMed]
12. Duan, M.; Ulibarri, J.; Liu, K J.; Mao, P. Role of Nucleotide Excision Repair in Cisplatin Resistance. *Int. J. Mol. Sci.* **2020**, *21*, 9248. [CrossRef] [PubMed]
13. Fidrus, E.; Hegedűs, C.; Janka, E.A.; Paragh, G.; Emri, G.; Remenyik, É. Inhibitors of Nucleotide Excision Repair Decrease UVB-Induced Mutagenesis—An In Vitro Study. *Int. J. Mol. Sci.* **2021**, *22*, 1638. [CrossRef] [PubMed]
14. Boichuk, S.; Bikinieva, F.; Nurgatina, I.; Dunaev, P.; Valeeva, E.; Aukhadieva, A.; Sabirov, A.; Galembikova, A. Inhibition of AKT-Signaling Sensitizes Soft Tissue Sarcomas (STS) and Gastrointestinal Stromal Tumors (GIST) to Doxorubicin via Targeting of Homology-Mediated DNA Repair. *Int. J. Mol. Sci.* **2020**, *21*, 8842. [CrossRef] [PubMed]
15. Mechetin, G.V.; Endutkin, A.V.; Diatlova, E.A.; Zharkov, D.O. Inhibitors of DNA Glycosylases as Prospective Drugs. *Int. J. Mol. Sci.* **2020**, *21*, 3118. [CrossRef] [PubMed]

International Journal of *Molecular Sciences*

MDPI

Article

New Hybrid Compounds Combining Fragments of Usnic Acid and Thioether Are Inhibitors of Human Enzymes TDP1, TDP2 and PARP1

Nadezhda S. Dyrkheeva [1], Aleksandr S. Filimonov [2], Olga A. Luzina [2], Kristina A. Orlova [3], Irina A. Chernyshova [1], Tatyana E. Kornienko [1], Anastasia A. Malakhova [1,4], Sergey P. Medvedev [1,4], Alexandra L. Zakharenko [1,5], Ekaterina S. Ilina [1], Rashid O. Anarbaev [1], Konstantin N. Naumenko [1], Kristina V. Klabenkova [3,4], Ekaterina A. Burakova [3,4], Dmitry A. Stetsenko [3,4], Suren M. Zakian [1,4], Nariman F. Salakhutdinov [2] and Olga I. Lavrik [1,3,5,*]

1 Institute of Chemical Biology and Fundamental Medicine, Siberian Branch of the Russian Academy of Sciences, 630090 Novosibirsk, Russia; dyrkheeva.n.s@gmail.com (N.S.D.); chernyshova0305@gmail.com (I.A.C.); t.kornienko1995@gmail.com (T.E.K.); amal@bionet.nsc.ru (A.A.M.); medvedev@bionet.nsc.ru (S.P.M.); a.zakharenko73@gmail.com (A.L.Z.); katya.plekhanova@gmail.com (E.S.I.); anarbaev@niboch.nsc.ru (R.O.A.); k-naumenko@mail.ru (K.N.N.); zakian@bionet.nsc.ru (S.M.Z.)

2 N.N. Vorozhtsov Novosibirsk Institute of Organic Chemistry, Siberian Branch of the Russian Academy of Sciences, 630090 Novosibirsk, Russia; alfil@nioch.nsc.ru (A.S.F.); luzina@nioch.nsc.ru (O.A.L.); anvar@nioch.nsc.ru (N.F.S.)

3 Department of Natural Sciences, Novosibirsk State University, 630090 Novosibirsk, Russia; kristina-orlova1999@mail.ru (K.A.O.); k.klabenkova@g.nsu.ru (K.V.K.); e.burakova1@nsu.ru (E.A.B.); d.stetsenko@nsu.ru (D.A.S.)

4 Federal Research Centre Institute of Cytology and Genetics, Siberian Branch of the Russian Academy of Sciences, 630090 Novosibirsk, Russia

5 Department of Physical and Chemical Biology and Biotechnology, Altai State University, 656049 Barnaul, Russia

* Correspondence: lavrik@niboch.nsc.ru; Tel.: +7-383-363-51-95

Citation: Dyrkheeva, N.S.; Filimonov, A.S.; Luzina, O.A.; Orlova, K.A.; Chernyshova, I.A.; Kornienko, T.E.; Malakhova, A.A.; Medvedev, S.P.; Zakharenko, A.L.; Ilina, E.S.; et al. New Hybrid Compounds Combining Fragments of Usnic Acid and Thioether Are Inhibitors of Human Enzymes TDP1, TDP2 and PARP1. *Int. J. Mol. Sci.* **2021**, *22*, 11336. https://doi.org/10.3390/ijms222111336

Academic Editor: Maria Luisa Balestrieri

Received: 30 September 2021
Accepted: 17 October 2021
Published: 20 October 2021

Abstract: Tyrosyl-DNA phosphodiesterase 1 (TDP1) catalyzes the cleavage of the phosphodiester bond between the tyrosine residue of topoisomerase 1 (TOP1) and the 3′ phosphate of DNA in the single-strand break generated by TOP1. TDP1 promotes the cleavage of the stable DNA–TOP1 complexes with the TOP1 inhibitor topotecan, which is a clinically used anticancer drug. This article reports the synthesis and study of usnic acid thioether and sulfoxide derivatives that efficiently suppress TDP1 activity, with IC_{50} values in the 1.4–25.2 μM range. The structure of the heterocyclic substituent introduced into the dibenzofuran core affects the TDP1 inhibitory efficiency of the compounds. A five-membered heterocyclic fragment was shown to be most pharmacophoric among the others. Sulfoxide derivatives were less cytotoxic than their thioester analogs. We observed an uncompetitive type of inhibition for the four most effective inhibitors of TDP1. The anticancer effect of TOP1 inhibitors can be enhanced by the simultaneous inhibition of PARP1, TDP1, and TDP2. Some of the compounds inhibited not only TDP1 but also TDP2 and/or PARP1, but at significantly higher concentration ranges than TDP1. Leader compound **10a** showed promising synergy on HeLa cells in conjunction with the TOP1 inhibitor topotecan.

Keywords: usnic acid; thioether; tyrosyl-DNA phosphodiesterase 1; TDP1 inhibitor; inhibiting activity; TDP2; PARP1; topotecan; synergy; HEK293 knockout cell line

1. Introduction

Genomic instability is one of the major driving forces of carcinogenesis. Studies of DNA repair mechanisms and their regulation are directly related to the search for the optimal ways of treating oncological and other human diseases. Chemotherapy is one of the main methods of treating various malignancy types. Anticancer chemotherapeutic drugs, by their principle of action, damage DNA in a targeted manner. They are powerful cell poisons that have a detrimental effect on the rapidly dividing cells of malignant tumors with a comparatively less negative damaging effect on the healthy, normally dividing cells and tissues of the organism. Although DNA repair is essential for a healthy cell, during anticancer therapy repair enzymes of the cancer cells counteract the efficacy of anticancer agents. Thus, DNA repair leads to a decrease in the effectiveness of therapy and contributes to the resistance of malignancies to chemotherapy. In recent years, much focus has been put on the DNA repair enzymes as targets for drug development. Researchers are actively searching for new compounds that suppress the DNA repair enzymes activity to increase the efficiency of anticancer therapy. Currently, tyrosyl-DNA phosphodiesterase 1 (TDP1) and poly(ADP-ribose) polymerase 1 (PARP1) are considered as promising target DNA repair enzymes for creating drugs [1,2].

TDP1 is involved in repairing stalled topoisomerase 1–DNA complexes by catalyzing the hydrolysis of the phosphodiester bond between the tyrosine residue of topoisomerase 1 (TOP1) and the 3′ phosphate of DNA in the single-strand break generated by TOP1. TDP1 also catalyzes the cleavage of phosphodiester bonds in other DNA–protein adducts and a number of different lesions at the 3′ end of DNA [3]. TDP1 plays a key role in removing the damage from DNA caused by the anticancer drugs used in clinical practice, such as topotecan (Tpc) and irinotecan, which are derivatives of the natural compound camptothecin [4,5]. Consequently, TDP1 activity may be a possible cause of tumor resistance to TOP1 inhibitors.

Tyrosyl-DNA phosphodiesterase 2 (TDP2) is a DNA repair enzyme that catalyzes the hydrolysis of dead-end complexes between DNA and the topoisomerase 2 (TOP2) active site tyrosine residue. TDP2 can remove a variety of covalent adducts from DNA through hydrolysis of a 5′ phosphodiester bond, giving rise to DNA with a free 5′ phosphate. TOP2 inhibitors stabilize the TOP2–DNA covalent complex and induce cell death [6]. TOP2 inhibitors (etoposide, doxorubicin) are widely used in clinical practice as antineoplastic drugs. TDP2 activity reduces the effectiveness of these drugs and, vice versa, TDP2 deficiency leads to a significant increase in sensitivity to TOP2 inhibitors [7,8]. The same as TDP1 inhibitors, TDP2 inhibitors can significantly increase the effectiveness of chemotherapy by synergizing with TOP2 inhibitors. The most effective TDP2 inhibitors today are deazaflavins [9,10]. Deazaflavin has been shown to synergize in vitro with etoposide at non-toxic concentrations [9]. It should be noted that deazaflavins have unsatisfactory pharmacokinetic characteristics, which makes it necessary to search for inhibitors of new structural types. Several TDP2 inhibitors of different chemical groups have already been proposed, but most of them had moderate efficacy [11]. Recently, though, some more efficient TDP2 inhibitors have been found [12–14]

TDP1 and TDP2 have little overlapping activity because TDP1 has a weak activity for 5′-phosphotyrosyl bonds, and TDP2 has a weak activity for 3′-phosphotyrosyl bonds [5]. Nevertheless, the ability of TDP1 and TDP2 to take on the functions of each other makes it highly promising to use the selective inhibitors of these two enzymes together, or to create agents capable of simultaneously inhibiting both TDP1 and TDP2. Simultaneous suppression of the activity of these two enzymes can be used to increase the effectiveness of a large set of clinically important anticancer drugs, TOP1 and TOP2 inhibitors. Recently, the first triple inhibitors of TOP1/TDP1/TDP2 have been discovered by Pommier's group [15], which exhibit only moderate activity against TDP1 and weak activity against TDP2.

PARP1 catalyzes the synthesis of poly(ADP-ribose) (PAR), which is responsible for the post-translational modification of proteins, as an immediate DNA damage response of the cell. PARP1 is a member of various DNA repair pathways [16,17]. Inhibition of PARP1 activity makes it possible to sensitize tumor cells to the action of chemotherapeutic drugs. Olaparib was the first example of therapeutic synthetic lethality in oncology approved by the FDA for the treatment of advanced ovarian cancers associated with defective BRCA1/2 [18]. To date, the FDA has approved four different PARP inhibitors, olaparib (2014), rucaparib (2016), niraparib (2017), and talazoparib (2018), for the treatment of ovarian, fallopian tube, breast, and peritoneal cancers [19]. More inhibitors are in various stages of development and preclinical testing. In addition to cancer, PARP inhibitors are also promising in the treatment of cardiovascular diseases [20]. The anticancer effect of TOP1 inhibitors can be significantly enhanced by the simultaneous inhibition of PARP1 and TDP1. PARP1–TDP1 interplay was shown in a series of publications. PARylation of TDP1 enhances its recruitment to sites of DNA damage without interfering with the catalytic activity of TDP1 [21]. Briefly, it was shown [21] that the N-terminal domain of TDP1 directly binds the C-terminal domain of PARP1, and TDP1 is PARylated but not inactivated by PARP1. PARP1 is known to interact with TDP1 directly with K_D 120 nM [22]. PARP1 stimulates the enzymatic activity of TDP1 on AP sites [23]. TDP1 can repair the covalent DNA–PARP1 crosslink at the apurinic/apyrimidinic (AP) site in double-stranded DNA [24,25]. Other authors also noticed that TDP1 together with PARP1 inhibition could be a successful cancer treatment strategy [26,27]. TDP1 expression is correlated with other DNA repair genes, including PARP1, BRCA2, and BRCA1 [28]. The combination of 70% to 90% Tdp1 knockdown and 10 mmol/L of the PARP1 inhibitor rucaparib was found to reduce the proliferation of A204, Birch, RH30, and CW9019 cells more than either of these treatments alone [29]. These facts make urgent the search for dual inhibitors of TDP1 and PARP1. In the other work [30], TDP1, together with PARP1, were shown to be essential cellular proteins in cancerogenic [31] human papillomavirus HPV18 replication, thus making those proteins good targets for developing HPV inhibitors. The authors also noticed that TDP1 and PARP1 inhibitors might also be effective against HPV-induced cancer.

We have previously discovered highly effective inhibitors of TDP1 based on the secondary metabolite of lichens usnic acid **1** (UA)—compounds **2**–**5** (Figure 1) [32–36]—the synergistic action of which, when combined with the TOP1 inhibitor Tpc, was confirmed in experiments on cell cultures and in animal models [36–38].

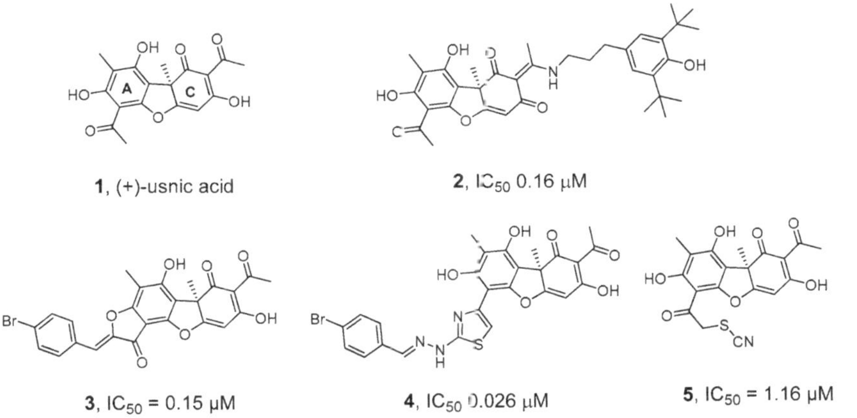

Figure 1. Structures of usnic acid (**1**) and its derivatives (**2**–**5**) that inhibit TDP1 [32–36].

We also previously studied the PARP1-inhibiting activity of various UA derivatives [39]. Most of the UA derivatives modified by the substituents in the C ring and the introduction of a fused ring or heterocycles into the A ring had low PARP1 affinity. The PARP1 inhibitory activity was facilitated by the introduction of aromatic substituents into the acyl fragment of the A ring of UA. Analysis of the literature showed that the pharmacophore fragments with respect to PARP1 are heterocycles containing one or several heteroatoms, primarily nitrogen (Figure 2) [40].

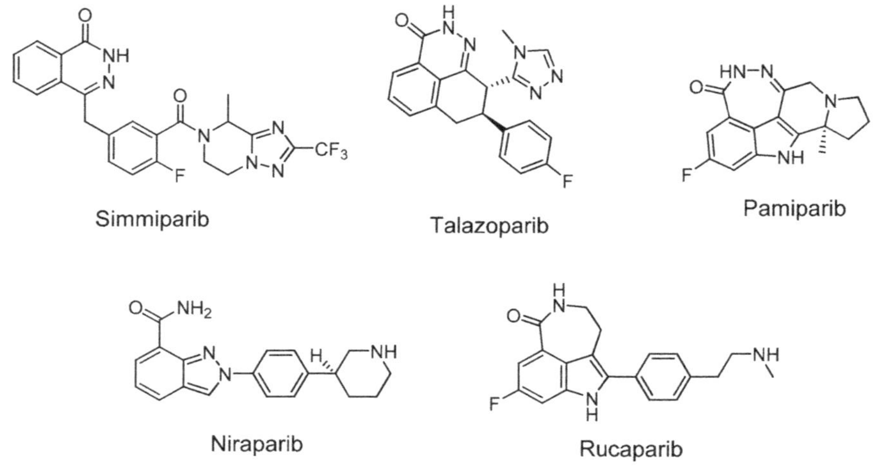

Figure 2. Some known inhibitors of PARP1.

We chose UA derivative **7** for the design of the joint TDP1 and PARP1 action inhibitors, which allow easy introduction of heteroatom-rich fragments into the acyl fragment of the ring A of UA (Figure 3). The activity of this type of UA derivative against DNA repair enzymes, including TDP1 and PARP1, has not been studied previously.

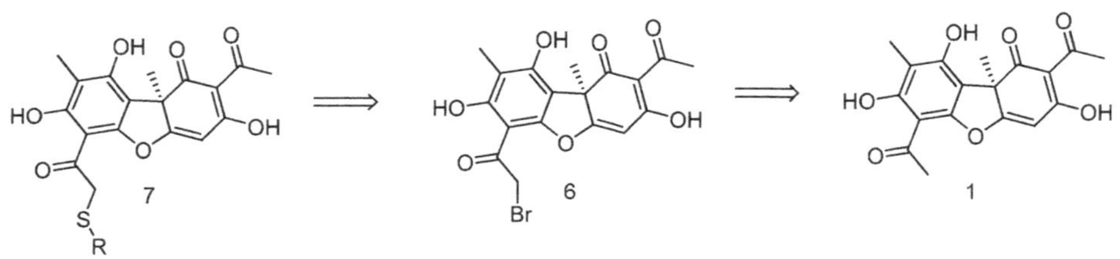

Figure 3. Retrosynthetic scheme for the synthesis of compound **7**.

In this work, we propose to use an approach for the synthesis of potential TDP1 inhibitors or dual inhibitors based on the construction of a derivative from a UA backbone, a flexible linker, and a heterocycle to synthesize a set of compounds derivatized through cycle A. The aim of this study was to assess the inhibitory ability of the UA derivatives against three DNA repair enzymes, search for selective and multitarget agents, study the inhibition mechanism assessment of toxicity, and study the prospects of using these compounds as potential pharmacological agents of in anticancer therapy.

2. Results and Discussion

2.1. Chemistry

In order to find out the structure–activity relationship (SAR), we synthesized the set of usnic acid (UA) derivatives of **7**, with five- and hexa-membered mono- and bicyclic heterocyclic along with non-heterocyclic and acyclic substituents.

The desired novel and known UA thioethers **7a–k** were synthesized using the procedure described early [41], by reaction of bromousnic acid **6** with the corresponding thiols in presence of NaOH. The target thioethers were obtained in 49–94% yield (Scheme 1).

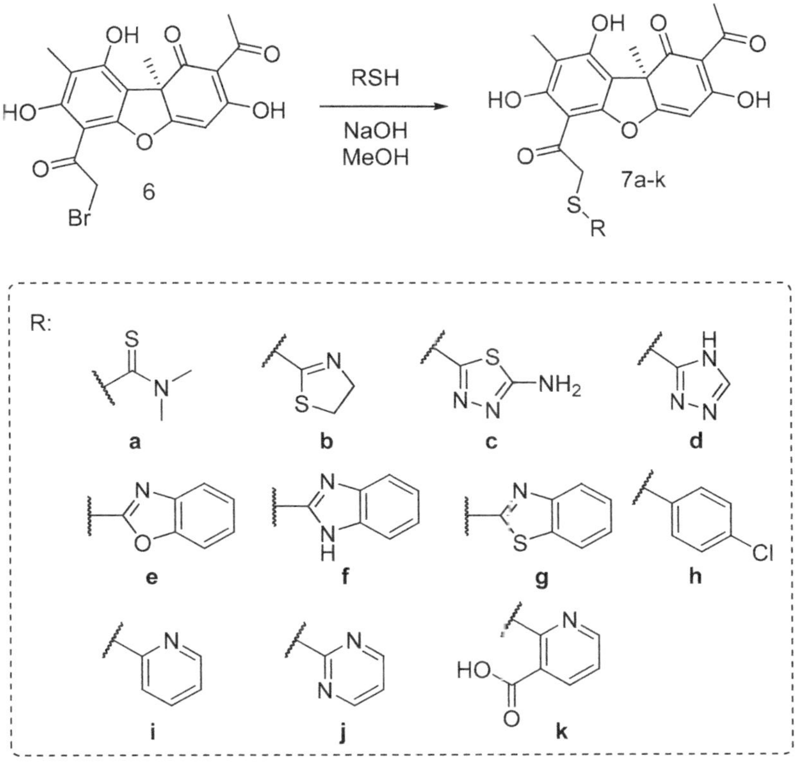

Scheme 1. Synthesis of the thioether usnic acid derivatives.

UA derivative **9** was synthesized in two steps (Scheme 2). At first, the reaction of the bromousnic acid with sodium thiocyanate in acetone resulted in obtaining compound **8**, with a 95% yield. Then, UA thiocyanate derivative **8** was hydrolyzed. The hydrolysis was performed with (33%) sulfuric acid in glacial acetic acid to obtain the desired compound **9**, with a 38% yield.

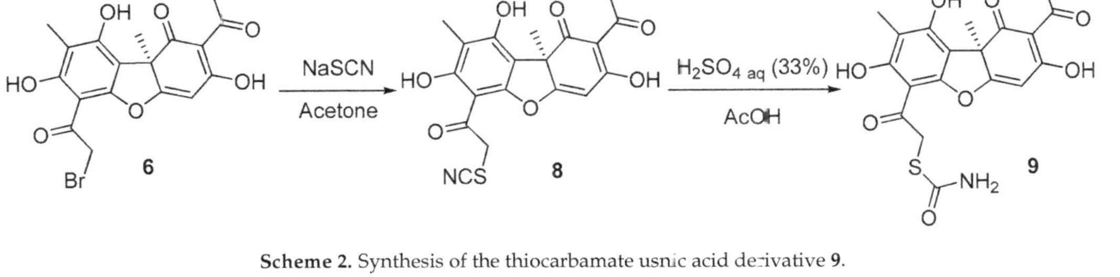

Scheme 2. Synthesis of the thiocarbamate usnic acid derivative **9**.

The target sulfoxide derivatives of **10** were obtained using the procedure described in [42], which is by a peroxidation reaction of the corresponding thioethers **7g**,**h** with mCPBA in CH_2Cl_2 at 0 °C (Scheme 3). The desired compounds **10a**,**b** were obtained, with 70–75% yields.

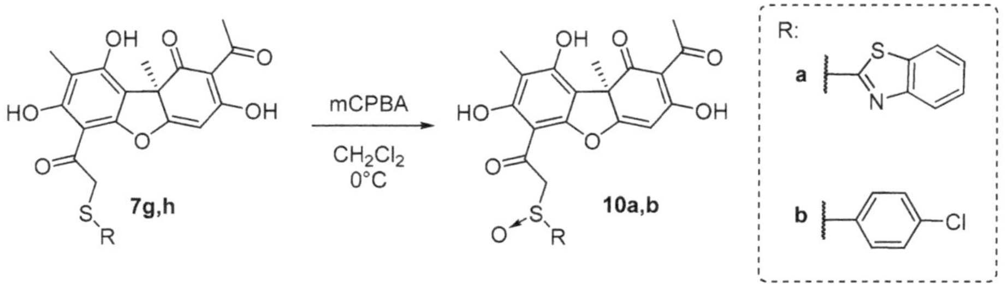

Scheme 3. Synthesis of the sulfoxide usnic acid derivatives **10a,b**.

2.2. Biology

2.2.1. Real-Time Fluorescence Assay of TDP1 Activity

We tested UA and all 14 newly synthesized UA derivatives for their TDP1 inhibitory properties by measuring their IC_{50} values using a real-time fluorescent assay [43]. All 14 obtained compounds (Table 1) were shown to inhibit TDP1 in the low-enough concentration range (IC_{50} 1.7–25.2 μM). UA does not inhibit TDP1 in these concentrations. We can assume that the structure of the heterocyclic fragment bound to the sulfur atom affects the TDP1 inhibitory activity of the thioether derivatives. There is a five-membered heterocyclic fragment in the effective compounds (**7b**,**c**,**e–h** and **10a**,**b**), with an IC_{50} of 1.4–4.4 μM, except for **7d** (IC_{50} 25.2 μM), whereas the compounds containing a six-membered heterocycle (**7i–k**) inhibit TDP1 at higher concentrations ($IC_{50} > 11$ μM). It is possible that the presence of a 4-halophenyl substituent in the six-membered heterocycle (**7h** and **10b**, IC_{50} 2.2 μM and 1.4 μM, respectively) enhances the inhibitory properties of the compounds. Four of the most effective inhibitors of TDP1 were compounds **7g**,**h** and **10a**,**b** (IC_{50} of 1.7 μM, 2.2 μM, 2.1 μM, and 1.4 μM, respectively). Compounds **10a**,**b** are sulfoxide analogs of **7g**,**h**.

Table 1. TDP1, TDP2, and PARP1 inhibition ability and cytotoxicity of the thioether UA derivatives.

		R	IC_{50} (TDP1), μM	HEK293A CC_{50}, μM	HEK293FT CC_{50}, μM	HeLa CC_{50}, μM	PARP1, 1 mM	TDP2, 1 mM
1	UA	-	>50	ND	ND	20 ± 10	-	-
2	**9**	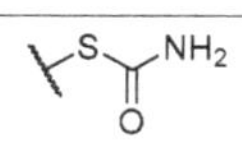	6.6 ± 1.0	>50	7 ± 2	13 ± 1	+	+
3	**7a**	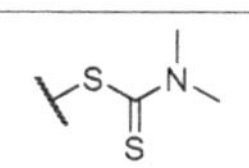	5.4 ± 2.9	>50	>50	13.0 ± 0.5	+	+
4	**7b**	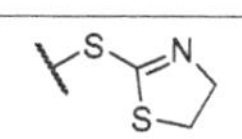	4.4 ± 1.0	>50	16 ± 5	11.0 ± 0.5	+	+
5	**7c**	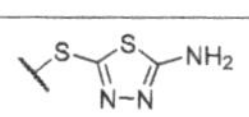	4.3 ± 0.5	>50	35 ± 2	33 ± 2	+	+
6	**7d**	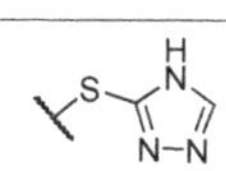	25.2 ± 6.5	ND	ND	ND	+	+
7	**7e**	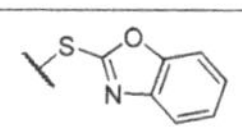	3.2 ± 0.5	26 ± 2	4.5 ± 1.0	2.9 ± 0.2	-	+
8	**7f**	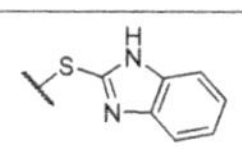	2.4 ± 1.0	>50	10 ± 2	13 ±	-	+

Table 1. *Cont.*

		R	IC_{50} (TDP1), μM	HEK293A CC_{50}, μM	HEK293FT CC_{50}, μM	HeLa CC_{50}, μM	PARP1, 1 mM	TDP2, 1 mM
9	**7g**	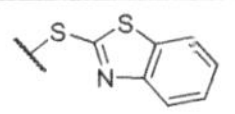	1.7 ± 0.6	3.5 ± 0.3	3 ± 1	2.0 ± 0.6	-	+
10	**10a**	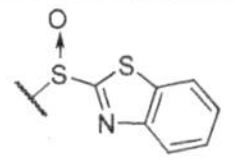	2.1 ± 0.2	20 ± 2	15 ± 2	15 ± 1	-	+
11	**7h**	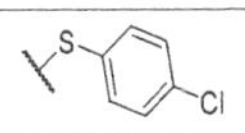	2.2 ± 0.5	10 ± 2	4 ± 1	2.5 ± 0.5	+	+
12	**10b**	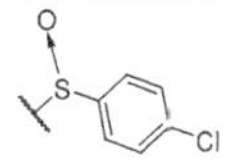	1.4 ± 0.2	>50	20 ± 3	27 ± 2	-	+
13	**7i**	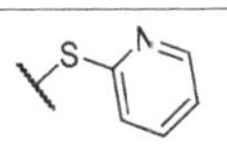	11.9 ± 0.4	ND	ND	ND	-	+
14	**7j**	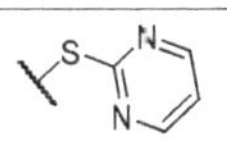	16.9 ± 2.4	ND	ND	ND	+	+
15	**7k**	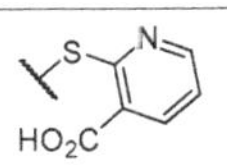	19.6 ± 2.8	ND	ND	ND	+	+

2.2.2. Type of Inhibition of TDP1 Enzyme Reaction for the Most Effective Compounds

The reversible enzyme inhibitors were classified as competitive, uncompetitive, non-competitive, or mixed, depending on the binding of the enzyme, to the enzyme–substrate complex (ES) or to the enzyme–substrate–inhibitor triple complex (ESI). The enzyme could bind the inhibitor at the different steps of catalysis (before or after the substrate) and at the substrate-binding site or another site. The effect of the inhibitor on enzymatic activity can be analyzed using graphical representations of the Michaelis–Menten equation. We defined the type of inhibition for the four most effective inhibitors of TDP1: **7g,h** and **10a,b** (IC_{50}~1.4–2.2 μM, Table 1) by obtaining the dependence of the reaction rate on the substrate concentration and dependence of V_{max} and K_M on the concentration of the inhibitor under steady-state reaction conditions. For all four TDP1 inhibitors, we observed a fall in both the V_{max} and K_M values with an increasing concentration of the inhibitor, which is typical for the uncompetitive type of inhibition (Table S1, Figure S25) when the inhibitor binds the enzyme–substrate complex but does not bind free enzyme. Thus, it was determined that no studied inhibitors demonstrate the competitive mechanism of inhibition. TDP1 has a catalytic pocket with a narrow positively charged cleft where it binds DNA and with a relatively large cleft that contains a mixed charge distribution for the TOP1 fragment binding [44–46]. It is known that TDP1 hydrolyzes the DNA–TOP1 adducts via two coordinated nucleophilic attacks of two His residues (His263 and His493) with the covalent intermediate complex formation [45,46]. The enzyme reaction catalyzed by TDP1 could be inhibited in two steps: the first one is the inhibition of the TDP1 binding to the DNA substrate (the nucleophilic attack of His263). The second step is releasing TDP1 from the transition complex (nucleophilic attack by a water molecule activated by His493). Inhibition of the second step prevents the release of the DNA 3′-phosphate product from the DNA–TDP1 complex. Spinocerebellar ataxia with axonal neuropathy (SCAN1) is a neurological disease that is caused by a TDP1 His493Arg mutation. This mutation prevents the second step of hydrolysis of the intermediate complex and the enzyme remains covalently bound to DNA [47]. The uncompetitive inhibitors also prevent the second step of the reaction, stabilizing the enzyme–DNA covalent complex, which can lead to the accumulation of the single-strand breaks in the cell.

2.2.3. The Effect of TDP1 Inhibitors on Cells Viability

The low intrinsic cytotoxicity of the additional drug component in the clinically used anticancer cocktails with traditional drugs is very important for treatment. This approach in chemotherapy can reduce the potential resulting toxic load on the body in the case of a combined action of these inhibitors on the activity of several targets. The aim of this part of the study was to demonstrate the ability of the found TDP1 inhibitors to enhance the cytotoxicity of the clinically used camptothecin derivative Tpc on the cancer cells. First, we analyzed the intrinsic cytotoxicity for 10 of the most effective TDP1 inhibitors (IC_{50} 1.4–6.6 μM, Table 1) on the HEK293A, HEK293FT, and HeLa cell lines by colorimetric tests. The effects of the tested compounds on cell survival are shown on Figure 4 and the CC_{50} values for three cell lines are shown in Table 1. The intrinsic cytotoxicity of a number of the studied compounds is quite high, namely, they have close IC_{50} values against the TDP1 and CC_{50} values on one or several studied cell lines. Most of the compounds generally were more toxic for the HEK293FT and HeLa cell lines than for the HEK293A cell line (Table 1 and Figure 4). The cytotoxicity of the three leader compounds, **7g**, **10a**, and **7h**, was quite high. Interestingly, the sulfoxide analogs **10a**,**b** were less cytotoxic than their thioester analogs **7g**,**h**. Compound **10b** was one of the least toxic among the ten tested compounds.

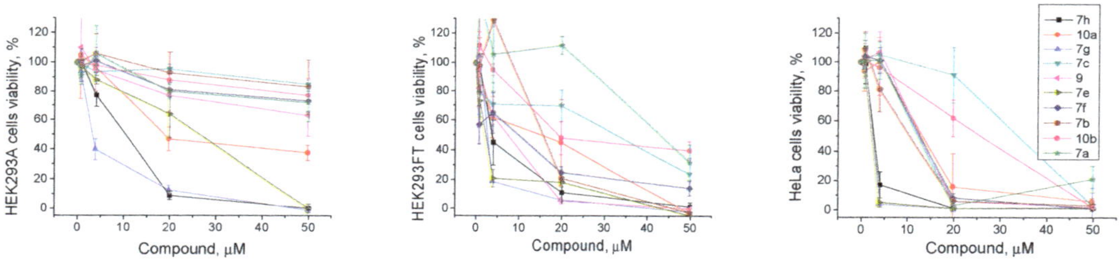

Figure 4. TDP1 inhibitors' intrinsic cytotoxicity on HEK293A, HEK293FT, and HeLa cells, and the dose-dependent action of the most effective compounds. The values were normalized on the control wells with the cells in 1% DMSO.

2.2.4. PARP1 and PARP2 Activity in the Presence of TDP1 Inhibitors

The next step of our study was to test the inhibitors for their ability to inhibit PARP1 and PARP2. UA and the 14 synthesized compounds were tested for the suppression of the ability of PARP1 and PARP2 to include a radioactive label into the acid-insoluble reaction product, poly(ADP-ribose) (PAR). None of the tested compounds inhibited PARP2 in the applied reaction conditions. All the results obtained for PARP1 inhibition are shown in Table 1. Eight compounds exhibited a weak inhibitory activity against PARP1 (40–80% of residual activity) at a 1 mM concentration, working at significantly higher concentration ranges than against TDP1. The best effect we observed was for **9**; the residual activity was $40 \pm 5\%$ under the reaction conditions (0.5 mM concentration of the compound), indicated in Materials and Methods. For PARP1, compounds with smaller substituents, an aliphatic or benzene moiety, had a higher inhibitory effect.

2.2.5. TDP2 Activity in the Presence of TDP1 Inhibitors

To study the ability of the investigated compounds to inhibit TDP2 activity, we separated the TDP2 reaction products under denaturing conditions in polyacrylamide gel (PAGE). We tested the ability of TDP2 to eliminate tyrosine residues from the 5′ end of the oligonucleotide substrate in the absence and presence of TDP1 inhibitors. UA did not inhibit TDP2 (Table 1, Figure S26). All 14 newly synthesized UA derivatives inhibited TDP2 at a 1 μM concentration, but almost did not inhibit its activity at a 0.200 mM concentration.

The residual TDP2 activity for the best four TDP1 inhibitors, **7g,h** and **10a,b**, was from 30 to 60%, depending on the tested compound, at a 1 mM concentration of the inhibitor. We tested the ability of compounds **7g,h** and **10a,b** to inhibit TDP2 activity at 0.2 mM and 0.5 mM concentrations (Figure S26). All four tested compounds inhibited TDP2 insignificantly, at higher concentration ranges than TDP1. Interestingly, the best TDP1 inhibitors, **7g,h** and **10a**, exhibited the highest effect on TDP2.

2.2.6. The Effect of TDP1 Inhibitors on Cell Viability in Combination with Topotecan

TDP1 is able to cleave the TOP1–DNA complex, thus preventing the action of Tpc and decreasing the efficiency of this clinically used drug. The inhibition of TDP1 activity can increase the efficiency of Tpc. Next, we checked the cytotoxic effect of the combination of Tpc and examined the TDP1 inhibitors compared to Tpc alone on HeLa cells and HEK293 cells.

Only one compound, **10a**, of the most effective TDP1 inhibitors (IC_{50} 2.1 μM, Table 1) showed promising synergy on HeLa cells in conjunction with Tpc. We observed the suppressed cell growth in the joint presence of the TDP1 inhibitor and Tpc on HeLa cells compared to Tpc alone (Figure 5). We observed the synergistic effect both when we titrated **10a** with addition of fixed Tpc (1 μM or 2 μM) and vice versa when we titrated Tpc with addition of fixed **10a** (5 μM). The CC_{50} value of **10a** decreased from 15 μM (only **10a**) to 3 μM (**10a** with 2 μM Tpc). The CC_{50} value of Tpc decreased from 2 μM (only Tpc) to 0.7 μM (Tpc with 5 μM **10a**).

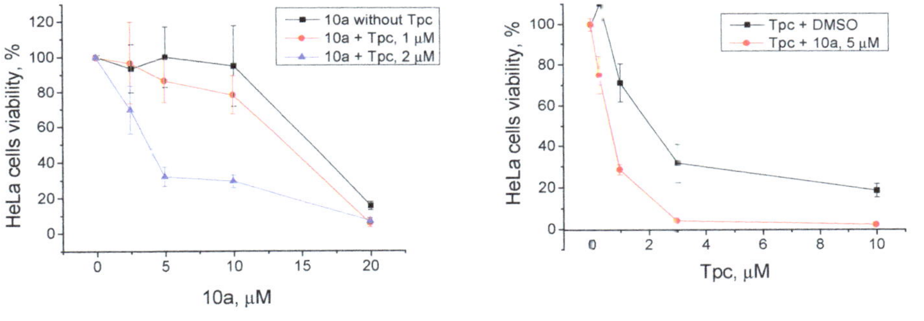

Figure 5. Left panel: The influence of Tpc at 1 μM and 2 μM concentrations on **10a** cytotoxicity. B: The influence of the **10a** derivative at a 5 μM concentration on Table 1. DMSO, for the red curve on the cells treated with 1 μM topotecan and 1% DMSO, and for the blue curve on the cells treated with 2 μM topotecan and 1% DMSO, in order to exclude the effect of topotecan itself and see the effect of the drug combination. Right panel: cells treated with 1% DMSO (black line) or 5 μM **10a** (red line) were taken as 100% in order to evaluate the effect of the combination rather than the individual substance **10a**.

In our previous works, we checked the cytotoxic effect of Tpc and the TDP1 inhibitors —monoterpene 3-carene-derived compounds [48,49] and UA combined with monoterpenoids [49], measured separately and jointly with Tpc—using a panel of HEK293FT and HEK293A *Tdp1* knockout isogenic clones. We showed that *Tdp1* knockout cells were more sensitive to Tpc compared to WT cells. The data on the HEK293FT mutants were of low reproducibility [48], which is why we decided to change the basic cell line to HEK293A [49]. In this work, we created new *PARP1-/-* HEK293A cells using the CRISPR–Cas9 approach (Figure S27). The sensitivity of the *PARP1-/-* cell line to Tpc was lower than WT HEK293A cells (CC_{50} 50 ± 5 nM and 27 ± 4 nM respectively, Figure S28). We checked the intrinsic cytotoxicity of the leader compound (**10a**) on the HEK293 wild type and *TDP1-/-* and *PARP1-/-* cell lines using a colorimetric test (Figure 6). The cytotoxicity of **10a** was nearly the same for all the cell lines, with a CC_{50} of 15–20 μM (Figure 6). We did not observe any effect of the joint presence of **10a** and Tpc on any of these cells (data not shown). There was no difference between the WT and mutant cells.

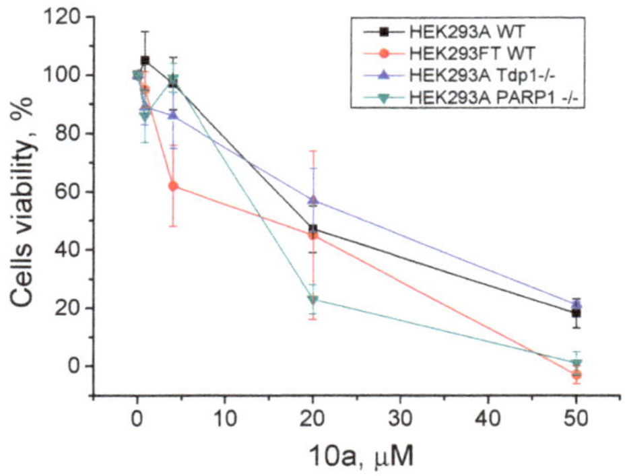

Figure 6. Intrinsic cytotoxicity of **10a** on HEK293FT and HEK293A cells, as well as on the Tdp1 and *PARP1* knockout HEK293A cell lines (HEK293A *Tdp1-/-* and HEK293A *PARP1-/-*); the dose-dependent action. The values were normalized to the control wells with the cells in 1% DMSO.

3. Materials and Methods

3.1. Chemistry

The analytical and spectral studies were conducted in the Chemical Service Center for the collective use of the Siberian Branch of the Russian Academy of Sciences. PMR and ^{13}C NMR spectra were recorded in $CDCl_3$ or DMSO-d_6 using solvent resonances (^{1}H 7.24 ppm, ^{13}C 76.90 ppm, and ^{1}H 2.50 ppm, ^{13}C 39.50, respectively) as the standards on a Bruker AV-400 spectrometer (Bruker Corporation, Germany; operating frequencies 400.13 MHz for ^{1}H and 100.61, for ^{13}C). Mass spectra (ionizing-electron energy 70 eV) were measured with a DFS Thermo Scientific high-resolution mass spectrometer (Thermo Fisher Scientific, Waltham, MA, USA). Macherey–Nagel silica gel (63–200 μ) was used for the column chromatography. Thin-layer chromatography was performed on TLC Silica gel 60 plates (UV-254, Merck, Darmstadt, Germany) (Figures S1–S24).

(+)-UA {**1**, $[\alpha]_D$ +478° (*c* 0.1, $CHCl_3$)} was isolated from a mixture of *Usnea* lichen species using the literature method [50]. Bromousnic acid **2** was synthesized according to the literature method [51]. Usnic acid derivative **9** was synthesized according to the literature [35].

Synthetic starting materials, reagents, and solvents were purchased from Sigma-Aldrich (St. Louis, MO, USA), Acros Organics (Geel, Belgium), and AlfaAesar (Heysham, UK) (95–99% pure). All chemicals were used as described unless otherwise noted. Reagent-grade solvents were redistilled prior to use.

3.1.1. Reaction of 2 with Thiols (General Method)

A weighed portion of KOH (1.1 mmol), MeOH (6 mL) and the appropriate thiol (1.1 mmol) were placed into a flask, stirred at room temperature for 10–15 min, treated with a solution of **2** (1 mmol) in CH_2Cl_2 (2 mL), stirred at room temperature for 2–3 h until the reaction was finished (TLC monitoring), washed with distilled H2O (two times the volume), dried over $MgSO_4$, and concentrated. If necessary, the solid was chromatographed over silica gel using a CH_2Cl_2 eluent.

(2R)-4-acetyl-10-{2-[(dimethylcarbamothioyl)sulfanyl]acetyl}-5,11,13-trihydroxy-2,12-dimethyl-8-oxatricyclo [7.4.0.0^{2,7}]trideca-1(13),4,6,9,11-pentaen-3-one 7a

Yellow amorphous powder, yield 94%. ^{1}H NMR ($CDCl_3$, δ) 1.77 (3H, s), 2.09 (3H, s), 2.65 (3H, s), 3.46 (3H, s), 3.54 (3H, s), 4.92 (2H, m), 6.02 (1H, s), 11.11 (OH, s), 12.81 (OH, s), 18.84 (OH, s). ^{13}C NMR ($CDCl_3$, δ): 7.4, 27.7, 31.9, 41.5, 45.6, 47.9, 58.8, 98.5, 100.9, 104.1, 105.0, 109.3, 154.7, 157.7, 163.5, 178.8, 191.4, 194.5, 195.2, 197.8, 201.6. HRMS m/z 463.0750 (calcd for $C_{21}H_{21}O_7N_2{}^{32}S_2$, 463.0754).

(2R)-4-acetyl-10-[2-(4,5-dihydro-1,3-thiazol-2-ylsulfanyl)acetyl]-5,11,13-trihydroxy-2,12-dimethyl-8-oxatricyclo[7.4.0.0^{2,7}]trideca-1(13),4,6,9,11-pentaen-3-one 7b

Yellow amorphous powder, yield 49%. ^{1}H NMR ($CDCl_3$, δ) 1.73 (3H, s), 2.06 (3H, s), 2.63 (3H, s), 3.41 (2H, bt), 4.15 (2H, bt), 4.43 (2H, m), 5.99 (1H, s), 11.09 (OH, s), 12.71 (OH, s), 18.81 (OH, s). ^{13}C NMR ($CDCl_3$, δ): 7.5, 27.8, 31.9, 43.0, 58.8, 63.7, 75.7, 98.7, 100.2, 104.1, 105.1, 109.5, 154.6, 157.9, 163.7, 167.7, 178.7, 191.5, 194.2, 197.8, 201.7. HRMS m/z 461.0592 (calcd for $C_{21}H_{19}O_7N^{32}S_2$, 461.0598).

(2R)-4-acetyl-10-{2-[(5-amino-1,3,4-thiadiazol-2-yl)sulfanyl]acetyl}-5,11,13-trihydroxy-2,12-dimethyl-8-oxatricyclo[7.4.0.0^{2,7}]trideca-1(13),4,6,9,11-pentaen-3-one 7c

Yellow amorphous powder, yield 81% ^{1}H NMR (DMSO-d_6, δ) 1.71 (3H, s), 1.99 (3H, s), 2.56 (3H, s), 4.62 (2H, s), 6.89 (1H, s), 7.32 (NH2, s), 11.56 (OH, bs), 12.75 (OH, s). ^{13}C NMR (DMSO-d_6, δ): 7.6, 27.9, 31.5, 44.2, 58.3, 98.7, 100.0, 105.1, 105.6, 107.5, 107.6, 148.7, 154.8, 157.2, 157.4, 162.3, 162.6, 170.2, 178.3, 190.9, 195.4, 197.4, 201.0. HRMS m/z 475.0507 (calcd for $C_{20}H_{17}O_7N_3{}^{32}S_2$, 475.0502).

(2R)-4-acetyl-10-[2-(1H-1,3-benzodiazol-2-ylsulfanyl)acetyl]-5,11,13-trihydroxy-2,12-dimethyl-8-oxatricyclo[7.4.0.0^{2,7}]trideca-1(13),4,6,9,11-pentaen-3-one 7f

Yellow amorphous powder, yield 54%. ^{1}H NMR ($CDCl_3$, δ) 1.70 (3H, s), 2.07 (3H, s), 2.64 (3H, s), 4.67 (2H, m), 5.94 (1H, s), 7.16 (2H, m), 7.47 (2H, m), 11.18 (OH, s), 12.64 (OH, bs), 18.78 (OH, bs). ^{13}C NMR ($CDCl_3$, δ): 7.3, 27.7, 31.7, 42.9, 58.5, 98.6, 100.0, 104.1, 104.9, 109.1, 113.8, 122.1, 138.4, 148.5, 154.5, 157.9, 163.3, 178.3, 191.3, 194.5, 197.6, 201.5. HRMS m/z 492.0981 (calcd for $C_{25}H_{20}O_7N_2{}^{32}S$, 492.0986).

(2R)-4-acetyl-5,11,13-trihydroxy-2,12-dimethyl-10-[2-(pyridin-2-ylsulfanyl)acetyl]-8-oxatricyclo[7.4.0.0^{2,7}]trideca-1(13),4,6,9,11-pentaen-3-one 7i

Yellow amorphous powder, yield 90% ^{1}H NMR ($CDCl_3$, δ) 1.76 (3H, s), 2.09 (3H, s), 2.65 (3H, s), 4.70 (2H, m), 5.92 (1H, s), 6.98 (1H, dd, J_1 = 7.3 Hz, J_2 = 5.0 Hz), 7.26 (1H, d, J = 7.3 Hz), 7.49 (1H, dt, J_1 = 7.3 Hz, J_2 = 1.5 Hz), 8.33 (1H, d, J = 5.0 Hz), 11.09 (OH, s), 12.86 (OH, s), 18.83 (OH, s). ^{13}C NMR ($CDCl_3$, δ): 7.5, 27.8, 32.0, 40.3, 58.9, 98.5, 100.5, 104.1, 105.1, 109.5, 119.8, 122.0, 136.1, 149.3, 154.7, 156.8, 157.7, 163.8, 179.0, 191.6, 195.9, 197.9, 201.7. HRMS m/z 453.0874 (calcd for $C_{23}H_{19}O_7N^{32}S$, 453.0877).

(2R)-4-acetyl-5,11,13-trihydroxy-2,12-dimethyl-10-[2-(pyrimidin-2-ylsulfanyl)acetyl]-8-oxatricyclo[7.4.0.0^{2,7}]trideca-1(13),4,6,9,11-pentaen-3-one 7j

Yellow amorphous powder, yield 92% ^{1}H NMR ($CDCl_3$, δ) 1.77 (3H, s), 2.09 (3H, s), 2.65 (3H, s), 4.65 (2H, m), 5.98 (1H, s), 6.96 (1H, t, J = 6.4 Hz), 8.45 (2H, d, J = 6.4 Hz), 11.09 (OH, s), 12.82 (OH, s), 18.83 (OH, s). ^{13}C NMR ($CDCl_3$, δ): 7.5, 27.8, 32.0, 41.4, 58.9, 98.5, 100.5, 104.1, 105.1, 109.5, 116.77, 154.7, 157.2, 157.4, 163.7. 170.7, 178.9, 191.5, 195.2, 197.8, 201.7. HRMS m/z 454.0823 (calcd for $C_{22}H_{18}O_7N_2{}^{32}S$, 454.0829).

3.1.2. Hydrolysis of Compound **8**

Compound **8** (1 mmol) was dissolved in glacial acetic acid (15 mL). The sulfuric acid water solution (30%, 0.5 mL) was added and the mixture was stirred for 4 h. After that, the resulted mixture was treated with water and the precipitate that formed was filtered off and dried in air. The solid was chromatographed over silica gel using a CH_2Cl_2 eluent with an MeOH gradient.

(2R)-4-acetyl-10-[2-(carbamoylsulfanyl)acetyl]-5,11,13-trihydroxy-2,12-dimethyl-8-oxatricyclo[7.4.0.0^{2,7}]trideca-1(13),4,6,9,11-pentaen-3-one 8

Yellow amorphous powder, yield 38%. ^{1}H NMR ($CDCl_3$, δ) 1.74 (3H, s), 2.07 (3H, s), 2.64 (3H, s), 4.43 (2H, m), 5.80 (NH2, bs) 6.00 (1H, s), 11.12 (OH, s), 12.76 (OH, s), 18.82 (OH, s). ^{13}C NMR ($CDCl_3$, δ): 7.5, 27.8, 31.9, 40.8, 58.8, 98.7, 100.2, 104.2, 105.1, 109.5, 154.7, 158.0, 163.7, 167.7, 178.7, 191.5, 194.9, 197.8, 201.7. HRMS m/z 419.0671 (calcd for $C_{19}H_{17}O_8N^{32}S$, 419.0669).

3.1.3. Oxidation of Compounds **7g,h**

Thioether **7g** or **7h** (1 mmol) was dissolved in methylene chloride (5 mL) on an ice bath. Meta-chloroperoxybenzoic acid (3 mmol) was added and the resulted solution was stirred for 30 min at 0 °C. After that, the mixture was treated with a saturated sodium sulfite solution (3 mL) and the resulted mixture was stirred for 1 h at room temperature. Finally, the mixture was extracted with methylene chloride, dried over magnesium sulfate, and evaporated.

(2R)-4-acetyl-10-[2-(1,3-benzothiazole-2-sulfinyl)acetyl]-5,11,13-trihydroxy-2,12-dimethyl-8-oxatricyclo[7.4.0.0^{2,7}]trideca-1(13),4,6,9,11-pentaen-3-one 10a

Yellow amorphous powder, yield 75%. ^{1}H NMR ($CDCl_3$, δ) 1.64 and 1.69 (3H, s), 2.06 and 2.09 (3H, s), 2.63 and 2.63 (3H, s), 4.72 and 5.01 (2H, m), 5.93 and 5.98 (1H, s), 7.46 (2H, m), 8.00 (2H, m), 11.18 and 11.21 (OH, s), 12.52 and 12.59 (OH, s), 18.78 (OH, bs). ^{13}C NMR ($CDCl_3$, δ): 7.4 and 7.4, 27.7, 31.7 and 31.8, 58.5 and 58.6, 68.0 and 68.6, 99.0, 101.3 and 101.4, 104.4 and 104.5, 105.0 and 105.1, 109.5 and 109.6, 122.1 and 122.2, 123.9 and 124.0, 126.4 and 127.0, 136.0 and 136.1, 153.5, 154.5 and 154.5, 158.8 and 158.9, 163.8 and 164.0, 176.1 and 176.2, 178.1, 190.4 and 190.6, 191.4, 197.6, 201.6 and 201.7.

3.2. Biology

3.2.1. Preparation of Labeled Oligonucleotides

A single-stranded oligodeoxynucleotide 5′-FAM-AAC GTC AGG GTC TTC C-BHQ-1-3′ containing a 6-carboxamido fluorescein (6-FAM) fluorophore at the 5′-end and a Black Hole Quencher 1 (BHQ-1) at the 3′-end was employed as an internally-quenched probe for real-time detection of TDP1 activity [43]. Another single-stranded oligonucleotide 5′-Tyr-AAC GTC AGG GTC TTC C-FAM-3′ containing a 6-FAM label at the 3′-end and an L-tyrosine residue attached via the phenolic OH group to the 5′-terminal phosphate was used as a substrate for detection of TDP2 activity (Tyr = HOOC-CH(NH_2)-CH_2-C_6H_4-*p*-O~).

Oligonucleotides were synthesized essentially as described in a previous paper [33]. Briefly, both oligonucleotides were assembled on a Biosset ASM-800 automated DNA/RNA synthesizer (Novosibirsk, Russia) on 200 nmol scale by β-cyanoethyl phosphoramidite chemistry. The doubly labeled probe was prepared from standard 5′-DMTr-deoxynucleoside 3′-phosphoramidites, 6-FAM phosphoramidite for 5′-labeling, and a BHQ-1 CPG support for attachment of a Black Hole Quencher^{TM} BHQ-1 (all from Glen Research, Sterling, VA, USA), deprotected and purified under standard conditions as described above [33].

Synthesis of a 5′-tyrosinyl oligodeoxynucleotide substrate 5′-Tyr-AAC GTC AGG GTC TTC C-FAM-3′ was carried out on a specially prepared CPG 500 Å support containing N^{α}-Fmoc-L-tyrosine with an unprotected phenol group esterified via the carboxy group to the 5′-OH group of a thymidine residue linked to the support through the 3′-OH by a succinate linker. Synthesis of the tyrosine support was accomplished as described previously [33]. In contradistinction to the previous paper, in this work the solid-phase DNA assembly was carried out with 'reversed' 3′-DMTr-deoxynucleoside 5′-phosphoramidites (ChemGenes, Wilmington, MA, USA) in the 5′→3′ direction using a free phenolic OH of the support-bound tyrosine as an anchoring point to introduce the tyrosine residue onto the 5′-end. 6-FAM phosphoramidite (Glen Research, Sterling, VA, USA) was coupled at the last cycle of the synthesis to place the label at the 3′-end. The oligonucleotide with a phosphotyrosine residue was cleaved from support by alkaline hydrolysis of the ester linkages (100 µL of 0.1 M NaOH, 1 h at 25 °C). Excess NaOH was quenched by ca. 50 µL of Amberlyst cation exchange resin beads in NH_4^+ form. The solid support was discarded, the supernatant and aqueous washings (2 × 100 µL) were combined, evaporated to dryness, and treated with 200 µL of a conc. (ca. 28%) aqueous NH_3 solution at 55 °C for 16 h to remove all the remaining *N*-protecting groups from oligonucleotide. The DNA substrate was isolated, purified, and analyzed as described previously [33]: [M-H]$^-$ calc. 5666.83, found 5666.90.

Human recombinant tyrosyl-DNA phosphodiesterase 1 (TDP1) and human recombinant poly(ADP-ribose)-polymerase1 (PARP1) were expressed in the *E. coli* system and purified as described [52,53].

3.2.2. Expression and Purification of Human Recombinant Tyrosyl-DNA Phosphodiesterase 2 (TDP2)

The recombinant N-terminally His-tagged TDP2 was expressed in *E. coli* BL21 (DE3) cells. The plasmid pLATE-31-TDP2 expression vector was constructed as follows. cDNA encoding the TDP2 protein was cloned in the pLATE-31 expression vector using a LICator LIC Cloning and Expression System (Thermo Scientific, Waltham, MA, USA). Using specific primers and total cDNA of HeLa cells, the TDP2 coding sequence was amplified by PCR and annealed with a linearized pLATE-31 vector. The sequence of the cloned cDNA was confirmed at the SB RAS Genomics Core Facility (ICBFM SB RAS, Novosibirsk, Russia). Plasmids were transformed into BL21 cells by electroporation, and the cells were grown in LB medium at pH 7.5 with 100 mg/mL ampicillin at 30 °C. Two hours after induction with 1 mM IPTG, cells were harvested. Cell pellets were thawed on ice, resuspended in binding buffer (0.5 M NaCl, 5% glycerol, 50 mM Tris-HCl, pH 8.0, and a mixture of protease inhibitors), and broken by sonication. After centrifugation, 10 mM imidazole was added to the supernatant. The Ni sepharose column (GE Healthcare, UK) was washed with binding buffer (0.5 M NaCl, 10 mM imidazole, 50 mM Tris-HCl, and pH 8.0). Elution of the proteins was carried out with an elution buffer (0.5 M NaCl, 500 mM imidazole, 50 mM Tris-HCl, pH 8.0, and protease inhibitors), and the eluate was loaded into the heparin sepharose column (GE Healthcare, Pollards Wood, UK). Elution of the proteins was carried out with a NaCl gradient of 0.1–1 M in 50 mM Tris-HCl, pH 8.0, with protease inhibitors. The protein TDP2 was stored in 50 mM NaCl, 50 mM Tris-HCl pH 8.0, 1 mM EDTA, 2 mM DTT, and 50% glycerol at −20 °C. The enzyme samples were estimated to be more than 95% pure. Enzyme concentrations were estimated by Bradford assay. The Coomassie-stained protein gel is shown in Supplementary Material (Figure S29).

3.2.3. Real-Time Detection of TDP1 Activity

The biosensor (5′-[FAM] AAC GTC AGGGTC TTC C [BHQ]-3′) was synthesized in the Laboratory of Nucleic Acid Chemistry at the Institute of Chemical Biology and Fundamental Medicine (Novosibirsk, Russia) and was used for TDP1 enzyme activity real-time fluorescence detection [43]. The reaction mixture (200 μL) contained a TDP1 reaction buffer (50 mM Tris-HCl, pH 8.0, 50 mM NaCl, and 7 mM β-mercaptoethanol), 50 nM oligonucleotide, varied concentrations of the tested compounds, and purified TDP1 in a final concentration 1.5 nM. The reaction mixtures were incubated at a constant temperature of 26 °C in a POLARstar OPTIMA fluorimeter (BMG LABTECH, GmbH, Ortenberg, Germany). Fluorescence intensity was measured (Ex485/Em520 nm) every 1 min for 10 min. The average values of the half maximal inhibitory concentration (IC_{50}) were determined using an eleven-point concentration response curve and calculated using MARS Data Analysis 2.0 (BMG LABTECH). The 50% inhibitory concentration (IC_{50}) was defined as the concentration of the compound that inhibited 50% of the enzyme activity when compared to the untreated controls. At least three independent experiments were carried out to obtain the IC_{50} values. To determine the kinetic parameters of the TDP1 enzymatic reaction, the apparent maximum rate of enzymatic reaction (V_{max}), Michaelis constant (K_M), possible inhibition mechanism, and steady-state kinetic experiments were carried out at 5 fixed concentrations of the substrate, with variation in the inhibitor concentrations [54]. The standard reaction mixtures (200 μL) contained reaction buffer components; 50 nM, 100 nM, 200 nM, 500 nM, or 1000 nM substrate; an inhibitor, and 1.5 nM recombinant human TDP1. The initial kinetic curves were obtained in three independent experiments and statistically processed in OriginPro 8.6.0 (OriginLab, Northampton, MA, USA).

3.2.4. Gel-Based TDP2 Activity Assay

Oligonucleotide 5′- tyrosine -AAC GTC AGG GTC TTC C- FAM -3′ was synthesized as described above and used for the indication of TDP2 enzyme activity in polyacrylamide gel. TDP2 gel-based assays were performed to a final volume 20 μL using 100 nM substrate incubated with 200 nM recombinant human TDP2 in the absence or presence of an inhibitor for 10 min at 37 °C in a buffer containing 50 mM Tris-HCl, pH 8.0, 50 mM NaCl, 7 mM β-mercaptoethanol, and BSA. Reactions were terminated by the addition of a gel loading buffer (TBE, 10% formamide, 7 M carbamide, 0.1% xylene cyanol, and 0.1% bromophenol blue, and 20 mM EDTA). The samples were heated before loading at 90 °C for 7 min. The reaction products were separated by electrophoresis in a 20% denaturing PAGE with 7 M carbamide at a ratio of acrylamide to bisacrylamide of 19:1. A Typhoon FLA 9500 phosphorimager (GE Healthcare, Boston, MA, USA) was used for gel scanning and imaging, and the data were analyzed with QuantityOne 4.6.7 software (Bio-Rad Laboratories, Inc., Hercules, CA, USA).

3.2.5. PARP1 and PARP2 Enzyme Assay

The radioactive labeled [32P]-NAD+ was synthesized from α–[32P]-ATP according to [55]. The reaction of autopoly(ADP-ribosyl)ation was carried out as follows: for PARP1, 50 mM Tris-HCl, pH 8.0, 10 mM $MgCl_2$, 150 mM NaCl, and 7 mM β-mercaptoethanol, as well as activated DNA 2 oe/mL, 0.3 mM [^{32}P]-NAD+ at 37 °C. The reaction was initiated by adding PARP1 to 200 nM and the reaction mixtures were incubated for 2 min. For PARP2: 50 mM Tris-HCl, pH 8.0, 3 mM spermin, 150 mM NaCl, and 7 mM β-mercaptoethanol, as well as activated DNA 2 oe/mL, 0.6 mM [^{32}P] NAD+ at 37 °C. The reaction was initiated by adding PARP2 to 800 nM and the reaction mixtures were incubated for 5 min. The tested compounds were added at a final concentration 500 nM for reactions with PARP1 and 1 μM for reactions with PARP2. The reaction was stopped by placing 10 μL aliquots onto Whatman 1 paper filters soaked with 5% TCA. The filters were washed with 5% TCA four times and dried in the air after the removal of TCA with 90% ethanol. The incorporation of radioactivity into the product was calculated using a Typhoon FLA 9500 scanner (GE Healthcare, Chicago, IL, USA). Measurements were done in at least two independent experiments.

3.2.6. Cell Culture Cytotoxicity Assay

Cytotoxicity of the compounds was examined against human cell lines HEK293A (human embryonic kidney)—WT, TDP1 deficient (*Tdp1-/-*), PARP1 deficient (*PARP1-/-*), and HeLa (cervical cancer) using an EZ4U colorimetric test (Biomedica, Vienna, Austria). The HEK293A cell line was obtained from Thermo Fisher Scientific (Waltham, MA, USA), and the HeLa cell line was obtained from the Russian Cell Culture Collection (RCCC) Institute of Cytology RAS, St. Petersburg, Russia. The cells were grown in DMEM/F12 medium (Gibco, Thermo Fisher Scientific, Waltham, MA, USA), with 1x GlutaMAX (Gibco, Thermo Fisher Scientific, Waltham, MA, USA), 50 IU/mL penicillin, 50 μg/mL streptomycin (MP Biomedicals), and in the presence of 10% fetal bovine serum (Biolot, Saint-Petersburg, Russia) in a 5% CO_2 atmosphere. Cells were grown in the presence of 1% DMSO in the control wells. After the formation of a 30–50% monolayer, the tested compounds were added to the medium, and the cell culture was monitored for 3 days. The values were normalized to their own control in each case. At least three independent experiments were carried out. The 50% cytotoxic concentration (CC_{50}) was defined as the compound concentration that reduced the cell viability by 50% when compared to the untreated controls. The compound concentration that caused 50% cell growth inhibition was determined using OriginPro 8.6.0 software (OriginLab, Northampton, MA, USA). The measurements were carried out in three independent experiments.

3.2.7. Plasmid Construction for Human *PARP1* Gene Knockout

sgRNAs design was performed using the Benchling CRISPR tool (https://www.benchling.com/, accessed on 9 November 2019). Two protospacers (PAM sequences in brackets) were selected for the DNA sequence deletion that includes 3–5 exons of the *PARP1* gene: PARP1-gRNA1 CTAGAACCTCCAATACCATG (TGG) and PARP1-gRNA2 GCAAGTGACCACAAAGGTGC (AGG). Corresponding oligonucleotides were cloned in plasmid pSpCas9(BB)-2A-GFP (PX458) (the plasmid was a gift from Feng Zhang (Addgene plasmid #48138; http://n2t.net/addgene:48138; RRID:Addgene_48138)) as previously described [48]. Transfection-grade plasmid DNA was isolated using the Plasmid Plus Midi Kit (QIAGEN, Hilden, Germany).

3.2.8. Knockout HEK293A Clones' Generation

5×10^5 HEK293A cells were plated into each well of a 12-well plate and co-transfected with the constructed plasmids PARP1-gRNA1 and PARP1-gRNA2 (0.25 µg of each) using a Lipofectamine 3000 Reagent (Thermo Fisher Scientific, Waltham, MA, USA). The growth medium contained DMEM/F12 (Gibco) 1:1, 10% FBS (Gibco), 100 U/mL penicillin–streptomycin (Gibco, Thermo Fisher Scientific, Waltham, MA, USA), and 1× GlutaMAX (Gibco, Thermo Fisher Scientific, Waltham, MA, USA). A total of 48 h after transfection the cells were detached using TrypLE Express (TrypLE, Gibco, Thermo Fisher Scientific, Waltham, MA, USA), and the GFP-positive cell population was enriched by cell sorting using the BD FACSAria III Cell Sorter (BD Biosciences, Franklin Lakes, NJ, USA). Transfected cells were plated onto a 96-well plate, one cell per well. Single-cell clones grew for two weeks before they were replicated to another 96-well plate, so we obtained two equal 96-well plates with cell clones: one plate was used for PCR analysis of the deletion in the *PARP1* gene, and the other plate was used for the mutant cell clone multiplying.

3.2.9. Analysis of CRISPR/Cas9-Mediated Deletions in *PARP1* Gene

Genome DNA was extracted from cells on one of two 96-well plates using 50 µL of QuickExtract™ DNA Extraction Solution (Lucigen, Madison, WI, USA) per well. The DNA extracts were diluted with 200 µL of mQ water. Two microliters of the diluted DNA extract were used to PCR amplify the target region with primers to detect the presence of deletions (PARP1-Del-F 5′-AGTGTGCCCTGCGTATTTGC-3′ and PARP1-Del-R 5′- CACAGGGATGAATCTTTCTGGTC-3′) and wild-type alleles (PARP1-In-F 5′- CGCTCCCTTGGTACCACATATG -3′ and PARP1-In-R 5′-GGCTTACTGACAGTCAGCCAAG-3′). Both reactions were run on an S1000 Thermal Cycler (Bio-Rad, Singapore) using BioMaster HS-Taq PCR-Color (2×) (Biolabmix, Novosibirsk, Russia) with the following program: 95 °C for 3 min; 35 cycles: 95 °C for 30 s; 60 °C for 30 s; 72 °C for 30 s; and 72 °C for 3 min. The products of the reactions were resolved in 1% agarose gel stained with ethidium bromide.

4. Conclusions

TDP1 promotes the cleavage of the stable DNA–TOP1 complexes with the clinically used anticancer drug topotecan (Tpc), which is a TOP1 inhibitor. TDP1 activity may be a possible cause of tumor resistance to TOP1 inhibitors. A series of new UA thioether and sulfoxide derivatives were synthesized.

The usnic acid thioether and sulfoxide derivatives efficiently suppressed TDP1 activity with the IC_{50} values in the 1.4–25.2 µM range. The structure of the heterocyclic substituent affects the TDP1 inhibitory efficiency of these compounds. Derivatives containing a five-membered heterocyclic fragment itself or fused to a benzene ring (**7b**,**c**,**e–g** and **10a**) inhibit TDP1 in the low micromolar concentration range (IC_{50} of 1.4–4.4 µM). Compounds containing a six-membered heterocycle (**7i–k**) inhibit TDP1 at higher concentrations (IC_{50} > 11 µM). The presence of a halogen in the *para*-position of the benzene substituent enhances the inhibitory properties of the compounds. For the most effective inhibitors of TDP1 **7g**,**h** and their sulfoxide analogs **10a**,**b**, we observed the uncompetitive type of inhibition. The uncompetitive inhibitors prevent the second step of the reaction stabilizing the enzyme–DNA covalent complex. Thus, uncompetitive TDP1 inhibitors could lead to the accumulation of the single-strand breaks in cancer cells.

The anticancer effect of the TOP1 inhibitors can be significantly enhanced by the simultaneous inhibition of PARP1, TDP1, and TDP2. We tested the ability of the synthesized compounds to inhibit the TDP1, TDP2, and PARP1 activities. We found the compounds act as dual or triple inhibitors of TDP1, TDP2, and PARP1. Some of the compounds inhibited not only TDP1 but also TDP2 and PARP1, but at significantly higher concentration ranges than TDP1.

Interestingly, the sulfoxide analogs **10a,b** were less cytotoxic than their thioester analogs **7g,h**. Compound **10b** was one of the less toxic among the ten tested compounds. One of the most effective TDP1 inhibitors, **10a** (IC_{50} 2.1 μM, Table 1), also moderately inhibit TDP2, and showed promising synergy on HeLa cells in conjunction with topotecan. That is of great importance for further development of sensitizers to topotecan and other clinically used TOP1 inhibitors.

Supplementary Materials: The following are available online at https://www.mdpi.com/article/10.3390/ijms222111336/s1.

Author Contributions: Chemistry investigation, A.S.F., O.A.L., E.A.B., K.V.K. and D.A.S.; in vitro investigation, N.S.D., A.L.Z., E.S.I., K.A.O., I.A.C. and T.E.K.; recombinant proteins purification: R.O.A. and K.N.N.; *PARP1* knockout clones' generation, A.A.M., S.P.M. and S.M.Z.; methodology, N.F.S. and O.I.L.; writing—original draft, N.S.D., A.L.Z., A.S.F. and O.A.L.; writing—review and editing, O.A.L., N.F.S. and O.I.L. All authors have read and agreed to the published version of the manuscript.

Funding: This research was funded by a grant from the Ministry of Science and Higher Education Russian Federation (agreement no. 075-15-2020-773).

Institutional Review Board Statement: Not applicable.

Informed Consent Statement: Not applicable.

Acknowledgments: The authors would like to acknowledge the Multi-Access Chemical Research Center SB RAS, Novosibirsk, Russia, for their assistance with the analytical and spectroscopic measurements. Cell lines were obtained from the Russian Cell Culture Collection (RCCC) Institute of Cytology RAS, St. Petersburg, Russia.

Conflicts of Interest: The authors declare no conflict of interest.

References

1. Curtin, N.J.; Szabo, C. Poly(ADP-ribose) polymerase inhibition: Past, present and future. *Nat. Rev. Drug Discov.* **2020**, *19*, 711–736. [CrossRef]
2. Zakharenko, A.; Dyrkheeva, N.; Lavrik, O. Dual DNA topoisomerase 1 and tyrosyl-DNA phosphodiesterase 1 inhibition for improved anticancer activity. *Med. Res. Rev.* **2019**, *39*, 1427–1441. [CrossRef] [PubMed]
3. Kawale, A.S.; Povirk, L.F. Tyrosyl-DNA phosphodiesterases: Rescuing the genome from the risks of relaxation. *Nucleic Acids Res.* **2018**, *46*, 520–537. [CrossRef] [PubMed]
4. Alagoz, M.; Gilbert, D.C.; El-Khamisy, S.; Chalmers, A.J. DNA repair and resistance to topoisomerase I inhibitors: Mechanisms, biomarkers and therapeutic targets. *Curr. Med. Chem.* **2012**, *19*, 3874–3885. [CrossRef] [PubMed]
5. Pommier, Y.; Huang, S.Y.; Gao, R.; Das, B.B.; Murai, J.; Marchand, C. Tyrosyl-DNA-phosphodiesterases (TDP1 and TDP2). *DNA Repair* **2014**, *19*, 114–129. [CrossRef] [PubMed]
6. Pommier, Y.; Leo, E.; Zhang, H.; Marchand, C. DNA topoisomerases and their poisoning by anticancer and antibacterial drugs. *Chem. Biol.* **2010**, *17*, 421–433. [CrossRef]
7. Zeng, Z.; Cortés-Ledesma, F.; El Khamisy, S.F.; Caldecott, K.W. TDP2/TTRAP is the major 5′-tyrosyl DNA phosphodiesterase activity in vertebrate cells and is critical for cellular resistance to topoisomerase II-induced DNA damage. *J. Biol. Chem.* **2011**, *286*, 403–409. [CrossRef]
8. Gómez-Herreros, F.; Romero-Granados, R.; Zeng, Z.; Alvarez-Quilón, A.; Quintero, C.; Ju, L.; Umans, L.; Vermeire, L.; Huylebroeck, D.; Caldecott, K.W.; et al. TDP2-dependent non-homologous end-joining protects against topoisomerase II-induced DNA breaks and genome instability in cells and in vivo. *PLoS Genet.* **2013**, *9*, e1003226. [CrossRef]
9. Marchand, C.; Abdelmalak, M.; Kankanala, J.; Huang, S.Y.; Kiselev, E.; Fesen, K.; Kurahashi, K.; Sasanuma, H.; Takeda, S.; Aihara, H.; et al. Deazaflavin Inhibitors of Tyrosyl-DNA Phosphodiesterase 2 (TDP2) Specific for the Human Enzyme and Active against Cellular TDP2. *ACS Chem. Biol.* **2016**, *11*, 1925–1933. [CrossRef]
10. Raoof, A.; Depledge, P.; Hamilton, N.M.; Hamilton, N.S.; Hitchin, J.R.; Hopkins, G.V.; Jordan, A.M.; Maguire, L.A.; McGonagle, A.E.; Mould, D.P.; et al. Toxoflavins and deazaflavins as the first reported selective small molecule inhibitors of tyrosyl-DNA phosphodiesterase II. *J. Med. Chem.* **2013**, *56*, 6352–6370. [CrossRef]
11. Laev, S.S.; Salakhutdinov, N.F.; Lavrik, O.I. Tyrosyl-DNA phosphodiesterase inhibitors: Progress and potential. *Bioorg. Med. Chem.* **2016**, *24*, 5017–5027. [CrossRef]

12. Zhang, Y.; He, X.Z.; Yang, H.; Liu, H.Y.; An, L.K. Robustadial A and B from Eucalyptus globulus Labill. and their anticancer activity as selective tyrosyl-DNA phosphodiesterase 2 inhibitors. *Phytother. Res.* **2021**, *35*, 5282–5289. [CrossRef]
13. Yang, H.; Zhu, X.Q.; Wang, W.; Chen, Y.; Hu, Z.; Zhang, Y.; Hu, D.X.; Yu, L.M.; Agama, K.; Pommier, Y.; et al. The synthesis of furoquinolinedione and isoxazoloquinolinedione derivatives as selective Tyrosyl-DNA phosphodiesterase 2 (TDP2) inhibitors. *Bioorg. Chem.* **2021**, *111*, 104881. [CrossRef]
14. Senaweera, S.; He, T.; Cui, H.; Aihara, H.; Wang, Z. 4-Benzylideneisoquinoline-1,3(2H,4H)-diones as tyrosyl DNA phosphodiesterase 2 (TDP2) inhibitors. *Med. Chem. Res.* **2021**, *30*, 371–386. [CrossRef]
15. Wang, P.; Elsayed, M.S.A.; Plescia, C.B.; Ravji, A.; Redon, C.E.; Kiselev, E.; Marchand, C.; Zeleznik, O.; Agama, K.; Pommier, Y.; et al. Synthesis and Biological Evaluation of the First Triple Inhibitors of Human Topoisomerase 1, Tyrosyl-DNA Phosphodiesterase 1 (Tdp1), and Tyrosyl-DNA Phosphodiesterase 2 (Tdp2). *J. Med. Chem.* **2017**, *60*, 3275–3288. [CrossRef] [PubMed]
16. Lavrik, O.I. PARPs' impact on base excision DNA repair. *DNA Repair* **2020**, *93*, 102911. [CrossRef] [PubMed]
17. Ray Chaudhuri, A.; Nussenzweig, A. The multifaceted roles of PARP1 in DNA repair and chromatin remodelling. *Nat. Rev. Mol. Cell Biol.* **2017**, *18*, 610–621. [CrossRef]
18. Lord, C.J.; Tutt, A.N.; Ashworth, A. Synthetic lethality and cancer therapy: Lessons learned from the development of PARP inhibitors. *Annu. Rev. Med.* **2015**, *66*, 455–470. [CrossRef] [PubMed]
19. Kim, D.S.; Camacho, C.V.; Kraus, W.L. Alternate therapeutic pathways for PARP inhibitors and potential mechanisms of resistance. *Exp. Mol. Med.* **2021**, *53*, 42–51. [CrossRef]
20. Pacher, P.; Szabó, C. Role of poly(ADP-ribose) polymerase 1 (PARP-1) in cardiovascular diseases: The therapeutic potential of PARP inhibitors. *Cardiovasc. Drug Rev.* **2007**, *25*, 235–260. [CrossRef]
21. Das, B.B.; Huang, S.Y.; Murai, J.; Rehman, I.; Amé, J.C.; Sengupta, S.; Das, S.K.; Majumdar, P.; Zhang, H.; Biard, D.; et al. PARP1-TDP1 coupling for the repair of topoisomerase I-induced DNA damage. *Nucleic Acids Res.* **2014**, *42*, 4435–4449. [CrossRef] [PubMed]
22. Moor, N.A.; Vasil'eva, I.A.; Anarbaev, R.O.; Antson, A.A.; Lavrik, O.I. Quantitative characterization of protein-protein complexes involved in base excision DNA repair. *Nucleic Acids Res.* **2015**, *43*, 6009–6022. [CrossRef] [PubMed]
23. Lebedeva, N.A.; Anarbaev, R.O.; Sukhanova, M.; Vasil'eva, I.A.; Rechkunova, N.I.; Lavrik, O.I. Poly(ADP-ribose)polymerase 1 stimulates the AP-site cleavage activity of tyrosyl-DNA phosphodiesterase 1. *Biosci. Rep.* **2015**, *35*, e00230. [CrossRef] [PubMed]
24. Horton, J.K.; Stefanick, D.F.; Zhao, M.L.; Janoshazi, A.K.; Gassman, N.R.; Seddon, H.J.; Wilson, S.H. XRCC1-mediated repair of strand breaks independent of PNKP binding. *DNA Repair* **2017**, *60*, 52–63. [CrossRef]
25. Prasad, R.; Horton, J.K.; Dai, D.P.; Wilson, S.H. Repair pathway for PARP-1 DNA-protein crosslinks. *DNA Repair* **2019**, *73*, 71–77. [CrossRef]
26. Brettrager, E.J.; van Waardenburg, R.C.A.M. Targeting Tyrosyl-DNA phosphodiesterase I to enhance toxicity of phosphodiester linked DNA-adducts. *Cancer Drug Resist.* **2019**, *2*, 1153–1163. [CrossRef]
27. Elsayed, W.; El-Shafie, L.; Hassan, M.K.; Farag, M.A.; El-Khamisy, S.F. Isoeugenol is a selective potentiator of camptothecin cytotoxicity in vertebrate cells lacking TDP1. *Sci. Rep.* **2016**, *6*, 26626. [CrossRef]
28. Al Abo, M.; Sasanuma, H.; Liu, X.; Rajapakse, V.N.; Huang, S.Y.; Kiselev, E.; Takeda, S.; Plunkett, W.; Pommier, Y. TDP1 is Critical for the Repair of DNA Breaks Induced by Sapacitabine, a Nucleoside also Targeting ATM- and BRCA-Deficient Tumors. *Mol. Cancer Ther.* **2017**, *16*, 2543–2551. [CrossRef]
29. Fam, H.K.; Walton, C.; Mitra, S.A.; Chowdhury, M.; Osborne, N.; Choi, K.; Sun, G.; Wong, P.C.; O'Sullivan, M.J.; Turashvili, G.; et al. TDP1 and PARP1 deficiency are cytotoxic to rhabdomyosarcoma cells. *Mol. Cancer Res.* **2013**, *11*, 1179–1192. [CrossRef] [PubMed]
30. Toots, M.; Ustav, M., Jr.; Männik, A.; Mumm, K.; Tämm, K.; Tamm, T.; Ustav, E.; Ustav, M. Identification of several high-risk HPV inhibitors and drug targets with a novel high-throughput screening assay. *PLoS Pathog.* **2017**, *13*, e1006168. [CrossRef] [PubMed]
31. Lei, J.; Ploner, A.; Elfström, K.M.; Wang, J.; Roth, A.; Fang, F.; Sundström, K.; Dillner, J.; Sparén, P. HPV Vaccination and the Risk of Invasive Cervical Cancer. *N. Engl. J. Med.* **2020**, *383*, 1340–1348. [CrossRef]
32. Zakharenko, A.; Luzina, O.; Koval, O.; Nilov, D.; Gushchina, I.; Dyrkheeva, N.; Švedas, V.; Salakhutdinov, N.; Lavrik, O. Tyrosyl-DNA Phosphodiesterase 1 Inhibitors: Usnic Acid Enamines Enhance the Cytotoxic Effect of Camptothecin. *J. Nat. Prod.* **2016**, *79*, 2961–2967. [CrossRef]
33. Zakharova, O.; Luzina, O.; Zakharenko, A.; Sokolov, D.; Filimonov, A.; Dyrkheeva, N.; Chepanova, A.; Ilina, E.; Ilyina, A.; Klabenkova, K.; et al. Synthesis and evaluation of aryliden- and hetarylidenfuranone derivatives of usnic acid as highly potent Tdp1 inhibitors. *Bioorg. Med. Chem.* **2018**, *26*, 4470–4480. [CrossRef]
34. Filimonov, A.S.; Chepanova, A.A.; Luzina, O.A.; Zakharenko, A.L.; Zakharova, O.D.; Ilina, E.S.; Dyrkheeva, N.S.; Kuprushkin, M.S.; Kolotaev, A.V.; Khachatryan, D.S.; et al. New Hydrazinothiazole Derivatives of Usnic Acid as Potent Tdp1 Inhibitors. *Molecules* **2019**, *24*, 3711. [CrossRef]
35. Zakharenko, O.; Luzina, O.; Sokolov, D.; Zakharova, O.; Rakhmanova, M.; Chepanova, A.; Dyrkheeva, N.; Lavrik, O.; Salakhutdinov, N. Usnic acid derivatives are effective inhibitors of tyrosyl-DNA phosphodiesterase 1//Rus. *J. Bioorg. Chem.* **2017**, *43*, 84–90. [CrossRef]
36. Zakharenko, A.L.; Luzina, O.A.; Sokolov, D.N.; Kaledin, V.I.; Nikolin, V.P.; Popova, N.A.; Patel, J.; Zakharova, O.D.; Chepanova, A.A.; Zafar, A.; et al. Novel tyrosyl-DNA phosphodiesterase 1 inhibitors enhance the therapeutic impact of topotecan on in vivo tumor models. *Eur. J. Med. Chem.* **2019**, *161*, 581–593. [CrossRef]

37. Nikolin, V.P.; Popova, N.A.; Kaledin, V.I.; Luzina, O.A.; Zakharenko, A.L.; Salakhutdinov, N.F.; Lavrik, O.I. The influence of an enamine usnic acid derivative (a tyrosyl-DNA phosphodiesterase 1 inhibitor) on the therapeutic effect of topotecan against transplanted tumors in vivo. *Clin. Exp. Metastasis.* **2021**, *38*, 431–440. [CrossRef] [PubMed]
38. Dyrkheeva, N.S.; Zakharenko, A.L.; Novoselova, E.S.; Chepanova, A.A.; Popova, N.A.; Nikolin, V.P.; Luzina, O.A.; Salakhutdinov, N.F.; Ryabchikova, E.I.; Lavrik, O.I. Antitumor Activity of the Combination of Topotecan and Tyrosyl-DNA-Phosphodiesterase 1 Inhibitor on Model Krebs-2 Mouse Ascite Carcinoma. *Mol. Biol.* **2021**, *55*, 312–317. [CrossRef]
39. Zakharenko, A.; Sokolov, D.; Luzina, O.; Sukhanova, M.; Khodyreva, S.; Zakharova, O.; Salakhutdinov, N.; Lavrik, O. Influence of usnic acid and its derivatives on the activity of mammalian poly(ADP-ribose)polymerase 1 and DNA polymerase beta. *Med. Chem.* **2012**, *8*, 883–893. [CrossRef] [PubMed]
40. Zhao, Y.; Zhang, L.X.; Jiang, T.; Long, J.; Ma, Z.Y.; Lu, A.P.; Cheng, Y.; Cao, D.S. The ups and downs of Poly(ADP-ribose) Polymerase-1 inhibitors in cancer therapy-Current progress and future direction. *Eur. J. Med. Chem.* **2020**, *203*, 112570. [CrossRef]
41. Shtro, A.A.; Zarubaev, V.V.; Luzina, O.A.; Sokolov, D.N.; Kiselev, O.I.; Salakhutdinov, N.F. Novel derivatives of usnic acid effectively inhibiting reproduction of influenza A virus. *Bioorg Med. Chem.* **2014**, *22*, 6826–6836. [CrossRef]
42. Sokolov, D.N.; Luzina, O.A.; Salakhutdinov, N.F. Synthesis of Sulfones and Sulfoxides Based on (+)-usnic Acid. *Chem. Nat. Compd.* **2018**, *54*, 46–49. [CrossRef]
43. Zakharenko, A.; Khomenko, T.; Zhukova, S.; Koval, O.; Zakharova, O.; Anarbaev, R.; Lebedeva, N.; Korchagina, D.; Komarova, N.; Vasiliev, V.; et al. Synthesis and biological evaluation of novel tyrosyl-DNA phosphodiesterase 1 inhibitors with a benzopentathiepine moiety. *Bioorg. Med. Chem.* **2015**, *23*, 2044–2052. [CrossRef]
44. Davies, D.R.; Interthal, H.; Champoux, J.J.; Hol, W.G. Insights into substrate binding and catalytic mechanism of human tyrosyl-DNA phosphodiesterase (Tdp1) from vanadate and tungstate-inhibited structures. *J. Mol. Biol.* **2002**, *324*, 917–932. [CrossRef]
45. Davies, D.R.; Interthal, H.; Champoux, J.J.; Hol, W.G. Crystal structure of a transition state mimic for Tdp1 assembled from vanadate, DNA, and a topoisomerase I-derived peptide. *Chem. Biol.* **2003**, *10*, 139–147. [CrossRef]
46. Davies, D.R.; Interthal, H.; Champoux, J.J.; Hol, W.G. The crystal structure of human tyrosyl-DNA phosphodiesterase, Tdp1. *Structure* **2002**, *10*, 237–248. [CrossRef]
47. Interthal, H.; Chen, H.J.; Kehl-Fie, T.E.; Zotzmann, J.; Leppard, J.B.; Champoux, J.J. SCAN1 mutant Tdp1 accumulates the enzyme–DNA intermediate and causes camptothecin hypersensitivity. *EMBO J.* **2005**, *24*, 2224–2233. [CrossRef]
48. Il'ina, I.V.; Dyrkheeva, N.S.; Zakharenko, A.L.; Sidorenko, A.Y.; Li-Zhulanov, N.S.; Korchagina, D.V.; Chand, R.; Ayine-Tora, D.M.; Chepanova, A.A.; Zakharova, O.D.; et al. Design, Synthesis, and Biological Investigation of Novel Classes of 3-Carene-Derived Potent Inhibitors of TDP1. *Molecules* **2020**, *25*, 3496. [CrossRef]
49. Dyrkheeva, N.S.; Filimonov, A.S.; Luzina, O.A.; Zakharenko, A.L.; Ilina, E.S.; Malakhova, A.A.; Medvedev, S.P.; Reynisson, J.; Volcho, K.P.; Zakian, S.M.; et al. New Hybrid Compounds Combining Fragments of Usnic Acid and Monoterpenoids for Effective Tyrosyl-DNA Phosphodiesterase 1 Inhibition. *Biomolecules* **2021**, *11*, 973. [CrossRef]
50. Polovinka, M.P.; Salakhutdinov, N.F.; Panchenko, M.Y. Method for Preparing Usnic Acid. Patent RU2317076, 20 February 2006.
51. Luzina, O.; Filimonov, A.; Zakharenko, A.; Chepanova, A.; Zakharova, O.; Ilina, E.; Dyrkheeva, N.; Likhatskaya, G.; Salakhutdinov, N.; Lavrink, O. Usnic Acid Conjugates with Monoterpenoids as Potent Tyrosyl-DNA Phosphodiesterase 1 Inhibitors. *J. Nat. Prod.* **2020**, *83*, 2320–2329. [CrossRef]
52. Dyrkheeva, N.; Anarbaev, R.; Lebedeva, N.; Kuprushkin, M.; Kuznetsova, A.; Kuznetsov, N.; Rechkunova, N.; Lavrik, O. Human Tyrosyl-DNA Phosphodiesterase 1 Possesses Transphosphooligonucleotidation Activity With Primary Alcohols. *Front. Cell Dev. Biol.* **2020**, *8*, 604732. [CrossRef] [PubMed]
53. Sukhanova, M.V.; Khodyreva, S.N.; Lavrik, O.I. Poly(ADP-ribose) polymerase-1 inhibits strand-displacement synthesis of DNA catalyzed by DNA polymerase beta. *Biochemistry* **2004**, *69*, 558–568. [CrossRef]
54. Komarova, A.O.; Drenichev, M.S.; Dyrkheeva, N.S.; Kulikova, I.V.; Oslovsky, V.E.; Zakharova, O.D.; Zakharenko, A.L.; Mikhailov, S.N.; Lavrik, O.I. Novel group of tyrosyl-DNA-phosphodiesterase 1 inhibitors based on disaccharide nucleosides as drug prototypes for anti-cancer therapy. *J. Enzym. Inhib. Med. Chem.* **2018**, *33*, 1415–1429. [CrossRef] [PubMed]
55. Sherstyuk, Y.V.; Ivanisenko, N.V.; Zakharenko, A.L.; Sukhanova, M.V.; Peshkov, R.Y.; Eltsov, I.V.; Kutuzov, M.M.; Kurgina, T.A.; Belousova, E.A.; Ivanisenko, V.A.; et al. Design, Synthesis and Molecular Modeling Study of Conjugates of ADP and Morpholino Nucleosides as A Novel Class of Inhibitors of PARP-1, PARP-2 and PARP-3. *Int. J. Mol. Sci.* **2019**, *21*, 214. [CrossRef] [PubMed]

International Journal of *Molecular Sciences*

MDPI

Article

Inhibitors of Nucleotide Excision Repair Decrease UVB-Induced Mutagenesis—An In Vitro Study

Eszter Fidrus [1,2], Csaba Hegedűs [1,2], Eszter Anna Janka [1], György Paragh [3,4], Gabriella Emri [1] and Éva Remenyik [1,*]

1 Department of Dermatology, Faculty of Medicine, University of Debrecen, 98 Nagyerdei Krt, 4032 Debrecen, Hungary; fidrus.eszter@med.unideb.hu (E.F.); hegeduscsaba88@gmail.com (C.H.); janka.eszter.a@gmail.com (E.A.J.); gemri@med.unideb.hu (G.E.)
2 Doctoral School of Health Sciences, University of Debrecen, 4032 Debrecen, Hungary
3 Department of Dermatology, Roswell Park Comprehensive Cancer Center, 665 Elm St, Buffalo, NY 14203, USA; gyorgy.paragh@roswellpark.org
4 Department of Cell Stress Biology, Roswell Park Comprehensive Cancer Center, 665 Elm St, Buffalo, NY 14203, USA
* Correspondence: remenyik@med.unideb.hu; Tel.: +36-52-412-345

Abstract: The high incidence of skin cancers in the Caucasian population is primarily due to the accumulation of DNA damage in epidermal cells induced by chronic ultraviolet B (UVB) exposure. UVB-induced DNA photolesions, including cyclobutane–pyrimidine dimers (CPDs), promote mutations in skin cancer driver genes. In humans, CPDs are repaired by nucleotide excision repair (NER). Several commonly used and investigational medications negatively influence NER in experimental systems. Despite these molecules' ability to decrease NER activity in vitro, the role of these drugs in enhancing skin cancer risk is unclear. In this study, we investigated four molecules (veliparib, resveratrol, spironolactone, and arsenic trioxide) with well-known NER-inhibitory potential in vitro, using UVB-irradiated CHO epithelial and HaCaT immortalized keratinocyte cell lines. Relative CPD levels, hypoxanthine phosphoribosyltransferase gene mutation frequency, cell viability, cell cycle progression, and protein expression were assessed. All four molecules significantly elevated CPD levels in the genome 24 h after UVB irradiation. However, veliparib, spironolactone, and arsenic trioxide reduced the mutagenic potential of UVB, while resveratrol did not alter UVB-induced mutation formation. UVB-induced apoptosis was enhanced by spironolactone and arsenic-trioxide treatment, while veliparib caused significantly prolonged cell cycle arrest and increased autophagy. Spironolactone also enhanced the phosphorylation level of mammalian target of rapamycin (mTOR), while arsenic trioxide modified UVB-driven mitochondrial fission. Resveratrol induced only mild changes in the cellular UVB response. Our results show that chemically inhibited NER does not result in increased mutagenic effects. Furthermore, the UVB-induced mutagenic potential can be paradoxically mitigated by NER-inhibitor molecules. We identified molecular changes in the cellular UVB response after NER-inhibitor treatment, which may compensate for the mitigated DNA repair. Our findings show that metabolic cellular response pathways are essential to consider in evaluating the skin cancer risk–modifying effects of pharmacological compounds.

Keywords: UVB radiation; UVB mutagenesis; nucleotide excision repair (NER); cyclobutane–pyrimidine dimer (CPD) photolesion; veliparib; resveratrol; arsenic trioxide; spironolactone

Citation: Fidrus, E.; Hegedűs, C.; Janka, E.A.; Paragh, G.; Emri, G.; Remenyik, É. Inhibitors of Nucleotide Excision Repair Decrease UVB-Induced Mutagenesis—An In Vitro Study. *Int. J. Mol. Sci.* **2021**, *22*, 1638. https://doi.org/10.3390/ijms22041638

Academic Editors: Konstantin Volcho Olga Lavrik and Ennio Prosperi
Received: 8 December 2020
Accepted: 2 February 2021
Published: 6 February 2021

Publisher's Note: MDPI stays neutral with regard to jurisdictional claims in published maps and institutional affiliations.

1. Introduction

The incidence of melanoma [1–4] and nonmelanoma skin cancers [4,5] is increasing in lighter skin types and is attributed to enhanced exposure of the skin to ultraviolet-B (UVB) [6–8]. UVB radiation induces DNA damage in epidermal cells [9]. The most common UVB-induced DNA changes are pyrimidine–pyrimidone photoproducts (6-4PPs) and cyclobutane–pyrimidine dimers (CPDs) [9,10]. These photolesions disrupt DNA structure by forming stable bonds between two adjacent pyrimidine bases [10,11]. CPDs form up to five times more frequently after UVB radiation than 6-4PPs [12,13], and CPDs are the leading cause of UV-signature mutations, specific markers for UV-induced DNA damage [11]. Wei et al. showed that CPDs show different accumulation throughout the genome, as enrichment of UV-signature mutations on specific genetic locations (mutational hotspots) can be detected [14,15].

In human cells, UVB-induced photolesions are repaired by nucleotide excision repair (NER) [16]. NER is highly effective in the repair of 6-4PPs, but less effective in repairing CPDs. Nakagawa et al. showed that 6 h after UVB irradiation, 6-4PPs are completely removed [17], while more than 40% of UVB-induced CPDs are left unrepaired even 24 h after UVB injury [17]. Furthermore, mutation data from epidermal cancer suggest that UVB-induced molecular changes are mainly attributed to CPDs [9,10,18]. Defects in the repair process cause rare genetic disorders, including xeroderma pigmentosum (XP), Cockayne syndrome (CS), and trichothiodystrophy (TTH). Patients with these diseases are extremely sensitive to sunlight and, in the case of XP, the risk of skin cancer at an early age is very high [19].

NER function can be both positively and negatively influenced by various chemical agents. Nicotinamide [20,21] and some plant derivatives [22–25] were shown to enhance NER activity in vitro. However, some chemical agents impair DNA repair [26–32], raising the possibility of enhanced skin cancer risk in exposed individuals. Among the repair-inhibiting drugs, veliparib, an inhibitor of poly [ADP-ribose] polymerase 1 (PARP1), is currently in clinical trials to target different malignancies [33–36]. Spironolactone [37,38] and arsenic trioxide [39–41] are used in clinical practice for their diuretic and chemotherapeutic properties, respectively. Resveratrol, a natural phytophenol, is a promising compound in UV protection via its anti-inflammatory [42–45], anti-oxidant [45,46], and anti-carcinogenic [47–49] effects. Although these chemicals are widely used, there are no in vivo data examining the risk of UVB-induced tumorigenesis in treated individuals or the role of NER inhibition in this risk. NER functionality may have no linear and obligate relationship with UVB-driven mutagenesis; other factors may also influence photocarcinogenesis.

In this study, we aimed to investigate the mutagenic effects of four chemical agents—veliparib, resveratrol, spironolactone, and arsenic trioxide—with known in vitro NER-inhibitory properties [26,29,31,50]. Since the ability of these molecules to impair NER function is well proven, our aim was to assess whether decreased repair function and increased CPD accumulation by the treatment of the tested chemicals lead to enhanced mutagenesis of epithelial cells in vitro. In addition to their impact on CPD formation and UV-induced mutation burden, we identified other molecular pathways (apoptosis, cell cycle progression, or autophagy) modified by these molecules, which have significant role in cellular UVB response.

2. Results

2.1. All Tested Chemicals Enhance CPD Formation after UVB Irradiation

To verify the NER-inhibitory effect of veliparib, resveratrol, arsenic trioxide, and spironolactone, CHO cells were pretreated with the chemicals and irradiated with 20 mJ/cm^2 UVB. In our previous study, we evaluated the kinetics of CPD removal after UVB irradiation, and we found that most of the UV-induced lesions (~60%) are eliminated from the DNA in the first 24 h [51]. According to this and other studies aiming to assess repair efficacy [17,52], we chose to measure the relative CPD content of the cells 24 h after UVB.

Twenty-four hours post-UVB, a large number of CPDs remained in the cellular DNA reflecting the slow repair of CPD lesions by NER [17]. The relative amount of CPDs was significantly higher in the treated groups compared to that in the nontreated counterparts. In many cases, the number of CPD lesions showed more than a 50% increase, e.g., the detectable CPD amount was 88% higher after 4 μg/mL arsenic-trioxide treatment compared with that in the sham-treated group (Figure 1). These findings are consistent with other studies showing that these molecules impair the removal of CPDs [26,28,29,31,32,50]. To validate these data, we repeated the experiments in HaCaT human keratinocyte cell line using the most effective concentrations. In our previous study, we have already presented that veliparib treatment reduces CPD repair in this cell line [29], and we also found similar results by resveratrol, spironolactone, and arsenic-trioxide treatment, too (Figure S1).

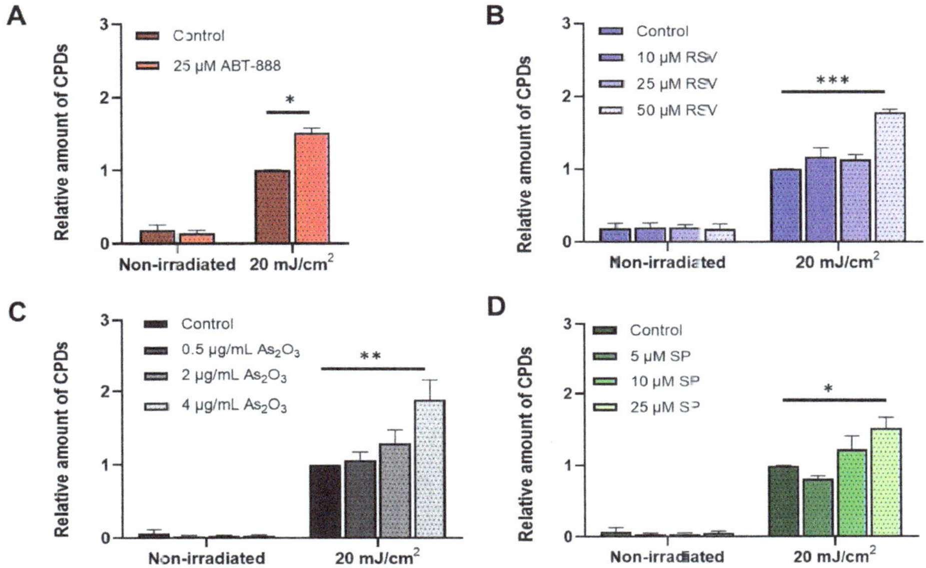

Figure 1. Relative number of cyclobutane–pyrimidine dimer (CPD) photolesions after ultraviolet B (UVB) irradiation. CHO cells were pretreated with (**A**) veliparib (ABT-888), (**B**) resveratrol (RSV), (**C**) arsenic trioxide (As_2O_3), or (**D**) spironolactone (SP), and then irradiated with 20 mJ/cm^2 UVB. CPD lesions were detected by CPD-specific ELISA 24 h after the irradiation. CPD amounts were normalized to UVB-irradiated vehicle controls. We present the mean ± SEM of at least three independent experiments. *, **, and *** indicate statistically significant difference at $p < 0.05$, $p < 0.01$, and $p < 0.001$.

2.2. CPD Accumulation and UVB-Induced Mutagenesis Show a Nonlinear Relationship

To evaluate the mutagenic effect of UVB radiation, *HPRT* gene mutation assays were performed [53]. This assay detects cells carrying heritable mutations in the *HPRT* gene. First, we assessed the dose dependence of CPD accumulation. Our results show that CPD levels, detected 24 h after UVB irradiation, increased linearly with the UVB dose (Figure 2A) [54]. Interestingly, UVB-induced mutagenesis did not exhibit a linear dose–response relationship with CPD accumulation. UVB-induced *HPRT* mutation frequency increased from 0 to 10 mJ/cm^2 UVB, then the mutational rate dropped at 15 mJ/cm^2 (Figure 2B), suggesting that UVB doses with lower cytotoxic effects are more mutagenic [55]. Based on this observation, we chose 10 mJ/cm^2 UVB in our experimental system for *HPRT* mutation detection. Figure S2 presents the linear reduction of viability after increasing UVB doses.

2.3. Veliparib, Arsenic-Trioxide, and Spironolactone Treatments Prevent UVB-Induced Mutagenesis

To assess the effects of chemically induced NER inhibition on UVB mutagenesis (Figure 3), we examined whether the compounds increased the mutation frequency of the HPRT gene after UVB irradiation. Since HaCaT cells were extremely intolerant to 6-thioguanine (6-TG) selection medium and formed a very low number of colonies after UVB radiation, HPRT assay was only carried out on CHO cells, which is the cell line conventionally used for measuring HPRT mutagenesis [53]. Contrary to expectations, we found that veliparib, arsenic-trioxide, and spironolactone treatment decreased the number of cells carrying nonfunctional mutations in their HPRT gene after UVB exposure (Figure 3A,B,D,E). Higher concentrations of the treatment compounds decreased the mutation rates almost to baseline, which was surprising in light of the increased CPD content with the same agents 24 h after the UVB exposure. The fourth molecule, resveratrol, caused nonsignificant increases in UVB-induced mutagenesis (Figure 3C).

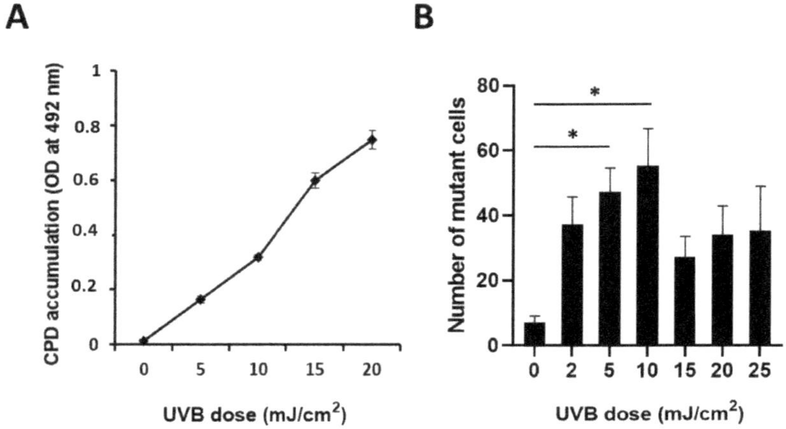

Figure 2. CPD accumulation and UVB-induced mutagenesis following different doses of UVB exposure. (**A**) Accumulation of CPD photolesions with increasing doses of UVB radiation in CHO cells. CPDs were detected 24 h after UVB exposure. Mean ± SEM ($n = 3$). (**B**) Number of mutated cells in response to increasing doses of UVB radiation in CHO cells after a 10-day culture in selective medium containing 6-thioguanine. Mean ± SEM, $n = 6$. * indicates statistically significant difference at $p < 0.05$.

2.4. Resveratrol, Arsenic Trioxide, and Spironolactone Enhanced UVB-Induced Apoptosis

While chemically induced attenuation of NER was triggered by the test compounds, some of them also showed marked anti-mutagenic activity. Therefore, we aimed to investigate other mechanisms that can affect UVB-induced mutations. Arsenic derivatives have well-known cytotoxic effects [56,57], especially in combination with other mutagens, such as UVB [58]. Apoptosis serves as a protective mechanism to diminish mutant clone formation [59]. Therefore, we hypothesized that the applied chemicals may induce an elevated photosensitive response in the UV-exposed cells, thereby inducing the clearance of cells with unrepaired DNA damage, resulting in decreased UVB-induced mutagenesis.

Because 10 mJ/cm^2 UVB (used for HPRT mutagenesis assay) caused a very moderate decrease in cell viability (more than 80% of the cells left viable) (Figure S2), we decided to choose 20 mJ/cm^2 for apoptosis measurements. To verify that the UVB dose does not influence the effects of a chemical treatment on HPRT mutagenesis, we repeated HPRT mutation assays with two of the most anti-mutagenic treatments, 25 μM ABT-888 and 25 μM SP. Although 20 mJ/cm^2 was less mutagenic than 10 mJ/cm^2, the direction of the changes after inhibitor treatments remained the same (Figure S3).

CHO cells were pretreated with the drugs and exposed to UVB. Subsequently, i.e., 48 h after irradiation, cells were stained with Alexa Fluor 488–conjugated Annexin V (AV) and propidium iodide (PI). Viable, apoptotic, and necrotic subpopulations were determined by flow cytometry. Both arsenic-trioxide and spironolactone treatments increased the proportion of dead cells in response to UVB. At higher treatment concentrations (4 μg/mL arsenic trioxide or 25 μM spironolactone), the mean percentage of living cells was between 11% and 23%. In the group exposed to UVB only, more than 50% of the cells were viable 48 h after the irradiation (Figure 4E–H). Resveratrol, the only tested molecule without detectable anti-mutagenic properties, induced only a mild increase in UV-induced apoptosis at high concentrations (Figure 4C,D); this is probably one of the reasons for the unaltered mutagenic response. Surprisingly, veliparib, a molecule with high DNA-repair-inhibitory properties and anti-mutagenic effects, caused no alterations in cell viability after UVB (Figure 4A,B). In HaCaT cells, we observed very similar alterations by the treatments (Figure S4); however, veliparib caused a mild decrease in cell viability in our previous study [29].

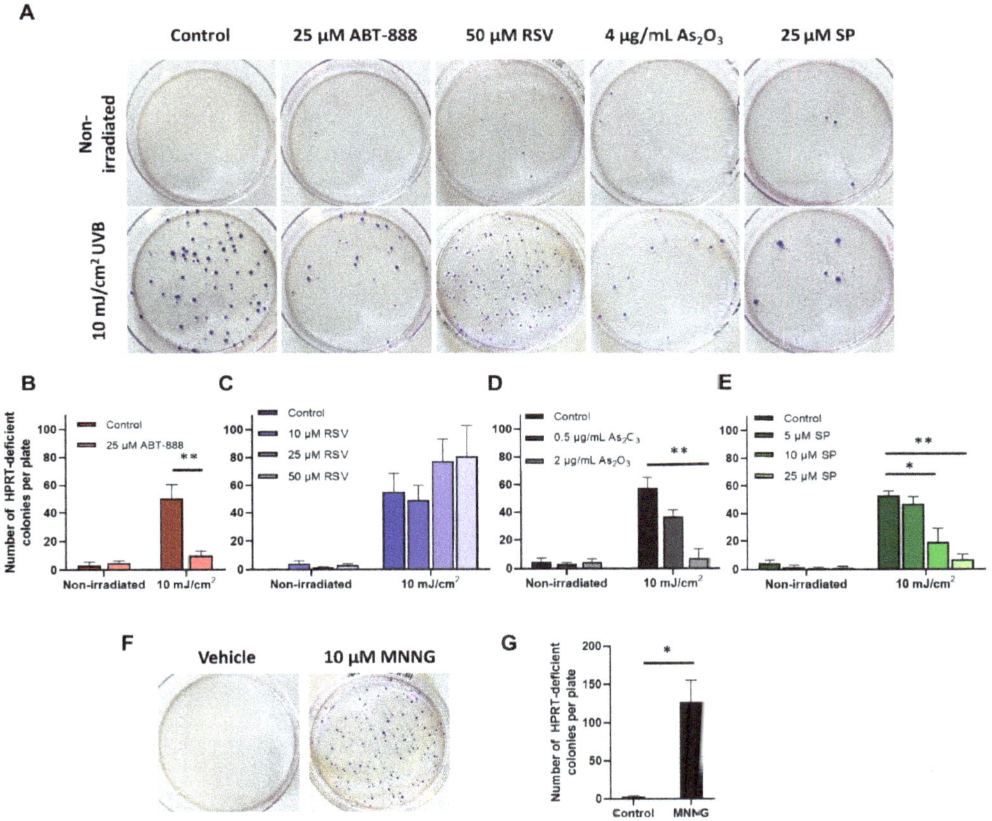

Figure 3. Effects of nucleotide excision repair (NER) inhibitors on UVB-induced HPRT gene mutation burden. (**A**) HPRT gene mutation assay after UVB and NER-inhibitor treatments. Cells were pretreated and exposed to 10 mJ/cm^2 UVB radiation. HPRT mutant cells were selected in Dulbecco's modified Eagle media (DMEM) containing 5 µM 6-thioguanine. One representative experiment from at least three independent measurements is presented. (**B–E**) Mean ± SEM of ≥3 HPRT gene mutation assays after treatment with (**B**) veliparib (ABT-888), (**C**) resveratrol (RSV), (**D**) arsenic trioxide (As_2O_3), and (**E**) spironolactone (SP). (**F**,**G**) 10 µM 1-methyl-3-nitro-1-nitrosoguanidine (MNNG) was used as a positive control for the assays. * and ** indicate statistically significant differences at $p < 0.05$ and $p < 0.01$.

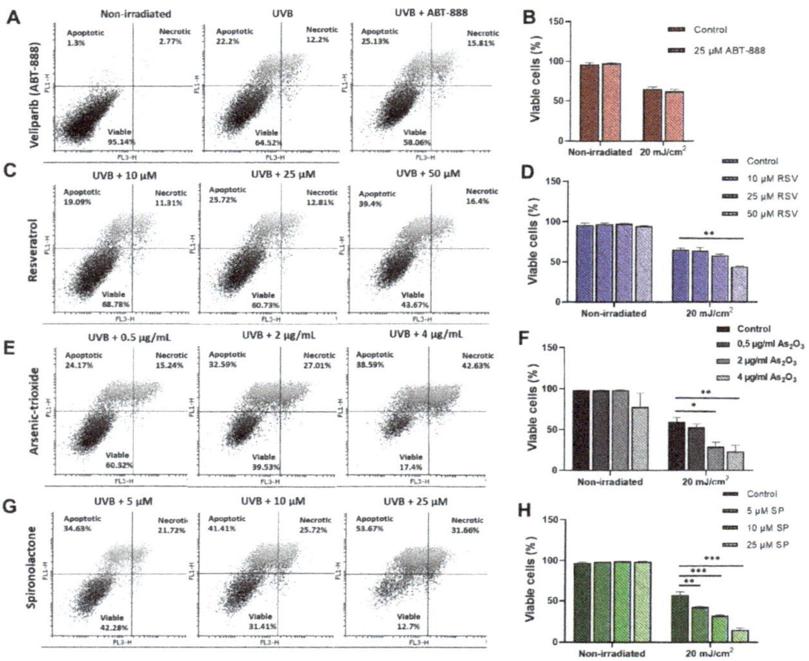

Figure 4. Changes in UVB-induced alterations in cell viability induced by pretreatment with NER-inhibitory molecules. Cells were pretreated with the indicated concentrations of (**A**,**B**) veliparib (ABT-888), (**C**,**D**) resveratrol (RSV), (**E**,**F**) arsenic trioxide (As_2O_3), or (**G**,**F**) spironolactone (SP) and exposed to 20 mJ/cm^2 UVB or left unexposed (nonirradiated). After 48 h of irradiation, apoptotic (Annexin V+/propidium iodide (PI−), necrotic (Annexin V+/PI+), and viable (Annexin V−/PI−) cells were detected by flow cytometry. (**B**,**D**,**F**,**H**) Bars represent the percentage of living cells 48 h after UVB exposure after inhibitor treatments. We calculated the mean ± SEM of three independent experiments, where *, **, and *** denote statistically significant differences at $p < 0.05$, $p < 0.01$, and $p < 0.001$.

2.5. Veliparib and Resveratrol Augment UVB-Induced Cell Cycle Arrest

Besides apoptosis, cell cycle arrest is one of the main cellular mechanisms that can attenuate the long-term effects of mutagenic exposure. During this process, cell division is halted in cells with unrepaired DNA lesions, extending the time for repair [60,61]. To assess whether the tested molecules can modify UVB-induced cell cycle arrest, we analyzed cell cycle progression of CHO cells 1, 3, and 6 days after 20 mJ/cm^2 UVB exposure. One day post-UVB, a large number of cells was detected in the G_2/M phases in every UV-irradiated group, consistent with previous findings showing that UVB-radiation-induced cell cycle arrest is mainly manifested at G_2/M [62]. Restoration of the cell cycle began at 3 days after UVB exposure and was nearly indistinguishable from the nonirradiated group 6 days after UVB radiation. When cells were pretreated with 25 µM veliparib, the percentage of cells in the G_2/M phase showed a mild but statistically significant increase compared to that in the vehicle control. The increase lasted up to 6 days after the exposure (Figure 5A,B), suggesting that veliparib treatment extends the recovery time of cells from DNA damage, which possibly contributes to its anti-mutagenic effect. The increase of UVB-induced cell cycle block and decreased proliferative capacity after veliparib treatment were also observed in HaCaT cells [29]. Resveratrol also caused a moderate elevation in the G_2/M block 3 days after the UVB, but this difference was only short-term and was not detectable

6 days after UVB exposure (Figure 5C). HaCaT cells did not show change in cell cycle progression after resveratrol treatment (Figure S5). Arsenic-trioxide and spironolactone treatments did not affect cell cycle progression after UVB either in CHO (Figure S6A,B) or in HaCaT cells (Figure S6C,D).

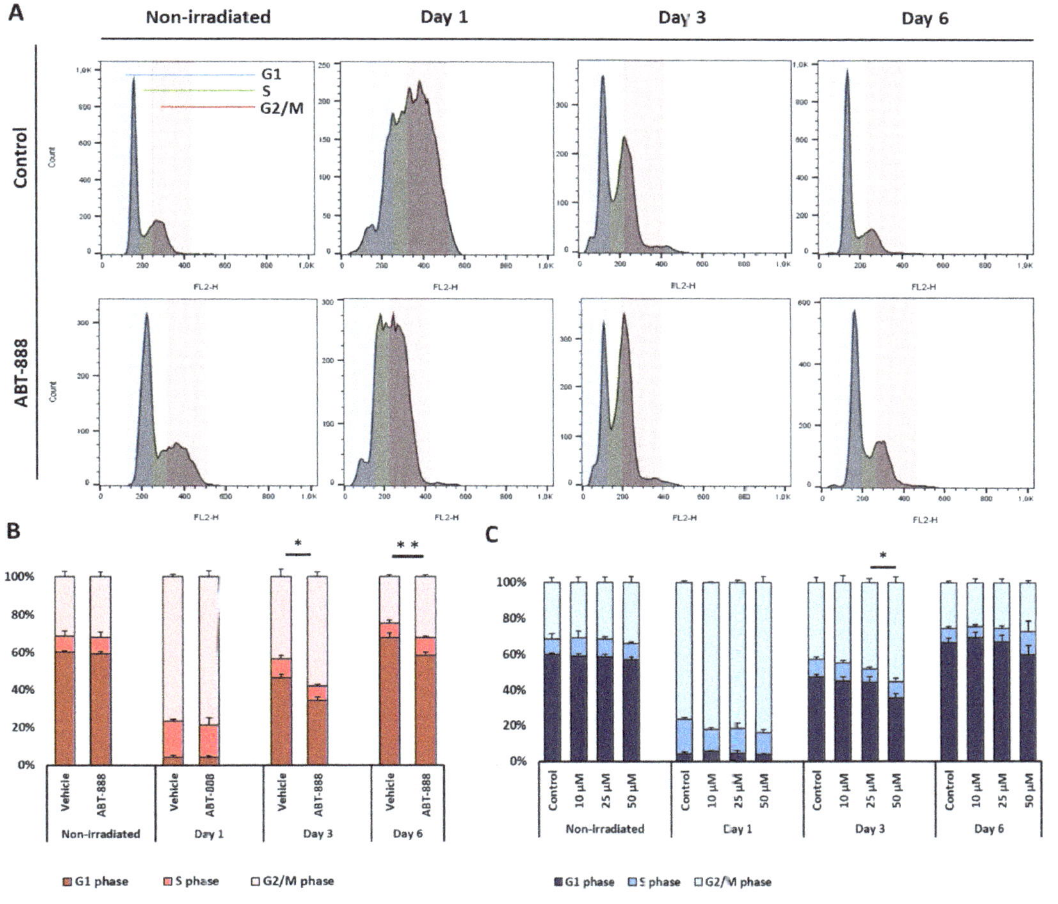

Figure 5. Cell cycle progression after veliparib and resveratrol treatment followed by UVB irradiation. (**A**) Cells were treated with 25 µM veliparib (ABT-888) and exposed to 20 mJ/cm^2 UVB or left unexposed (nonirradiated). Cell cycle progression was analyzed 1, 3, and 6 days after exposure using propidium-iodide staining followed by flow cytometry. G1 (blue), S (green), and G2/M (red) phase cell populations were distinguished, as indicated. (**B**) The distribution of cells in different phases was calculated after veliparib and (**C**) different doses of resveratrol treatment (mean ± SEM; $n = 3$). * and ** denote statistically significant differences at $p < 0.05$ and $p < 0.01$, respectively.

2.6. Altered Protein Expression in Diverse Stress–Response Pathways May Orchestrate UVB-Induced Mutagenesis

We aimed to identify general upstream regulators at the protein level that can shed light on the anti-mutagenic nature of the compounds. Thus, we looked for changes in the expression or activation levels of proteins involved in key pathways linked to DNA damage with UVB-induced mutagenesis. First, we measured the phosphorylation level of p53 protein and the expression of the phosphatidylethanolamine conjugated form of microtubule-associated protein 1A/1B-light chain 3 protein (LC3-I-II). Phospho-p53 and LC3-I and II are widely accepted markers of DNA damage sensing and cellular autophagy, respectively [63,64]. Western blot analysis revealed that LC3-II expression was increased after UVB exposure, and veliparib treatment enhanced these effects 20 h post-UVB (Figure 6A,B). Thus, PARP1 inhibition by veliparib promotes UVB-induced autophagy [29]. Similar to LC3 expression, p53 phosphorylation levels increased after UVB irradiation and this effect was augmented by PARP inhibition (Figure 6A,C). Total p53 expression was also enhanced by UVB exposure, but veliparib treatment did not affect it (Figure 6A). No differences in LC3 expression (Figure S7A–C) or p53 phosphorylation levels (Figure S7D–F) were detected after resveratrol, spironolactone, or arsenic-trioxide treatments, suggesting that the observed cellular changes were mediated by p53-independent mechanisms in these cases.

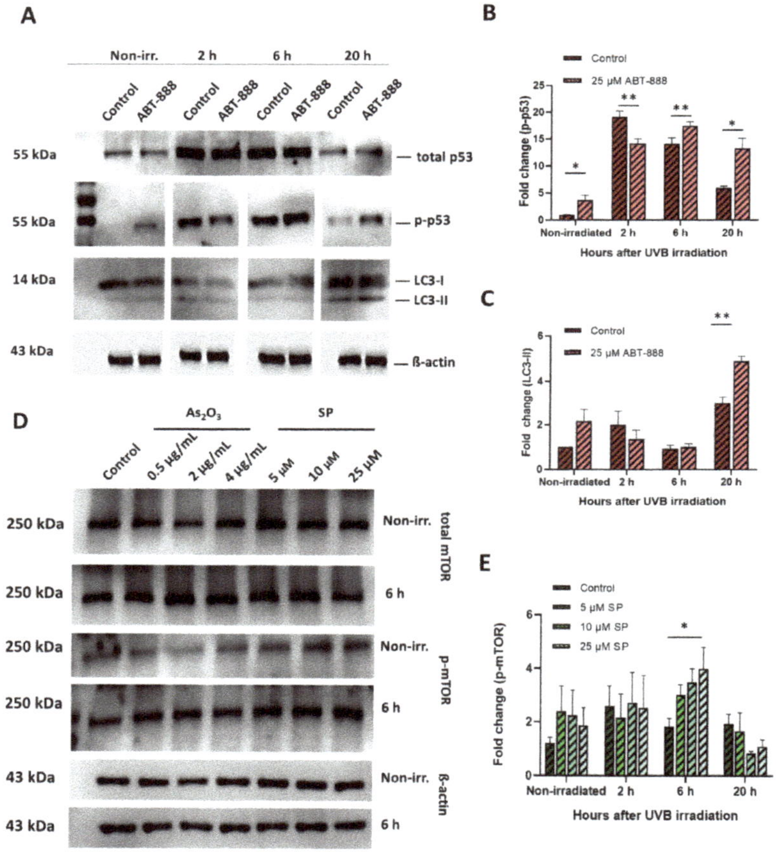

Figure 6. Protein expression and activation changes after veliparib (ABT-888), arsenic-trioxide (As_2O_3), or spironolactone (SP) treatment followed by UVB irradiation. Cells were treated with 25 μM veliparib (ABT-888) or different concentrations of As_2O_3 or SP and exposed to 20 mJ/cm^2 UVB or left unexposed (nonirradiated). Control cells were UVB irradiated without any additional treatment. (**A**) Level of total tumor protein 53 (p53), phospho-p53 (p-p53) and microtubule-associated protein 1A/1B-light chain 3 protein (LC3-I and LC3-II) following ABT-888 treatment and (**D**) total mammalian target of rapamycin (mTOR) and phosphorylated mTOR (p-mTOR) (following SP or As_2O_3 treatment) were detected by Western blotting. Bars represent the mean of (**B**) $n = 6$ for p53 phosphorylation; (**C**) $n = 3$ for LC3-II expression after ABT-888 treatment and (**E**) $n = 4$ for mTOR phosphorylation after SP or As_2O_3 treatment. Data were normalized to ß-actin. * and ** denote statistically significant differences at $p < 0.05$ and $p < 0.01$, respectively.

In addition to the p53 pathway, activation of the mammalian target of rapamycin (mTOR) signaling is another key mechanism for the regulation of diverse stress–response pathways, including apoptosis, senescence, and autophagy. In contrast to p53, phosphorylated mTOR promotes the survival and proliferation of UVB-exposed cells [65]. We found, that arsenic-trioxide and spironolactone treatments increased mTOR phosphorylation 6 h after UVB, but this increase was only statistically significant after spironolactone treatment. Total mTOR expression was not affected (Figure 6D,E). The other two chemicals did not affect mTOR phosphorylation after UVB (Figure S8A–C).

3. Discussion

Gross NER defects in XP patients result in accelerated photocarcinogenesis. Several human medications are effective inhibitors of NER [26,28,29,31], raising the possibility that the clinical application of these molecules enhances photocarcinogenesis. Although some of these molecules are widely used in clinical practice in vivo therapeutic application of these compounds is not associated with increased risk of photocarcinogenesis [35,66,67].

We confirmed that all these test molecules impair the elimination of CPD photolesions from DNA after UVB irradiation. Three molecules—veliparib, spironolactone, and arsenic trioxide—exerted strong anti-mutagenic effects (*HPRT* gene mutation assay), while resveratrol not affected UVB-induced mutation formation. Spironolactone and arsenic trioxide induced a marked loss in cell viability upon UVB exposure, while inhibition of PARP1 by veliparib caused a prolongation of UVB-induced cell cycle arrest.

The detected anti-mutagenic effects of the tested compounds could be interesting in the aspect of synthetic lethality, which is a promising therapeutical approach for the treatment of various cancers [68]. Inhibitors of PARP1 are already investigated due to their synthetic lethal effects with the combination of pre-existing *BRCA1/2* (Breast Cancer gene 1/2) loss-of-function mutations [69,70]. Our findings that these inhibitors are able to reduce the survival of cells carrying mutations, also support the hypothesis that selective elimination of cancerous cells can be achieved by inducing defects in cellular repair pathways, supplementing the deleterious effects of other genetic deficiencies. The possibility of testing the other anti-mutagenic chemicals (besides veliparib) as synthetic lethal compounds should be also considered—especially in the case of spironolactone, which has markedly less harmful side effects compared to arsenic trioxide.

These results suggest, that possible synthetic lethal properties of the chemicals can originate from their ability to inhibit specific subunits of the NER complex. Although PARP1 is involved in the recognition of CPDs and thus in the initiation of the NER process through the activation of DDB2 (DNA damage-binding protein 2) [71,72] and XPC (Xeroderma pigmentosum, complementation group C) [73], this multi-faceted protein also plays a role in the regulation of other repair pathways. For example, PARP1 has been shown to interact with CSB (Cockayne syndrome group B), a protein involved in both transcription-coupled NER and base excision repair (BER) [71,74]. Thus, its synthetic lethality is may be linked to a complex dysregulation of DNA repair. In contrast, spironolactone was found to inhibit NER by inducing the rapid proteosomal degradation of the XPB (Xeroderma pigmentosum, complementation group B) subunit [31,32], which suggests a more specific interaction between pre-existing mutations and defective NER repair resulting in the elimination of genetically damaged cells. Arsenic trioxide was also shown to inhibit the XPC subunit [28,50], but this molecule has other versatile effects. For instance, arsenic trioxide regulates the survival of damaged cells through the induction of Bax/caspase-3 pathway [57,75], ROS (relative oxygen species) production [57], and the downregulation of survivin [75]. Thus, its lethal effects on UV-exposed cells cannot be merely explained by XPC inhibition. The underlying mechanism of resveratrol-induced defect in CPD removal is less known, but Keuser et al. found a significant compaction of the chromatin structure after resveratrol treatment. Resveratrol induces the activation of sirtuins (SIRTs), which are members of the class III histone deacetylases (HDACs), and regulates various cellular pathways including DNA damage recognition and repair through the deacetylation of target proteins. Furthermore, deacetylation-mediated chromosome remodeling via SIRTs may influence the accessibility of repair proteins to the damaged DNA and result in the accumulation of CPDs and other types of DNA lesions, as well [26].

P53 is considered to be the guardian of the genome; it regulates the balance between pro- and anti-apoptotic signals during the stress response [64]. In our study, veliparib significantly enhanced UV-induced p53 phosphorylation. This is intriguing in the view of other studies, which found that the application of PARP1 inhibitors in cancer therapy is more effective in p53-deficient cells [76,77]. This is a result of direct interaction between PARP1 and p53 protein, where the status of p53 (gain-of function/loss-of-function) dictates the outcome of cell survival. For example, loss-of-function mutations of p53 were associated with higher sensitivity to PARPi (inhibitor of poly-(adenosine diphosphate [ADP]) ribose polymerase) [78–80]. These results suggest that increased activation or gain-of-function mutation of p53 following chemically induced PARP1 inhibition may serve to protect cells from intensive genome instability due to the loss of PARP1 function.

Except veliparib, none of the other molecules caused significant alterations in p53 phosphorylation, suggesting that changes are at least partly mediated by p53-independent pathways in these cases. The phosphorylation of mTOR, involved in cell survival, was increased by spironolactone, in contradiction to the strong apoptotic response. We hypothesize that increased mTOR phosphorylation is a cellular strategy to counteract the cytotoxic effect of SP.

Autophagy represents another key strategy in cell survival [64], which was enhanced by ABT-888 treatment. However, the role of autophagy in tumorigenesis is quite controversial. While autophagy helps remove damaged cellular organelles and thereby prevents tumor formation, it can also fuel metabolism by recycling damaged molecules to promote the survival of pre-cancerous cells during metabolic stress in a nutrient-deficient environment [81].

We observed that the mutagenic effect of UVB and CPD accumulation did not exhibit parallel increases, as UVB-induced mutagenesis decreased at higher UVB doses. This shows that there is no obligate linear relationship between repair activity and UVB mutagenesis, but other factors should be also considered while assessing the possible risk of skin cancer induction by a chemical treatment. Alterations in the cellular UVB response, such as elevated apoptosis, cell cycle arrest, and autophagy, may play a significant role in counterbalancing the negative effect of repair inhibition on mutagenesis.

In conclusion, we demonstrated that UVB-induced mutagenesis is a highly complex process even in an in vitro model, and decreased cellular repair activity does not necessarily result in elevated mutagenicity. These results suggest that, if a molecule inhibits DNA repair in vitro, its effects on other cellular processes also need to be assessed before its mutagenic potential can be predicted. Furthermore, DNA repair inhibitors need not be considered necessarily mutagenic. We also showed that three out of four compounds reduced UVB-induced mutagenesis in vitro, which suggest that further in vivo studies are warranted to establish the safety of these molecules.

4. Materials and Methods

4.1. Cell Culture

CHO-K1 (Chinese hamster ovary; ATCC, Manassas, VA, USA) and HaCaT (immortalized human keratinocyte; ATCC, Manassas, VA, USA) cells were cultured in 4500 mg/L glucose containing Dulbecco's modified Eagle media (DMEM, Biosera, Budapest, Hungary) supplemented with L-glutamine (Biosera, Budapest, Hungary), 10% heat-inactivated fetal bovine serum (FBS; Biosera, Budapest, Hungary), and 0.5% antibiotic/antimycotic solution (Biosera, Budapest, Hungary). Cells were maintained in a humidified incubator at 37 °C with 5% CO_2 atmosphere.

4.2. Cell Treatment

Cells were harvested with trypsin-EDTA (Biosera, Budapest, Hungary) and then seeded in six-well plates for the hypoxanthine phosphoribosyltransferase (*HPRT*) gene mutation assay or 24-well plates for all other experiments. At ~80% confluence, cells were pretreated with 25 µM ABT-888 (PARP1 inhibitor, veliparib, Selleckchem, Houston, TX, USA), 10–50 µM resveratrol (Abcam, Cambridge, UK), 5–25 µM spironolactone (Selleckchem, Houston, TX, USA), or 0.5–4 µg/mL As_2O_3 (Sigma-Aldrich, St. Louis, MO, USA) solution. In the case of veliparib treatment, we chose the concentration that caused marked inhibition of PARP1 protein, according to our previous [29] and current experiments (Figure S9). For the other chemicals, we identified three different concentrations due to their more complex and not fully understood mode of action—based on prior published data [26,27,30–32]. As_2O_3 was dissolved in 1 M NaOH (Sigma-Aldrich, St. Louis, MO, USA) and diluted in Dulbecco's phosphate-buffered saline (DPBS; Biosera, Budapest, Hungary). Other chemicals were dissolved in dimethyl sulfoxide (DMSO, Sigma-Aldrich, St. Louis, MO, USA). Pretreated cells were incubated for 120 min at 37 °C before UVB irradiation.

4.3. UVB Irradiation

After pretreatment and incubation, the culture medium was removed, and cells were covered with a thin layer of DPBS. Cells were irradiated with 20 mJ/cm^2 UVB, using two UVB broadband tubes (TL-20W/12 RS; Philips, Eindhoven, The Netherlands). For the *HPRT* gene mutation assay, the UVB dose was reduced to 10 mJ/cm^2, as it was found to be the most mutagenic for CHO cells (Figure 2). The proper dosage of UVB was determined by a UVX Digital Radiometer (UVP Inc., San Gabriel, CA, USA). After irradiation, the DPBS was replaced by DMEM supplemented as described above.

4.4. CPD-Specific Enzyme-Linked Immunosorbent Assay (ELISA)

A CPD-specific ELISA was performed as previously described by Boros et al. [82]. Genomic DNA was extracted by an Invitrogen™ PureLink™ Genomic DNA Mini Kit (Thermo Fisher Scientific, Waltham, MA, USA) 24 h after the UVA irradiation. Flat-bottomed 96-well plates were coated with 0.003% protamine-sulfate and incubated at 37 °C to completely dry. DNA was denatured at 100 °C for 10 min, then immediately chilled on ice for 15 min. Denatured DNA was distributed to wells in triplicate (15 ng DNA to each well) and incubated at 37 °C overnight. Plates were washed with PBS (Biosera, Budapest, Hungary) containing 0.05% Tween-20 (Amresco, Solon, OH, USA) (PBS-T) and incubated with 150 µL/well 5% FBS at 37 °C for 30 min to prevent nonspecific antibody binding. After washing plates three times with PBS-T, anti-CPD monoclonal antibody (clone TDM-2, dilution 1:1500, Cosmo Bio Co., Ltd., Tokyo, Japan) was added to each well. Plates were incubated at 37 °C for 60 min. After washing three times, HRP-conjugated anti-mouse IgG secondary antibody (dilution 1:3000, Bio-Rad, Hercules, CA, USA) was added and plates were incubated at 37 °C for 30 min. Plates were washed three times with PBS-T and once with 150 µL/well citrate-phosphate buffer (0.51% $C_6H_8O_7.H_2O$ (Sigma-Aldrich, St. Louis, MO, USA) and 0.73% Na_2HPO_4 (Sigma-Aldrich, St. Louis, MO, USA) in distilled water; pH 5.0). Substrate solution (0.04% o-phenylenediamine (Sigma-Aldrich, St. Louis, MO, USA) and 0.006% H_2O_2 (Sigma-Aldrich, St. Louis, MO, USA) dissolved in citrate-phosphate buffer with H_2O_2 added to the solution, when o-phenylenediamine was completely dissolved) was added to each well. Plates were incubated until the appropriate color intensity appeared. To stop the enzyme reaction, 50 µL/well 2 N H_2SO_4 (Sigma-Aldrich, St. Louis, MO, USA) was added to each well. Absorbance was measured at 492 nm using an Epoch Microplate Spectrophotometer (BioTek, Budapest, Hungary).

4.5. HPRT Gene Mutation Assay

CHO cells were cultured in DMEM containing HAT (hypoxanthine–aminopterin–thymidine; HAT Media Supplement (50×) Hybri-Max™; Sigma-Aldrich, St. Louis, MO, USA) for a week to eliminate preexisting HPRT-mutant cells from the culture. CHO cells were treated with the previously specified inhibitor molecules and exposed to 0–25 mJ/cm^2 UVB. Cells were cultured for one more week and then harvested with trypsin-EDTA (Biosera, Budapest, Hungary). In the case of each sample, an equal number of cells (1 × 10^6) were seeded into 100 mm Petri dishes in selective DMEM containing 5 μM 6-thioguanine (6-TG; Sigma-Aldrich, St. Louis, MO, USA). The 6-TG-resistant cells were allowed to form visible clones for 10 days. Clones were washed with PBS, fixed with 100% methanol (Sigma-Aldrich, St. Louis, MO, USA) for 10 min, and stained with May–Grünwald–Giemsa (Molar Chemicals, Halásztelek, HU, Hungary). HPRT-mutant colonies were counted. For the positive control, 10 μM 1-methyl-3-nitro-1-nitrosoguanidine (MNNG; TCI Europe N.V., Zwijndrecht, Belgium) was used.

4.6. *Apoptosis Assay*

Cell viability was measured 48 h after UVB irradiation using Alexa Fluor 488–conjugated Annexin V/propidium iodide (PI) dual staining (apoptosis assay, Alexa Fluor^{TM} 488 Annexin V/Dead Cell Apoptosis Kit, Thermo Fisher Scientific, Waltham, MA, USA). The supernatant of the cells was collected, living cells were harvested with trypsin-EDTA and added to the supernatant. To avoid the loss of apoptotic cells, cell culture media was not changed between UVB exposure and viability measurement. Cells were labeled according to the manufacturer's instructions. Cells were analyzed by flow cytometry using a FACS Calibur (Becton Dickinson, San Jose, CA, USA) flow cytometer and CellQuestPro software 5.2 (Becton Dickinson, San Jose, CA, USA). Fluorescence intensity was measured in the FL-1 (for Annexin V) and FL-3 (for PI) channels. For data evaluation, FlowJo 10.6.2. (Becton Dickinson, San Jose, CA, USA) flow cytometry software was used.

4.7. *Cell Cycle Analysis*

Cell cycle progression was quantified 1, 3, and 6 days after UVB irradiation. Cells were harvested with trypsin-EDTA, washed with DPBS, and fixed with ice-cold 80% ethanol (VWR, Radnor, PA, USA). Equal numbers of cells were centrifuged at 3500 rpm, for 5 min and re-suspended in 50 μL DPBS containing 0.2 mg/mL RNase A (Sigma-Aldrich, St. Louis, MO, USA), 0.1 μL Triton-X 100 (Amresco, Solon, OH, USA), and 5 mg/mL PI (Thermo Fisher Scientific, Waltham, MA, USA). Samples were incubated at 37 °C for 45 min and then supplemented with 0.5% bovine serum albumin (BSA; VWR, Radnor, PA, USA). Cell cycle progression was analyzed by flow cytometry with an FACS Calibur and fluorescence was measured on the x-axis in the FL2-A channel. Doublet discrimination was performed for single-cell analysis. FlowJo software was used for analyzing the data.

4.8. *Western Blot*

Cells were lysed in RIPA (Radioimmunoprecipitation assay) buffer containing protease-inhibitor cocktail (dilution 1:1000) 2, 6, or 24 h after UVB irradiation. Lysates were centrifuged at 15,000 rpm for 5 min at 4 °C. Protein concentration in the supernatants was measured using a Pierce BCA (Bicinchoninic acid) assay kit (Thermo Fisher Scientific, Waltham, MA, USA). Protein samples were mixed with 5× loading buffer (bromophenol blue (0.25%), β-mercaptoethanol (5%; Sigma-Aldrich, St. Louis, MO, USA), glycerol (50%; Sigma-Aldrich, St. Louis, MO, USA), SDS (sodium dodecyl sulfate; 10%; Duchefa Biochemie, Haarlem, The Neatherlands), Tris-HCl (0.25 M, pH 6.8; Sigma-Aldrich, St. Louis, MO, USA)), then boiled at 100 °C for 10 min. Proteins were separated on 7.5%, 10%, or 12.5% polyacrylamide gels, then transferred to nitrocellulose membranes (Bio-Rad, Hercules, CA, USA). Membranes were washed in TBS-T (TBS buffer containing 0.05% Tween-20), blocked in 5% nonfat dry milk for 1 h, and incubated with the primary antibody overnight at 4 °C. Antibodies used for Western blotting are listed in Supplementary Table S1. Antibodies were diluted in TBS-T containing 5% BSA. After washing with TBS-T, membranes were incubated with horseradish peroxidase (HRP)–conjugated goat anti-mouse/anti-rabbit IgG secondary antibodies (Bio-Rad, Hercules, CA, USA; dilution 1:2000) for 1 h with gentle shaking. After washing, protein bands were visualized using Pierce™ ECL Western

Blotting Substrate (Thermo Fisher Scientific, Waltham, MA, USA) or SuperSignal West Femto Maximum Sensitivity Substrate (Thermo Fisher Scientific, Waltham, MA, USA). For band quantification, ImageJ 1.8.0 software (Research Services Branch, National Institute of Mental Health, Bethesda, MD, USA) was used.

4.9. Statistical Analysis

The normality of the population was determined using the Shapiro–Wilk test. If two groups were compared, we used independent t-test (two tailed), as the Shapiro–Wilk test showed normal distribution. When we compared three or more groups, one-way ANOVA complemented by Dunnett's post-hoc test was used, if the data showed normal distribution. Kruskal–Wallis test complemented with Dunn's post hoc test was performed, if the distribution of the data was not normal. Statistical calculations were performed using GraphPad Prism 7 (GraphPad Software Inc., San Diego, CA, USA) and SPSS 25 software. (SPSS package for Windows, Release 25.; SPSS, Chicago, IL, USA). Data are presented as mean ± SEM. Statistically significant differences are denoted by *, **, and *** for $p < 0.05$, $p < 0.01$, and $p < 0.001$.

Supplementary Materials: Supplementary Materials can be found at https://www.mdpi.com/1422-0067/22/4/1638/s1.

Author Contributions: Conceptualization: É.R., E.F., and G.P.; methodology: E.F. and C.H.; formal analysis: E.A.J.; writing—original draft preparation: E.F.; writing, reviewing, and editing: C.H., É.R., G.E., and G.P.; funding acquisition: É.R.; supervision: É.R. All authors have read and agreed to the published version of the manuscript.

Funding: This work was supported by the European Union and the European Regional Development Fund GINOP-2.3.2-15-2016-00005; the Hungarian National Research Development and Innovation Fund NKFIH K120206; and the ÚNKP-20-4-I New National Excellence Program of the Ministry for Innovation and Technology from the source of the National Research, Development, and Innovation Fund.

Institutional Review Board Statement: Not applicable.

Informed Consent Statement: Not applicable.

Data Availability Statement: Not applicable.

Conflicts of Interest: The authors declare no conflict of interest. G.P. is a consultant for ADC Therapeutics and Buffalo Biolabs.

Abbreviations

6-4PP	Pyrimidine (6-4) pyrimidone photoproduct
ABT-888	Veliparib
As_2O_3	Arsenic trioxide
CHO	Chinese hamster ovary cell line
CPD	Cyclobutane–pyrimidine dimer
HPRT	Hypoxanthine phosphoribosyltransferase 1
LC3	Microtubule-associated protein 1A/1B-light chain 3
mTOR	mammalian target of rapamycin
NER	Nucleotide excision repair
p53	Cellular tumor antigen p53
PARP1	Poly [ADP-ribose] polymerase 1
RSV	Resveratrol
SP	Spironolactone
XP	Xeroderma pigmentosum
UVB	Ultraviolet B radiation

References

1. Olsen, C.; Green, A.C.; Pandeya, N.; Whiteman, D. Trends in Melanoma Incidence Rates in Eight Susceptible Populations through 2015. *J. Investig. Dermatol.* **2019**, *139*, 1392–1395. [CrossRef]
2. Glazer, A.M.; Winkelmann, R.R.; Farberg, A.S.; Rigel, D.S. Analysis of Trends in US Melanoma Incidence and Mortality. *JAMA Dermatol.* **2017**, *153*, 225–226. [CrossRef]
3. Sacchetto, L.; Zanetti, R.; Comber, H.; Bouchardy, C.; Brewster, D.; Broganelli, P.; López, M.D.C.; Coza, D.; Galceran, J.; Gavin, A.; et al. Trends in incidence of thick, thin and in situ melanoma in Europe. *Eur. J. Cancer* **2018**, *92*, 108–118. [CrossRef] [PubMed]
4. Apalla, Z.; Lallas, A.; Sotiriou, E.; Lazaridou, E.; Ioannides, D. Epidemiological trends in skin cancer. *Dermatol. Pract. Concept.* **2017**, *7*, 1–6. [CrossRef] [PubMed]
5. Griffin, L.L.; Ali, F.R.; Lear, J. Non-melanoma skin cancer. *Clin. Med.* **2016**, *16*, 62–65. [CrossRef]
6. Boukamp, P. Non-melanoma skin cancer: What drives tumor development and progression? *Carcinogenesis* **2005**, *26*, 1657–1667. [CrossRef]
7. de Zwaan, S.E.; Haass, N.K. Genetics of basal cell carcinoma. *Australas J. Dermatol.* **2010**, *51*, 81–92. [CrossRef]
8. Emri, G.; Paragh, G.; Tósaki, Á.; Janka, E.; Kollár, S.; Hegedűs, C.; Gellén, E.; Horkay, I.; Koncz, G.; Remenyik, É. Ultraviolet radiation-mediated development of cutaneous melanoma: An update. *J. Photochem. Photobiol. B Biol.* **2018**, *185*, 169–175. [CrossRef] [PubMed]
9. de Lima-Bessa, K.M.; Armelini, M.G.; Chiganças, V.; Jacysyn, J.F.; Amarante-Mendes, G.P.; Sarasin, A.; Menck, C.F.M. CPDs and 6-4PPs play different roles in UV-induced cell death in normal and NER-deficient human cells. *DNA Repair* **2008**, *7*, 303–312. [CrossRef]
10. Pfeifer, G.P.; You, Y.-H.; Besaratinia, A. Mutations induced by ultraviolet light. *Mutat. Res.* **2005**, *571*, 19–31. [CrossRef]
11. Brash, D.E. UV signature mutations. *Photochem. Photobiol.* **2015**, *91*, 15–26. [CrossRef] [PubMed]
12. Lo, H.-L.; Nakajima, S.; Ma, L.; Walter, B.; Yasui, A.; Ethell, D.W.; Owen, L.B. Differential biologic effects of CPD and 6-4PP UV-induced DNA damage on the induction of apoptosis and cell-cycle arrest. *BMC Cancer* **2005**, *5*, 135. [CrossRef]
13. Meador, J.A.; Walter, R.B.; Mitchell, D.L. Induction, Distribution and Repair of UV Photodamage in the Platyfish, *Xiphophorus signum*. *Photochem. Photobiol.* **2000**, *72*, 260–266. [CrossRef]
14. Wei, L.; Christensen, S.R.; Fitzgerald, M.E.; Graham, J.; Hutson, N.D.; Zhang, C.; Huang, Z.; Hu, Q.; Zhan, F.; Xie, J.; et al. Ultradeep sequencing differentiates patterns of skin clonal mutations associated with sun-exposure status and skin cancer burden. *Sci. Adv.* **2021**, *7*, eabd7703. [CrossRef]
15. You, Y.H.; Szabó, P.E.; Pfeifer, G.P. Cyclobutane pyrimidine dimers form preferentially at the major p53 mutational hotspot in UVB-induced mouse skin tumors. *Carcinogenesis.* **2000**, *21*, 2113–2117. [CrossRef] [PubMed]
16. Scharer, O.D. Nucleotide excision repair in eukaryotes. *Cold Spring Harb. Perspect. Biol.* **2013**, *5*, a012609. [CrossRef] [PubMed]
17. Nakagawa, A.; Kobayashi, N.; Muramatsu, T.; Yamashina, Y.; Shirai, T.; Hashimoto, M.W.; Ikenaga, M.; Mori, T. Three-dimensional visualization of ultraviolet-induced DNA damage and its repair in human cell nuclei. *J. Investig. Dermatol.* **1998**, *110*, 143–148. [CrossRef] [PubMed]
18. Jans, J.; Schul, W.; Sert, Y.-G.; Rijksen, Y.; Rebel, H.; Eker, A.P.; Nakajima, S.; Van Steeg, H.; De Gruijl, F.R.; Yasui, A.; et al. Powerful skin cancer protection by a CPD-photolyase transgene. *Curr. Biol.* **2005**, *15*, 105–115. [CrossRef] [PubMed]
19. de Boer, J.; Hoeijmakers, J.H. Nucleotide excision repair and human syndromes. *Carcinogenesis* **2000**, *21*, 453–460. [CrossRef] [PubMed]
20. Chhabra, G.; Garvey, D.R.; Singh, C.K.; Mintie, C.A.; Ahmad, N. Effects and Mechanism of Nicotinamide Against UVA- and/or UVB-mediated DNA Damages in Normal Melanocytes. *Photochem. Photobiol.* **2019**, *95*, 331–337. [CrossRef]
21. Surjana, D.; Halliday, G.M.; Damian, D.L. Nicotinamide enhances repair of ultraviolet radiation-induced DNA damage in human keratinocytes and ex vivo skin. *Carcinogenesis* **2013**, *34*, 1144–1149. [CrossRef] [PubMed]
22. Katiyar, S.K.; Vaid, M.; Van Steeg, H.; Meeran, S.M. Green tea polyphenols prevent UV-induced immunosuppression by rapid repair of DNA damage and enhancement of nucleotide excision repair genes. *Cancer Prev. Res.* **2010**, *3*, 179–189. [CrossRef] [PubMed]
23. Chang, L.; Sheu, H.; Huang, Y.S.; Tsai, T.R.; Kuo, K. A novel function of emodin: Enhancement of the nucleotide excision repair of UV- and cisplatin-induced DNA damage in human cells. *Biochem. Pharmacol.* **1999**, *58*, 49–57. [CrossRef]
24. Vaid, M.; Sharma, S.D.; Katiyar, S.K. Proanthocyanidins inhibit photocarcinogenesis through enhancement of DNA repair and xeroderma pigmentosum group A-dependent mechanism. *Cancer Prev. Res.* **2010**, *3*, 1621–1629. [CrossRef] [PubMed]
25. Guillermo-Lagae, R.; Deep, G.; Ting, H.; Agarwal, C.; Agarwal, R. Silibinin enhances the repair of ultraviolet B-induced DNA damage by activating p53-dependent nucleotide excision repair mechanism in human dermal fibroblasts. *Oncotarget* **2015**, *6*, 39594–39606. [CrossRef]
26. Keuser, B.; Khobta, A.; Galle, K.; Anderhub, S.; Schulz, I.; Pauly, K.; Epe, B. Influences of histone deacetylase inhibitors and resveratrol on DNA repair and chromatin compaction. *Mutagenesis* **2013**, *28*, 569–576. [CrossRef]
27. Okui, T.; Fujiwara, Y. Inhibition of human excision DNA repair by inorganic arsenic and the co-mutagenic effect in V79 Chinese hamster cells. *Mutat. Res.* **1986**, *172*, 69–76. [CrossRef]
28. Holcomb, N.; Goswami, M.; Han, S.G.; Scott, T.; D'Orazio, J.; Orren, D.K.; Gairola, C.G.; Mellon, I. Inorganic arsenic inhibits the nucleotide excision repair pathway and reduces the expression of XPC. *DNA Repair* **2017**, *52*, 70–80. [CrossRef]

29. Hegedűs, C.; Boros, G.; Fidrus, E.; Kis, G.N.; Antal, M.; Juhász, T.; Janka, E.A.; Jankó, L.; Paragh, G.; Emri, G.; et al. PARP1 Inhibition Augments UVB-Mediated Mitochondrial Changes-Implications for UV-Induced DNA Repair and Photocarcinogenesis. *Cancers* **2019**, *12*, 5. [CrossRef]
30. Sergey, A.; Mériam, A.; Brino, L.; Marc-Egly, J.; Larsen, A.K.; Coin, F. A small molecule screen identifies an inhibitor of DNA repair inducing the degradation of TFIIH and the chemosensitization of tumor cells to platinum. *Chem. Biol.* **2014**, *21*, 398–407.
31. Kemp, M.G.; Krishnamurthy, S.; Kent, M.N.; Schumacher, D.L.; Sharma, P.; Excoffon, K.J.; Travers, J.B. Spironolactone Depletes the XPB Protein and Inhibits DNA Damage Responses in UVB-Irradiated Human Skin. *J. Investig. Dermatol.* **2019**, *139*, 448–454. [CrossRef]
32. Ueda, M.; Matsuura, K.; Kawai, H.; Wakasugi, M.; Matsunaga, T. Spironolactone-induced XPB degradation depends on CDK7 kinase and SCF(FBXL18) E3 ligase. *Genes Cells* **2019**, *24*, 284–296. [CrossRef] [PubMed]
33. Hendrickson, A.E.W.; Menefee, M.E.; Hartmann, L.C.; Long, H.J.; Northfelt, D.W.; Reid, J.M.; Boakye-Agyeman, F.; Kayode, O.; Flatten, K.S.; Harrell, M.I.; et al. A Phase I Clinical Trial of the Poly(ADP-ribose) Polymerase Inhibitor Veliparib and Weekly Topotecan in Patients with Solid Tumors. *Clin. Cancer Res.* **2018**, *24*, 744–752. [CrossRef]
34. Kummar, S.; Ji, J.; Morgan, R.; Lenz, H.-J.; Puhalla, S.L.; Belani, C.P.; Gandara, D.R.; Allen, D.; Kiesel, B.; Beumer, J.H.; et al. A phase I study of veliparib in combination with metronomic cyclophosphamide in adults with refractory solid tumors and lymphomas. *Clin. Cancer Res.* **2012**, *18*, 1726–1734. [CrossRef] [PubMed]
35. Lowery, M.A.; Kelsen, D.P.; Capanu, M.; Smith, S.C.; Lee, J.W.; Stadler, Z.K.; Moore, M.J.; Kindler, H.L.; Golan, T.; Segal, A.; et al. Phase II trial of veliparib in patients with previously treated BRCA-mutated pancreas ductal adenocarcinoma. *Eur. J. Cancer* **2018**, *89*, 19–26. [CrossRef] [PubMed]
36. Ramalingam, S.S.; Blais, N.; Mazieres, J.; Martin Reck, C.; Jones, M.; Juhasz, E.; Urban, L.; Orlov, S.; Barlesi, F.; Kio, E.; et al. Randomized, Placebo-Controlled, Phase II Study of Veliparib in Combination with Carboplatin and Paclitaxel for Advanced/Metastatic Non-Small Cell Lung Cancer. *Clin. Cancer Res.* **2017**, *23*, 1937–1944. [CrossRef] [PubMed]
37. Struthers, A.D.; Krum, H.; Williams, G.H. A comparison of the aldosterone-blocking agents eplerenone and spironolactone. *Clin. Cardiol.* **2008**, *31*, 153–158. [CrossRef]
38. Lainscak, M.; Pelliccia, F.; Rosano, G.; Vitale, C.; Schiariti, M.; Greco, C.; Speziale, G.; Gaudio, C. Safety profile of mineralocorticoid receptor antagonists: Spironolactone and eplerenone. *Int. J. Cardiol.* **2015**, *200*, 25–29. [CrossRef]
39. Soignet, S.L.; Frankel, S.R.; Douer, D.; Tallman, M.S.; Kantarjian, H.; Calleja, E.; Stone, R.M.; Kalaycio, M.; Scheinberg, D.A.; Steinherz, P.; et al. United States multicenter study of arsenic trioxide in relapsed acute promyelocytic leukemia. *J. Clin. Oncol.* **2001**, *19*, 3852–3860. [CrossRef] [PubMed]
40. Cicconi, L.; Fenaux, P.; Kantarjian, H.; Tallman, M.; Sanz, M.A.; Lo-Coco, F. Molecular remission as a therapeutic objective in acute promyelocytic leukemia. *Leukemia* **2018**, *32*, 1671–1678. [CrossRef]
41. Gill, H.; Yim, R.; Lee, H.K.K.; Mak, V.; Lin, S.-Y.; Kho, B.; Yip, S.-F.; Lau, J.S.M.; Li, W.; Ip, H.-W.; et al. Long-term outcome of relapsed acute promyelocytic leukemia treated with oral arsenic trioxide-based reinduction and maintenance regimens: A 15-year prospective study. *Cancer* **2018**, *124*, 2316–2326. [CrossRef] [PubMed]
42. Sirerol, J.A.; Feddi, F.; Mena, S.; Rodriguez, M.L.; Sirera, P.; Aupí, M.; Pérez, S.; Asensi, M.; Ortega, A.; Estrela, J.M. Topical treatment with pterostilbene, a natural phytoalexin, effectively protects hairless mice against UVB radiation-induced skin damage and carcinogenesis. *Free Radic. Biol Med.* **2015**, *85*, 1–11. [CrossRef]
43. Fujimura, A.T.; Martinez, R.M.; Pinho-Ribeiro, F.A.; Silva, A.M.; Baracat, M.M.; Georgetti, S.R.; Verri, W.A.; Chorilli, M.; Casagrande, R. Resveratrol-Loaded Liquid-Crystalline System Inhibits UVB-Induced Skin Inflammation and Oxidative Stress in Mice. *J. Nat. Prod.* **2016**, *79*, 1329–1338. [CrossRef] [PubMed]
44. Adhami, V.M.; Afaq, F.; Ahmad, N. Suppression of ultraviolet B exposure-mediated activation of NF-kappaB in normal human keratinocytes by resveratrol. *Neoplasia* **2003**, *5*, 74–82. [CrossRef]
45. Subedi, L.; Lee, T.H.; Wahedi, H.M.; Baek, S.; Kim, S.Y. Corrigendum to "Resveratrol-Enriched Rice Attenuates UVB-ROS-Induced Skin Aging via Downregulation of Inflammatory Cascades". *Oxid. Med. Cell. Longev.* **2018**, *2018*, 6052623. [CrossRef] [PubMed]
46. Afaq, F.; Adhami, V.M.; Ahmad, N. Prevention of short-term ultraviolet B radiation-mediated damages by resveratrol in SKH-1 hairless mice. *Toxicol. Appl. Pharmacol.* **2003**, *186*, 28–37. [CrossRef]
47. Vitale, N.; Kisslinger, A.; Paladino, S.; Procaccini, C.; Matarese, G.; Pierantoni, G.M.; Mancini, F.P.; Tramontano, D. Resveratrol couples apoptosis with autophagy in UVB-irradiated HaCaT cells. *PLoS ONE* **2013**, *8*, e80728. [CrossRef]
48. Aziz, M.H.; Afaq, F.; Ahmad, N. Prevention of ultraviolet-B radiation damage by resveratrol in mouse skin is mediated via modulation in survivin. *Photochem. Photobiol.* **2005**, *81*, 25–31. [CrossRef]
49. Back, J.H.; Zhu, Y.; Calabro, A.; Queenan, C.; Kim, A.S.; Arbesman, J.; Kim, A.L. Resveratrol-mediated downregulation of Rictor attenuates autophagic process and suppresses UV-induced skin carcinogenesis. *Photochem. Photobiol.* **2012**, *88*, 1165–1172. [CrossRef]
50. Nollen, M.; Ebert, F.; Moser, J.; Mullenders, L.H.F.; Hartwig, A.; Schwerdtle, T. Impact of arsenic on nucleotide excision repair: XPC function, protein level, and gene expression. *Mol. Nutr. Food Res.* **2009**, *53*, 572–582. [CrossRef] [PubMed]
51. Boros, G. Investigation of UVB-Induced Cellular Mechanisms in Human Keratinocytes using a Novel Approach of Delivering In Vitro Synthesized mRNA Encoding Cyclobutane Pyrimidine Dimer-Specific Photolyase. Ph.D. Thesis, University of Debrecen, Debrecen, Hungary, 2015.

52. Budden, T.; Davey, R.J.; Vilain, R.E.; Ashton, K.A.; Braye, S.G.; Beveridge, N.J.; Bowden, N.A. Repair of UVB-induced DNA damage is reduced in melanoma due to low XPC and global genome repair. *Oncotarget* **2016**, *7*, 60940–60953. [CrossRef] [PubMed]
53. Johnson, G.E. Mammalian cell HPRT gene mutation assay: Test methods. *Methods Mol. Biol.* **2012**, *817*, 55–67.
54. Greinert, R.; Boguhn, O.; Harder, D.; Breitbart, E.W.; Mitchell, D.L.; Volkmer, B. The dose dependence of cyclobutane dimer induction and repair in UVB-irradiated human keratinocytes. *Photochem. Photobiol.* **2000**, *72*, 701. [CrossRef]
55. Ikehata, H.; Mori, T.; Douki, T.; Cadet, J.; Yamamoto, M. Quantitative analysis of UV photolesions suggests that cyclobutane pyrimidine dimers produced in mouse skin by UVB are more mutagenic than those produced by UVC. *Photochem. Photobiol. Sci.* **2018**, *17*, 404–413. [CrossRef] [PubMed]
56. Qian, W.; Liu, J.; Jin, J.; Ni, W.; Xu, W. Arsenic trioxide induces not only apoptosis but also autophagic cell death in leukemia cell lines via up-regulation of Beclin-1. *Leuk. Res.* **2007**, *31*, 329–339. [CrossRef]
57. Kumar, S.; Yedjou, C.; Tchounwou, P.B. Arsenic trioxide induces oxidative stress, DNA damage, and mitochondrial pathway of apoptosis in human leukemia (HL-60) cells. *J. Exp. Clin. Cancer Res.* **2014**, *33*, 42. [CrossRef]
58. Lee, C.-H.; Yu, C.-L.; Liao, W.-T.; Kao, Y.-H.; Chai, C.-Y.; Chen, G.-S.; Yu, H.-S. Effects and interactions of low doses of arsenic and UVB on keratinocyte apoptosis. *Chem. Res. Toxicol.* **2004**, *17*, 1199–1205. [CrossRef]
59. Zhang, W.; Hanks, A.N.; Boucher, K.M.; Florell, S.R.; Allen, S.M.; Alexander, A.; Brash, D.E.; Grossman, U. UVB-induced apoptosis drives clonal expansion during skin tumor development. *Carcinogenesis* **2004**, *26*, 249–257. [CrossRef] [PubMed]
60. Eastman, A. Cell cycle checkpoints and their impact on anticancer therapeutic strategies. *J. Cell. Biochem.* **2004**, *91*, 223–231. [CrossRef]
61. Paulovich, A.G.; Toczyski, D.P.; Hartwell, L.H. When Checkpoints Fail. *Cell* **1997**, *88*, 315–321. [CrossRef]
62. Pavey, S.; Russell, T.; Gabrielli, B. G2 phase cell cycle arrest in human skin following UV irradiation. *Oncogene* **2001**, *20*, 6103–6110. [CrossRef]
63. Tanida, I.; Ueno, T.; Kominami, E. LC3 and Autophagy. *Methods Mol. Biol.* **2008**, *445*, 77–88.
64. Ryan, K.M. p53 and autophagy in cancer: Guardian of the genome meets guardian of the proteome. *Eur. J. Cancer* **2011**, *47*, 44–50. [CrossRef]
65. Strozyk, E.; Kulms, D. The role of AKT/mTOR pathway in stress response to UV-irradiation: Implication in skin carcinogenesis by regulation of apoptosis, autophagy and senescence. *Int. J. Mol. Sci.* **2013**, *14*, 15260–15285. [CrossRef]
66. Thomas-Schoemann, A.; Batteux, F.; Mongaret, C.; Nicco, C.; Chéreau, C.; Annereau, M.; Dauphin, A.; Goldwasser, F.; Weill, B.; Lemare, F.; et al. Arsenic trioxide exerts antitumor activity through regulatory T cell depletion mediated by oxidative stress in a murine model of colon cancer. *J. Immunol.* **2012**, *189*, 5171–5177. [CrossRef]
67. Carone, L.; Oxberry, S.G.; Twycross, R.; Charlesworth, S.; Mihalyo, M.; Wilcock, A. Spironolactone. *J. Pain Symptom Manag.* **2017**, *53*, 288–292. [CrossRef]
68. Li, S.; Topatana, W.; Juengpanich, S.; Cao, J.; Hu, J.; Zhang, B.; Ma, D.; Cai, X.; Chen, M. Development of synthetic lethality in cancer: Molecular and cellular classification. *Signal Transduct. Target Ther.* **2020**, *5*, 1–14. [CrossRef] [PubMed]
69. Helleday, T. The underlying mechanism for the PARP and BRCA synthetic lethality: Clearing up the misunderstandings. *Mol. Oncol.* **2011**, *5*, 387–393. [CrossRef] [PubMed]
70. Lord, C.J.; Tutt, A.N.; Ashworth, A. Synthetic Lethality and Cancer Therapy: Lessons Learned from the Development of PARP Inhibitors. *Annu. Rev. Med.* **2015**, *66*, 455–470. [CrossRef]
71. Chaudhuri, A.R.; Nussenzweig, A.R.C.A. The multifaceted roles of PARP1 in DNA repair and chromatin remodelling. *Nat. Rev. Mol. Cell Biol.* **2017**, *18*, 610–621. [CrossRef] [PubMed]
72. Pines, A.; Vrouwe, M.G.; Marteijn, J.A.; Typas, D.; Luijsterburg, M.S.; Cansoy, M.; Mullenders, L. PARP1 promotes nucleotide excision repair through DDB2 stabilization and recruitment of ALC1. *J. Cell Biol.* **2012**, *199*, 235–249. [CrossRef]
73. Robu, M.; Shah, R.G.; Purohit, N.K.; Zhou, P.; Naegeli, H.; Shah, G.M. Poly(ADP-ribose) polymerase 1 escorts XPC to UV-induced DNA lesions during nucleotide excision repair. *Proc. Natl. Acad. Sci. USA* **2017**, *114*, E6847–E6856. [CrossRef]
74. Thorslund, T.; Von Kobbe, C.; Harrigan, J.A.; Indig, F.E.; Christiansen, M.; Stevnsner, T.; Bohr, V.A. Cooperation of the Cockayne syndrome group B protein and poly(ADP-ribose) polymerase 1 in the response to oxidative stress. *Mol. Cell. Biol.* **2005**, *25*, 7625–7636. [CrossRef] [PubMed]
75. Chiu, H.; Ho, Y.-S.; Wang, Y.-J. Arsenic trioxide induces autophagy and apoptosis in human glioma cells in vitro and in vivo through downregulation of survivin. *J. Mol. Med.* **2011**, *89*, 927–941. [CrossRef] [PubMed]
76. Liu, Q.; Gheorghiu, L.; Drumm, M.; Clayman, R.; Eidelman, A.; Wszolek, M.F.; Olumi, A.; Feldman, A.; Wang, M.; Marcar, L.; et al. PARP-1 inhibition with or without ionizing radiation confers reactive oxygen species-mediated cytotoxicity preferentially to cancer cells with mutant TP53. *Oncogene* **2018**, *37*, 2793–2805. [CrossRef] [PubMed]
77. Vance, S.M.; Liu, E.; Zhao, L.; Parsels, J.D.; Parsels, L.A.; Brown, J.L.; Maybaum, J.; Lawrence, T.S.; Morgan, M.A. Selective radiosensitization of p53 mutant pancreatic cancer cells by combined inhibition of Chk1 and PARP1. *Cell Cycle* **2011**, *10*, 4321–4329. [CrossRef]
78. Valenzuela, M.T.; Guerrero, R.; Nuñez, M.I.; De Almodóvar, J.M.R.; Sarker, M.; De Murcia, G.; Oliver, F.J. PARP-1 modifies the effectiveness of p53-mediated DNA damage response. *Oncogene* **2002**, *21*, 1108–1116. [CrossRef]
79. Kubota, E.; Williamson, C.T.; Ye, R.; Elegbede, A.; Peterson, L.; Lees-Miller, S.P.; Bebb, D.G. Low ATM protein expression and depletion of p53 correlates with olaparib sensitivity in gastric cancer cell lines. *Cell Cycle* **2014**, *13*, 2129–2137. [CrossRef]

80. Kanai, M.; Hanashiro, K.; Kim, S.-H.; Hanai, S.; Boulares, A.H.; Miwa, M.; Fukasawa, K. Inhibition of Crm1-p53 interaction and nuclear export of p53 by poly(ADP-ribosyl)ation. *Nat. Cell Biol.* **2007**, *9*, 1175–1183. [CrossRef]
81. Lorin, S.; Hamaï, A.; Mehrpour, M.; Codogno, P. Autophagy regulation and its role in cancer. *Semin. Cancer Biol.* **2013**, *23*, 361–379. [CrossRef]
82. Boros, G.; Miko, E.; Muramatsu, H.; Weissman, D.; Emri, E.; Rózsa, D.; Nagy, G.; Juhász, A.; Juhász, I.; Van Der Horst, G.; et al. Transfection of pseudouridine-modified mRNA encoding CPD-photolyase leads to repair of DNA damage in human keratinocytes: A new approach with future therapeutic potential. *J. Photochem. Photobiol. B Biol.* **2013**, *129*, 93–99. [CrossRef] [PubMed]

International Journal of
Molecular Sciences

MDPI

Article

Inhibition of AKT-Signaling Sensitizes Soft Tissue Sarcomas (STS) and Gastrointestinal Stromal Tumors (GIST) to Doxorubicin via Targeting of Homology-Mediated DNA Repair

Sergei Boichuk [1,2,*], Firuza Bikinieva [1], Ilmira Nurgatina [1], Pavel Dunaev [1], Elena Valeeva [2], Aida Aukhadieva [1], Alexey Sabirov [3] and Aigul Galembikova [1]

[1] Department of Pathology, Kazan State Medical University, 420012 Kazan, Russia; firuza1995@mail.ru (F.B.); ilmiranurgatina@gmail.com (I.N.); dunaevpavel@mail.ru (P.D.); arom.1705@mail.ru (A.A.); ailuk000@mail.ru (A.G.)

[2] Central Research Laboratory, Kazan State Medical University, 420012 Kazan, Russia; vevaleeva@ya.ru

[3] Department of Pathology, Tatarstan Cancer Center, 420029 Kazan, Russia; a-sabirov@yandex.ru

* Correspondence: boichuksergei@mail.ru; Tel.: +7-917-397-80-93; Fax. +7-843-236-06-52

Received: 9 October 2020; Accepted: 20 November 2020; Published: 22 November 2020

Abstract: Activation of the phosphoinositide 3-kinase (PI3K)/Akt/mTOR pathway is well documented for a broad spectrum of human malignancies supporting their growth and progression. Accumulating evidence has also implicated AKT as a potent modulator of anti-cancer therapies via regulation of DNA damage response and repair (DDR) induced by certain chemotherapeutic agents and ionizing radiation (IR). In the present study, we examined the role of AKT signaling in regulating of Rad51 turnover and cytotoxic effects of topoisomerase II inhibitor, doxorubicin (Dox) in soft tissue sarcomas (STS) and gastrointestinal stromal tumors (GIST) in vitro. Blocking of AKT signaling (MK-2206) enhanced cytotoxic and pro-apoptotic effects of Dox in vast majority of STS and GIST cell lines. The phosphorylated form of Akt co-immunoprecipitates with Rad51 after Dox-induced DNA damage, whereas Akt inhibition interrupts this interaction and decreases Rad51 protein level by enhancing protein instability via proteasome-dependent degradation. Inhibition of Akt signaling in Dox-treated cells was associated with the increased number of γ-H2AX-positive cells, decrease of Rad51 foci formation and its colocalization with γ-H2AX foci, thereby revealing unsuccessful DDR events. This was also in consistency with an increase of tail moment (TM) and olive tail moment (OTM) in Dox-treated GIST and STS cells cultured in presence of Akt inhibitor after Dox washout. Altogether, our data illustrates that inhibition of AKT signaling is STS and GIST might potentiate the cytotoxic effect of topoisomerase II inhibitors via attenuating the homology-mediated DNA repair.

Keywords: soft tissue sarcomas (STS); gastrointestinal stromal tumors (GIST); AKT signaling; Rad51 recombinase; homology-mediated DNA repair; apoptosis; sensitization

1. Introduction

Activation of the phosphoinositide 3-kinase (PI3K)/Akt/mTOR pathway is frequently detected in human malignancies, including gastrointestinal stromal tumors (GIST) and soft tissue sarcomas (STS). This pathway supports tumor growth and progression and is thought to be linked with resistance to the current therapeutic regimens, including certain chemotherapeutic agents, ionizing radiation (IR) and targeted-based therapies. In particular, activation of PI3K/Akt/mTOR pathway in GIST is due to the constitutive activation of c-KIT/PDGFRA autophosphorylation and involved in oncogenesis and tumor progression at various disease stages [1], and is also thought to be linked with GIST resistance to imatinib mesylate (IM) [2,3]. Moreover, the preclinical experiments illustrated that a long-term exposure of IM- and sunitinib (SU)-resistant cell lines to SU induces epigenetic silencing of *PTEN* with a consequent overactivation of the PI3K/AKT pathway [4]. This also correlates with a clinical data illustrating that low/negative IHC-staining for PTEN was associated with aggressive disease [5], thereby suggesting that *PTEN* depletion and/or silencing is also associated with aggressive phenotype and resistance to RTK inhibition. Important, molecular and genomic changes in IM-resistant GISTs illustrated that the PI3K/Akt/mTOR pathway has a greater importance in IM-resistant GIST than other pathways downstream of c-KIT or PDGFRA (e.g., MEK/MAPK pathway), therefore illustrating a rationale for targeting the PI3K/Akt/mTOR pathway in GIST [6]. Indeed, inhibition of PI3K [7], AKT [8] and mTOR [1] has been shown promising results in vitro and in xenograft models and led to the clinical trials to examine an efficiency against IM-resistant GIST. However, some of them have not succeeded to date [9,10], whereas the other trials to examine the novel drugs targeting the elements of this pathway are currently ongoing (NCT01991379, NCT01735968 and NCT01468688).

Activation of the PI3K/AKT/mTOR pathway is also well-documented for soft tissue sarcomas (STS). In particular, for leiomyosarcoma (LMS) the most common genetic abnormalities include the loss of function mutations in *p53* and/or *PTEN* or activating mutations in the genes encoding the signaling molecules of the PI3K/AKT/mTOR pathway [11–13]. Similarly, to LMS, the PI3K/AKT/mTOR signaling pathway is often aberrantly activated in Ewing's sarcoma (ES) [14,15], rhabdomyosarcoma, a well-known pediatric sarcoma of soft tissues displaying very similar histology and therapeutic options with ES [16], osteosarcoma (OS) [17,18], thereby illustrating that the PI3K/AKT/mTOR pathway is a suitable therapeutic target for STS as well as for other human cancers.

Besides, the PI3K/AKT/mTOR pathway is considered as a perspective molecular therapeutic target for STS and GIST, accumulating evidence also illustrates the tight connection between this pathway and checkpoint responses and repair of DNA damage, induced by certain chemotherapeutic agents and/or ionizing radiation. This reflects, predominantly, the regulatory role of AKT in DNA double-stand break (DSB) repair, including non-homology end-joining (NHEJ) and homologous recombination (HR)—mediated DNA repair, which in turn also illustrates the AKT-mediated pathway, a perspective target to sensitize STS and GIST to DNA-damaging agents, including topoisomerase II (Topo II) inhibitors. Given that Topo II inhibitors are currently used for therapy of STS [19–22] and taking into account that GIST were also previously shown to be sensitive to the Topo II inhibitors [23,24], we thought to examine whether inhibition of the PI3K/AKT/mTOR pathway might enhance their sensitivity to Topo II inhibitor, doxorubicin (DOX) via targeting of the molecular pathways involved in DNA DSB repair.

In the present study we characterized the importance of AKT-pathway for HR-mediated repair of DSBs in STS and GIST in vitro and their relevance to the tumor cell sensitivity to topoisomerase II inhibitor, doxorubicin (Dox). We found that inhibition of AKT-signaling in GIST and STS cell lines results in a significantly decreased expression of Rad51 recombinase and number of residual Rad51/BRCA1 foci in Dox-treated tumor cells. This was due to the decreased stability of Rad51 as a consequence of the enhanced proteasomal degradation. Finally, as a result of the impaired homology-mediated DNA repair, we observed a substantial decrease of viability in AKT-inhibited tumor cells after Dox treatment, which was evidenced by MTS-based colorimetric assay and increased expression of apoptotic markers (cleaved forms of caspase-3 and poly-(ADP)-ribose-polymerase (PARP) and the numbers of hypodiploid cells).

Therefore, overactivation of AKT-signaling pathway in STS and GIST might serve as a prospective molecular target to enhance cytotoxic effects of DNA-topoisomerase II inhibitors inducing DNA DSBs in STS and GIST.

2. Results

2.1. Inhibition of AKT-Signaling Enhances Cytotoxicity of Topo II Inhibitors in STS and GIST

To examine whether inhibition of AKT signaling potentiates the cytotoxic activities of Dox in STS and GIST, we performed MTS-based survival assay with a broad spectrum of cancer cell lines, including SK-LMS-1 leiomyosarcoma, RD rhabdomyosarcoma, HT-1080 fibrosarcoma, A673 Ewing's sarcoma, U2-OS osteosarcoma, IM-sensitive and resistant gastrointestinal stromal tumors (e.g., GIST T-1 vs. GIST T-1R and GIST 430, respectively). Cells indicated above were treated with Dox for 72 h alone or in combination with MK-2206, a selective AKT-inhibitor. We observed that inhibition of AKT signaling potentiated cytotoxic activity of Dox and inhibited growth of the vast majority of cancer cell lines included in present study (Figure 1). The IC50 values for MK-2206, Dox used alone or in combination are shown on Table 1. In particular, we observed an approximately two-fold decrease of IC50 for all types of cancer cell lines treated with combination of Dox and MK-2206, when compared to cells treated with Dox alone. Synergism of Dox and MK-2206 was also calculated by combination index (CI) values for each molar ratio of Dox and MK-2206 (Table 2) and was depicted as a heat-map shown on Figure 1B. In addition, the average synergistic effects of Dox and MK-2206 were calculated by using of R-package of computational tool Synergy Finder. Indeed, we observed a prominent synergism for Dox and MK-2206 in RD and SK-LMS-1 cells (Figure S1) and other STS cell lines, including U2-OS osteosarcoma and HT-1080 fibrosarcoma cells (Figure S2).

2.2. Inhibition of AKT-Signaling Enhances Doxorubicin-Induced Apoptosis of STS and GIST

To further corroborate these findings, we examined whether inhibition of AKT signaling can potentiate pro-apoptotic effect of Dox in STS and GIST. For this purpose, cancer cell lines were treated with the low dose of Dox (0.25 g/mL) in absence (control) or presence of MK-2206 (5 M) and further subjected for western blotting to examine expression of apoptotic markers (cleaved forms of PARP and caspase-3). Strikingly, a substantial increase of cleaved caspase-3 and PARP was observed in the majority of STS and GIST cells treated with combination of AKT and Topo II inhibitors (Figure 2). As expected, MK-2206 used alone has no cytotoxic effect on the tumor cells, whereas pro-apoptotic effect of Dox used alone was much less when compared to the effect of combination of these inhibitors. The most sensitive for AKT inhibition were Dox-treated RD rhabdomyosarcoma cells (Figure 2A), U2-OS osteosarcoma (Figure 2B), GIST T-1R (Figure 2D) and 430 (Figure 2E) which was evidenced by a substantial cell death after the treatment with Dox and MK-2206. Quantification by mean pixel density revealed that PARP and caspase-3 cleavage was substantially increased in all types of cancer cell lines treated with Dox in combination of MK-2206 (Figure S3.1). The inhibitory effect of MK-2206 on AKT pathway in tumor cells was confirmed by a substantial decrease of AKT phosphorylation (Figure 2). Similarly, changes in phospho-AKT (Ser473) expression were quantified and normalized to actin (Figure S3.1), thereby illustrating that decreased AKT phosphorylation in MK-2206-inhibited cells was specific and was not due to the changes in expression of the total AKT.

In addition to increased levels of cleaved caspase-3 and PARP in cancer cells treated with Dox in presence of AKT inhibitor, we also observed the increased numbers of apoptotic (e.g., Annexin V-positive cells) cells in RD rhabdomyosarcoma and GIST T-1R cells treated with combination of Dox and MK-2206 when compared to non-treated cells or cells treated with Dox alone (Figure S4), thereby revealing that AKT inhibition effectively sensitized STS and GIST cells to Dox treatment and induced apoptotic cell death.

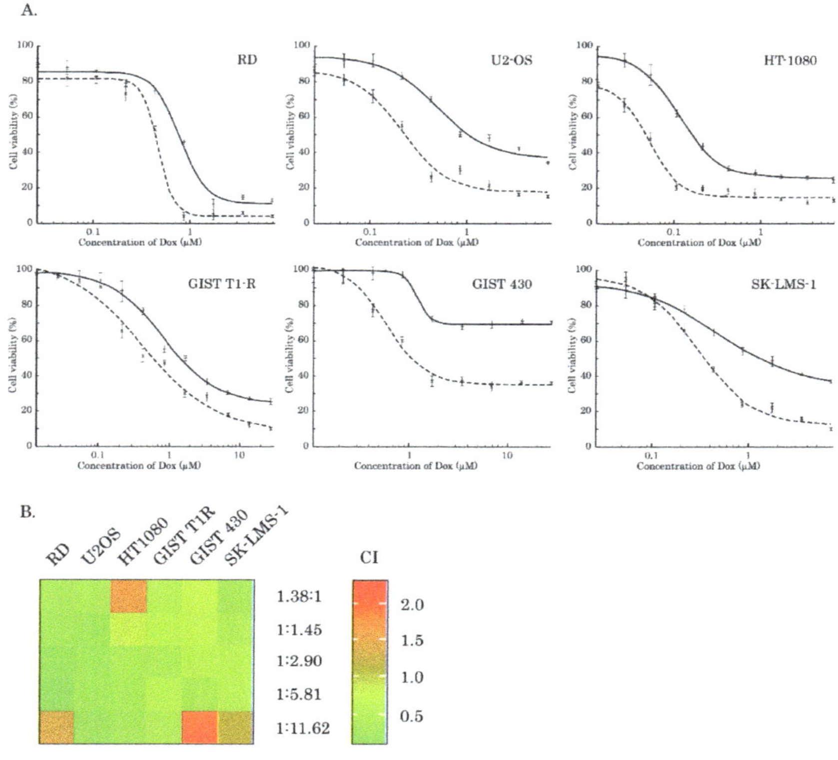

Figure 1. An impact of AKT inhibition on cytotoxic activity of doxorubicin (Dox) in soft tissue sarcomas (STS) and gastrointestinal stromal tumors (GIST). (**A**) MTS-based viability assay in RD rhabdomyosarcoma, U2-OS osteosarcoma, HT-1080 fibrosarcoma, GIST T-1R, GIST 430 and SK-LMS-1 leiomyosarcoma cells treated with Dox alone (solid lines) or in combination of MK-2206 (dashed lines) for 72 h. The data was normalized to DMSO-treated controls. Values are the means ± standard deviation ($n = 3$). (**B**) the heat-map illustrating the combination index (CI) values for Dox and MK2206 combinations at various molar ratios for cancer cell lines indicated above. CI values were calculated with CompuSyn Software (Version 1.0), based on the Chou–Talalay algorithm. CI values < 1 indicate the synergy between Dox and MK-2206.

Table 1. IC50 values for Dox in STS and GIST treated in absence or presence of MK-2206, an AKT inhibitor.

IC50	MK-2206 (μM)	Dox (μM)	Dox + MK-2206 (μM)	Fold Increase
GIST T1-R	32.9 ± 4.3	0.77 ± 0.13	0.42 ± 0.1	1.8
GIST 430	17 ± 0.2	1.21 ± 0.08	0.59 ± 0.07	2.1
SK-LMS-1	13.5 ± 1.02	0.49 ± 0.05	0.3 ± 0.03	1.6
RD	26.4 ± 1	0.79 ± 0.08	0.47 ± 0.03	1.7
U2-OS	27.8 ± 1.2	0.49 ± 0.07	0.22 ± 0.04	2.2
HT-1080	19.7 ± 0.4	0.116 ± 0.005	0.05 ± 0.006	2.3

Table 2. CI values for each molar ratio of Dox and MK-2206 in STS and GIST.

Cell Line	Molar Ratio of DOX: MK-2206	CI Value ± SD
RD	1:11.62	1.41 ± 0.23
	1:5.81	0.07 ± 0.01
	1:2.90	0.15 ± 0.02
	1:1.45	0.31 ± 0.04
	1.38:1	0.34 ± 0.04
SK-LMS-1	1:11.62	1.18 ± 0.10
	1:5.81	0.53 ± 0.06
	1:2.90	0.60 ± 0.02
	1:1.45	0.45 ± 0.05
	1.38:1	0.33 ± 0.03
HT-1080	1:11.62	0.32 ± 0.06
	1:5.81	0.35 ± 0.05
	1:2.90	0.41 ± 0.06
	1:1.45	0.83 ± 0.09
	1.38:1	1.58 ± 0.60
U2-OS	1:11.62	0.19 ± 0.02
	1:5.81	0.30 ± 0.02
	1:2.90	0.26 ± 0.01
	1:1.45	0.26 ± 0.02
	1.38:1	0.40 ± 0.10
GIST T1-R	1:11.62	0.45 ± 0.05
	1:5.81	0.60 ± 0.06
	1:2.90	0.42 ± 0.05
	1:1.45	0.64 ± 0.04
	1.38:1	0.57 ± 0.06
GIST 430	1:11.62	2.32 ± 0.42
	1:5.81	0.43 ± 0.07
	1:2.90	0.61 ± 0.09
	1:1.45	0.69 ± 0.11
	1.38:1	0.75 ± 0.11

2.3. Inhibition of AKT Signaling Attenuates Repair of Dox-Induced DNA Damage

Next, we performed alkaline-based DNA comet assay to examine whether enhanced apoptosis of cancer cells treated with Dox and MK-2206 was due to DNA damage which remained unrepairable in presence of AKT inhibitor. For this purpose, STS or GIST cells were initially treated with Dox for 2 h to induce DNA damage, and followed by the wash-out of chemotherapeutic agent and further cultured for 8 h in absence or presence of MK-2206. Strikingly, we observed a substantial increase of tail moment (TM) and olive tail moment (OTM) in GIST cells, thereby illustrating unrepairable DNA damage in AKT-inhibited tumor cells (Figure 3A,B). Similar data was obtained in GIST cells that were not washed from Dox. Again, we observed that MK-2206 substantially attenuated DNA damage repair which was evidenced by increased TM (Figure S5A,B). As expected, we observed a substantial increase of TM in Dox-treated cells when compared to the previous experimental settings, thereby illustrating the substantial DNA damage due to the permanent exposure to Dox. Given that alkaline-based DNA comet assay aimed to detect all the types of DNA lesions, we performed neutral version of single-cell electrophoresis to examine whether inhibition of AKT-signaling attenuates repair of Dox-induced DNA DSBs. Similar to alkaline version, we observed an increase of both TM and OTM in GIST cells treated with Dox in presence of MK-2206, thereby revealing that AKT-signaling plays an important role in regulating of repair of Dox-induced DNA lesions, including DNA DSBs (Figure S5C,D).

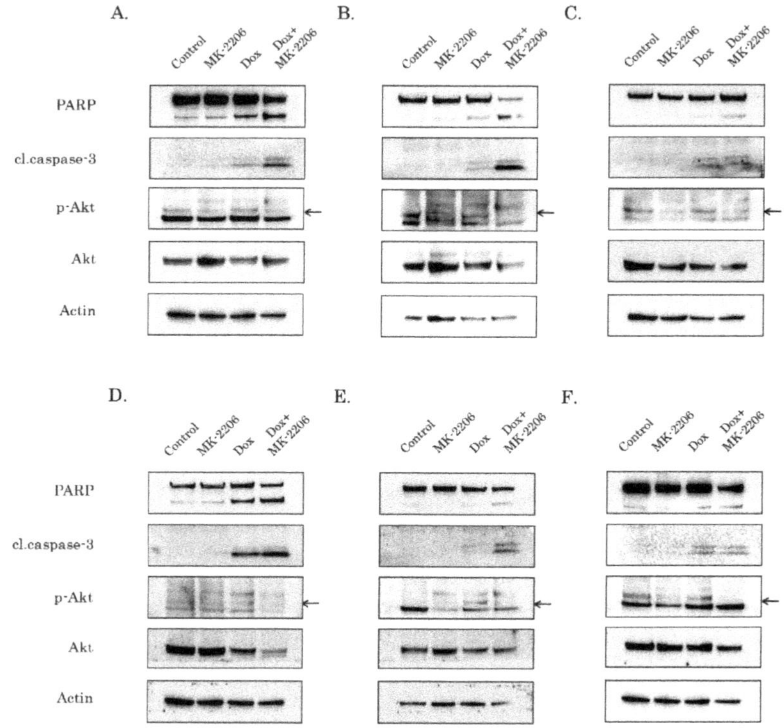

Figure 2. AKT inhibition potentiates pro-apoptotic activity of Dox in STS and GIST. Immunoblot analysis for apoptosis markers (cleaved forms of poly-(ADP)-ribose-polymerase (PARP) and caspase-3) in RD rhabdomyosarcoma (**A**), U2-OS osteosarcoma (**B**), HT-1080 fibrosarcoma (**C**), GIST T-1R (**D**), GIST 430 (**E**) and SK-LMS-1 leiomyosarcoma (**F**) cells after treatment with DMSO (control), Dox (0.25 g/mL) and MK-2206 (5 M) alone and in combination (e.g., Dox + MK-2206) for 72 h. The lysates were also probed for total and phosphorylated forms of AKT to illustrate AKT inhibition by MK-2206. pAKT expression is shown by arrows. Actin was used as a loading control.

Unrepairable DNA damage in Dox-treated tumor cells cultured in presence of AKT inhibitor was also confirmed by quantitative analysis of γ-H2AX-expression, a well-known marker of the DNA DSBs. As expected, MK-2206 used alone has no impact on γ-H2AX-expression, thereby revealing absence of DNA DSBs in GIST T-1R cells treated with AKT inhibitor (Figure 4B). In contrast, Dox-treated cells exhibited an increased pattern of γ-H2AX-expression. Of note, Dox was washed out from the cell culture 2 h post treatment to exclude permanent DNA damage, and hereby allowing to repair Dox-induced DNA lesions (Figure 4C). Strikingly, we observed a retention of γ-H2AX-expression after AKT inhibition in vast majority of Dox-treated GIST cells (Figure 4D). Similar findings were observed for various STS cell lines, including HT-1080 fibrosarcoma cells, as shown in Figure S6.1. In contrast to the AKT inhibitor, inhibition of MAPK-signaling pathway by selective inhibitor U0126, did not have the similar effects on DNA repair in Dox-treated GIST (Figure S6.2). WB data also revealed a substantial increase of γ-H2AX-expression in all tumor cell lines treated with combination of Dox and MK-2206 when compared to the cells treated with Dox alone (Figure 5). Strikingly, our WB data also illustrated a substantial decrease of Rad51 expression in most of Dox-treated cells (Figure 4).

All together this data illustrates that inhibition of AKT-signaling in STS and GIST sensitizes them to Dox due to the inhibition of DNA damage repair.

2.4. Rad51 Expression Is Substantially Reduced in AKT-Inhibited Cancer Cells

Given that attenuation of DNA damage repair in AKT-inhibited cancer cells might be due to the decreased efficacy of homology-mediated repair of DNA DSBs, we analyzed the expression of Rad51 recombinase, known as a key protein involved in DSB repair. For this purpose, we performed a quantitative analysis of Rad51 expression in cancer cells treated with Dox and MK-2206 by a similar way as shown on Figure 4. As expected, no difference in Rad51 expression was found between control (Figure 6A) and MK-2206-treated cells (Figure 6B). When Dox-treated cells were washed out (to exclude a permanent DNA damage), we found a substantial increase of Rad51 intensity in cancer cells (Figure 6C). Strikingly, intensity of Rad51-mediated fluorescence substantially decreased when MK-2206 was introduced in cell culture after Dox washout (Figure 6D). This was observed for GIST T-1R cells (Figure 6) and STS (data not shown), thereby illustrating the possible mechanism of MK-2206-induced sensitization of cancer cells to Topo II inhibitors. These findings were in consistency with WB data illustrating a substantial decrease of Rad51 expression in AKT-inhibited cancer cells (Figure 5).

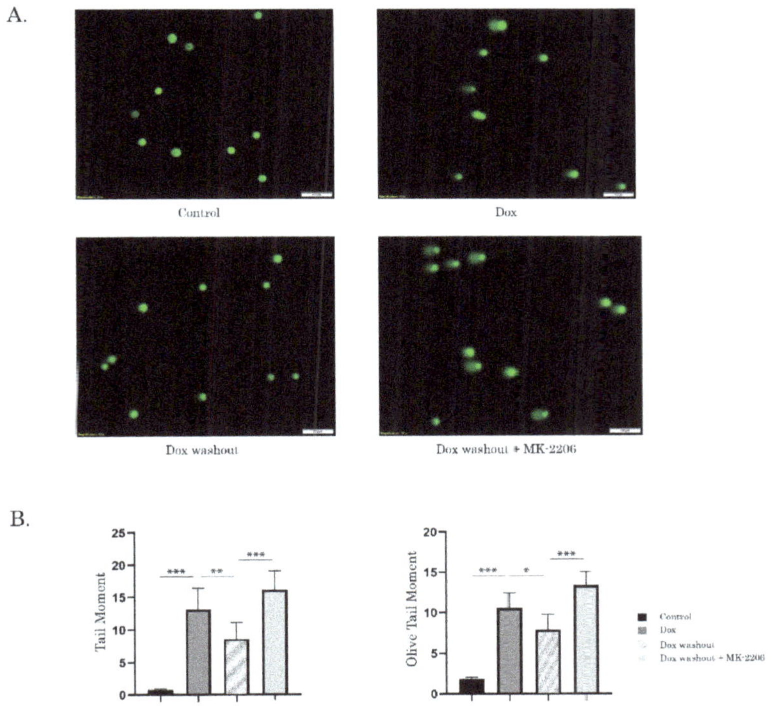

Figure 3. Inhibition of AKT-signaling attenuates DNA double-stand break (DSB) repair. GIST T-1R cells treated with DMSO (negative control) and Dox 0.5 µg/mL for 2 h. After the drugs were washed out, cancer cells were further cultured in absence (Dox washout) or presence of MK-2206 (5 µM), an AKT inhibitor (Dox washout + MK-2206) for 8 h. (**A**) representative images of comets from the experimental settings shown above (Scale bars = 100 µm). (**B**) graphic depiction of the calculated tail moment (TM) and olive tail moment (OTM) from alkaline comet assay shown in Figure 3A. Columns, mean of at least three independent experiments with a minimum of 50 cells counted per each experiment; bars, SE. * $p < 0.05$; ** $p < 0.01$; *** $p < 0.001$.

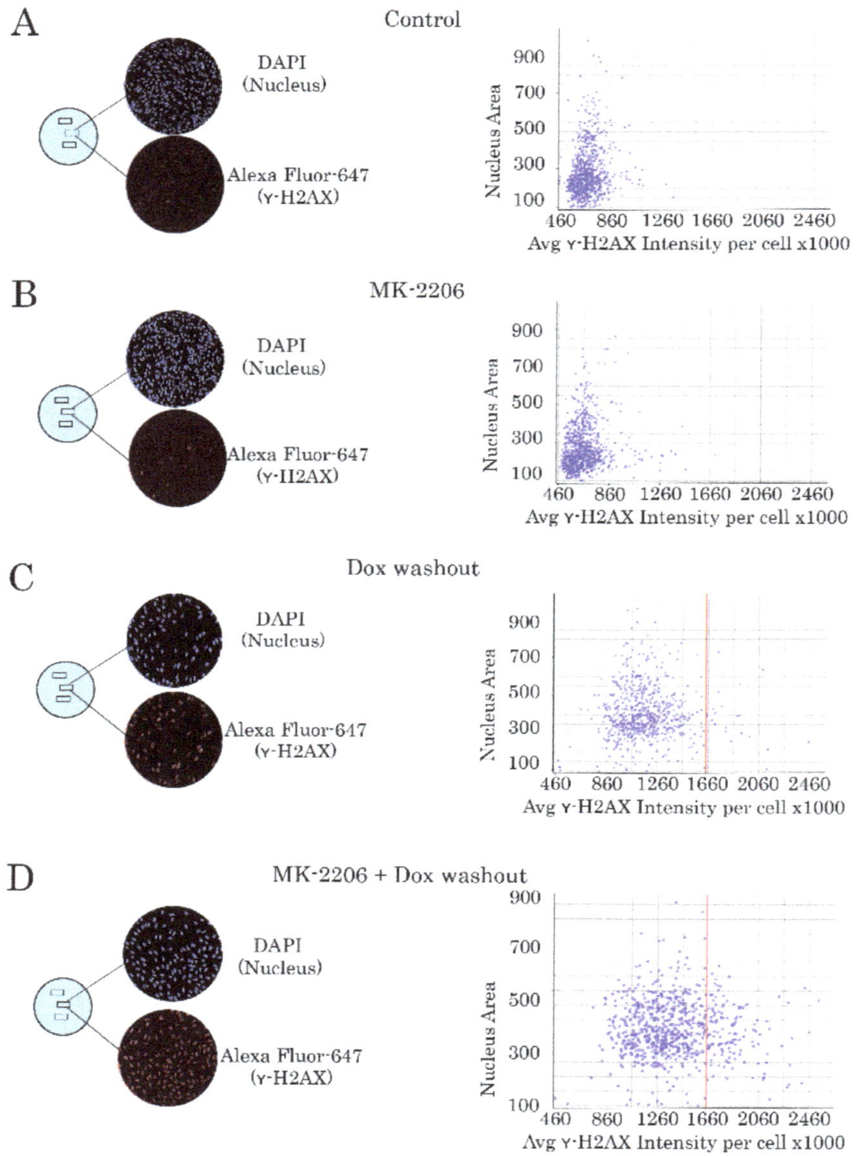

Figure 4. Inhibition of AKT-signaling delays the kinetics of γ-H2AX decline in Dox-treated GIST T-1R cells (representative experiment). Left panel-cancer cells were treated with DMSO (negative control) (**A**), MK-2206, a selective AKT inhibitor (5 µM) (**B**), and Dox (0.5 µg/mL) for 2 h. The cells were further washed from the compounds indicated above and cultured for 8 h in absence (Dox washout) (**C**) or presence of MK-2206 (Dox washout + MK-2206) (**D**). Right panel-histograms illustrating the intensity of γ-H2AX-specific fluorescence at the single-nucleus level. GIST T-1R cells were grown on slides for 24 h and treated with Dox and MK-2206 as indicated above. Cells were fixed with paraformaldehyde and stained with DAPI (blue) and γ-H2AX-specific antibody (red). The intensity of γ-H2AX-specific fluorescence was measured for each nucleus (DAPI) and calculated automatically. All images were acquired by GE Cytell imager as shown in Section 4.10.

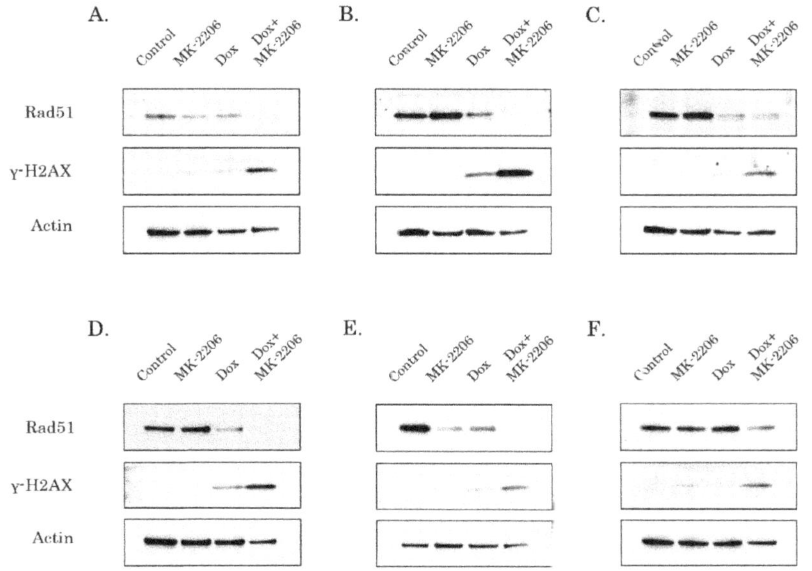

Figure 5. An impact of AKT inhibition on γ-H2AX and Rad51 recombinase in cancer cells. Immunoblot analysis was performed on RD rhabdomyosarcoma (**A**), U2-OS osteosarcoma (**B**), HT-1080 fibrosarcoma (**C**), GIST T-1R (**D**), GIST 430 (**E**), SK-LMS-1 leiomyosarcoma and (**F**) treated with DMSO (control), Dox (0.25 g/mL). MK-2206 (5 M) alone and in combination (e.g., Dox + MK-2206) for 48–72 h. Actin was used as a loading control.

The impact of AKT-signaling on Rad51 expression was also assessed by immunofluorescence staining. Non-treated GIST T-1R cells exhibited a low level of γ-H2AX/Rad51 foci, whereas vast majority of tumor cells became γ-H2AX/Rad51-foci positive after Dox treatment (Figure 7A).

Strikingly, a substantial decrease of Rad51 foci-positive cells was found when cancer cells were treated with Dox in combination with MK-2206 (Figure 7A,B). Moreover, inhibition of AKT-signaling attenuated co-localization between Rad51 and γ-H2AX foci after Dox treatment (Figure 7C, bottom panel), thus suggesting a failure in the recruitment of Rad51 to DSBs in AKT-inhibited tumor cells after Dox treatment. Similar results were observed for STS cells lines, including RD rhabdomyosarcoma cells, as shown in Figure S7. Indeed, the majority of Dox-treated cells exhibited high co-localization between Rad51 and γ-H2AX foci, whereas inhibition of AKT-signaling dramatically reduced the number of cells with co-localized proteins indicated above.

Of note, increased number of Rad51 foci in Dox-treated cells was in contrast to the substantial decrease of Rad51 expression observed in most of cancer cells utilized in present study, as shown in Figure 4. These might be due to the differences in the methodologies utilized for these assays. Indeed, RIPA buffer which is commonly utilized for WB analysis might be not effective in extracting of the chromatin-bound proteins, whereas the increased number of Rad51 nuclear foci observed during immunofluorescence staining mainly represents Rad51 associated with chromatin. To confirm this possibility, we examined the distribution of Rad51 in different subcellular fractions in GIST/STS cells treated with Dox alone or in presence of AKT inhibitor, as well. The results showed that while in control cells Rad51 was evenly distributed between Triton X-100 soluble (i.e., cytoplasmic and soluble nuclear) and DNAase-soluble fractions (i.e., chromatin-associated), inhibition of AKT-signaling with MK-2206 induced its re-distribution from the chromatin, whereas Dox brought a pronounced immobilization of the protein in the DNAase-soluble fraction (Figure S8). Strikingly, inhibition of AKT signaling in Dox-treated cells attenuated recruitment of Rad51 in chromatin-associated fraction, thereby revealing the attenuation of Rad51-mediated repair mechanisms. In contrast to Rad51, expression of pRPA (Ser 4/8) in DNA-soluble fraction did not differ between GIST T-1R cells treated with Dox alone or in presence with AKT inhibitor, thereby suggesting that processing of DSBs in AKT-inhibited cells remained unchanged.

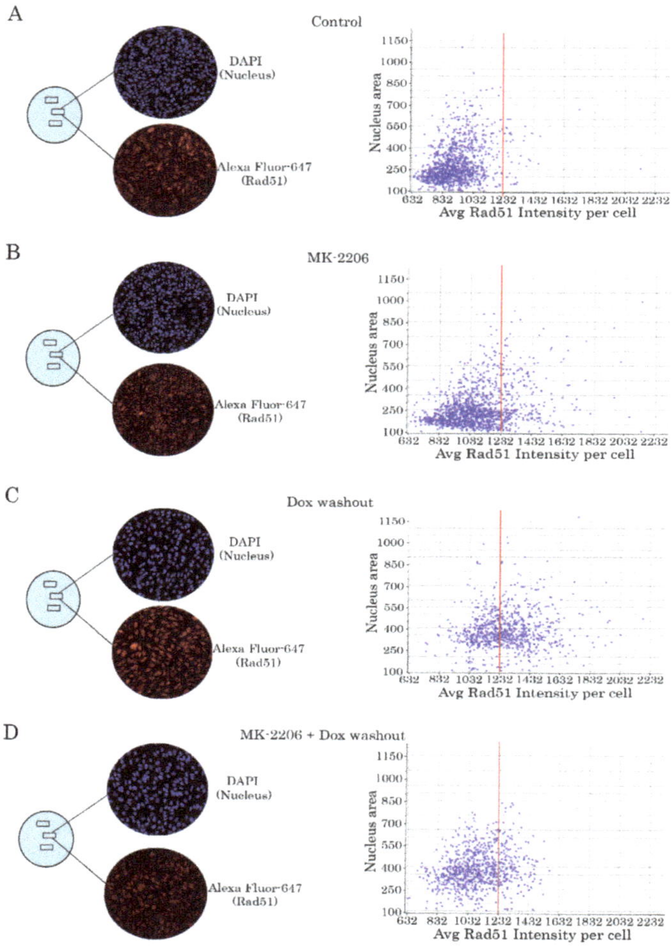

Figure 6. Inhibition of AKT-signaling attenuates Rad51 expression in Dox-treated cells (representative experiment). (**A–D**) GIST T-1R cells were treated similarly as shown for Figure 4. Left panel—cells were fixed with paraformaldehyde and stained with DAPI (blue) to outline the nucleus and Rad51-specific antibody (red). The intensity of Rad51-specific fluorescence was assayed for each nucleus (DAPI) and quantified automatically. Right panel—histograms illustrating the intensity of Rad51-specific fluorescence at the single-nucleus level. All images were acquired by GE Cytell imager as shown in "Materials and methods".

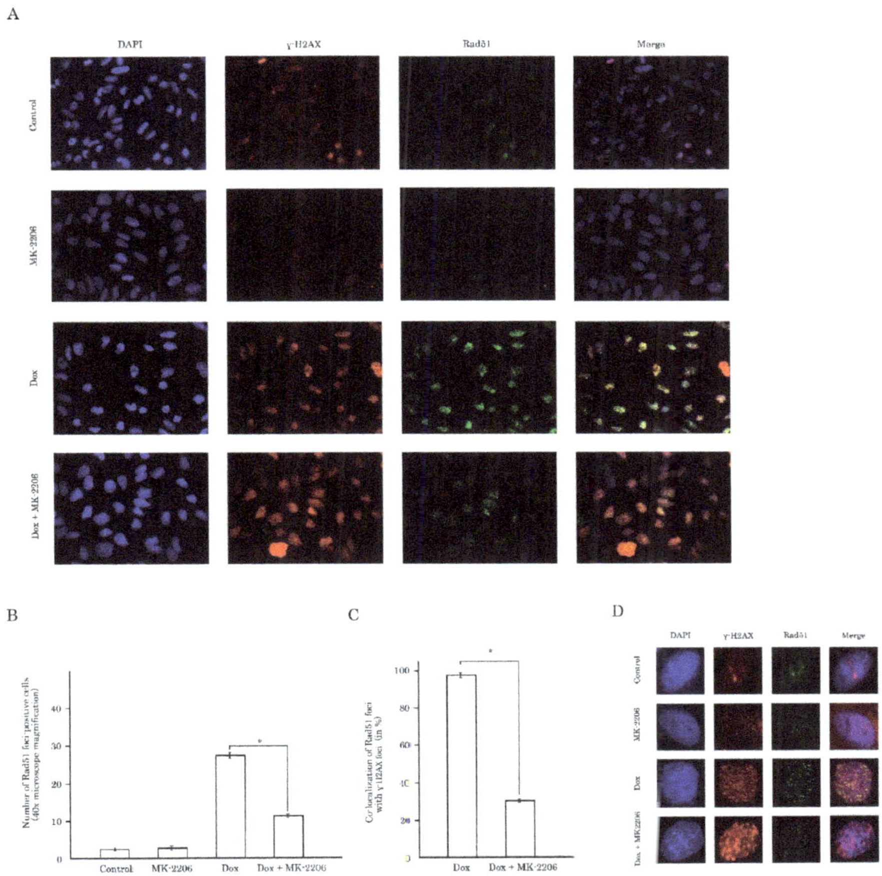

Figure 7. Blockage of Akt signaling pathway disrupts involvement of the Rad51 protein in the repair of DNA DSBs in GIST T-1R cells. (**A**) immunofluorescence staining of GIST T-1R cells for γ-H2AX and Rad51. The cells were cultured in the presence of DMSO (control), MK-2206 (5 μM for 48 h), Dox (0.25 μg/mL for 4 h) or pretreated with MK-2206 (5 μM) for 48 h prior Dox treatment. DAPI staining (blue) was used to outline the nucleus. Scale bars = 20 μm. (**B**) graph depicting the number of GIST T-1R cells positive for Rad51 foci after Dox treatment alone or in presence of MK-2206 from three independent experiments. Cells treated with DMSO (control) and MK-2206 were used as the negative controls. * $p < 0.05$. (**C**) graph showing co-localization of Rad51 foci with γ-H2AX foci in GIST T1-R cells after Dox treatment alone or in presence of MK-2206 from three independent experiments. * $p < 0.05$. (**D**) distribution of γ-H2AX and Rad51 foci in the nucleus in a single-cell level. GIST T-1R cells were pretreated with DMSO (control) or MK-2206 (5 μM) for 48 h prior Dox treatment (0.25 μg/mL for four hours). DAPI staining (blue) was used to outline the nucleus. Scale bars = 10 μm.

2.5. AKT Impacts Rad51 Stability in GIST and STS Cell Lines

Given that AKT inhibition has a negative impact of Rad51 levels in the number of Rad51 foci in Dox-treated GIST and STS, we further examined the molecular mechanisms responsible of this phenomenon.

First, RT-PCR data illustrated very small, and statistically insignificant differences between Rad51 mRNA transcripts in control and MK-2206 treated GIST T-1R cells. Similarly, no differences in Rad51 *mRNA* levels were observed between non-treated and MK-2206-treated SK-LMS-1 leiomyosarcoma and RD rhabdomyosarcoma cell lines, thereby revealing that inhibition of AKT-signaling has no impact on *Rad51* transcription in GIST and STS (Figure S9).

Next, we examined whether low levels of Rad51 in AKT-inhibited cells were due to increased turnover of the protein via ubiquitin-mediated proteasomal pathway. To test this possibility, we initially inhibited synthesis of the new protein by cycloheximide (CHX) and compared Rad51 levels in control and MK-2206-treated cells. Indeed, we observed the substantial decrease of half-life of Rad51 in GIST cells treated with CHX in presence of MK-2206, thereby suggesting about the rapid turnover of protein in AKT-inhibited cancer cells (Figure 8, upper and middle panels). Next, we found that MG-132, a 26S proteasome inhibitor effectively restored Rad51 levels in AKT-inhibited cells, thereby revealing an increase proteasome-mediated degradation of Rad51 in MK-2006-treated cells (Figure 8, bottom panel).

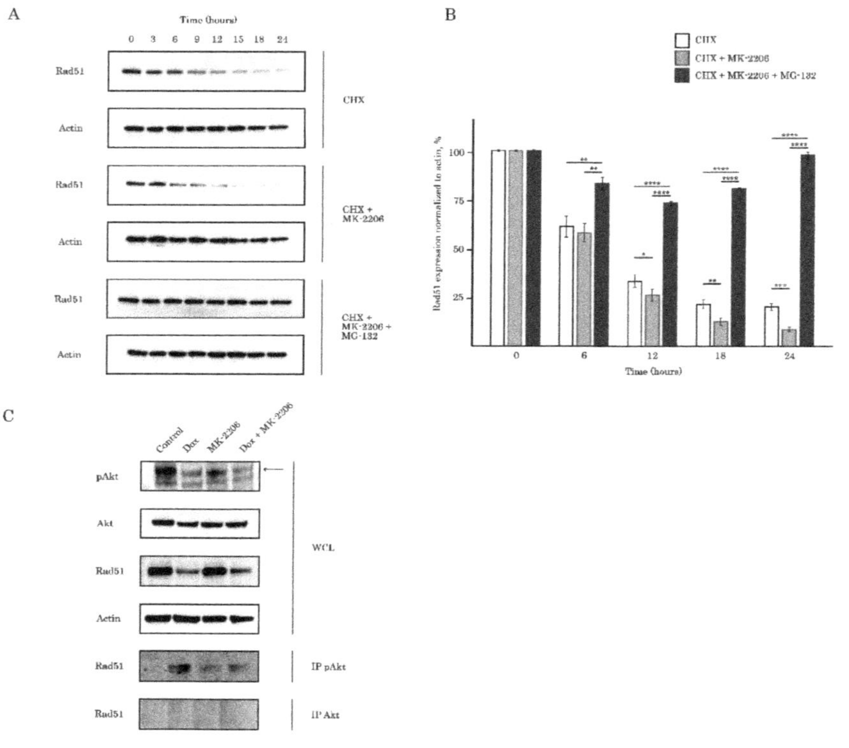

Figure 8. AKT is required for regulation of Rad51 stability. (**A**) the half-life of Rad51 was determined by cycloheximide (CHX) chase analysis. GIST T-1R cells were treated with MK-2206 (5 μM) or DMSO (control), followed by treatment with 10 μg/mL cycloheximide for 0, 3, 6, 9, 12, 15, 18 and 24 h. Immunoblotting for Rad51 and actin was performed on samples from all time points. To examine an impact of AKT-inhibition on proteasomal degradation of Rad51, cells were treated with MK-2206 and CHX in presence of MG-132 (2 μM) and were subjected for immunoblotting for Rad51 and actin as a loading control. (**B**) densitometric analysis of Rad51 in CHX chase analysis indicated above; bars, SD. * $p < 0.05$; ** $p < 0.01$; *** $p < 0.001$; **** $p < 0.0001$. (**C**) GIST T-1R cell lysates were immunoprecipitated with total (left) or phosphorylated at Ser473 (right) AKT Abs and immunoblotted with Rad51 Abs to demonstrate endogenous complex formation. A whole cell lysate (WCL) was included. pAKT expression is shown by arrows. Actin was used as a loading control. For immunoprecipitation experiments cells were treated with Dox (1 μg/mL) and MK-2206 (10 μM).

Given the functional link between Akt and Rad51, we hypothesized that they might be in a complex together. For this purpose, we examined the interaction between endogenous Akt/pAkt and Rad51. Lysates from GIST T-1R cells were immunoprecipitated for Akt or pAkt Ser473, followed by Western blotting with Rad51 antibody to detect the complexes. As shown in Figure 8B, specific co-immunoprecipitation was detected between Rad51 and pAkt (but not total Akt) after Dox treatment, thus confirming the interaction with endogenous proteins and reveling physical interaction between activated Akt and Rad51 after DNA damage. Strikingly, inhibition of AKT signaling in Dox-treated cells abolished this interaction. As expected, negative results were also observed in control (non-treated cells) and cells exposed to MK-2206.

All together, these results suggest that AKT inhibition in tumor cells leads to Rad51 down-regulation via ubiquitin-mediated proteasome pathway.

Collectively, our data illustrates that inhibition of AKT pathway in GIST and STS attenuates homology-mediated DNA repair and sensitizes tumor cells to the DNA-damaging agents, such as doxorubicin.

3. Discussion

It is well-known that an efficiency of most commonly used anticancer agents, including the classic genotoxic chemotherapeutic drugs and ionizing radiation, is due to their DNA-damaging properties which in turn triggers apoptotic cell death of cancer cells harboring unrepairable DNA lesions. Apoptosis induced by DNA lesions mainly results from double-strand breaks and stalled replication forks, which activates DNA damage checkpoints networks consisting of DNA damage sensors, signal transducers and effector pathways. The central sensors are composed of the PI3K-related kinases and include ataxia telangiectasia mutated (ATM), ataxia telangiectasia and Rad3-related (ATR) and DNA-dependent protein kinase (DNA-PK). Upon activation by DSBs and DNA replication blocks, PI3Ks indicated above activate specific substrates that mediate replication fork stability, cell cycle arrest, DNA repair and apoptosis.

The accumulating evidence implicated PI3Ks and AKT are tightly coregulated in both checkpoint response and DNA DSB repair. This reflected, predominantly, the regulatory role of AKT in NHEJ-mediated DNA repair. Indeed, AKT and DNA-PK were shown to activate each other to induce and maintain an effective NHEJ-mediated DNA repair. For example, Stronach et al. demonstrated that AKT relocates to the nucleus of cisplatin-resistant cancer cells where it is phosphorylated specifically on S473 by DNA-PK, and this activation inhibits cisplatin-mediated apoptosis. Importantly, direct interaction between DNA-PK and AKT in cisplatin-resistant but not sensitive cells was revealed by immunoprecipitation [25]. The regulatory role of DNA-PK in Akt-S473 phosphorylation was also shown for glioblastoma cells in response to IR [26]. The role of AKT in DNA-PKcs-dependent DNA DSB repair has been extensively studied by Toulany M. with co-authors. In particular, they showed that inhibition of AKT1 effectively sensitized non-small cell lung cancer cell lines to IR by inhibiting DNA-PKcs-dependent DNA DSB repair [27]. They also found that AKT (predominantly Akt1) mediated DNA-PKcs autophosphorylation at S2056 that is required for efficient DNA-DSB repair and release of DNA-PKcs from the DNA damage sites induced by IR. Akt1 played a critical role in formation of the functional complex to DNA duplex ends marked by Ku dimers [28]. Lastly, they found that Akt1 and Akt3, but not Akt2, interact with DNA-PKcs in K-Ras mutant cells and stimulate DSB repair, thereby protecting cancer cell against IR [29].

In contrast to NHEJ, the data illustrating the regulatory role of AKT in homology-mediated DNA repair (HR) is controversial. For example, Mueck K. with co-authors demonstrated that regulatory role of Akt1 in Rad51 foci formation in IR-treated NSCLC cells and recruitment of Rad51 recombinase to γ-H2AX foci, a well-known marker of DNA DSBs. This was in a concordance with a significant decrease of Rad51 protein in the nucleus of irradiated cells exhibiting AKT knockdown (AKT-KD). Lastly, the increased number of BRCA1 foci in AKT-KD cells exposed to IR also illustrated the impaired HR repair which was revealed by using SceI-based GFP-reporter assay [30]. In contrast,

Plo I. with co-authors observed the opposite effect of AKT in homology-mediated DNA repair. For example, AKT silencing restored formation of IR-induced BRCA1 foci in breast cancer cells, whereas HR-related proteins (e.g., BRCA1 and Rad51) sequestered in the cytoplasm upon activation of AKT1. Important, this was observed in tumor cell lines and biopsies from AKT-high sporadic breast cancers, thereby illustrating that AKT1 inhibits homologous recombination in breast cancer cells in vitro and in vivo [31]. Similarly, negative regulatory role of Akt on homology-mediated repair was shown for BRCA1-deficient breast cancer cells. In particular, Akt1 promoted chromosome instability in Brca1-deficent cells by impairing nuclear localization of Chk1 and disrupting its interaction with Rad51, thereby leading to attenuation of homology-mediated repair [32].

Activation of AKT-mediated pathway is well-documented for the soft tissue sarcomas and can be used as independent prognostic factor for tumor recurrence, overall survival (OS) and disease-free survival (DFS) [12,33,34]. Similarly, activation of PI3-kinase/AKT pathway was well-documented for GIST cell lines and patient samples and played an important role for was critical to survival in IM-resistant GIST [1–3,7], thereby illustrating a rationale for combination therapy (e.g., of imatinib mesylate and AKT inhibitor) for patients with GIST [8].

Despite the sensitivity of STS and (in less extent) GIST to DNA-damaging agents was shown in multiple reports [20–23], however, to date, a little is known about the role of AKT in DNA DSB repair (in particular HR) in STS and GIST. Given that inhibition of FGF-signaling in GIST effectively sensitized them to Topo II inhibitors via attenuating HR-mediated DNA DSB repair [35], we sought to examine the downstream signaling pathways responsible for this phenomenon. We present here the novel data illustrating that AKT- but not a MEK-signaling pathway regulates an efficiency of homology-mediated DNA damage repair in STS and GIST. In particular, phosphorylated (i.e., activated) form of endogenous Akt physically interacts with Rad51 recombinase after DNA damage (Figure 8C). This was in consistency with previous data illustrating that Akt might be physically associated with various DNA DSBs repair proteins. For example, Toulany M. with the colleagues demonstrated that Akt1 phosphorylated at Ser472/Ser473 interacts with activated (i.e., phosphorylated at T2609) DNA-PKcs in various cell lines including NSCLC cells A549 and H460. Of note, this interaction was found shortly (3–5 min) of post IR exposure (4Gy), thereby illustrating an important regulatory role of Akt1 in DDR signaling [27]. The detailed analysis of this interaction demonstrated that Akt1 interacts with DNA-PKs through its C-terminal domain, stimulates autophosphorylation of DNA-PKcs at S2056, promotes its kinase activity and accumulation of DNA-PKcs at DNA-DSBs [28]. Similarly, by utilizing the proximity ligation assay, Sahlberg S. with co-authors demonstrated that phosphorylated forms of Akt and DNA-PKcs also interact with each other in IR-exposed colon cancer cell lines, whereas Akt deficiency significantly impaired the rejoining of radiation-induced DNA double strand breaks and sensitized colon cancer cells to IR [36]. Next, we found that AKT regulates Rad51 protein stability via proteasome-dependent pathway (Figure 8A,B) and MK-2206-induced inhibition of AKT-signaling has a strong impact of Rad51 expression, especially after Dox-induced DNA damage. Again, our data is in a close consistency with the data published recently by Mueck C. with co-authors illustrating that knockdown of Akt1 significantly reduced the amount of Rad51 protein in the nuclear fraction of irradiated non-small ling cancer cells [30]. We also observed that AKT signaling plays a regulatory role in the recruitment of Rad51 to DNA damage sites which was evidenced by a substantial decrease of number of residual Rad51 foci in Dox-treated tumor cells (Figure 7). As a result, inhibition of AKT-signaling pathway effectively sensitized STS and GIST to DNA-damaging agent, doxorubicin, and decreased tumor cell viability due by enhancing of apoptotic cell death (Figures 1 and 2, Figures S3 and S4).

Collectively, our data illustrates that overactivation of AKT-signaling pathway in STS and GIST might serve as a prospective molecular target to enhance cytotoxicity of DNA-topoisomerase II inhibitors against STS and GIST. This is in consistency with several reports illustrating a perspective role of PI3K/AKT/mTOR inhibitors to sensitize STS to standard chemotherapeutic regimens, including Topo II inhibitors. For example, Babichev Y. with colleagues demonstrated a high potency of PI3K vs.

mTOR inhibitors used in combination of doxorubicin against LMS in vitro and in the leiomyosarcoma xenografts [37]. Similarly, AKT inhibition by pentacyclic triterpene ursolic acid (UA) effectively sensitized human STS cell lines in vitro to Dox treatment and induced their apoptosis [38]. Moreover, phosphorylation of AKT which was observed in STS cell lines after Dox treatment (10 M) illustrates that activation of AKT signaling represents a compensatory mechanism counteracting the cytotoxic and anti-proliferative effects of Dox [38]. This, in turn, illustrates a rationale of AKT inhibition in cancer cells treated with Dox. Our study effectively supplements these findings and illustrates the molecular mechanism responsible for this phenomenon. We show here for the first time that inhibition of AKT pathway in Dox-treated STS and GIST reduces Rad51 levels and inhibits recruitment of Rad51 recombinase to sites of DNA DSBs after Dox treatment, thereby attenuating the homology-mediated repair mechanisms. This was evidenced by substantial decrease of Rad51 foci that were co-localized with γ-H2AX when cancer cells were treated with Dox in presence of AKT inhibitor (Figure 7 and Figure S7). Moreover, data obtained from fractionation experiment revealed the substantially decrease in Rad51 level in chromatin fraction when cancer cells were treated with Dox in presence of MK-2206 but not with Dox alone (Figure S8). Thus, AKT inhibition may have a synergistic effect with Dox, standard chemotherapy for several types of STS due to inhibition of DNA DSB repair via attenuation of recruitment of Rad51 recombinase to the DNA damage lesions. Of note, we did not observe the enhanced phosphorylation of AKT in STS and GIST after Dox-induced DNA damage (Figure 2). This might be due to enhanced basal level of expression of phosphorylated form of AKT in STS and GIST cells and also in consistency with previous findings, illustrating that Dox treatment enhances AKT phosphorylation in sarcoma cell lines when was used in a high dose (10M), whereas no increase of AKT phosphorylation was observed when the cells were treated with 1M of Dox [38], which was similar to our experimental conditions. Of note, the effect of inhibition of AKT signaling might be not limited to sensitization of STS to the Topo II inhibitors. For example, MK-2206 and eribulin, a microtubule dynamics inhibitor, synergistically inhibited STS cell growth in vitro and in vivo, also revealing a rationale for the development of an AKT inhibitor in combination with eribulin for therapy of patients with STS [39]. Similarly to Dox-treated STS, expression of phosphorylated AKT significantly increased in eribulin-treated STS, thereby revealing AKT phosphorylation a common compensatory mechanism counteracting the cytotoxic and anti-proliferative effects of chemotherapeutic agents used for therapy of patients with STS. Thus, future studies with inhibitors targeting these pathways are warranted.

4. Materials and Methods

4.1. Chemical Compounds

Doxorubicin (Dox) were obtained from Sigma (St. Louis, MO, USA), MK-2206 and U0126 were purchased from Selleck Chem (Houston, TX, USA).

4.2. Antibodies

The following primary antibodies were used for immunoblotting and immunofluorescence: Cleaved form of caspase-3 (#9662S), phospho-AKT S473 (#4060P), AKT (#4691P), phospho-histone H2A.X Ser139 (#9718S) (Cell Signaling, Danvers, MA, USA), PARP (#436400) (Life Technologies, Carlsbad, CA, USA), γ-H2AX S139 (sc-517348), Rad51 (sc-8349) (Santa Cruz Biotechnology, Santa Cruz, CA, USA), Rad51 (ab-133534, Abcam, Burlingame, CA, USA), phosphor-RPA32 Ser4/8 (A300-245)(Bethyl Laboratories, Montgomery, TX, USA), beta-actin (A00730-200, Gene Script, Piscataway, NJ, USA), HRP-conjugated secondary antibodies for Western blotting were purchased from Santa Cruz Biotechnology (Santa Cruz, CA, USA) and Alexa Fluor 488- or Texas Red-conjugated secondary antibodies for immunofluorescence staining were obtained from Invitrogen (Carlsbad, CA, USA).

4.3. Cell lines and Culture Conditions

SK-LMS-1 leiomyosarcoma, RD rhabdomyosarcoma, HT-1080 fibrosarcoma and U2-OS osteosarcoma cell lines were purchased from American Type Culture Collection (ATCC). GIST T-1 cell line was generated from a metastatic pleural tumor from stomach GIST. This cell line exhibited the heterozygous 57-base pair deletion (V570-Y578) in *KIT* exon 11 [40]. IM-resistant GIST T-1R subline was generated in our laboratory. For this purpose, GIST T-1 cell line was continuously cultured with IM with a gradually increased concentration of the drug from 0.4 nM to 1000 Nm [41]. The IM-resistant GIST 430 cell line was generated from human GIST after developing resistance to IM therapy. This cell line has a heterozygous primary *KIT* exon 11 deletion (V560_L576del) and a secondary *KIT* exon 13-point mutation (V654A) [1]. GIST cell lines were cultured in a humidified atmosphere of 5% CO_2 at 37 °C (LamSystems, Myass, Russia).

4.4. Cellular Survival MTS-Based Assay

STS and GIST cells were seeded in 96-well flat-bottomed plates (Corning Inc., Corning, NY, USA) and were cultured for 24 h. The cells were further incubated for 48–72 h with indicated concentrations of MK-2206, doxorubicin or DMSO (control). To assess the number of living cells, the MTS reagent (Promega, Madison, WI, USA) was added to the culture medium for at least 1 h. Cellular viability was determined using a MultiScan FC plate reader (Thermo Fisher Scientific, Waltham, MA, USA) at 492 nm. Calculated IC50 data were defined as the concentration of compound that inhibits cell growth by 50% at 48–72 h. Data were normalized to the DMSO-treated control group.

4.5. Western Blotting and Co-Immunoprecipitation (Co-IP)

For Western blotting analysis, cells were washed twice with PBS and lysed on ice for 20 min by using a RIPA buffer (25 mM Tris-HCl pH 7.6, 150 mM NaCl, 5 mM EDTA, 1% NP-40, 1% sodium deoxycholate and 0.1% SDS), supplemented with protease and phosphatase inhibitors. Cellular extracts were further centrifuged for 30 min at 13,000 rpm at 4°C. The pellet was removed, and the protein concentration in the supernatant (whole-cell lysate) was determined by the Bradford assay. The samples containing 30 µg of protein were resolved on 4% to 12% Bis-Tris or 3% to 8% Tris-acetate NuPAGE gels (Invitrogen, Carlsbad, CA, USA), transferred to a nitrocellulose membrane (Bio-Rad, Hercules, CA, USA), probed with specific antibody, and visualized by enhanced chemiluminescence (Western Lightning Plus-ECL reagent, Perkin Elmer, Waltham, MA, USA) on an gel-documenting system "Fusion Solo WL.2M" (Vilber Lourmat, Collégien, France). Densitometric analysis was performed in FIJI Software (Laboratory for Optical and Computational Instrumentation, University of Wisconsin, Madison, WI, USA).

For Co-IP, cells were washed with PBS and lysed on ice for 15 min using by TEB buffer (50 mM Tris-HCl pH 7.5, 150 mM NaCl, 1% NP-40 and 10% glycerol), supplemented with protease and phosphatase inhibitors. The lysates were cleared by centrifugation and further incubated with the corresponding precipitating Abs overnight (rotating device at 4 °C). Then protein A and G Sepharose beads (Santa Cruz Biotechnology, Santa Cruz, CA, USA) were added to the samples and incubated for 1 h (rotating device at 4 °C). The beads were washed 3 times with TEB buffer and resolved by SDS-PAGE. Then 1X sample buffer was added to the resolved precipitates and the mixture was boiled for 5 min. Subsequently, Western blotting was performed as indicated above.

4.6. Immunofluorescence Staining

Cells were seeded on glass coverslips coated with poly-L-lysine (Sigma-Aldrich, St. Louis, Missouri, USA) and allowed to attach for 48 h before treatment. After washing with PBS, cells were fixed in 4% paraformaldehyde solution (in PBS) for 30 min at 4 °C and further permeabilized with 0.5% Triton X-100 for 5 min at 4 °C. Then the cells were incubated for 30 min in blocking solution containing 10% normal goat serum and 0.5% bovine serum albumin (in PBS). The cells were further incubated for 30 min with the blocking solution containing 10% goat serum and 0.5% bovine serum albumin. After blocking procedure, the cells were incubated with primary antibodies for overnight at 4 °C. Next day the cells were washed twice with PBS, incubated with Alexa Fluor 488, or TexRed-conjugated secondary antibodies (Invitrogen, Carlsbad, CA, USA) for 30 min at room temperature in the dark. After brief DAPI (Sigma-Aldrich, St. Louis, MI, USA) staining, the coverslips were mounted on glass slides and visualized on fluorescence microscope "Olympus BX63" (Tokyo, Japan). Finally, the images were captured by using a Spot advanced imaging system.

4.7. RNA Extraction and Real-Time Quantitative PCR

Total RNA was extracted from cancer cells according to the standard protocol as described elsewhere [41]. PCR reaction mix for real-time quantitative PCR (qPCR) has consisted of the following components: 1 µL synthesized cDNA, 5x qPCRmix-HS SYBR (PB025, Evrogen, Moscow, Russia), 10 mM each forward and reverse PCR primers for *RAD51* or control genes. Real-time qPCR was carried according to the manufacturer's protocol by the CFX96 Real-Time detection system (Bio-Rad, Hercules, CA, USA). The absolute levels of each mRNA were normalized relative to *GAPDH* as a control gene. The production of quantitative data based on the number of cycles required for the fluorescent detection amplifying of target genes (the Ct value). The relative level of expression of the target genes was based on the following formula $2^{-\Delta\Delta Ct}$.

4.8. Single-Cell Electrophoresis (Comet Assay)

Formation of DNA DSBs was assayed by using the alkaline version of single-cell electrophoresis. Comet slides were stained by SYBR Green (Trevigen, Gaithersburg, MD, USA) and visualized by fluorescent microscopy (Olympus BX63, Tokyo, Japan). Comet assay parameters (TM and OTM) were evaluated by the ImageJ software. 50 nuclei were evaluated per treatment. Differences between control and treated cells in TM and OTM were analyzed by Kruskal–Wallis test followed by Dunn's test with Benjamini–Hochberg adjustment in R software (R Foundation for Statistical Computing, Vienna, Austria; URL https://www.R-project.org/).

4.9. Sub-Cellular Fractionation

For sub-cellular fractionation experiment cancer cells were lysed according to the protocol by Mladenov et al. [42]. Briefly, cells were collected, washed in PBS and Triton X-100-soluble fraction was extracted by using CSK-buffer. The pellet was washed with buffer A and then extracted with buffer A containing Benzonase® Nuclease (Sigma, St. Louis, MO, USA). The supernatant was considered as a DNase-soluble fraction. The pellet was washed and extracted with buffers containing increasing concentrations of ammonium sulfate. The remaining pellet was labeled as the nuclear matrix fraction.

4.10. Images Quantification and Registration

Cells were seeded into 6-well plates and allowed to grow for 48 h before treatment. The cells were fixed, permeabilized and stained for γ-H2AX or Rad51 by using specific antibodies. Plates were imaged with a × 10 objective using a Cytell Cell Imaging System (GE Healthcare, Chicago, Illinois, USA). MyBioApp Protocol with specified parameters was created to acquire the data and to quantify the signal intensity in nuclei, cytoplasm, and in whole cells. DAPI (blue channel) was used for the nuclear masks, whereas Alexa Fluor 647 Mouse (red channel) was used for γ-H2AX-staining and Alexa Fluor 647 Rabbit (red channel) for Rad51-staining. Graphics illustrating the average intensity of the nuclear γ-H2AX-staining and Rad51-staining were automatically generated by using MyBioApp Protocol and then passed to MS Excel for further processing and analysis.

4.11. Statistics

All the experiments were performed for a minimum of three times. Data for each group are indicated as the mean ± standard deviation (SD). The results were considered statistically significant at $p < 0.05$. CompuSyn Version 1.0 (CompuSyn Inc., New York, NY, USA), based on the Chou–Talalay algorithm [43] was used to calculate the values of CI at each molar ratio of Dox and MK-2206. The combinations of the drugs indicated above that yielded CI values < 1 were considered as synergistic [44,45]. The degree of combination effects were also quantified by using R-package of the computational tool SynergyFinder (https://bioconductor.org/packages/release/bioc/html/synergyfinder.html) [46]. Zero potency interaction (ZIP) reference model was used to calculate synergy. Average delta score over a dose–response matrix was used as a summary interaction score for a drug combination. Value of delta score >0, =0 or <0 corresponds to the synergy, zero interaction and antagonism, respectively [46]. Means of normalized protein levels were compared by using the analysis of variance (ANOVA) with subsequent pairwise comparisons (Tukey HSD test) in R software (R Foundation for Statistical Computing, Vienna, Austria; URL https://www.R-project.org/).

5. Conclusions

Collectively, our data illustrates that inhibition of AKT-signaling pathway STS and GIST might be considered as a potent tool to enhance their sensitivity to doxorubicin, a topoisomerase II inhibitor. This might be due to the attenuation of homology-mediated DNA repair in AKT-inhibited tumor cells. In particular, we observed the decrease of Rad51 on the protein level in AKT-inhibited cells and attenuated recruitment of Rad51 recombinase to the chromatin and the sites of DSBs after Dox treatment. As a consequence of attenuation of homology-mediated repair of DNA DSBs, AKT-inhibited tumor cells underwent the apoptotic cell death after Dox treatment, which was evidenced by increases expression of well-known apoptotic markers—cleaved forms of caspase-3 and PARP.

Supplementary Materials: Supplementary materials can be found at http://www.mdpi.com/1422-0067/21/22/8842/s1.

Author Contributions: Conception and design: S.B.; Development of methodology: S.B., A.G., F.B., A.A., P.D., E.V., I.N., A.S.; Acquisition of data (provided animals, acquired and managed patients, provided facilities, etc.): S.B., A.G., F.B., A.A., P.D., E.V., I.N., A.S.; Analysis and interpretation of data (e.g., statistical analysis, biostatistics, computational analysis): S.B., A.G., F.B., A.A., P.D., I.N.; Writing, review and/or revision of the manuscript: S.B.; Administrative, technical, or material support (i.e., reporting or organizing data, constructing databases): S.B.; Study supervision: S.B. All authors have read and agreed to the published version of the manuscript.

Funding: This study was supported by grant from Russian Science Foundation (RSF) (grant No. 20-15-00001).

Acknowledgments: We thank Kirill Syuzov for assistance in the statistical analysis.

Conflicts of Interest: The authors declare no conflict of interest.

References

1. Bauer, S.; Duensing, A.; Demetri, G.D.; Fletcher, J.A. KIT oncogenic signaling mechanisms in imatinib-resistant gastrointestinal stromal tumor: PI3-kinase/AKT is a crucial survival pathway. *Oncogene* **2007**, *26*, 7560–7568. [CrossRef] [PubMed]
2. Heinrich, M.C.; Corless, C.L.; Blanke, C.D.; Demetri, G.D.; Joensuu, H.; Roberts, P.J.; Eisenberg, B.L.; von Mehren, M.; Fletcher, C.D.; McDougall, K.; et al. Molecular correlates of imatinib resistance in gastrointestinal stromal tumors. *J. Clin. Oncol.* **2006**, *24*, 4764–4774. [CrossRef] [PubMed]
3. Wang, C.M.; Huang, K.; Zhou, Y.; Du, C.Y.; Ye, Y.W.; Fu, H.; Zhou, X.Y.; Shi, Y.Q. Molecular mechanisms of secondary imatinib resistance in patients with gastrointestinal stromal tumors. *J. Cancer Res. Clin. Oncol.* **2010**, *136*, 1065–1071. [CrossRef] [PubMed]
4. Yang, J.; Ikezoe, T.; Nishioka, C.; Takezaki, Y.; Hanazaki, K.; Taguchi, T.; Yokoyama, A. Long-term exposure of gastrointestinal stromal tumor cells to sunitinib induces epigenetic silencing of the PTEN gene. *Int. J. Cancer* **2012**, *130*, 959–966. [CrossRef] [PubMed]
5. Ricci, R.; Maggiano, N.; Castri, F.; Rinelli, A.; Murazio, M.; Pacelli, F.; Potenza, A.E.; Vecchio, F.M.; Larocca, L.M. Role of PTEN in gastrointestinal stromal tumor progression. *Arch. Pathol. Lab. Med.* **2004**, *128*, 421–425.

6. Patel, S. Exploring novel therapeutic targets in GIST: Focus on the PI3K/Akt/mTOR pathway. *Curr. Oncol. Rep.* **2013**, *15*, 386–395. [CrossRef]
7. Van Looy, T.; Wozniak, A.; Floris, G.; Sciot, R.; Li, H.; Wellens, J.; Vanleeuw, U.; Fletcher, J.A.; Manley, P.W.; Schöffski, P.; et al. Phosphoinositide 3-kinase inhibitors combined with imatinib in patient-derived xenograft models of gastrointestinal stromal tumors: Rationale and efficacy. *Clin. Cancer Res.* **2014**, *20*, 6071–6082. [CrossRef]
8. Zook, P.; Pathak, H.B.; Belinsky, M.G.; Gersz, L.; Devarajan, K.; Zhou, Y.; Godwin, A.K.; von Mehren, M.; Rink, L. Combination of imatinibmesylate and AKT inhibitor provides synergistic effects in preclinical study of gastrointestinal stromal tumor. *Clin. Cancer Res.* **2016**, *23*, 171–180. [CrossRef]
9. Conley, A.P.; Araujo, D.; Ludwig, J.; Ravi, V.; Samuels, B.L.; Choi, H.; Thall, P.F.; Patel, S.; Benjamin, R.; Trent, J. A randomized phase II study of perifosine (P) plus imatinib for patients with imatinib-resistant gastrointestinal stromal tumor (GIST). *J. Clin. Oncol.* **2009**, *27*, 10563. [CrossRef]
10. Schoffski, P.; Reichardt, P.; Blay, J.Y.; Dumez, H.; Morgan, J.A.; Ray-Coquard, I.; Hollaender, N.; Jappe, A.; Demetri, G.D. A phase I-II study of everolimus (RAD001) in combination with imatinib in patients with imatinib-resistant gastrointestinal stromal tumors. *Ann. Oncol.* **2010**, *21*, 1990–1998. [CrossRef]
11. Yang, J.; Du, X.; Chen, K.; Ylipää, A.; Lazar, A.J.; Trent, J.; Lev, D.; Pollock, R.; Hao, X.; Hunt, K.; et al. Genetic aberrations in soft tissue leiomyosarcoma. *Cancer Lett.* **2009**, *275*, 1–8. [CrossRef] [PubMed]
12. Hernando, E.; Charytonowicz, E.; Dudas, M.E.; Mendez, S.; Matushansky, I.; Mills, J.; Socci, N.D.; Behrendt, N.; Ma, L.; Maki, R.G.; et al. The AKT-mTOR pathway plays a critical role in the development of leiomyosarcomas. *Nat. Med.* **2007**, *13*, 748–753. [CrossRef] [PubMed]
13. Bathan, A.J.; Constantinidou, A.; Pollack, S.M.; Jones, R.J. Diagnosis, prognosis, and management of leiomyosarcoma: Recognition of anatomic variants. *Curr. Opin.* **2013**, *25*, 384–389. [CrossRef] [PubMed]
14. Mateo-Lozano, S.; Tirado, O.M.; Notario, V. Rapamycin induces the fusion-type independent downregulation of the EWS/FLI-1 proteins and inhibits Ewing's sarcoma cell proliferation. *Oncogene* **2003**, *22*, 9282–9287. [CrossRef] [PubMed]
15. Manara, M.C.; Nicoletti, G.; Zambelli, D.; Ventura, S.; Guerzoni, C.; Landuzzi, L.; Lollini, P.L.; Maira, S.M.; García-Echeverría, C.; Mercuri, M.; et al. NVP-BEZ235 as a new therapeutic option for sarcomas. *Clin. Cancer Res.* **2010**, *16*, 530–540. [CrossRef] [PubMed]
16. Petricoin, E.F., 3rd; Espina, V.; Araujo, R.P.; Midura, B.; Yeung, C.; Wan, X.; Eichler, G.S.; Johann, D.J., Jr.; Qualman, S.; Tsokos, M.; et al. Phosphoprotein pathway mapping: Akt/mammalian target of rapamycin activation is negatively associated with childhood rhabdomyosarcoma survival. *Cancer Res.* **2007**, *67*, 3431–3440. [PubMed]
17. Zhang, J.; Yu, X.H.; Yan, Y.G.; Wang, C.; Wang, W.J. PI3K/Akt signaling in osteosarcoma. *Clin. Chim. Acta* **2015**, *444*, 182–192. [CrossRef]
18. Zhou, Y.; Zhu, L.B.; Peng, A.F.; Wang, T.F.; Long, X.H.; Gao, S.; Zhou, R.P.; Liu, Z.L. LY294002 inhibits the malignant phenotype of osteosarcoma cells by modulating the phosphatidylinositol 3-kinase/Akt/fatty acid synthase signaling pathway in vitro. *Mol. Med. Rep.* **2015**, *11*, 1352–1357. [CrossRef]
19. Schöffski, P.; Cornillie, J.; Wozniak, A.; Li, H.; Hompes, D. Soft tissue sarcoma: An update on systemic treatment options for patients with advanced disease. *Oncol. Res. Treat.* **2014**, *37*, 355–362. [CrossRef]
20. Judson, I.; Verweij, J.; Gelderblom, H.; Hartmann, J.T.; Schöffski, P.; Blay, J.Y.; Kerst, J.M.; Sufliarsky, J.; Whelan, J.; Hohenberger, P.; et al. Doxorubicin alone versus intensified doxorubicin plus ifosfamide for first-line treatment of advanced or metastatic soft-tissue sarcoma: A randomised controlled phase 3 trial. *Lancet Oncol* **2014**, *15*, 415–423. [CrossRef]
21. Dickson, M.A.; D'Adamo, D.R.; Keohan, M.L.; D'Angelo, S.P.; Carvajal, R.D.; Gounder, M.M.; Maki, R.G.; Qin, L.X.; Lefkowitz, R.A.; McKennon, O.R.; et al. Phase II Trial of Gemcitabine and Docetaxel with Bevacizumab in Soft Tissue Sarcoma. *Sarcoma* **2015**, *2015*, 532478. [CrossRef] [PubMed]
22. García-Del-Muro, X.; López-Pousa, A.; Maurel, J.; Martín, J.; Martínez-Trufero, J.; Casado, A.; Gómez-España, A.; Fra, J.; Cruz, J.; Poveda, A.; et al. Randomized phase II study comparing gemcitabine plus dacarbazine versus dacarbazine alone in patients with previously treated soft tissue sarcoma: A Spanish Group for Research on Sarcomas study. *J. Clin. Oncol.* **2011**, *29*, 2528–2533. [CrossRef] [PubMed]
23. Boichuk, S.; Lee, D.J.; Mehalek, K.R.; Makielski, K.R.; Wozniak, A.; Seneviratne, D.S.; Korzeniewski, N.; Cuevas, R.; Parry, J.A.; Brown, M.F.; et al. Unbiased compound screening identifies unexpected drug sensitivities and novel treatment options for gastrointestinal stromal tumors. *Cancer Res.* **2014**, *74*, 1200–1213. [CrossRef] [PubMed]

24. Pessetto, Z.Y.; Ma, Y.; Hirst, J.J.; von Mehren, M.; Weir, S.J.; Godwin, A.K. Drug repurposing identifies a synergistic combination therapy with imatinibmesylate for gastrointestinal stromal tumor. *Mol. Cancer Ther.* **2014**, *13*, 2276–2287. [CrossRef] [PubMed]
25. Stronach, E.A.; Chen, M.; Maginn, E.N.; Agarwal, R.; Mills, G.B.; Wasan, G.H. DNA-PK mediates AKT activation and apoptosis inhibition in clinically acquired platinum resistance. *Neoplasia* **2011**, *13*, 1069–1080. [CrossRef] [PubMed]
26. Szymonowicz, K.; Oeck, S.; Krysztofiak, A.; van der Linden, J.; Iliakis, G.; Jendrossek, V. Restraining AKT1 phosphorylation attenuates the repair of radiation-induced DNA double-strand breaks and reduces the survival of irradiated cancer cells. *Int. J. Mol. Sci.* **2018**, *19*, 2233. [CrossRef]
27. Toulany, M.; Kehlbach, R.; Florczak, U.; Sak, A.; Wang, S.; Chen, J.; Lobrich, M.; Rodemann, H.P. Targeting of AKT1 enhances radiation toxicity of human tumor cells by inhibiting DNA-PKcs-dependent DNA double-strand break repair. *Mol. Cancer Ther.* **2008**, *7*, 1772–1781. [CrossRef]
28. Toulany, M.; Lee, K.J.; Fattah, K.R.; Lin, Y.F.; Fehrenbacher, B.; Schaller, M.; Chen, B.P.; Chen, D.J.; Rodemann, H.P. Akt promotes post-irradiation survival of human tumor cells through initiation, progression, and termination of DNA-PKcs-dependent DNA double-strand break repair. *Mol. Cancer Res.* **2012**, *10*, 945–957. [CrossRef]
29. Toulany, M.; Maier, J.; Iida, M.; Rebholz, S.; Holler, M.; Grottke, A.; Juker, M.; Wheeler, D.L.; Rothbauer, U.; Rodemann, H.P. Akt1 and Akt3 but not Akt2 through interaction with DNA-PKcs stimulate proliferation and post-irradiation cell survival of K-RAS-mutated cancer cells. *Cell Death Discov.* **2017**, *3*, 17072. [CrossRef]
30. Mueck, K.; Rebholz, S.; Harati, M.; Rodemann, H.; Toulany, M. AKT1 stimulates homologous recombination repair of DNA double-strand breaks in a Rad51-dependent manner. *Int. J. Mol. Sci.* **2017**, *18*, 2473. [CrossRef]
31. Plo, I.; Laulier, C.; Gauthier, L.; Lebrun, F.; Calvo, F.; Lopez, B.S. AKT1 inhibits homologous recombination by inducing cytoplasmic retention of BRCA1 and RAD51. *Cancer Res.* **2008**, *68*, 9404–9412. [CrossRef] [PubMed]
32. Jia, Y.; Song, W.; Zhang, F.; Yan, J.; Yang, Q. Akt1 inhibits homologous recombination in Brca1-deficient cells by blocking the Chk1-Rad51 pathway. *Oncogene* **2013**, *32*, 1943–1949. [CrossRef] [PubMed]
33. Valkov, A.; Kilvaer, T.K.; Sorbye, S.W.; Donnem, T.; Smeland, E.; Bremnes, R.M.; Busund, L.T. The prognostic impact of Akt isoforms, PI3K and PTEN related to female steroid hormone receptors in soft tissue sarcomas. *J. Transl. Med.* **2011**, *9*, 200. [CrossRef]
34. Tomita, Y.; Morooka, T.; Hoshida, Y.; Zhang, B.; Qiu, Y.; Nakamichi, I.; Hamada, K.; Ueda, T.; Naka, N.; Kudawara, I.; et al. Prognostic significance of activated AKT expression in soft-tissue sarcoma. *Clin. Cancer Res.* **2006**, *12*, 3070–3077. [CrossRef] [PubMed]
35. Boichuk, S.; Dunaev, P.; Galembikova, A.; Bikinieva, F.; Nurgatina, I.; Mustafin, I.; Aukhadieva, A.; Kurtasanov, R.; Andriutsa, N.; Shagimardanova, E.; et al. Inhibition of FGFR2-Signaling Attenuates a Homology-Mediated DNA Repair in GIST and Sensitizes Them to DNA-Topoisomerase II Inhibitors. *Int. J. Mol. Sci.* **2020**, *21*, 352. [CrossRef]
36. Sahlberg, S.H.; Gustafsson, A.S.; Pendekanti, P.N.; Glimelius, B.; Stenerlow, B. The influence of akt isoforms on radiation sensitivity and DNA repair in colon cancer cell lines. *Tumour. Biol.* **2014**, *35*, 3525–3534. [CrossRef] [PubMed]
37. Babichev, Y.; Kabaroff, L.; Datti, A.; Uehling, D.; Isaac, M.; Al-Awar, R.; Prakesch, M.; Sun, R.; Boutros, P.; Venier, R.; et al. PI3K/AKT/mTOR inhibition in combination with doxorubicin is an effective therapy for leiomyosarcoma. *J. Transl. Med.* **2016**, *8*, 67. [CrossRef] [PubMed]
38. Villar, V.; Vögler, O.; Barceló, F.; Martín-Broto, J.; Martínez-Serra, J.; Ruiz-Gutiérrez, V.; Alemany, R. Down-regulation of AKT signaling by ursolic acid induces intrinsic apoptosis and sensitization to doxorubicin in soft tissue sarcoma. *PLoS ONE* **2016**, *11*, e0155946. [CrossRef]
39. Hayasaka, N.; Takada, K.; Nakamura, H.; Arihara, Y.; Kawano, Y.; Osuga, T.; Murase, K.; Kikuchi, S.; Iyama, S.; Emori, M.; et al. Combination of eribulin plus AKT inhibitor evokes synergistic cytotoxicity in soft tissue sarcoma cells. *Sci. Rep.* **2019**, *9*, 5759. [CrossRef]
40. Taguchi, T.; Sonobe, H.; Toyonaga, S.; Yamasaki, I.; Shuin, T.; Takano, A.; Araki, K.; Akimaru, K.; Yuri, K. Conventional and molecular cytogenetic characterization of a new human cell line, GIST-T1, established from gastrointestinal stromal tumor. *Lab. Investig.* **2002**, *82*, 663–665. [CrossRef]
41. Boichuk, S.; Galembikova, A.; Dunaev, P.; Valeeva, E.; Shagimardanova, E.; Gusev, O.; Khaiboullina, S. A novel receptor tyrosine kinase switch promotes gastrointestinal stromal tumor drug resistance. *Molecules* **2017**, *22*, 2152. [CrossRef] [PubMed]

42. Mladenov, E.; Anachkova, B.; Tsaneva, I. Sub-nuclear localization of Rad51 in response to DNA damage. *Genes Cells* **2006**, *11*, 513–524. [CrossRef] [PubMed]
43. Chou, T.C.; Talalay, P. Quantitative analysis of dose-effect relationships: The combined effects of multiple drugs or enzyme inhibitors. *Adv. Enzyme Regul.* **1984**, *22*, 27–55. [CrossRef]
44. Chang, T.T.; Chou, T.C. Rational approach to the clinical protocol design for drug combinations: A review. *Acta Paediatr. Taiwan* **2000**, *41*, 294–302. [PubMed]
45. Chou, T.C. Preclinical versus clinical drug combination studies. *Leuk. Lymphoma* **2008**, *49*, 2059–2080. [CrossRef] [PubMed]
46. Yadav, B.; Wennerberg, K.; Aittokallio, T.; Tang, J. Searching for drug synergy in complex dose–response landscapes using an interaction potency model. *Comput. Struct. Biotechnol. J.* **2015**, *13*, 504–513. [CrossRef]

International Journal of *Molecular Sciences*

Article

PARP Inhibitor Olaparib Causes No Potentiation of the Bleomycin Effect in VERO Cells, Even in the Presence of Pooled ATM, DNA-PK, and LigIV Inhibitors

Valentina Perini [1], Michelle Schacke [1], Pablo Liddle [1], Salomé Vilchez-Larrea [2], Deborah J. Keszenman [3,*] and Laura Lafon-Hughes [1,*]

1 Instituto de Investigaciones Biológicas Clemente Estable (IIBCE), Departamento de Genética, Montevideo 11.600, Uruguay; valeperini95@gmail.com (V.P.); michimefuz@gmail.com (M.S.); pabloliddle@googlemail.com (P.L.)

2 Instituto de Investigaciones en Ingeniería Genética y Biología Molecular "Dr. Héctor N. Torres", Consejo Nacional de Investigaciones Científicas y Técnicas, Ciudad Autónoma de Buenos Aires 1428, Argentina; vilchez.ingebi@gmail.com

3 Laboratorio de Radiobiología Médica y Ambiental, Grupo de Biofisicoquímica, Centro Universitario Regional Litoral Norte, Universidad de la República (UdelaR), Salto 50.000, Uruguay

* Correspondence: dkeszen@gmail.com (D.J.K.); lauralafon2010@gmail.com (L.L.-H.)

Received: 30 September 2020; Accepted: 19 October 2020; Published: 5 November 2020

Abstract: Poly(ADP-ribosyl)polymerase (PARP) synthesizes poly(ADP-ribose) (PAR), which is anchored to proteins. PAR facilitates multiprotein complexes' assembly. Nuclear PAR affects chromatin's structure and functions, including transcriptional regulation. In response to stress, particularly genotoxic stress, PARP activation facilitates DNA damage repair. The PARP inhibitor Olaparib (OLA) displays synthetic lethality with mutated homologous recombination proteins (BRCA-1/2), base excision repair proteins (XRCC1, Polβ), and canonical nonhomologous end joining (LigIV). However, the limits of synthetic lethality are not clear. On one hand, it is unknown whether any limiting factor of homologous recombination can be a synthetic PARP lethality partner. On the other hand, some BRCA-mutated patients are not responsive to OLA for still unknown reasons. In an effort to help delineate the boundaries of synthetic lethality, we have induced DNA damage in VERO cells with the radiomimetic chemotherapeutic agent bleomycin (BLEO). A VERO subpopulation was resistant to BLEO, BLEO + OLA, and BLEO + OLA + ATM inhibitor KU55933 + DNA-PK inhibitor KU-0060648 + LigIV inhibitor SCR7 pyrazine. Regarding the mechanism(s) behind the resistance and lack of synthetic lethality, some hypotheses have been discarded and alternative hypotheses are suggested.

Keywords: poly(ADP-ribosylation); PARP-1; Olaparib; KU55933; KU-0060648; SCR7 pyrazine; VERO cells; synergism; resistance; CDKN2A

1. Introduction

Poly(ADP-ribosylation) or PARylation is a post-translational protein modification that is catalyzed by poly(ADP-ribosyl)polymerases (PARPs). PARPs consume nicotinamide adenine dinucleotide (NAD+) and release nicotinamide (Nam) every time they add ADP-ribose monomers to originate a lineal or ramified chain of up to 400 residues that is covalently anchored to a single amino acid. Such a nucleic-acid-like polymer, called poly(ADP-ribose) or PAR, is strongly negatively charged since it harbors two phosphate groups per residue. PAR is recognized by PAR-binding protein domains (e.g., a macrodomain) and facilitates the assembly of multiprotein complexes through non-covalent interactions. For this reason, PAR has been called a "glue" [1]. PAR synthesis is stimulated by genotoxic insults, increasing up to 500 times overall or 7 to 8 times on specific proteins [2,3].

Both endogenous metabolism and exogenous genotoxic agents induce different types of DNA damage, such as oxidative lesions, single-base modifications, crosslinks, local changes in DNA structure, single-strand breaks (SSBs), and double-strand breaks (DSBs), which can be repaired by different DNA repair systems while the cell cycle is coordinately slowed down. Directly or indirectly generated DSBs are the most challenging type of DNA damage for the cell because they can lead to mutations, carcinogenesis, or cell death. DSBs are marked by the spread of the phosphorylation of the histone H2AX on Ser139 by the canonical kinases ATM, ATR, or DNA-PK [4] through a feed-forward signaling loop to give γH2AX foci. A less studied chromatin-bound kinase called VRK1 can also give rise to γH2AX foci, at least in response to DNA damage induced by γ-rays [5]. Lesions are then repaired either after DNA replication by homologous recombination (HR) or throughout the cell cycle by canonical nonhomologous end-joining (c-NHEJ) or alternative NHEJ, also called microhomology-mediated end joining (MMEJ). Unrepaired damage may lead to replication fork stalling and single-stranded DNA exposure, activating ATR [6]. Regarding genomic stability, it is accepted that HR is the DSBs repair pathway that allows for the highest fidelity, MMEJ is the one promoting the highest genomic instability, and c-NHEJ is in between. C-NHEJ repair occurs with a half-time of 10–30 min after damage, while MMEJ processing has a half-time of 30 min to 20 h, being considered a rescue pathway [7]. Some molecular actors have been identified in each pathway. C-NHEJ is dependent on KU70/80, DNA-PK, and LigIV. C-NHEJ is promoted by 53BP1 and antagonized by BRCA1. In contrast, DNA end-resection by MRE11 exonuclease activity prevents c-NHEJ and promotes HR with BRCA-1 and BRCA-2 recruitment (in the presence of an undamaged template) or allows for MMEJ. The latter does not involve DNA-PK, 53BP1, or LigIV but is dependent on LigIII instead. These pathways are quite complex and MMEJ is still poorly understood [7–9].

The PARP family has 18 members in humans. Some of them are enzymatically inactive; others act just as mono-(ADP-ribosyl) transferases and, to our knowledge, only four of them synthesize PAR [10], namely, the canonical ancient PARP-1, PARP-2, and two tankyrases (TNKS-1 and TNKS-2). PARP-1 is the most studied member of the family and is involved in the PARP response to stress, inflammation, or genotoxic insults [11,12].

The interpreted role for PARP-1 in the DNA damage response (DDR) has changed over time in the literature. In the 20 years between 1985 and 2005, there were enthusiastic positive results with PARP inhibitors (PARPis). In different human and rodent normal or transformed cell types, diverse PARPis (3AB, INO1001, E7016, AG14361, 4-ANI) applied before, during, or after DNA damage induction sensitized cells to different genotoxic agents, including ultraviolet-C radiation (UV-C), methylmethanosulfonate (MMS), ionizing radiation (IR), and the radiomimetic agent bleomycin (BLEO) [13–24]. The variable sensitization effects, i.e., additive or synergistic, suggested different biological interpretations and putative clinical implications [25]. Still, this data was promising at least as a way to reduce chemotherapy doses and consequently diminish side effects. Such results did also push the field to go on and try to understand the mechanisms involved.

Interestingly, nowadays, a paradigm shift has occurred. The current synthetic lethality paradigm states that PARPis do not appreciably affect normal cells but do potentiate DNA damage effects in cells that have certain DNA repair pathways that are blocked by the alteration (by mutation or enzymatic inhibition) of a crucial actor of the pathway. This paradigm has huge clinical implications, opening the possibility that PARPis could be used to selectively kill aberrant cells while keeping normal cells. In fact, it was under this paradigm that Olaparib (OLA, Lynparza, AZD-2281) reached U.S. Food and Drug Administration (FDA) approval to treat BRCA-1 mutated cancers (ucm572143, ucm592357, and NCT02987543). Interestingly, BRCA-1 is not only crucial for HR but also regulates BER [26,27], and OLA displays synthetic lethality with mutated BER proteins, such as XRCC1 or Polβ [28,29] and with mutated LigIV in c-NHEJ [30]. Last but not least, PARPis can affect not only DNA repair but also modulate gene transcription [31]. In fact, the potential clinical impact of

PARPis is not limited to cancer therapy but also reaches many pathologies that are characterized by inflammation [11,32–34]. Interestingly, the major constraint of anti-cancer BLEO therapy is the early inflammation of lung parenchyma, leading to pulmonary fibrosis [35]. Thus, we reasoned that PARPis might concomitantly potentiate BLEO-directed effects in DNA-repair-hampered cells while diminishing inflammation-related side effects in normal cells.

Not only PARP-1 but also PARP-2 [36–38], the mono(ADP-ribosyl)transferase PARP-3 [39–41], and TNKS [42] are involved in the cellular response to DNA damage, particularly DSBs. OLA was initially considered a potent and specific PARP-1 inhibitor with an in vitro IC_{50} (ivitr-IC_{50}) of 5 nM [43]. Later, an even lower IC_{50} was determined using full-length PARP-1 (ivitr-IC_{50}: 1.4 nM), while it was demonstrated that it is also a potent PARP-2 (ivitr-IC_{50}: 12 nM) [44] and PARP-3 inhibitor (ivitr-IC_{50}: 4 nM) [41] but is not a potent TNKS-1 or TNKS-2 inhibitor (ivitr-IC_{50}: 1230 nM and 2340 nM, respectively) [44]. As PARP-1, PARP-2, and PARP-3 are involved in DDR, the OLA effects include DNA repair hampering due to combined enzymatic inhibition/trapping of these enzymes [38,41,45,46].

VERO is a kidney epithelial cell line from a female African green monkey. It is a challenging cell culture model to use when analyzing resistance mechanisms because it preserves an epithelial morphology but can form colonies in agar and includes a subpopulation that is BLEO-resistant, according to Terasima et al. [47]. It is not tumorigenic, except at very high passages [48]. Moreover, its genome has been studied [49]. Finally, VERO cells are infection-prone [49–51], thus allowing studies of DNA damage, inflammation, PARylation, or epithelial-to-mesenchymal transition induction in the host cell by viral, bacterial, or parasitic infections.

In the present work, we addressed the question of whether OLA would potentiate DNA damage induced by BLEO in VERO cells. As a further step, we asked whether the simultaneous inhibition of ATM, DNA-PK, and LigIV (to block at least HR and c-NHEJ) would show synthetic lethality with the OLA effect in VERO cells. It did not. A hypothesis regarding the mechanisms involved is discussed as a guide for future research.

2. Results

2.1. OLA Did Not Potentiate BLEO Effects on VERO Cells

As a BLEO treatment has to be performed in the absence of FBS, in all experiments involving BLEO (regardless of the presence or absence of inhibitors), all the cells, including those with and without BLEO, were subjected to a 45 min serum depletion. Therefore, all differences could be attributed to BLEO itself rather than to this short serum depletion.

Previous experiments were carried out in order to choose appropriate experimental conditions (Figure A1A–D).

Cell viability dose-response curves with BLEO alone (45 min pulses) were evaluated immediately after treatment (t = 0 h), or 24 or 72 h later (Appendix A Figure A1A). The BLEO toxicity increased slightly above 40 µg/mL. Neither a BLEO concentration increase up to 500 µg/mL nor a treatment duration of 270 min (Figure A1B, evaluation 24 h post treatment) appreciably decreased the viability. In agreement with the report by Terasima et al. from 1972 [47], the VERO cell population was composed of a BLEO-sensitive and a BLEO-resistant subpopulation.

Experiments that involved adding the BLEO hydrolase inhibitor E-64 before (30 min) and during (45 min) BLEO treatment were carried out to exclude the possibility of the existence of this BLEO-degradative enzyme in our model (Figure A1C). The lack of an effect of E-64 indicated that the resistance was not due to the expression of BLEO hydrolase.

The toxicity of OLA alone was analyzed on VERO cells (Figure A1D). An approximately 70–80% cell viability was preserved, even after a 400 nM continuous 24 or 72 h treatment. As we wanted to analyze the specific effects on PARPs involved in DDR (PARP-1/2 and 3) and we knew that low concentrations were enough to inhibit at least PARP-1 in this cell line [43], we decided to regularly use

50 nM OLA in the combined experiments. Furthermore, a 72 h evaluation point was considered to be the most informative one in our MTT assays.

The cell viability was then assayed (Figure 1A) after a 45 min serum depletion, without (control; BLEO = 0 μg/mL) or with BLEO (BLEO = 40 or 160 μg/mL), without (violet bars) or with (light blue bars) a 50 nM OLA treatment. OLA exposure involved continuous co- and post-treatment that finished at the moment of evaluation with the MTT assay, 72 h post the BLEO treatment.

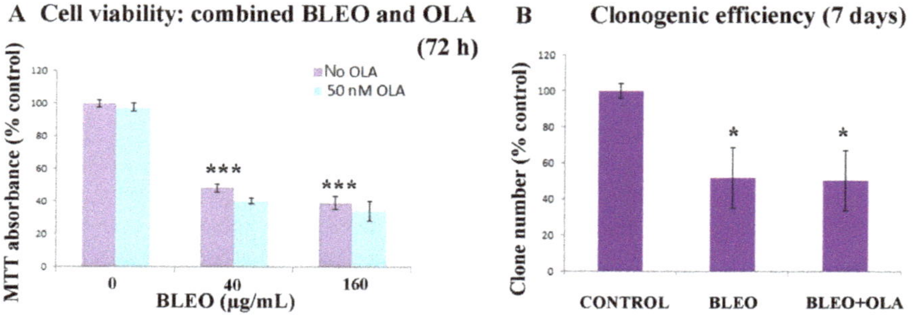

Figure 1. Neither the cell viability loss nor clonogenic efficiency loss induced by BLEO in VERO cells were potentiated by OLA. (**A**) Cell viability (MTT assay). Cells were exposed to BLEO and, when indicated, co- and post-exposed to 50 nM OLA. The result was evaluated after 72 h. Most experiments were carried out using 40 μg/mL BLEO and 50 nM OLA. The respective *n*'s were as follows. No OLA: 63, 72, and 22; 50 nM OLA: 75, 86, and 12 (see Figure A2 for 150 nM OLA). ANOVA ($p = 1.11 \times 10^{-16}$). Post-hoc tests: Tukey, Scheffe, and Bonferroni. ***: $p < 0.001$. (**B**) Clonogenic efficiency of VERO cells in the control condition or under a pulse treatment with 40 μg/mL BLEO (45 min) in the absence or presence of continuous treatment with 50 nM OLA. Data were from two independent experiments in triplicate. All results are expressed as mean ± SEM. Comparisons against control. ANOVA ($p = 0.0371$) and Holm *p*-value with only comparisons against the control considered. *: $p < 0.05$.

As shown in Figure 1A, there was no significant difference in cell viability that was attributable to 50 nM OLA (light blue bars) in basal (BLEO = 0) or BLEO-treated cells (40 or 160 μg/mL).

In order to distinguish between different possible scenarios, the clonogenic efficiency was also evaluated in cells treated with BLEO (40 μg/mL) or BLEO + OLA (50 nM) (Figure 1B). Two conclusions could be derived. First, taking into account the errors, cell viability results resembled clonogenic efficiencies (BLEO: 48 vs. 52%; BLEO + OLA: 41 vs. 50%), indicating that in the presence of 40 μg/mL BLEO, about one in every two cells was alive and cycling. Second, upon the OLA treatment, no difference was observed.

Although an even lower OLA concentration (25 nM) is known to have effects on VERO nuclear PARP-1 activity [43], and 50 nM OLA is enough to prevent or partially revert the epithelial-to-mesenchymal transition induced by TGF-β in NMuMG cells [52], a higher OLA concentration was assayed as well, just in case an unexpected shift occurred. As can be seen in Figure A2, the OLA concentration was tripled (to 150 nM) and still displayed no effect on the BLEO-treated cells.

OLA did not potentiate a BLEO lethal effect in VERO cells. The absence of potentiation of the BLEO effect was also evidenced with chemically different, less specific PARPis and with a PARG inhibitor, indicating that PAR metabolism was not crucially involved in the BLEO-induced DDR. The inhibitors used were 3-aminobenzamide (3AB), 5′-deoxy-5′-[4-[2-[(2,3-dihydro-1oxo-1H-isoindol-4-yl)amino]-2-oxoethyl]-1-piperazinyl]-5′-oxoadenosine dihydrochloride (EB), and 6,9-diamino-2-ethoxyacridine-DL-lactate monohydrate (DEA). Figure A3A represents PAR, its synthesis by PARPs, its degradation mainly by poly-ADP-glycohydrolase (PARG), and the inhibitors abbreviations associated with their targets. Figure A3B depicts the PAR quantification on the control untreated cells and cells treated with PARPis or the PARG inhibitor DEA. As the basal PAR was low and this was done once, these measurements did not have much sensitivity, but overall, they were a control to check that the inhibitors were active. The lack of potentiation [25] of BLEO effects by PARPis 3AB or EB was demonstrated (Figure A3C,D). Finally, PARG inhibition did not change the cell viability in the presence of BLEO (Figure A3E,F).

To sum up, despite being able to alter the PAR metabolism, neither PARP nor PARG inhibitors potentiated the toxic effects of BLEO in VERO cells.

2.2. Untreated VERO Cell Nuclei Harbor PARP, PARG, and PAR

Next, it was checked whether VERO cells were expressing some of the nuclear molecular actors of PARylation, as well as synthesizing basal PAR. As displayed in Figure 2A–D, the indirect immunocytofluorescence (ICF) and DAPI (blue) counterstain showed that nuclear PARP-1/2 (green) was distributed throughout the nucleus, while the PARG (red) distribution was punctuated and excluded the nucleolus. Relative intensity measurements (Figure 2E,F) following the lines drawn in Figure 2A,B, respectively (color-coded like the channels), also supported these observations. Regardless of the distribution, the important point is that VERO cells were expressing at least PARP-1/2 and PARG. Basal PAR was also detected, as demonstrated by the comparison of Figure 2H vs. Figure 2K and the respective relative intensity graphs (Figure 2L,M).

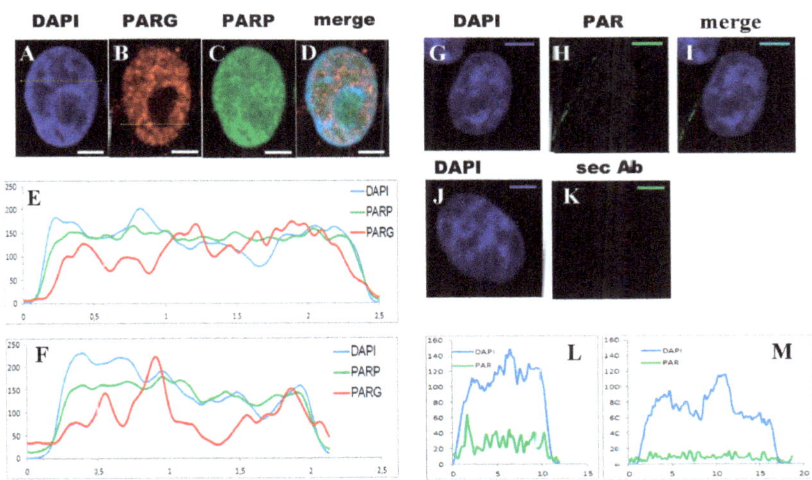

Figure 2. PAR, PARP, and PARG were detected in the VERO cell nuclei. (**A–D**) DAPI (blue), PARG (red), PARP (green), and the merged confocal images of representative nuclei. (**E**,**F**) Graphs displaying the fluorescence intensity measurements in the three channels of the correspondent nuclei images through two lines that are drawn in (**A**) or (**B**) respectively. Intensity in Relative units. Distance: 1 U ≈ 5 µm (**G–I**) Indirect immunocytofluorescence (ICF) with BD anti-PAR antibody. DAPI (blue), PAR (green), and merged channels. (**J**,**K**) Control of the anti-PAR ICF without the anti-PAR antibody with only the secondary antibody (sec Ab). (**L**,**M**) Blue and green channel intensities measured over a line in (**H**) (with anti-PAR) and (**I**) (without anti-PAR), respectively. Confocal images were obtained with the same settings and subject to identical processing adjustments. Relative intensity on the ordinates and distance in µm on the abscissas. PAR signal (green) was low but detectable in untreated VERO nuclei. Bar: 5 µm.

2.3. No Sharp PAR Increase Could Be Detected Immediately after the 45 min Pulse of BLEO

The first estimations in the literature suggested that PAR can increase up to 500-fold in response to a genotoxic insult [2]. Later, 50-fold increases under PARG inhibition and 7-fold increases on specific proteins have been reported [2,3]. To assess whether VERO cells respond to BLEO by increasing PAR levels, we performed inmunocytofluorescence experiments immediately after the end of the treatment (t = 0) with three different anti-PAR antibodies.

The 10H anti-PAR antibody has a known specificity for long PAR chains (above 20 residues) [53] and has been widely used to monitor the nuclear response to DNA damage [24,54]. Interestingly, some DDR proteins do not interact with short PAR chains (16-mer), while long PAR chains (55-mer) promote their integration into protein complexes [55]. Thus, the best antibody for detecting long-chain PAR induced by genotoxic stress is 10H anti-PAR. In DAPI-counterstained control or BLEO-treated (40 μg/mL, 45 min) cells (blue, Figure 3A–C), one in every several cells in the population displayed a strong nuclear PARylation signal, while the rest displayed no signal (Figure 3E; in Figure 3F, the red point under the calibration bar is a single H10-anti-PAR-positive nucleus). This observation would explain why the PAR increase in the cell population was not significant (Figure 3I, right-hand graph). As can be seen in Figure A4, a different cell type used as a positive control (CHO9 fibroblastic cell line) displayed nuclear PARylation more frequently in the same experimental conditions.

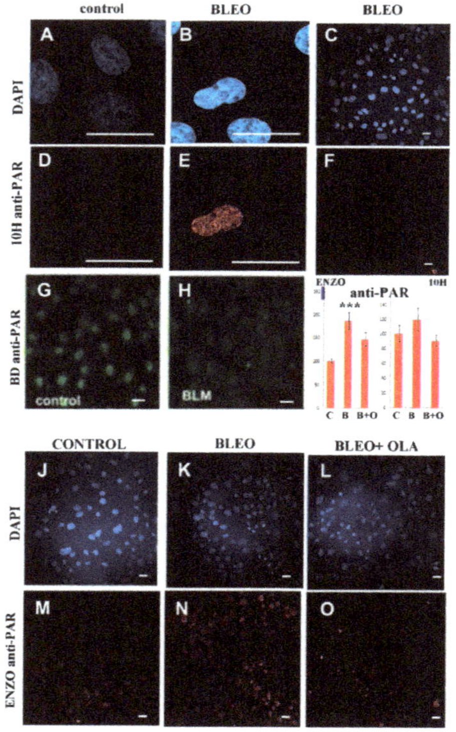

Figure 3. Neither of the three anti-PAR antibodies evidenced a sharp PAR increase using ICF immediately after the BLEO (45 min) treatment. This was observed in at least five independent experiments. (**A–C**) DAPI channel to see nuclei that were positive or negative for the (**D–F**) 10H anti-PAR signal. (**A,D**) control and (**B–F**) BLEO. (**G,H**) BD anti-PAR in (**G**) control or (**H**) BLEO-treated cells. (**I**) To estimate the signal increase, the PAR signal intensity was quantified in one of the experiments with two of the antibodies. The whole-field PAR intensity was adjusted by the DAPI intensity and expressed as a percentage of the control. Mean ± SEM. Left: ENZO anti-PAR. ANOVA ($p = 0.0010$). Post-hoc test: Holm against control, Tukey, or Scheffé. ***: $p < 0.001$. Right: 10H anti-PAR antibody showed no differences. (**J,M**) Control, (**K,N**) BLEO-, or (**L,O**) BLEO + OLA-treated cells. Panels **G** and **H** were extracted from Lafon-Hughes' unpublished Ph.D. thesis [56]. Bar: 25 μm.

It has formerly been reported that nuclear PAR in untreated VERO cells is detected with polyclonal rabbit BD anti-PAR or chicken Tulip anti-PAR antibodies but not with Tulip monoclonal 10H clone antibodies [57]. Thus, now we are reporting that a third antibody detected basal nuclear PAR in VERO cells. BD and ENZO anti-PAR antibodies are better suited to detecting short-chain PAR. According to the manufacturers, ENZO anti-PAR (BML-SA216) was specifically designed against "purified poly(ADP-ribose) polymer (chain length of 2–50 units)." This would explain why no signal was detected in the control cells with 10H anti-PAR (Figure 3D) but there was a PAR signal in control cells with BD (Figure 3G) or ENZO anti-PAR (Figure 3M).

No signal increase was detected with BD anti-PAR in BLEO-treated cells (Figure 3H, extracted from unpublished [56]). As a positive control, a slight signal increase (about ×1.5) was detected with BD anti-PAR only under extreme conditions leading to cell death (Figure A4). More recently, the anti-PAR reagent MABE1031 was also used to evaluate the PAR increase in response to H_2O_2, which was hampered in the presence of 50 nM OLA (Figure A4I).

Compared to control samples (Figure 3J,M), a slight but significant PAR increase (Figure 3I) was detected with ENZO anti-PAR due to the BLEO treatment (Figure 3K,N). Of notice, given the nature of BLEO treatment, some cells may have been damaged 45 min before and other ones just at the very last minute before fixation. Therefore, in this single fixation, we may have had many time points that were superimposed. Regarding the effect of OLA in BLEO-treated cells (Figure 3L,O), there was a tendency toward PAR diminution. In fact, unlike what happened in BLEO-treated cells, in BLEO + OLA treated cells, the PAR level was indistinguishable from the control ($p = 0.589$). Therefore, the next step was to check that OLA was not interfering with the DNA damage induction by BLEO.

2.4. OLA Did Not Hamper the DNA Damage Induction by BLEO

DNA damage induction was registered in the DAPI-counterstained cells (blue, Figure 4A–C) through γH2AX (red, Figure 4D–F) and 53BP1 (green, Figure 4G–I) detection using ICF, as well as in cells subjected to single-cell gel electrophoresis or a comet assay (Figure 4I–K). The percentage of cells with a γH2AX foci (Figure 4M), the percentage of cells with a pan-nuclear γH2AX signal (Figure 4N), and the relative DNA damage index (DDI, Figure 4O) were quantified from three independent experiments. Furthermore, in one of the experiments, the relative number of γH2AX, 53BP1, and mixed foci per cell was also evaluated (Figure A5). All our DNA damage induction measurements had the same stair-shape, from left to right: the control in the lower step, then BLEO, and then BLEO + OLA in the higher step. As this happened regarding the percentage of cells with γH2AX foci (Figure 4M), the percentage of pan-nuclear γH2AX cells (Figure 4N), the comet DDI relative to the control (Figure 4O), and even the relative abundance of γH2AX, 53BP1, and mixed foci (Figure A5), it was concluded that at t = 0, the BLEO cells were more damaged than the control cells. Moreover, the OLA undoubtedly did not interfere with BLEO to avoid the initial DNA damage induction.

2.5. OLA Did Not Potentiate BLEO, Even in the Presence of a Pool of DNA Repair Enzyme Inhibitors

All the data above suggested that VERO cells were repairing DSBs induced by BLEO via a mechanism that was independent of PARPs-1/2 and 3 since OLA can block their enzymatic activity or induce enzyme trapping. According to some authors [8,58], PARP-1 activation participates in MMEJ but not on HR or c-NHEJ. If this was the case, then the simultaneous block of HR and c-NHEJ should leave the cells dependent on MMEJ, thus dependent on PARP-1. Therefore, we hypothesized that the simultaneous inhibition of DNA-PK, LigIV, and ATM (3i) would either induce the death of remnant cells or force them to shift to MMEJ, thus becoming PARP-dependent and OLA-sensitive.

As depicted in the Figure 5 legend, 3i was toxic per se (viability = 66.82%) and BLEO + DMSO viability was 41.46%. According to Webb's equation, the expected viability if the combined effect of BLEO and 3i was additive was 27.70%. The real viability was 26.42 ± 1.25%. Thus, the 3i toxic effect was added to the BLEO effect. 3i did not potentiate the BLEO effect.

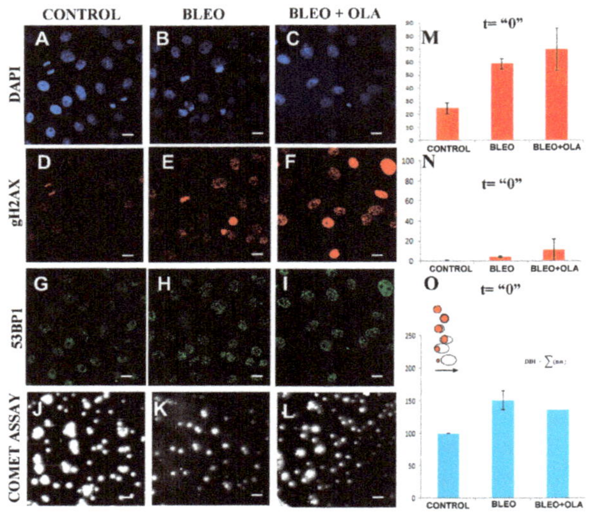

Figure 4. OLA did not hamper the DNA damage induction by BLEO. The initial (t = "0") BLEO-induced DNA damage in the presence or absence of the OLA co-treatment was evaluated All images and graphs follow the order: Control, BLEO and BLEO+OLA, from left to right. (**A–I**) Representative ICF images. (**A–C**): DAPI in blue; (**D–F**): gH2AX in red; (**G–I**): 53BP1 in green. (**J–L**) Representative single-cell electrophoresis or comet assay images (**M,** in red): Percentage of cells with a γH2AX foci. Data were from three independent experiments and $n \geq 200$ cells/condition. Two-tailed t-test. (**N,** in red) Pan-nuclear γH2AX cells were counted on the same set of images. (**O,** in blue) The comet assay was evaluated using the DNA damage index (DDI). In turn, the results were normalized to control the DDI to pool the experiments. Data were from three independent experiments, with $n \geq 140$ cells/condition. Two-tailed t-test. $p = 0.17$ (control vs. BLEO), $p = 0.099$ (control vs. BLEO + OLA), $p = 0.24$ (BLEO vs. BLEO + OLA). To calculate the DDI, each comet was assigned a degree from 1 to 5. The inset scheme represents comet heads (in red) and comet tails (in white) for successive degrees. 1: compact or with a simetric halo (first two drawings); 2: asimetric halo; 3: tail length = head length; 4: tail length > head length; 5: head separated from tail. Bar: 20 μm.

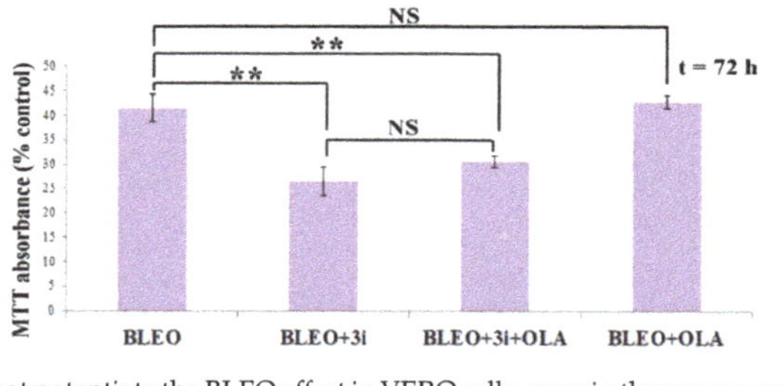

Figure 5. OLA did not potentiate the BLEO effect in VERO cells, even in the presence of certain inhibitors of DSB repair pathways. The cell viability relative to the control (MTT, 72 h) of BLEO-treated cells in the absence or presence of OLA and "3i" (0.1 μM DNA-PK inhibitor KU-0060648 + 0.1 μM LigIV inhibitor SCR7 pyrazine + 5 mM ATM inhibitor KU55933). DMSO, the inhibitors' vehicle, was maintained in a constant concentration in all conditions. Data were from six independent experiments, with $n = 32, 56, 54$, and 54. Mean ± SEM. According to ANOVA and Tukey, Schaffé, or Bonferroni a posteriori tests, OLA made no difference, whereas 3i or 3i + OLA differed from the BLEO treatment with **: $p < 0.01$. NS: statistical non-significance. Although not represented in the graph, 3i was toxic by itself (cell viability = 66.82 ± 2.9%).

To sum up, the unexpected results were that neither 3i nor OLA potentiated BLEO effects on the cell viability and that most (≈75%) of the cell subpopulation that resisted BLEO was still resistant in the presence of 3i and OLA.

3. Discussion

Resistance to BLEO itself in VERO cells can be due to several causes, including a low intake, high export, non-availability of free radicals (due to low Fe, hypoxia, or high antioxidant defenses) [59] or BLEO inactivation by BLEO hydrolase [60]. From these known putative causes, hypoxia was discarded since experiments were carried out under aerobic conditions. Furthermore, as BLEO sensitivity does not increase in the presence of the BLEO hydrolase inhibitor E-64, it can be inferred that the VERO cells did not express significant levels of this inactivating enzyme.

VERO cells have low basal PAR levels, which increase moderately in response to different genotoxic insults. Having tried three different anti-PAR antibodies, we did not detect a sharp increase in PAR levels in response to BLEO. ICF with a 10H antibody for long-chain PAR showed no basal long-chain PAR and a strong signal in very few BLEO-treated cells was observed. Although it is known that PAR levels can peak and then diminish, the nature of the BLEO treatment implied that at the fixing time, the damage induced at the beginning of the pulse coexisted with the damage induced at the end of the pulse (45 min later). Thus, if VERO cells had been good PAR synthesizers in response to BLEO, we should have detected different degrees of the PARylation signal regardless of its kinetics. PARylation in response to other agents used as positive controls, such as H_2O_2 or MMS, was also modest (a very high concentration of the latter, which induced cell death, was necessary to achieve a clear PAR increase). The VERO cells expressed PARP-1/2 and PARG, which were clearly detected by the ICF (Figure 2). Recent evidence shows that a low nuclear PARP-1 level is not the reason why VERO cells are not good PAR synthesizers. According to published Western Blots (WB) results, the level of PARP-1 expression in VERO cells is considerable, with it being similar to that in human epithelial cell lines, such as HEK293 (embryonic kidney), A549 (lung carcinoma) [61], or HEp-2 (laryngeal carcinoma) [62]. In fact, VERO has been chosen as one of the model epithelial cell lines for analyzing the effects of PARP-1 knockdown [61]. The VERO PARP-1 protein sequence (XP 007986456.1) harbors several point amino acid substitutions despite showing a good alignment with hPARP-1 (NP 001609.2). An effect of such mutations in PARP-1 structure and/or activity cannot be formally discarded since several hPARP-1 mutations causing diminished enzymatic activity have been identified [63]

The lack of the potentiation of BLEO-induced damage by OLA was not surprising given the cumulative evidence indicating that OLA can display a range of effects from pro-cell death to protective effects depending on the cellular context [11,13–24,28,29]. OLA promotes cell death differentially in cells that have hampered DSB repair (due to a BRCA mutation for example) rather than in cells that have a normal, fully functional, DNA repair systems. As VERO cells do not harbor known mutations in DNA repair genes, this was an expected result. However, the initial DNA damage was measured (γH2AX positive cells, γH2AX and 53BP1 foci number, and γH2AX pan-nuclear cells and DNA damage index) to exclude the possibility that OLA had interfered with the DNA damage induction. It was also confirmed that the 50 nM OLA treatment reached its targets and affected the PARylation. ENZO anti-PAR antibody, which was generated against a 50-unit PAR chain, depicted a measurable and significant PAR increase in the presence of BLEO, which decreased, as expected, under co-treatment with 50 nM OLA (Figure 3). Finally, the WB confirmed that the PARylation increase induced by H_2O_2 (detected using MABE1031 anti-PAR reagent) was inhibited by a co-treatment with 50 nM OLA (Figure A4, panel I). An interesting point is that, according to our preliminary data (Figure A4), OLA increased the 53BP1 foci number. These data regarding VERO cells agree with data in HeLa and PtK2 cell lines published by Saquilabon et al. [64], who determined that PARPis induce the increased recruitment of 53BP1 to DNA damage sites. It has been interpreted that PARP-1/2 activation is antagonistic to c-NHEJ, limiting the 53BP1 accumulation at damaged sites [64]. Antagonistic roles in the selection of DSB DNA repair pathways have been documented for PARP-1 vs. Ku [58,65] and BRCA vs. 53BP1 [66,67].

The lack of abundant PAR synthesis in response to DNA damage induction and the absence of potentiation by OLA could be due to the fact that VERO cells were repairing BLEO-induced DNA damage, mainly through PARP-independent pathways. In such a case, blocking these pathways would force cells to become PARP-dependent or die. For this reason, and as an effort to help delineate the boundaries of synthetic lethality, a pool of inhibitors (3i) targeting key enzymes of the two main pathways of DSB repair, namely, HR (ATM inhibitor) and C-NHEJ (DNA-PK and LigIV inhibitor), were used. Cells were exposed to BLEO and a co- and post-treatment with 3i alone or 3i + OLA.

The most important point was that since in VERO cells we did inhibit ATM, DNA-PK, and LigIV, and under that condition, a cell subpopulation survived, even in the presence of the PARP inhibitor OLA, no synergism occurred. In fact, our results indicated the presence of a VERO subpopulation that was resistant to BLEO, BLEO + OLA, and BLEO + OLA + ATM inhibitor KU55933 + DNA-PK inhibitor KU-0060648 + LigIV inhibitor SCR7 pyrazine (Figure 5).

An improved understanding of the synthetic lethality strategies will probably extend the use of PARP inhibitors beyond tumors with BRCA1/2 mutations [68]. Aside from drug–drug interactions, known causes for OLA resistance include disturbed PARylation metabolism, alterations in drug transporters, up-regulation of HR, and replication forks [69–71].

As discussed above, although the VERO cells expressed normal PARP levels, poor PARylation was observed and VERO cells harbored PARP mutations of unknown biological meaning. This deserves further study. PARPis resistance has been described in an ovarian cancer patient carrying a point mutation in PARP-1 [70]. Regarding transporters, it is known that P-glycoprotein (P-gp) transporters export OLA, pumping it out of the cell [41]. We have not evaluated whether VERO cells overexpress P-gp transporters. However, PAR diminution in the presence of OLA evidences that OLA is reaching PARP, making P-gp overexpression an unlikely explanation of OLA resistance in VERO cells.

Assuming that all the used inhibitors worked as in other models, one explanation regarding the resistance to BLEO + OLA + 3i would be that cells repaired DNA damage through a PARP-independent pathway rather than MMEJ (which is PARP-1 dependent) or HR (because ATM was inhibited) or c-NHEJ (since DNA-PK and LigIV were inhibited). Even under the HR defect (due to mutated or inhibited proteins), replication fork stabilization has been proposed as an alternative PARPis resistance mechanism [69,71]. According to Haynes et al. [72], checkpoint kinases ATR, CHK1, and WEE1 play different roles in replication fork stabilization, providing alternative mechanisms to be considered in combination therapy development in order to avoid drug resistance. For example, ATRis (VE-821, VE-822, or AZ20) overcome acquired and pre-existing PARPis resistance in multiple BRCA1-deficient cancer cell lines of distinct origins. Moreover, such cells are preferentially affected over BRCA-proficient or PARP-sensitive cells [73]. ATRi (AZD6738) and WEE1i (AZD1775) display differential activity in a subpopulation of non-germinal-center-B cell (non-GCB) diffuse large B-cell lymphoma (DLBCL) cell lines characterized by high MYC oncogene expression and cyclin-dependent kinase inhibitor 2A/B (CDNK2A/B) deletion [74]. Interestingly, CDKN2A codes alternatively spliced variants that are involved in cell cycle control, including the tumor suppressor p16INK4a. In turn, CHK1 inhibitor Prexasertib is effective against human head and neck squamous cell carcinoma cell lines with *CDKN2A* genetic losses. Conversely, restoration of p16 expression in hypersensitive cells prevents Prexasertib-induced cell proliferation drop [75].

Does the DDR response and coordination of DNA repair and cell cycling work properly in VERO cells? VERO cells have a homozygous ≈9 Mb deletion on chromosome 12, causing the loss of CDKN2A/B genes, besides the type I interferon gene cluster (which probably explains why they are so infection-prone). Although these mutations by themselves are not enough to transform cells into tumorigenic cells, the loss of CDKN2A/B may play a crucial role in the acquirement of immortality in the VERO cell lineage [49]. Interestingly, *CDKN2A* gene mutations are found in up to 40% of familial cases of melanoma, up to 25% of head and neck squamous cell carcinomas (HNSCC), some breast cancers and pancreatic cancers, and others. In some families, *CDKN2A* gene mutations facilitate the development of only one type of cancer, while in other families, they can lead to a cancer predisposition syndrome, increasing the risk of developing multiple types of cancer. Furthermore, somatic *CDKN2A* gene mutations have been found in some people with brain tumors and in children with acute lymphoblastic leukemia [76] (From this point of view, VERO cells could be considered a cancer-prone cell model. Furthermore, the effects of mono-treatments and combined treatments with checkpoint kinases inhibitors should be tested. Moreover, further research is needed to know whether CDKN2 mutations are mandatory over mutations regarding its synergism with PARPis. For example, does a BRCA mutated cell lose its synergism with PARPis in general, and with OLA in particular, if it is also mutated in *CDKN2A*? Could this be an explanation of why some BRCA-mutated breast cancers do not respond to OLA? Efforts to determine the synthetic lethality partners of PARPis [68], ATRis, CHK1is, and WEE1is are being done. Both positive and negative results will be useful for understanding the underlying mechanisms.

4. Materials and Methods

4.1. VERO Cell Culture

Cercopithecus aethiops (green monkey) VERO cells (ATCC CCL-81, [57]) were cultured in MEM-STA (Capricorn, Capricorn Scientific GmbH, 35085 Ebsdorfergrund, Germany) supplemented with 10% FBS (Capricorn # FBS-11A, collected in South America), penicillin/streptomycin (Capricorn PS-B, 100×) and 2 mM L-glutamine at 37 °C and 5% CO_2. Cells were seeded onto 96-well cell culture plates for MTT assay, round coverslips in 24-well plates for immunocytofluorescence (ICF), 30 mm dishes for the comet assay, and 50 mm dishes for the clonogenic assay.

4.2. Treatments with BLEO and Inhibitors

Bleomycin sulfate treatment (BLEO); from NOLVER (Montevideo, Uruguay), NIPPON KAYAKU (Japan), or LKM (Peru) consisted of a 45 min pulse in the absence of FBS. Controls and other experimental conditions were also subjected to 45 min FBS depletion. Initially, dose–response curves (0, 4, 10, 20, 40, 80, 160, 200, and 500 µg/mL) were found. The Olaparib (Tocris, Minneapolis, MN, USA) treatment was continuous. Dose–response curves (0, 50, 100, 150, and 200 nM) were also found. Then, 40 µg/mL BLEO and 50 nM OLA were selected for the following experiments: OLA effect on the PAR pool was checked and compared with other PARPis and with the PARG inhibitor DEA in untreated VERO cells using ICF with a BD anti-PAR antibody (Figure A3B). While DEA increased PAR, 25 nM OLA did induce an effect that was very similar to 250 nM OLA, diminishing the endogenous PAR by more than 100 nM EB or 5 mM 3AB. Like with OLA, the treatment with ATM, DNA-PK, and Lig IV inhibitors, namely, 10 µM KU55933 (SIGMA SML-1109, St. Louis, Missouri, USA [77]), 0.1 µM KU-0060648 (SIGMA SML-1257, St. Louis, MO, USA), and 0.1 µM SCR7 Pyrazine (SIGMA SML-1546, St. Louis, MO, USA [78]), was continuous. SIGMA is subsidiary of Merck KGaA (Darmstadt, Germany), and its headquarter is at St. Louis, Missouri, USA.Co-treatments were done with the correspondent inhibitors + BLEO for 45 min; then, for the MTT or clonogenic experiments, BLEO was removed and the inhibitors were added again in a fresh medium with FBS. For the comet assay or ICF, the experiment was stopped and the cells were fixed or lysed immediately after the 45 min treatment. All times in the graphs refer to the post-BLEO-treatment time. As such, t = "0" means "immediately after the 45 min BLEO treatment", t = 24 is 24 h later, and so on.

4.3. Verification of the PARPis and PARGi Effects on VERO Basal PAR Pool

ICF using an anti-PAR antibody on untreated VERO cells indicated that 25 nM OLA had an effect that was very similar to 250 nM OLA, diminishing the endogenous PAR more than 100 nM EB or 5 mM 3AB. Conversely, the PARG inhibitor DEA induced PAR accumulation.

4.4. Cell Viability Assay (MTT)

Cells were seeded in 96-well plates at a density according to the duration of the experiment (e.g., 15,000/well for 24 h experiments and 3000/well for 72 h experiments). The cell viability was determined by a 3-(4,5-dimethylthiazol-2-yl)-2,5-diphenyltetrazolium bromide (MTT) colorimetric assay [79], in which metabolically active cells reduced the dye to purple formazan. Cells were incubated for 1 h at 37 °C with MTT (0.5 mg/mL final concentration in 10 mM glucose in PBS). Formazan crystals were dissolved with DMSO. The absorbance was measured using a reference wavelength of 630 nm and a test wavelength of 570 nm on a Varioskan Flash microplate reader, Thermo Scientific, Waltham, MA, USA).

4.5. Clonogenic Assay

Cells were seeded in 50 mm cell culture dishes at a very low density (1000 cells for treatments and less (500 or 250) for the controls). After the attachment, cell cultures were treated for 45 min with BLEO in the absence or presence of inhibitors. Then, the culture media was changed and fresh inhibitors were added. The cells were incubated until control colonies reached at least 50 cells (counted under the microscope) for 8 days. Finally, colonies were fixed on 70% ethanol at room temperature (RT) for 10 min, briefly stained with 0.1% crystal violet (Fluka 61135), rinsed in abundant distilled water, dried at 37 °C, and manually counted by an observer under a magnifying glass. The plates were blind-coded. The clonogenic efficiency was expressed relative to the controls.

4.6. Comet Assay

The assay was carried out in alkaline conditions to detect SSB, DSB, and alkali-labile sites [80–82]. The slide's surface was pretreated with 1% normal melting point agarose (Sigma, St. Louis, MO, USA) in PBS. After a 45 min treatment, the cells were washed with PBS, incubated with trypsin-EDTA for 5 min at 37 °C, centrifuged, and resuspended in PBS; then, 20 µL of cell suspension was mixed with 80 µL of 0.75% low-melting agarose (Sigma, St. Louis, MO, USA) in PBS at 37 °C. Immediately, an 80 µL volume of the suspension was placed on a slide, covered with parafilm, and kept at 4 °C (10 min). The parafilm was removed and the slide was immersed in a cold lysis buffer (2.5 M NaCl, 100 mM EDTA, 10 mM Tris-HCl, and 8 g of NaOH/890 mL of water, adjusted to pH 10, to which was added 1% Triton-X-100 and 10% DMSO an hour before use) and kept at 4 °C (from 1 to 15 days). The slides were incubated in cold electrophoresis buffer (300 mM NaOH and 1 mM EDTA at pH 13) for 20 min at 4 °C to unwind the DNA strands and expose the alkali-labile sites. Electrophoresis was performed at 25 V for 20 min. The buffer volume was adjusted to achieve a current intensity in the range of 250 to 300 mA. After that, the slides were washed with neutralization buffer (0.4 M Tris-HCl, pH 7.5) three times for 5 min each and washed with distilled water. Subsequently, the slides were stained with 80 µL of DAPI (6 µg/mL) (10 min), washed with distilled water, and covered with coverslips.

Comets were blind-counted under epifluorescence and representative photographs were taken with the confocal microscope in non-confocal conditions (confocal aperture 5) under a 20× objective and classified according to the degree of damage in five categories: from 1 to 5 (α = degree of damage). The following criteria were used to assess the degree of damage: 1: no damage and intact or with a halo surrounding the core; 2: a little damage (the DNA was distributed in an oval); 3: the anterior–posterior axis measured twice the diameter; 4: the compact DNA was reduced and a large cloud of DNA (long tail of the comet) appeared; and 5: the tail was separated from the rest of compact DNA.

A double-blind count of 100 cells was performed. A damage index (DDI) was calculated using the equation: $DDI = \Sigma(n.\alpha)$, where α (which can range from 1 to 5) expresses the degree of damage and n is the number of cells with the degree of damage α.

4.7. Indirect Inmunocytofluorescence and Image Acquisition

Cells were washed with filtered PBS (fPBS, 0.22 µm pore size), fixed in 4% paraformaldehyde (PFA) in fPBS 15 min at 4 °C, washed in fPBS, permeabilized in 0.1% Triton-X100 in fPBS, and immersed in blocking buffer (0.2% Tween, 1% BSA in fPBS) for 30 min. Briefly, cells were incubated with the specific antibodies, namely, 1:300 rabbit anti-PARP-1/2 (Santa Cruz sc-7150, CA, USA), 1:500 rabbit anti-PAR (BD551813, Becton Dickinson (Franklin Lakes, NJ, USA)), 1:50 mouse anti-PAR (Enzo BML-SA216, Farmingdale, NY, USA), 1:100 10H-anti-PAR (Tulip #1020), 1:200 anti-hPARG (Abcam 16060, Cambridge, MA, USA), 1:400 mouse monoclonal anti-γH2AX (Abcam), and 1:300 anti-53BP1 (Abcam 36828), and diluted in blocking buffer for 2 h at 37 °C. After washing in fPBS/T (0.1% Tween), cells were incubated (1 h, RT) with the correspondent anti-antibodies mix (1:250 anti-mouse-Cy3 Jackson Immuno Research, 1:1000 goat-anti-rabbit 488 (#A-11034, Thermo Fisher Scientific, Waltham, MA, USA) in a blocking buffer. After washing in fPBS/T and fPBS, DAPI counterstaining (1.5 µg/mL in fPBS), and a final wash in fPBS, the coverslips were mounted in Prolong Gold (Molecular Probes P36930, Eugene, OR, USA) and sealed with nail polish. Controls without a primary antibody were run in parallel to check the specificity of the signals

Mainly an Olympus BX61/FV300 (Tokyo, Japan) and sometimes a Zeiss LSM800-Airyscan (Oberkochen, Germany) or a Leica microscope were used to take the confocal images/stacks. Fluorescence excitation was performed with the following lasers: diode 405 nm (DAPI), multiline argon 488 nm (Alexa Fluor 488), and helium-neon 561 nm (Cy3). Microscope settings were adjusted to register no signal in controls without primary antibodies. The scanning of optical sections was performed sequentially for the different fluorochromes. All images in each experimental series were taken with the same setting at the same confocal session. If modified, all were subject to the same degree of brightness/contrast adjustment and Gaussian blur filtering, including the control without a primary antibody. ImageJ free software [83] was used for the image processing.

4.8. Cell Counting Using Low-Magnification Fields

The ImageJ "Cell Counter" plug-in [84] (was employed to find out the percentage of nuclei that exhibited no γH2AX, well-defined γH2AX foci, or γH2AX pan-nuclear staining.

4.9. Relative PAR Quantification

The total field PAR intensity (RawIntDen, the sum of signal intensity in all the image pixels) was normalized using DAPI RawIntDen in each field (to account for putative differences in cell densities). Then, the data were expressed as a percentage of the control.

4.10. Statistical Analysis

The results were expressed as mean ± SEM. Differences between different experimental groups were tested for significance using a two-tailed unequal variances Student's *t*-test in Microsoft Excel 2007 [85] or a one-way analysis of variance (ANOVA), followed by post-hoc multiple comparisons tests (Tukey, Scheffe, Bonferroni, and Holm) in ASTATSA [86])

Author Contributions: Conceptualization, V.P., M.S., P.L., D.J.K., and L.L.-H.; methodology, V.P., M.S., P.L., S.V.-L., and L.L.-H.; formal analysis, V.P., M.S., P.L., S.V.-L., D.J.K., and L.L.-H.; investigation, V.P., M.S., P.L., S.V.-L., D.J.K., and L.L.-H.; resources, V.P., M.S., P.L., S.V.-L., D.J.K., and L.L.-H.; data curation, V.P., M.S., P.L., S.V.-L., D.J.K., and L.L.-H.; writing—original draft preparation, L.L.-H.; writing—review and editing, V.P., M.S., P.L., S.V.-L., D.J.K., and L.L.-H.; visualization, L.L.-H.; supervision, D.J.K; project administration, D.J.K.; funding acquisition, V.P., M.S., D.J.K.. and L.L.-H. All authors have read and agreed to the published version of the manuscript (CRediT taxonomy).

Funding: This research was partially funded by an undergraduate Student Research Support Program, call 2017: "Programa de Apoyo a la Investigación Estudiantil, Comisión Sectorial de Investigación Científica" (PAIE-CSIC UdelaR, Uruguay). Project title: Poly(ADP-ribosylation) and VERO cells' response to induced DNA damage. Responsible student: V.P.; responsible professor: D.J.K; co-responsible: L.L.H. Some reagents were bought through

Programa de Desarrollo de las Ciencias Básicas (PEDECIBA). APC contributors: PEDECIBA BIOLOGIA (aliquots given to researchers D.J.K., L.L.-H and R. Daniel Peluffo, CENUR); and Ministerio de Educación y Cultura.

Acknowledgments: Agencia Nacional de Investigación e Innovación (ANII-SNI). Bleomycin sulfate (Bleocris 15, LKM) was a gift from Laboratorios Teva Uruguay S.A., Grupo Biotoscana. The single-cell electrophoresis chamber and power supply were from the Epigenetics and Genomic Instability Lab, Instituto de Investigaciones Biológicas Clemente Estable (IIBCE). The alternative CO_2 incubators used came from the Neurochemistry Department, IIBCE, and the Microbiology Department, IIBCE. Part of the VERO cell stock was kept at Facultad de Ciencias, UdelaR. G. A. Folle kindly contributed to the text editing. Thanks to R.D. Peluffo and to Ministerio de Educación y Cultura, Dirección para el Desarrollo de la Ciencia y el Conocimimento, for financial contributions to pay APC.

Conflicts of Interest: The authors declare no conflict of interest. The funders had no role in the design of the study; in the collection, analyses, or interpretation of data; in the writing of the manuscript, or in the decision to publish the results.

Abbreviations

γH2AX	Histone H2AX phosphorylated on Ser 139
10H	Clone mouse monoclonal anti-PAR antibody (Tulip #1020)
3AB	3-Aminobenzamide (PARP inhibitor)
3i	Inhibition of DNA-PK, LigIV, and ATM
4-ANI	4-Amino-1,8-naphthalimide
53BP1	p53 binding protein 1
AG14361	2-[4-[(dimethylamino)methyl]phenyl]-1,3,10-triazatricyclo[6.4.1.$0^{4,13}$] trideca-2,4,6,8(13)-tetraen-9-one (PARP-1i)
ATM	Ataxia telangiectasia mutated
ATMi	ATM inhibitor
ATR	Ataxia telangiectasia and Rad3-related protein
ATRi	ATR inhibitor
AZ20	4{4-[(3R)3-Methylmorpholin-4-yl]-6-[1-(methylsulfonyl)cyclopropyl] pyrimidin-2-yl}-1*H*-indole
BER	Base excision repair
BLEO	Bleomycin
BRCA1/2	Breast cancer gene 1 or 2
CDKN2	Cyclin-dependent kinase inhibitor 2
CHK1	Checkpoint kinase 1
CHO9	Chinese hamster ovary 9 cell line
C-NHEJ	Canonic-non-homologous End Joining
DAPI	4′,6-diamidino-2-phenylindole (fluorescent DNA marker)
DDI	DNA damage index
DDR	DNA damage response
DEA	6,9-Diamino-2-ethoxyacridine-DL-lactate monohydrate
DMSO	Dimethyl sulfoxide
DNA-PK	DNA-dependent protein kinase
DNA-PKi	DNA-PK inhibitor
DSBs	Double-strand breaks
E-64	*trans*-Poxysuccinyl-l-leucylamido-(4-guanidino) butane
E7016	10-((4-Hydroxypiperidin-1-yl)methyl)chromeno[4,3,2-de]phthalazin-3(2H)-one (PARPi)
EB	5′-deoxy-5′-[4-[2-[(2,3-dihydro-1oxo-1H-isoindol-4-yl)amino]-2-oxoethyl]-1-piperazinyl]-5′-oxoadenosine dihydrochloride (PARPi)
FBS	Fetal bovine serum
HeLa	Henrietta Lacks human epithelial carcinoma cell line
HEp-2	Human epithelial type 2 cells
HNSCC	Head and neck squamous cell carcinomas
HR	Homologous recombination
IC_{50}	The half-maximal inhibitory concentration

ICF	Inmunocytofluorescence
INO1001	3-Aminobenzamide (PARP inhibitor)
IR	Ionizing radiation
KU-0060648	4-Ethyl-N-[4-[2-(4-morpholinyl)-4-oxo-4H-1-benzopyran-8-yl]-1-dibenzothienyl]-1-piperazineacetamide (DNA-PKi)
KU55933	2-Morpholin-4-yl-6-thianthren-1-yl-pyran-4-one
KU70/80	Ku heterodimer: Ku70 (XRCC6) and Ku80 (XRCC5)
LigIII	DNA ligase 3
LigIV	DNA ligase 4
LigIVi	LigIV inhibitor
MAPK	Mitogen-activated protein kinase
MEM	Minimum essential medium
MMEJ	Microhomology-mediated end joining
MMS	Methylmethanosulfonate
MRE11	Complex consisting of meiotic recombination 11
Mst-1	Macrophage-stimulating 1 (mitogen-activated protein kinase)
MTT	3-(4,5-dimethylthiazol-2-yl)-2,5-diphenyltetrazolium bromide
NAD+	Nicotinamide adenine dinucleotide
Nam	Nicotinamide
NMuMG	Nontransformed mouse mammary gland epithelial cell line
OLA	Olaparib (Lynparza, AZD-2281; PARP-1/2 & 3 inhibitor); IUPAC name:4-[[3-[4-(cyclopropanecarbonyl)piperazine-1-carbonyl]-4-fluorophenyl]methyl]-2*H*-phthalazin-1-one
p38	A class of MAPK
PAR	Poly(ADP-ribose)
PARG	Poly-ADP-glicohidrolase
PARP	Poly(ADP-ribosyl) polymerase
PARPis	PARP inhibitors
PARylation	Poly(ADP-ribosylation)
PBS	Phosphate-buffered saline
PFA	Paraformaldehyde
Polβ	DNA polymerase β
Prexacertib	5-[[5-[2-(3-aminopropoxy)-6-methoxyphenyl]-1*H*-pyrazol-3-yl]amino]pyrazine-2-carbonitrile (CHK1 inhibitor)
PtK2	Male rat kangaroo kidney epithelial cell
RT	Room temperature
SSBs	Single-strand breaks
TNKS	Tankyrase
UV-C	Ultraviolet C radiation
VE-821	3-amino-6-(4-methylsulfonylphenyl)-N-phenylpyrazine-2-carboxamide (ATRi)
VE-822	3-[3-[4-(methylaminomethyl)phenyl]-1,2-oxazol-5-yl]-5-(4-propan-2-ylsulfonylphenyl)pyrazin-2-amine (ATRi)
VERO	African green monkey kidney epithelial cell line
VRK1	Vaccinia-related kinase 1
WBWEE1	Western BlotsG2 checkpoint kinase
XRCC1	X-ray repair cross-complementing protein 1

Appendix A

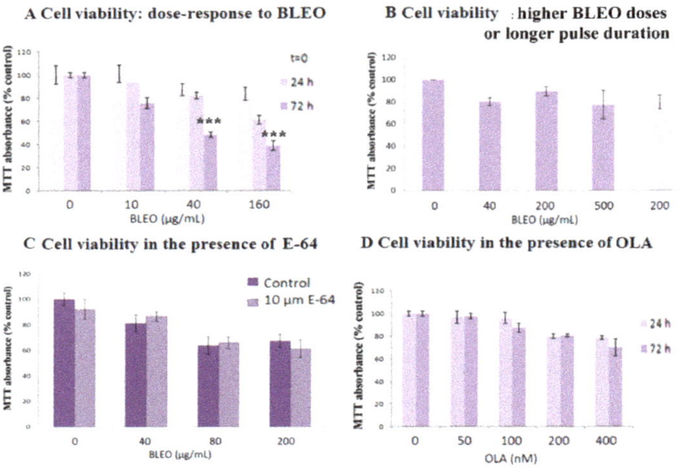

Figure A1. BLEO and OLA curves. Cell viability (MTT assay). (**A**) Cells were exposed to BLEO in the indicated doses (0–160 µg/mL) during a 45 min pulse. Then, cells were cultured during the indicated time before the MTT addition (0, 24, or 72 h, where "t = 0" refers to immediately after the treatment). The 2018–2020 dataset includes 20 MTT experiments with BLEO in VERO cells. At 72 h, n = 63, 6, 72, and 22 for increasing BLEO concentrations. Fewer data were recorded for shorter times and are depicted to facilitate understanding of the chosen schedule. ***: $p < 0.001$. (**B**) Cell viability (24 h) under higher BLEO doses or longer BLEO pulse durations. Cells were exposed to BLEO in the indicated doses (0–500 µg/mL) during a 45 min pulse or (last bar) during 270 min. (**C**) Cell viability (24 h) in the control and BLEO-exposed cells in the absence or presence of a co-treatment (45 min pulse) with bleomycin hydrolase inhibitor E64. (**D**) Cell viability after exposure to OLA (0–400 nM). The whole curve was done once at 24 h and once at 72 h. In contrast, for 50 nM Olaparib at 72 h, n = 75 from 12 independent experiments.

Appendix B

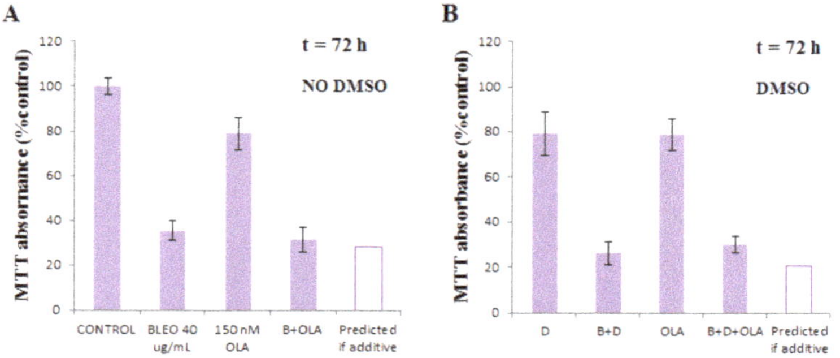

Figure A2. Concentrations of 150 nM OLA and BLEO. A single experiment was carried out to explore whether an increase in OLA concentration would lead to potentiation. Effect of combined 150 nM OLA and 40 µg/mL BLEO in the absence (**A**) or presence (**B**) of DMSO on VERO cell viability was evaluated using MTT (72 h). Mean ± SD. n = 6. Since the predicted additive effect according to Webb's equation [25] was equal or lower than the combined effect, synergism was discarded.

Appendix C

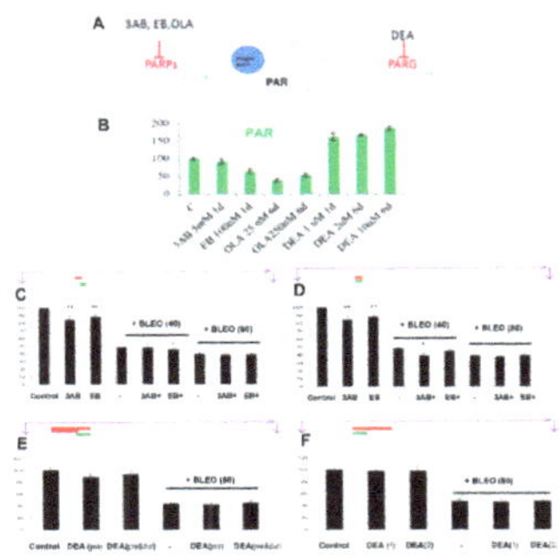

Figure A3. PARPis and BLEO in VERO. PARPis and BLEO in VERO. (**A**) Scheme representing PAR, its synthesis by PARPs, its degradation mainly by poly-ADP-glycohydrolase (PARG) activity, and the inhibitors abbreviations associated with the correspondent enzyme. 3AB, EB, and OLA are PARPis, while DEA inhibits PARG. (**B**) Test of the inhibitors' effects on PAR. The inhibitors used were 3-aminobenzamide (3AB, Sigma A-0788), 5′-deoxy-5′-[4-[2-[(2,3-dihydro-1oxo-1H-isoindol-4-yl)amino]-2-oxoethyl]-1-piperazinyl]-5′-oxoadenosine dihydrochloride (EB, Alexis Biochemicals), 6,9-diamino-2-ethoxyacridine-DL-lactate monohydrate (DEA, Trevigen 4680-096-03), and OLA. PAR was measured on confocal planes of monolayers using the Analyze/Measure ImageJ plugin, and the results are expressed as arbitrary fluorescence units/area. (**C–F**) Cell viability assays. Results are expressed as % MTT absorbance with respect to control. The violet arrows at the top of each graph represent the time in days (5 days) from seeding to MTT evaluation. The bleomycin treatment (40 to 80 µg/mL, 45 min) is represented in green and the inhibitors' treatment durations are in red. (**C**) Pre-treatment with 3AB or EB, (**D**) co-treatment with 3AB or EB, (**E**) pre- and co-treatment with the PARG inhibitor DEA, and (**F**) co- and post-treatment with DEA. The genotoxic treatment was always different from the control ($p < 0.001$). Asterisks refer to a comparison of a treatment vs. the control or with respect to the genotoxic treatment. *: $p < 0.05$, **: $p < 0.01$.

Appendix D

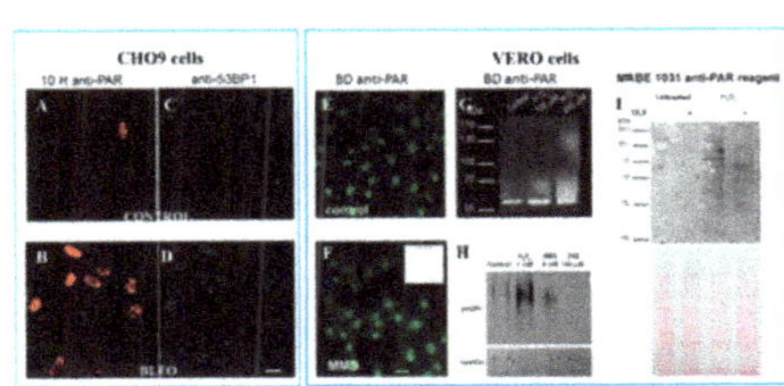

Figure A4. Positive controls for anti-PAR antibodies. (**A–D**) Comparison of the control (**A**,**C**) and BLEO-treated cells (**B**,**D**; 40 µg/mL, 45 min, like VERO) that were indirectly immunostained with 10H anti-PAR (**A**,**B**, red) and anti-53BP1 (**C**,**D**, green) Bar: 15 µm. PAR-positive nuclei were much more frequent in CHO9 than in VERO cells. (**E**,**F**) ICF with BD anti-PAR antibody (same experiment as in Figure 3G,H, extracted from Ph.D. thesis [53] Bar: 25 µm). Only after a very strong treatment with MMS (**F**; 6 mM, 45 min), which induced cell death of the whole cell population (insert), was a slight PAR increase noticed with BD anti-PAR. (**G**) WB detecting PAR with BD anti-PAR in control cells or cells exposed to 100 and 250 µM H_2O_2. (**H**) WB detecting PAR indirectly with the rabbit PAR-binding reagent MABE1031 (1:4000) and anti-rabbit-HRP in control VERO cells or under exposure to 1 mM H_2O_2 (10 min), 8 mM MMS (45 min), or the PARP inhibitor 3AB (2 h). (**I**) WB detecting PAR with MABE1031 in the control VERO cells or under exposure to OLA (1 h), H_2O_2 (last 10 min), or the combined treatment OLA + H_2O_2.

Appendix E

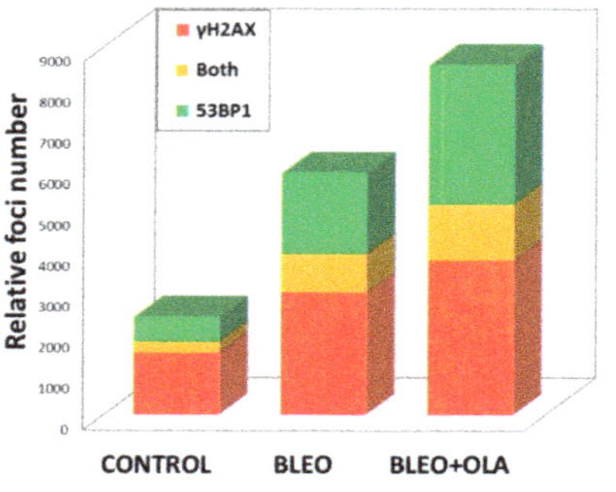

Figure A5. γH2AX vs. 53 BP1 foci distribution in one experiment. Foci counting required the following steps, which were done in batch: (1) Separate channels. (2) Segment images to obtain a mask using a fixed threshold for all images of the experiment in each channel. In this case, the following thresholds were used: 50 for the red channel (γH2AX), 62 for the green channel (53BP1), and 40 for the blue (DAPI) channel. (3) The Analyze Particles plug-in allowed for counting objects with certain properties, namely, circularity and size. Foci were defined as ovoid to circular objects (circularity > 0.5) and 5 to 200 pixels. Nuclei were defined as ovoid objects measuring 500 to 500,000 pixels. In each case, the program showed the masks, displayed results, and summarized the data. In this way, we obtained the number of γH2AX foci, 52 BP1 foci, and cell number. (4) In order to determine the number of mixed foci (positive both for γH2AX and 53 BP1), mask images (which are 0 or 1 matrixes) were multiplied. In this way, a signal remained in pixels positive for γH2AX and 53BP1 (1 × 1 = 1) while no signal was obtained in a pixel if at least one of the channels was black (1 × 0 = 0). After that, we had objects representing partial foci superimposition. In order to avoid artifacts or overcounting, we used the Analyze Particles Plugin again to count only whole foci. Again, foci were defined as ovoid to circular objects (circularity > 0.5) measuring 5 to 200 pixels. (5) Mixed foci were subtracted from the total γH2AX foci and from the total 53BP1 foci to obtain the numbers of foci with only one mark. (6) γH2AX alone, 53BP1 alone, and mixed foci numbers were normalized by cell number. Data were from a single experiment using 10 microscopic fields (40× zoom 2) per condition. Foci counting involved 247 control cells, 176 BLEO cells, and 119 (BLEO + OLA) cells.

References

1. Teloni, F.; Altmeyer, M. Survey and summary readers of poly(ADP-ribose): Designed to be fit for purpose. *Nucleic Acids Res.* **2016**, *44*, 993–1006. [CrossRef] [PubMed]
2. Althaus, F.R.; Kleczkowska, H.E.; Malanga, M.; Müntener, C.R.; Pleschke, J.M.; Ebner, M.; Auer, B. Poly ADP-ribosylation: A DNA break signal mechanism. *Proc. Mol. Cell. Biochem.* **1999**, *193*, 5–11. [CrossRef]
3. Gagné, J.P.; Pic, É.; Isabelle, M.; Krietsch, J.; Éthier, C.; Paquet, É.; Kelly, I.; Boutin, M.; Moon, K.M.; Foster, L.J.; et al. Quantitative proteomics profiling of the poly(ADP-ribose)-related response to genotoxic stress. *Nucleic Acids Res.* **2012**, *40*, 7788–7805. [CrossRef] [PubMed]
4. Blackford, A.N.; Jackson, S.P. ATM, ATR, and DNA-PK: The Trinity at the Heart of the DNA Damage Response. *Mol. Cell* **2017**, *66*, 801–817. [CrossRef]
5. Salzano, M.; Sanz-García, M.; Monsalve, D.M.; Moura, D.S.; Lazo, P.A. VRK1 chromatin kinase phosphorylates H2AX and is required for foci formation induced by DNA damage. *Epigenetics* **2015**, *10*, 373–383. [CrossRef]
6. Maréchal, A.; Zou, L. DNA damage sensing by the ATM and ATR kinases. *Cold Spring Harb. Perspect. Biol.* **2013**, *5*, a012716. [CrossRef]
7. Dueva, R.; Iliakis, G. Alternative pathways of non-homologous end joining (NHEJ) in genomic instability and cancer. *Transl. Cancer Res.* **2013**, *2*, 163–177. [CrossRef]

8. Mladenov, E.; Iliakis, G. The Pathways of Double-Strand Break Repair. In *DNA Repair—On the Pathways to Fixing DNA Damage and Errors*; Storici, F., Ed.; InTech: London, UK, 2011; pp. 143–168. ISBN 978-953-307-649-2.
9. Ray Chaudhuri, A.; Nussenzweig, A. The multifaceted roles of PARP1 in DNA repair and chromatin remodelling. *Nat. Rev. Mol. Cell Biol.* **2017**, *18*, 610–621. [CrossRef] [PubMed]
10. Hottiger, M.O. SnapShot: ADP-Ribosylation Signaling. *Mol. Cell* **2015**, *58*, 1134–1134.e1. [CrossRef]
11. Virag, L.; Szabo, C. The Therapeutic Potential of Poly(ADP-Ribose) Polymerase Inhibitors. *J. Natl. Cancer Inst.* **2011**, *103*, 334–346. [CrossRef]
12. Kraus, W.L. PARPs and ADP-Ribosylation: 50 Years ... and Counting. *Mol. Cell* **2015**, *58*, 902–910. [CrossRef]
13. Thraves, P.; Mossman, K.L.; Brennan, T.; Dritschilo, A. Radiosensitization of human fibroblasts by 3-aminobenzamide: An inhibitor of poly(ADP-ribosylation). *Radiat. Res.* **1985**, *104*, 119–127. [CrossRef]
14. Brock, W.A.; Milas, L.; Bergh, S.; Lo, R.; Szabo, C.; Mason, K.A. Radiosensitization of human and rodent cell lines by INO-1001, a novel inhibitor of poly(ADP-ribose) polymerase. *Cancer Lett.* **2004**, *205*, 155–160. [CrossRef]
15. Arundel-Suto, C.M.; Scavone, S.V.; Turner, W.R.; Suto, M.J.; Sebolt-Leopold, J.S. Effect of PD 128763, a new potent inhibitor of poly(ADP-ribose) polymerase, on X-ray-induced cellular recovery processes in Chinese hamster V79 cells. *Radiat. Res.* **1991**, *126*, 367–371. [CrossRef]
16. Oleinick, N.L.; Evans, H.H. Poly(ADP-ribose) and the response of cells to ionizing radiation. *Radiat. Res.* **1985**, *101*, 29–46. [CrossRef]
17. Russo, A.L.; Kwon, H.-C.; Burgan, W.E.; Carter, D.; Beam, K.; Weizheng, X.; Zhang, J.; Slusher, B.S.; Chakravarti, A.; Tofilon, P.J.; et al. In vitro and In vivo Radiosensitization of Glioblastoma Cells by the Poly (ADP-Ribose) Polymerase Inhibitor E7016. *Clin. Cancer Res.* **2009**, *15*, 607–612. [CrossRef]
18. Veuger, S.J.; Curtin, N.J.; Richardson, C.J.; Smith, G.C.M.; Durkacz, B.W. Radiosensitization and DNA repair inhibition by the combined use of novel inhibitors of DNA-dependent protein kinase and poly(ADP-ribose) polymerase-1. *Cancer Res.* **2003**, *63*, 6008–6015.
19. Noel, G.; Godon, C.; Fernet, M.; Giocanti, N.; Megnin-Chanet, F.; Favaudon, V. Radiosensitization by the poly(ADP-ribose) polymerase inhibitor 4-amino-1,8-naphthalimide is specific of the S phase of the cell cycle and involves arrest of DNA synthesis. *Mol. Cancer Ther.* **2006**, *5*, 564–574. [CrossRef] [PubMed]
20. Huet, J.; Laval, F. Potentiation of Cell Killing by Inhibitors of Poly (Adenosine Diphosphate-Ribose) Synthesis in Bleomycin-treated Chinese Hamster Ovary Cells. *Cancer Res.* **1985**, *45*, 987–991.
21. Kelland, L.R.; Burgess, L.; Steel, G.G. Radiation damage repair capacity of a human germ-cell tumour cell line: Inhibition by 3-aminobenzamide. *Int. J. Radiat. Biol. Relat. Stud. Phys. Chem. Med.* **1987**, *51*, 227–241. [CrossRef]
22. Horton, J.K.; Stefanick, D.F.; Naron, J.M.; Kedar, P.S.; Wilson, S.H. Poly(ADP-ribose) polymerase activity prevents signaling pathways for cell cycle arrest after DNA methylating agent exposure. *J. Biol. Chem.* **2005**, *280*, 15773–15785. [CrossRef]
23. Cleaver, J.E.; Bodello, W.J.; Morgan, W.F.; Zellell, B. Poly (ADP-ribose) of Repair of DNA Differences in the Regulation by and Ultraviolet Light According to Damage from Alkylating Agents Cell Type. *J. Biol. Chem.* **1983**, *258*, 9059–9068.
24. Vodenicharov, M.D.; Ghodgaonkar, M.M.; Halappanavar, S.S.; Shah, R.G.; Shah, G.M. Mechanism of early biphasic activation of poly(ADP-ribose) polymerase-1 in response to ultraviolet B radiation. *J. Cell Sci.* **2005**, *118*, 589–599. [CrossRef]
25. Chou, T.-C. Theoretical basis, experimental design, and computerized simulation of synergism and antagonism in drug combination studies. *Pharmacol. Rev.* **2006**, *58*, 621–681. [CrossRef] [PubMed]
26. Saha, T.; Smulson, M.; Rosen, E.M. BRCA1 regulation of base excision repair pathway. *Cell Cycle* **2010**, *9*, 2471–2472. [CrossRef]
27. Helleday, T. The underlying mechanism for the PARP and BRCA synthetic lethality: Clearing up the misunderstandings. *Mol. Oncol.* **2011**, *5*, 387–393. [CrossRef]
28. Ström, C.E.; Johansson, F.; Uhlén, M.; Szigyarto, C.A.K.; Erixon, K.; Helleday, T. Poly (ADP-ribose) polymerase (PARP) is not involved in base excision repair but PARP inhibition traps a single-strand intermediate. *Nucleic Acids Res.* **2011**, *39*, 3166–3175. [CrossRef]
29. Horton, J.K.; Stefanick, D.F.; Prasad, R.; Gassman, N.R.; Kedar, P.S.; Wilson, S.H. Base excision repair defects invoke hypersensitivity to PARP inhibition. *Mol. Cancer Res.* **2014**, *12*, 1128–1139. [CrossRef]

30. Czyż, M.; Toma, M.; Gajos-Michniewicz, A.; Majchrzak, K.; Hoser, G.; Szemraj, J.; Nieborowska-Skorska, M.; Cheng, P.; Gritsyuk, D.; Levesque, M.; et al. PARP1 inhibitor olaparib (Lynparza) exerts synthetic lethal effect against ligase 4-deficient melanomas. *Oncotarget* **2016**, *7*, 75551–75560. [CrossRef]
31. Ko, H.L.; Ren, E.C. Functional aspects of PARP1 in DNA repair and transcription. *Biomolecules* **2012**, *2*, 524–548. [CrossRef]
32. Feng, F.Y.; de Bono, J.S.; Rubin, M.A.; Knudsen, K.E. Chromatin to Clinic: The Molecular Rationale for PARP1 Inhibitor Function. *Mol. Cell* **2015**, *58*, 925–934. [CrossRef] [PubMed]
33. Bai, P. Review Biology of Poly (ADP-Ribose) Polymerases: The Factotums of Cell Maintenance. *Mol. Cell* **2015**, *58*, 947–958. [CrossRef]
34. Berger, N.A.; Besson, V.C.; Boulares, A.H.; Bürkle, A.; Chiarugi, A.; Clark, R.S.; Curtin, N.J.; Cuzzocrea, S.; Dawson, T.M.; Dawson, V.L.; et al. Opportunities for the repurposing of PARP inhibitors for the therapy of non-oncological diseases. *Br. J. Pharmacol.* **2018**, *175*, 192–222. [CrossRef]
35. Sleijfer, S. Bleomycin-induced pneumonitis. *Chest* **2001**, *120*, 617–624. [CrossRef]
36. Beck, C.; Robert, I.; Reina-San-Martin, B.; Schreiber, V.; Dantzer, F. Poly(ADP-ribose) polymerases in double-strand break repair: Focus on PARP1, PARP2 and PARP3. *Exp. Cell Res.* **2014**, *329*, 18–25. [CrossRef]
37. Sukhanova, M.V.; Abrakhi, S.; Joshi, V.; Pastre, D.; Kutuzov, M.M.; Anarbaev, R.O.; Curmi, P.A.; Hamon, L.; Lavrik, O.I. Single molecule detection of PARP1 and PARP2 interaction with DNA strand breaks and their poly(ADP-ribosyl)ation using high-resolution AFM imaging. *Nucleic Acids Res.* **2015**. [CrossRef]
38. Fouquin, A.; Guirouilh-Barbat, J.; Lopez, B.; Hall, J.; Amor-Guéret, M.; Pennaneach, V. PARP2 controls double-strand break repair pathway choice by limiting 53BP1 accumulation at DNA damage sites and promoting end-resection. *Nucleic Acids Res.* **2017**, *45*, 12325–12339. [CrossRef]
39. Rulten, S.L.; Fisher, A.E.O.; Robert, I.; Zuma, M.C.; Rouleau, M.; Ju, L.; Poirier, G.; Reina-San-Martin, B.; Caldecott, K.W. PARP-3 and APLF function together to accelerate nonhomologous end-joining. *Mol. Cell* **2011**, *41*, 33–45. [CrossRef] [PubMed]
40. Grundy, G.J.; Polo, L.M.; Zeng, Z.; Rulten, S.L.; Hoch, N.C.; Paomephan, P.; Xu, Y.; Sweet, S.M.; Thorne, A.W.; Oliver, A.W.; et al. PARP3 is a sensor of nicked nucleosomes and monoribosylates histone H2B Glu2. *Nat. Commun.* **2016**, *7*. [CrossRef] [PubMed]
41. Oplustil O'Connor, L.; Rulten, S.L.; Cranston, A.N.; Odedra, R.; Brown, H.; Jaspers, J.E.; Jones, L.; Knights, C.; Evers, B.; Ting, A.; et al. The PARP Inhibitor AZD2461 Provides Insights into the Role of PARP3 Inhibition for Both Synthetic Lethality and Tolerability with Chemotherapy in Preclinical Models. *Cancer Res.* **2016**, *76*, 6084–6094. [CrossRef]
42. Nagy, Z.; Kalousi, A.; Furst, A.; Koch, M.; Fischer, B.; Soutoglou, E. Tankyrases Promote Homologous Recombination and Check Point Activation in Response to DSBs. *PLoS Genet.* **2016**, *12*, 1–26. [CrossRef]
43. Vilchez Larrea, S.C.; Haikarainen, T.; Narwal, M.; Schlesinger, M.; Venkannagari, H.; Flawiá, M.M.; Villamil, S.H.F.; Lehtiö, L. Inhibition of poly(ADP-ribose) Polymerase Interferes with Trypanosoma cruzi Infection and Proliferation of the Parasite. *PLoS ONE* **2012**, *7*, e46063. [CrossRef]
44. Thorsell, A.G.; Ekblad, T.; Karlberg, T.; Löw, M.; Pinto, A.F.; Trésaugues, L.; Moche, M.; Cohen, M.S.; Schüler, H. Structural Basis for Potency and Promiscuity in Poly(ADP-ribose) Polymerase (PARP) and Tankyrase Inhibitors. *J. Med. Chem.* **2017**, *60*, 1262–1271. [CrossRef]
45. Sharif-Askari, B.; Amrein, L.; Aloyz, R.; Panasci, L. PARP3 inhibitors ME0328 and olaparib potentiate vinorelbine sensitization in breast cancer cell lines. *Breast Cancer Res. Treat.* **2018**, *172*, 23–32. [CrossRef]
46. Murai, J.; Huang, S.Y.N.; Das, B.B.; Renaud, A.; Zhang, Y.; Doroshow, J.H.; Ji, J.; Takeda, S.; Pommier, Y. Trapping of PARP1 and PARP2 by clinical PARP inhibitors. *Cancer Res.* **2012**, *72*, 5588–5599. [CrossRef]
47. Terasima, T.; Takabe, Y.; Katsumata, T.; Watanabe, M.; Umezawa, H. Effect of Bleomycin on Mammalian Cell Survival. *J. Natl. Cancer Inst.* **1972**, *49*, 1093–1100.
48. Sheets, R. History and Characterization of the Vero Cell Line. *Open Sess.* **2000**, 1–12.
49. Naoki, O.; Arihiro, K.; Toshiyuki, Y.; Noriko, H.; Fumio, K.; Suyoshi, S.; Makoto, K.; Kentaro, H.; Hattori, M. The genome landscape of the African Green Monkey kidney-derived vero cell line. *DNA Res.* **2014**, *21*, 673–683. [CrossRef]
50. Ammerman, A. NIH Public Access. *Curr. Protoc. Microbiol.* **2009**, 1–10. [CrossRef]
51. Duran-Rehbein, G.A.; Vargas-Zambrano, J.C.; Cuéllar, A.; Puerta, C.J.; Gonzalez, J.M. Mammalian cellular culture models of trypanosoma cruzi infection: A review of the published literature. *Parasite* **2014**, *21*. [CrossRef]

52. Schacke, M.; Kumar, J.; Colwell, N.; Hermanson, K.; Folle, G.A.; Nechaev, S.; Dhasarathy, A.; Lafon-Hughes, L. PARP-1/2 inhibitor olaparib prevents or partially reverts EMT induced by TGF-β in NMuMG cells. *Int. J. Mol. Sci.* **2019**, *20*, 518. [CrossRef]
53. Kawamitsu, H.; Hoshino, H.; Okada, H.; Miwa, M.; Momoi, H.; Sugimura, T. Monoclonal antibodies against poly(ADP-ribose) recognize different structures of poly(ADP-ribose). *Biochemistry* **1984**, *23*, 41–47. [CrossRef]
54. Gagné, J.P.; Isabelle, M.; Lo, K.S.; Bourassa, S.; Hendzel, M.J.; Dawson, V.L.; Dawson, T.M.; Poirier, G.G. Proteome-wide identification of poly(ADP-ribose) binding proteins and poly(ADP-ribose)-associated protein complexes. *Nucleic Acids Res.* **2008**, *36*, 6959–6976. [CrossRef]
55. Fahrer, J.; Kranaster, R.; Altmeyer, M.; Marx, A.; Bürkle, A. Quantitative analysis of the binding affinity of poly(ADP-ribose) to specific binding proteins as a function of chain length *Nucleic Acids Res.* **2007**, *35*. [CrossRef] [PubMed]
56. Lafon-Hughes, L. Localización Preferencial del Daño Genético en Regiones Eucromáticas y Replicantes de los Cromosomas y los Núcleos de Mamíferos. Roles de la Poli-ADP-Ribosa en Presencia y Ausencia de Daño Genético Inducido en Células VERO. Ph.D. Thesis, Facultad de Ciencias—PEDECIBA, IIBCE, INGEBI, Universidad de la República, Montevideo, Uruguay, 2014. Available online: http://www.bib.fcien.edu.uy/files/etd/biol/uy24-17142.pdf (accessed on 3 November 2020).
57. Lafon-Hughes, L.; Vilchez Larrea, S.C.; Kun, A.; Fernández Villamil, S.H. VERO cells harbor a poly-ADP-ribose belt partnering their epithelial adhesion belt. *PeerJ* **2014**, *2*, e617. [CrossRef] [PubMed]
58. Wang, M.; Wu, W.; Wu, W.; Rosidi, B.; Zhang, L.; Wang, H.; Iliakis, G. PARP-1 and Ku compete for repair of DNA double strand breaks by distinct NHEJ pathways. *Nucleic Acids Res.* **2006**, *34*, 6170–6182. [CrossRef] [PubMed]
59. Burger, R.M. Cleavage of nucleic acids by bleomycin. *Chem. Rev.* **1998**, *98*, 1153–1169. [CrossRef]
60. Lafon-Hughes, L.; Di Tomaso, M.V.; Liddle, P.; Toledo, A.; Reyes-Ábalos, A.L.; Folle, G.A. Preferential localization of γh2AX foci in euchromatin of retina rod cells after DNA damage induction. *Chromosome Res.* **2013**, *21*. [CrossRef]
61. Xia, C.; Wolf, J.J.; Sun, C.; Xu, M.; Studstill, C.J.; Chen, J.; Ngo, H.; Zhu, H.; Hahm, B. PARP1 Enhances Influenza A Virus Propagation by Facilitating Degradation of Host Type I Interferon Receptor. *J. Virol.* **2020**, *94*, e01572-19. [CrossRef]
62. Nguyen, M.L.; Kraft, R.M.; Blaho, J.A. African green monkey kidney Vero cells require de novo protein synthesis for efficient herpes simplex virus 1-dependent apoptosis. *Virology* **2005**, *336*, 274–290. [CrossRef]
63. The UniProt Consortium UniProt: A worldwide hub of protein knowledge. *Nucleic Acids Res.* **2019**, *47*, D506–D515. [CrossRef]
64. Cruz, G.M.S.; Kong, X.; Silva, B.A.; Khatibzadeh, N.; Thai, R.; Berns, M.W.; Yokomori, K. Femtosecond near-infrared laser microirradiation reveals a crucial role for PARP signaling on factor assemblies at DNA damage sites. *Nucleic Acids Res.* **2015**, *44*, 1–17. [CrossRef]
65. Hochegger, H.; Dejsuphong, D.; Fukushima, T.; Morrison, C.; Sonoda, E.; Schreiber, V.; Zhao, G.Y.; Saberi, A.; Masutani, M.; Adachi, N.; et al. Parp-1 protects homologous recombination from interference by Ku and Ligase IV in vertebrate cells. *EMBO J.* **2006**, *25*, 1305–1314. [CrossRef]
66. Scully, R.; Xie, A. Double strand break repair functions of histone H2AX. *Mutat. Res.-Fundam. Mol. Mech. Mutagen.* **2013**, *750*, 5–14. [CrossRef]
67. Orr, K.S.; Savage, K.I. The BRCA1 and BRCA2 Breast and Ovarian Cancer Susceptibility Genes—Implications for DNA Damage Response, DNA Repair and Cancer Therapy. In *Advances in DNA Repair*; InTech: London, UK, 2015.
68. Kawahara, N.; Ogawa, K.; Nagayasu, M.; Kimura, M.; Sasaki, Y.; Kobayashi, H. Candidate synthetic lethality partners to parp inhibitors in the treatment of ovarian clear cell cancer (Review). *Biomed. Rep.* **2017**, *7*, 391–399. [CrossRef] [PubMed]
69. D'Andrea, A.D. Mechanisms of PARP inhibitor sensitivity and resistance. *DNA Repair* **2018**, *71*, 172–176. [CrossRef]
70. Pettitt, S.J.; Krastev, D.B.; Brandsma, I.; Dréan, A.; Song, F.; Aleksandrov, R.; Harrell, M.I.; Menon, M.; Brough, R.; Campbell, J.; et al. Genome-wide and high-density CRISPR-Cas9 screens identify point mutations in PARP1 causing PARP inhibitor resistance. *Nat. Commun.* **2018**, *9*. [CrossRef]
71. Mweempwa, A.; Wilson, M.K. Mechanisms of resistance to PARP inhibitors—An evolving challenge in oncology. *Cancer Drug Resist.* **2019**, 608–617. [CrossRef]

72. Haynes, B.; Murai, J.; Lee, J.M. Restored replication fork stabilization, a mechanism of PARP inhibitor resistance, can be overcome by cell cycle checkpoint inhibition. *Cancer Treat. Rev.* **2018**, *71*, 1–7. [CrossRef]
73. Yazinski, S.A.; Comaills, V.; Buisson, R.; Genois, M.M.; Nguyen, H.D.; Ho, C.K.; Kwan, T.T.; Morris, R.; Lauffer, S.; Nussenzweig, A.; et al. ATR inhibition disrupts rewired homologous recombination and fork protection pathways in PARP inhibitor-resistant BRCA-deficient cancer cells. *Genes Dev.* **2017**, *31*, 318–332. [CrossRef]
74. Young, L.A.; O'Connor, L.O.; de Renty, C.; Veldman-Jones, M.H.; Dorval, T.; Wilson, Z.; Jones, D.R.; Lawson, D.; Odedra, R.; Maya-Mendoza, A.; et al. Differential activity of ATR and Wee1 inhibitors in a highly sensitive subpopulation of DLBCL linked to replication stress. *Cancer Res.* **2019**, *79*, 3762–3775. [CrossRef]
75. Gadhikar, M.A.; Zhang, J.; Shen, L.; Rao, X.; Wang, J.; Zhao, M.; Kalu, N.N.; Johnson, F.M.; Byers, L.A.; Heymach, J.; et al. CDKN2A/p16 Deletion in Head and Neck Cancer Cells Is Associated with CDK2 Activation, Replication Stress, and Vulnerability to CHK1 Inhibition. *Cancer Res.* **2018**, *78*, 781–797. [CrossRef] [PubMed]
76. MedlinePlus, National Library of Medicine. Available online: https://medlineplus.gov/genetics/gene/cdkn2a/ (accessed on 22 October 2020).
77. Bakr, A.; Oing, C.; Köcher, S.; Borgmann, K.; Dornreiter, I.; Petersen, C.; Dikomey, E.; Mansour, W.Y. Involvement of ATM in homologous recombination after end resection and RAD51 nucleofilament formation. *Nucleic Acids Res.* **2015**, *43*, 3154–3166. [CrossRef]
78. Srivastava, M.; Nambiar, M.; Sharma, S.; Karki, S.S.; Goldsmith, G.; Hegde, M.; Kumar, S.; Pandey, M.; Singh, R.K.; Ray, P.; et al. An inhibitor of nonhomologous end-joining abrogates double-strand break repair and impedes cancer progression. *Cell* **2012**, *151*, 1474–1487. [CrossRef]
79. Denizot, F.; Lang, R. Rapid colorimetric assay for cell growth and survival. Modifications to the tetrazolium dye procedure giving improved sensitivity and reliability. *J. Immunol. Methods* **1986**, *89*, 271–277. [CrossRef]
80. Klaude, M.; Eriksson, S.; Nygren, J.; Ahnstrom, G. The comet assay: Mechanisms and technical considerations. *Mutat. Res.* **1996**, *363*, 89–96. [CrossRef]
81. Tice, R.R.; Andrews, P.W.; Singh, N.P. The single cell gel assay: A sensitive technique for evaluating intercellular differences in DNA damage and repair. *Basic Life Sci.* **1990**, *53*, 291–301. [CrossRef]
82. Olive, P.L.; Banáth, J.P. The comet assay: A method to measure DNA damage in individual cells. *Nat. Protoc.* **2006**, *1*, 23–29. [CrossRef]
83. Rueden, C.T.; Schindelin, J.; Hiner, M.C.; DeZonia, B.E.; Walter, A.E.; Arena, E.T.; Eliceiri, K.W. ImageJ2: ImageJ for the next generation of scientific image data. *BMC Bioinform.* **2017**, *18*, 529. [CrossRef]
84. Kurt De Vos; University of Sheffield, Academic Neurology Image J, Cell Counter Plugin. Available online: https://imagej.nih.gov/ij/plugins/cell-counter.html (accessed on 22 October 2020).
85. Available online: https://worddisk.com/wiki/Microsoft_Excel/ (accessed on 4 November 2020).
86. Navendu, V. Online Web Statistical Calculators for Categorical Data Analysis. 2016. Available online: astatsa.com (accessed on 22 October 2020).

International Journal of
Molecular Sciences

MDPI

Article

The First Berberine-Based Inhibitors of Tyrosyl-DNA Phosphodiesterase 1 (Tdp1), an Important DNA Repair Enzyme

Elizaveta D. Gladkova [1,2], Ivan V. Nechepurenko [1], Roman A. Bredikhin [1], Arina A. Chepanova [3], Alexandra L. Zakharenko [3], Olga A. Luzina [1], Ekaterina S. Ilina [3], Nadezhda S. Dyrkheeva [3], Evgeniya M. Mamontova [3], Rashid O. Anarbaev [2,3], Jóhannes Reynisson [4], Konstantin P. Volcho [1,2,*], Nariman F. Salakhutdinov [1,2] and Olga I. Lavrik [2,3]

1 N. N. Vorozhtsov Novosibirsk Institute of Organic Chemistry, Siberian Branch of the Russian Academy of Sciences, 9, Akademika Lavrentieva Ave., Novosibirsk 630090, Russia; liza95@nioch.nsc.ru (E.D.G.); niv@nioch.nsc.ru (I.V.N.); bred@nioch.nsc.ru (R.A.B.); luzina@nioch.nsc.ru (O.A.L.); anvar@nioch.nsc.ru (N.F.S.)

2 Novosibirsk State University, Pirogova str. 1, Novosibirsk 630090, Russia; anarbaev@niboch.nsc.ru (R.O.A.); lavrik@niboch.nsc.ru (O.I.L.)

3 Novosibirsk Institute of Chemical Biology and Fundamental Medicine, Siberian Branch of the Russian Academy of Sciences, 8, Akademika Lavrentieva Ave., Novosibirsk 630090, Russia; arinachepanova@mail.ru (A.A.C.); sashaz@niboch.nsc.ru (A.L.Z.); katya.plekhanova@gmail.com (E.S.I.); elpida80@mail.ru (N.S.D.); evgeniya.mm.94@gmail.com (E.M.M.)

4 School of Pharmacy and Bioengineering, Keele University, Hornbeam Building, Staffordshire ST5 5BG, UK; j.reynisson@keele.ac.uk

* Correspondence: volcho@nioch.nsc.ru

Received: 31 August 2020; Accepted: 26 September 2020; Published: 28 September 2020

Abstract: A series of berberine and tetrahydroberberine sulfonate derivatives were prepared and tested against the tyrosyl-DNA phosphodiesterase 1 (Tdp1) DNA-repair enzyme. The berberine derivatives inhibit the Tdp1 enzyme in the low micromolar range; this is the first reported berberine based Tdp1 inhibitor. A structure–activity relationship analysis revealed the importance of bromine substitution in the 12-position on the tetrahydroberberine scaffold. Furthermore, it was shown that the addition of a sulfonate group containing a polyfluoroaromatic moiety at position 9 leads to increased potency, while most of the derivatives containing an alkyl fragment at the same position were not active. According to the molecular modeling, the bromine atom in position 12 forms a hydrogen bond to histidine 493, a key catalytic residue. The cytotoxic effect of topotecan, a clinically important topoisomerase 1 inhibitor, was doubled in the cervical cancer HeLa cell line by derivatives 11g and 12g; both displayed low toxicity without topotecan. Derivatives 11g and 12g can therefore be used for further development to sensitize the action of clinically relevant Topo1 inhibitors.

Keywords: berberine; tetrahydroberberine; Tdp1 inhibitor; cancer; molecular modeling; DNA repair enzyme; SAR

1. Introduction

A promising strategy to enhance the efficacy of anticancer therapy is the inhibition of various DNA repair enzymes, which counteract the effect of many anticancer drugs [1,2]. This is particularly important where resistance to chemotherapy is observed. An interesting example is the poly (ADP-ribose) polymerase (PARP), which inhibitors were studied both in combination with chemotherapeutic agents and as individual drugs. Now olaparib, rucaparib and niraparib are in clinical use for the treatment of ovarian cancers [3]. Another DNA repair enzyme, tyrosyl-DNA phosphodiesterase 1 (Tdp1) has attracted considerable interest in the last few years mainly due to its ability to repair DNA lesions caused topoisomerase 1 (Top1) poisons, a well-established class of anticancer drugs [4]. The anticancer activity of Top1 poisons, camptothecin and its clinically important derivatives, topotecan and irinotecan [5,6], is based on their ability to bind to the covalent intermediate complex Top1/DNA and prevent the restoring of DNA integrity. This leads to the stabilization of the covalent bond between catalytic tyrosine (Y723) of Top1 and the 3′- end of DNA (Figure 1). The Tdp1 mechanism of action is the phosphotyrosyl bond hydrolysis [7], resulting in the resumption of DNA replication and cell division.

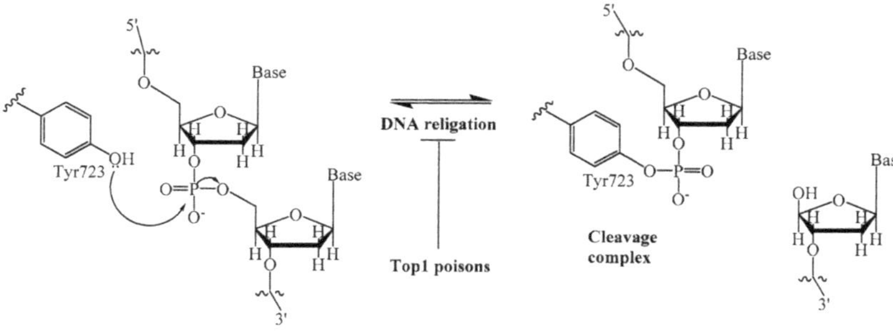

Figure 1. DNA single-strand cleavage by nucleophilic attack of Tyr723 of Top1, and covalent cleavage complex formation.

A few classes of Tdp1 inhibitors are known such as furamidines (compound 1, Figure 2), tetracyclines (compound **2**), aminoglycosides (compound **3**) [8,9]. Also, natural products of various types have been found to inhibit Tdp1, including derivatives of bile acids **4** [10,11], of lichen metabolite usnic acid **5a,b** [12,13] and **6** [14], monoterpenoid derivatives **7** [15–18] and oxinitidine **8** [19] with inhibitory activity in the micro- or submicromolar range. Importantly, the hydrazinothiazole derivative of usnic acid **5a** [13,20] and monoterpene-substituted 4-arylcoumarin **7a** [21] significantly increased topotecan efficacy in vivo.

The aim of this study was to establish the potency of a novel structural class of natural products, the derivatives of berberine **9** (Scheme 1), which like their usnic acid counterparts are phenolic compounds. It is known that the isoquinoline plant alkaloid berberines have many beneficial physiological effects, e.g., they are hypocholesterolemic, antibacterial, hypoglycemic agents, antioxidants and, finally, they suppress tumor growth [22–26]. Interestingly, the sulfonate derivatives of berberines and tetrahydroberberines have been reported as promising hypocholesterolemic agents [27,28]. Although berberine derivatives never used as Tdp1 inhibitors, a preliminary molecular modeling study indicated that berberines with 9-sulfonate group would bind to Tdp1. In the present study, 9-sulfonate-berberine and tetrahydroberberine derivatives **10–12** with aliphatic and aromatic substitutes were synthesized and their potency against Tdp1 was tested. To the best of our knowledge, this has not been done previously. Using the HeLa cervical cancer cell line, derivatives **11g** and **12g** were found to be nontoxic and sensitized the cancer cells to topotecan.

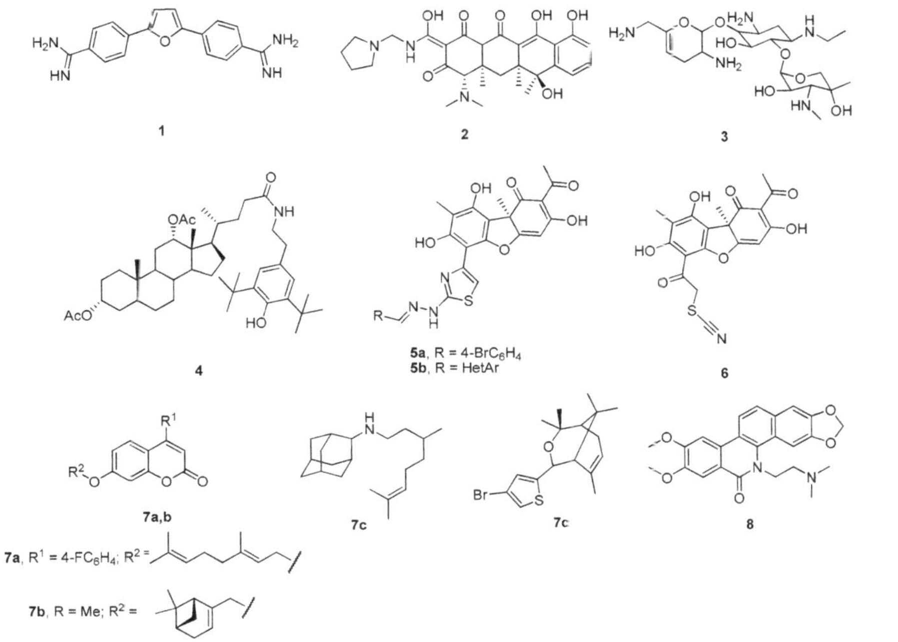

Figure 2. Examples of established Tdp1 inhibitors **1**–**8**.

2. Results and Discussion

2.1. Chemistry

Sulfonates **10**–**12** were synthesized in accordance with previously reported methods [27]. For this purpose, berberine **9** was selectively demethylated at 190 °C under vacuum as previously described [29]. After treatment with HBr, berberrubine hydrobromide **13** was isolated in 89% yield (Scheme 1). Compound **13** was reduced with sodium borohydride in methanol according to a procedure described previously [30] yielding tetrahydroberberrubine **14**. The bromination of compound **14** with a bromine in dioxane solution afforded 12-bromotetrahydroberberrubine **15** in 52% yield. The reaction of tetrahydroderivatives **14** and **15** with polyfluoroaryl and alkyl sulfonylchlorides as well as with tosylchloride in dichloromethane in the presence of triethylamine produced tetrahydroberberrubine 9-*O*-sulfonates **11a**–**h** (44–84% yields) and 12-bromotetrahydroberberrubine 9-*O*-sulfonates **12a**–**h** (49–93% yields). New berberine type **10** derivatives were synthesized by the reaction of berberrubine hydrobromide **13** with different alkyl sulfochlorides. The reactions were carried out in dichloromethane in the presence of triethylamine for 5 h at room temperature. Sulfonates **10** were isolated by precipitation from the reaction mixtures.

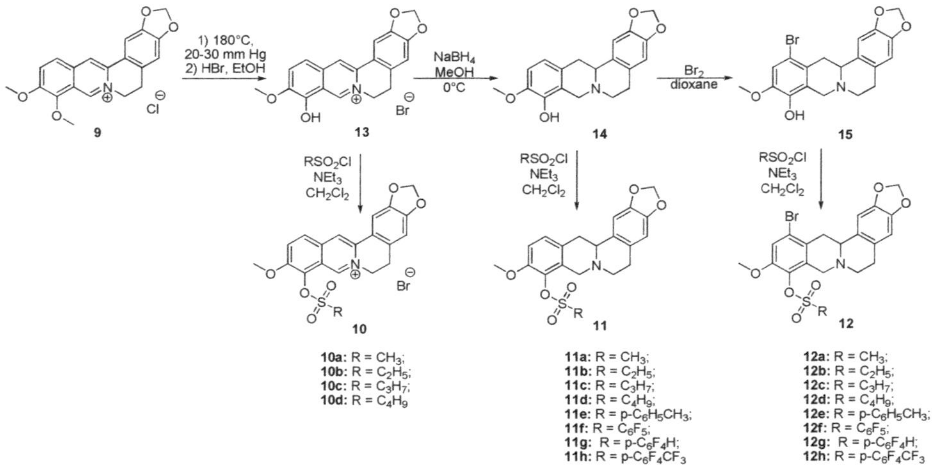

Scheme 1. The synthetic pathways to sulfonates **10**, **11** and **12**.

The structures of the new compounds were confirmed by ^{1}H-NMR,^{13}C-NMR, IR and HRMS methods; the results are shown in the experimental section and the supplementary information.

2.2. *Effects of the Berberine Sulfonates on Tdp1 Activity*

An oligonucleotide real-time biosensor was used based on the ability of Tdp1 to remove fluorophore quenchers from the 3′-end of DNA, as previously described [31]. The hexadecameric oligonucleotide carried 5(6)-carboxyfluorescein (FAM) at the 5′-end and the fluorophore quencher BHQ1 (Black Hole Quencher-1) at the 3′-end. Tdp1 inhibitors prevent removal of fluorophore quenchers, thus reducing fluorescence intensity. The results of the Tdp1 assay for derivatives **10–12** and cytotoxic effects are shown in Tables 1 and 2.

Table 1. Tdp1 inhibiting activity (IC_{50}—half maximal inhibitory concentration) of sulfonate berberine derivatives with aliphatic substituents and their cytotoxicity (CC_{50}—half maximal cytotoxic concentration) in HeLa cells. Furamidine was used as a positive control (IC_{50} = 1.2 ± 0.3 μM).

Code, Structure	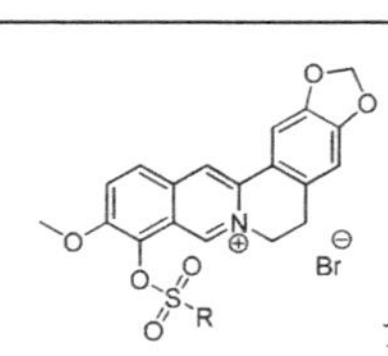	**10a–d**	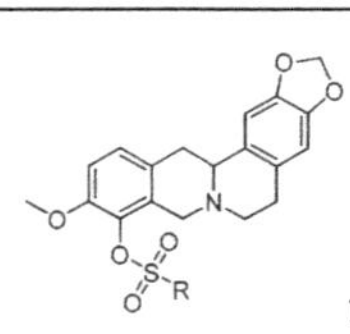	**11a–d**	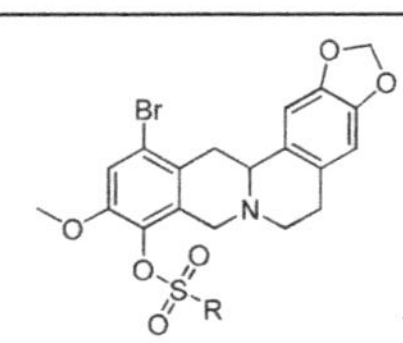	**12a–d**
R	IC_{50}, μM	CC_{50}, μM	IC_{50}, μM	CC_{50}, μM	IC_{50}, μM	CC_{50}, μM
a; Me	>15	ND *	>15	ND	>15	ND
b; Et	>15	ND	>15	ND	>15	ND
c; Pr	>15	ND	>15	ND	2.9 ± 1.3	>100
d; Bu	>15	ND	>15	ND	4 ± 1.0	>100

* ND—not determined.

Table 2. Tdp 1 inhibiting activity (IC_{50}) of sulfonate berberine derivatives with polyfluoroaromatic substituents and their cytotoxicity (CC_{50}) in HeLa cells. Furamidine was used as a positive control (IC_{50} = 1.2 ± 0.3 μM).

Code, structure	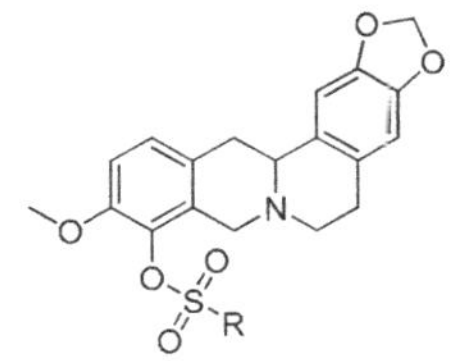 **11e–h**		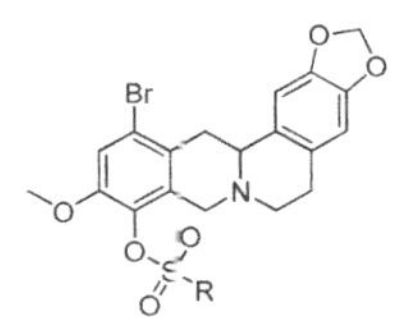 **12e–h**	
R	IC_{50}, μM	CC_{50}, μM	IC_{50}, μM	CC_{50}, μM
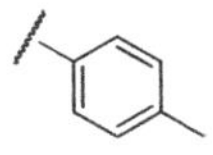 **e;**	>15	ND	>15	ND
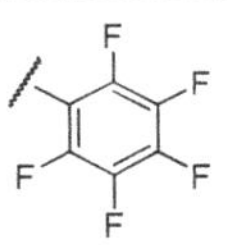 **f;**	1.0 ± 0.20	11 ± 2.0	0.53 ± 0.01	9.9 ± 4.5
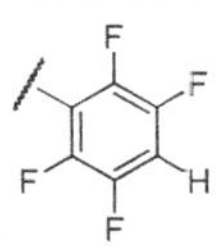 **g;**	1.05 ± 0.05	>100	1.3 ± 0.30	95 ± 5.0
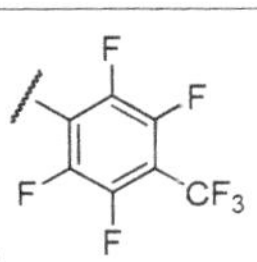 **h;**	0.9 ± 0.20	2.6 ± 0.1	1.4 ± 0.30	2.2 ± 1.5

It is clear from the data in Table 1 that both alkyl sulfonates of berberine and tetrahydroberberine are not very active with the exception of the tetrahydroberberine derivatives containing both sufficiently long alkyl substituent and bromine in a *para*-position to the sulfonate substituent, **12c** and **12d**. The favorable substitution pattern of these derivatives is confirmed by the results of the tetrahydroberberine sulfonates with aromatic and polyfluoroaromatic substituents as shown in Table 2. Both sulfonates with *para*-toluenesulfonyl substituent (**11e** and **12e**) are inactive, but all three fluorinated sulfonates (**11f**, **11g**, **11h**) show good inhibitory activity with IC_{50} values of ~1 μM. Bromine substitution is not important for **12e–h** activity as comparison with their non-brominated analogues **11e–h** shows.

Based on the data shown in Tables 1 and 2 it can be stated that some berberine derivatives inhibit Tdp1. It was found that the structure of the substituent in the sulfonate affects the inhibitory activity. Polyfluorinated arylsulfonates **11f–h** and **12f–h** exhibited inhibitory activity in the low micromolar range, while their non-fluorinated analogs (**11e**, **12e**) were inactive at these concentrations. In the series of alkylsulfonates, inhibitory activity was found only for compounds with a sufficiently long alkyl substituent (propyl-, butyl-) in the sulfonate group and in the presence of bromine substituent at position 12 (compounds **12c,d**). To explain the observed effects, a molecular modeling study was carried out.

2.3. Molecular Modeling of the Berberine Derivatives

Twenty-three berberine derivatives were docked into the binding site of Tdp1 (PDB ID: 6DIE, resolution 1.78 Å) [32] with three water molecules (HOH814, 821 and 1078). It is established that keeping these crystalline water molecules improves the prediction quality of the docking scaffold (for further information see the Methodology section) [13]. The binding predictions of the scoring functions used are given in Table S1, all the ligands show reasonable scores.

Compound **12f** is the ligand with the best IC_{50} value and according to the docking; **12f** fits neatly in the catalytic region as shown in Figure 3A. This region contains the catalytic histidine 263 and 493 amino acid residues thus the ligand blocs any activity of the enzyme. Indeed, **12f** forms a weak H-bond with the His493 imidazole site group via the bromine substituent as shown in Figure 3B as well as with Asn283's amide side chain. Finally, the amide group of Asn516 forms an H-bond with one of the oxygen atoms in the 1.3-benzodioxole moiety of the ligand. It is worth mentioning that the methoxy group on **12f** can potentially form a weak H-bond with the thiol on Cys205, if the flexibility of the protein would be accounted for, and the same methoxy group has lipophilic contacts with Ile285 aliphatic side chain, stabilizing the binding mode. Interestingly, the modeling of the **11f** ligand, which does not contain a bromine group, but also with a good IC_{50} value, did not give consistent results across the scoring functions used, i.e., different conformations were predicted. This strongly indicates that the bromine group with its weak H-bonds to Asn283 and His493 is essential for anchoring the ligand in the catalytic site.

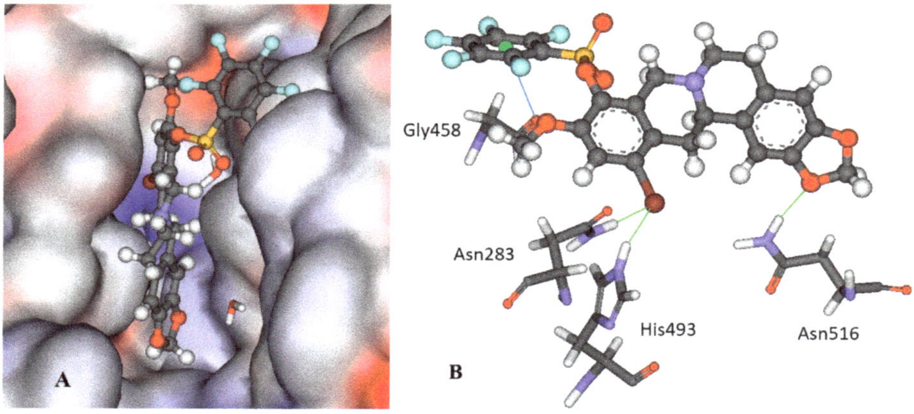

Figure 3. The docked configuration of **12f** in the binding site of Tdp1 as predicted by the ChemPLP scoring function. (**A**) The protein surface is rendered; blue depicts a hydrophilic region with a partial positive charge on the surface; red depicts hydrophobic region with a partial negative charge and grey shows neutral areas. The ligand occupies the catalytic pocket blocking access to it. (**B**) H-bonds are shown as green lines between **12f** and the amino acids Asn283, His493 and Asn516 side chains. A potential lone pair-π stacking interaction is shown as a blue line between the carboxylic backbone group in Gly458 and the centroid (green ball) of the fluorinated phenyl group (3.5 Å).

Interestingly, derivatives **11e** and **12e** are essentially inactive (IC_{50} >15 μM); they are structural analogues of the **11f/12f** pair with hydrogens on the phenyl ring as well as a *para* methyl substitution instead of fluorine groups. According to the modeling, the **11e/12e** pair has a different binding mode from **11f/12f** with the bromine moiety pointing into the aqueous phase for the former pair as shown in Figure S1 in the Supplementary Information. This indicates the importance of the fluorine substitution on the phenyl rings; the modeling suggests that the phenyl ring is leaning against the carboxyl moiety in the Gly458 backbone, which can form an interaction between the electron deficient fluoride substituted ring and the lone pairs of the oxygen atom in the carboxylic group as shown in Figure 3B.

To investigate the binding stability of the ligands molecular dynamic (MD) runs were conducted using the docked conformations of **11f**, **12f** and **12e** for 10 ps at 1000 K. In general, the ligands are stable within the binding pocket and not ejected; the most stable intramolecular bond being between the fluorinated phenyl ring and the Gly458 carboxyl group for **12f**. In contrast the phenyl group in **12e** is very mobile. The other H-bonding interactions predicted are often broken to be reestablished during the MD run.

2.4. Cytotoxicity

Top1 poisons are used as anticancer drugs for the treatment for various oncological diseases [33–35]. Since Tdp1 is involved in the removal of DNA damage caused by Top1 poisons, the activity of Tdp1 can lead to the development of drug resistance [36]. Thus, it is believed that Tdp1 inhibition can enhance the efficacy of Top1 poisons [37]. Tdp1 inhibitors should have the lowest possible intrinsic toxicity to minimize potential side effects. Therefore, we studied the intrinsic cytotoxicity of the compounds against HeLa cells (cervical carcinoma). EZ4U cell proliferation and cytotoxicity assay results are shown in Figure 4 for the ligands with Tdp1 inhibitory activity.

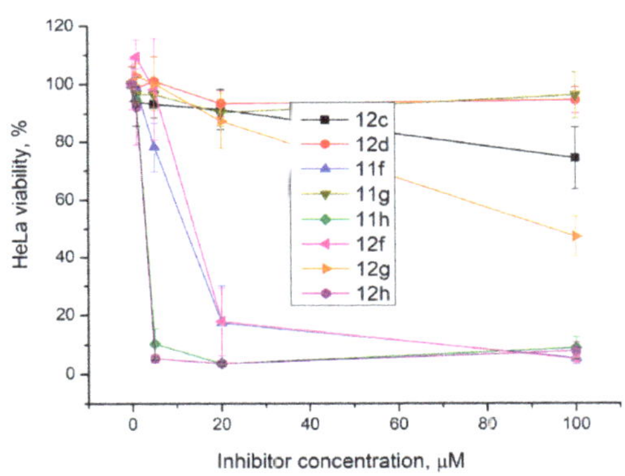

Figure 4. The berberine derivatives' cytotoxicity according to the EZ4U test. Error bars show standard deviations.

The results show that 9-O-sulfonates of 12-bromotetrahydoberberine with aliphatic substituents **12c** and **12d**, (Table 1 and Figure 4, black and red traces) are non-toxic up to 100 μM. The toxicity of 9-O-sulfonates strongly depends on the structure of the polyfluorinated fragment. Compounds with a CF_3-group in the *para*-position were the most toxic; CC_{50} values of 2.6 μM and 2.2 μM for **11h** and **12h,** respectively (Table 2 and Figure 4, green and brown traces). The replacement of the trifluoromethyl group with a fluorine atom in the *para*-position reduced toxicity with CC_{50} values of 10 μM for **11f** and **12f** (Table 2 and Figure 4, blue and magenta traces, respectively). The compounds with a hydrogen atom in the *para*-position were non-toxic (**11g**, Table 2 and Figure 4, violet trace) or moderately toxic (**12g**, Table 2 and Figure 4, orange trace).

2.5. Sensitizing Effects

The sensitizing effect of the berberine inhibitors on topotecan's cytotoxic potential was investigated. In order to determine the optimal concentration for the inhibitors to provide the maximum sensitizing effect, but remaining non-toxic, their concentrations were varied with topotecan concentration of 2 μM, its CC_{50} for HeLa cells. Topotecan significantly increased the cytotoxicity of compounds **11g**, **12g**, and **11f**, the reliability was confirmed by the Mann–Whitney U-test, $p = 0.05$ and the results are shown in Figure 5. The original data are given in Table S2.

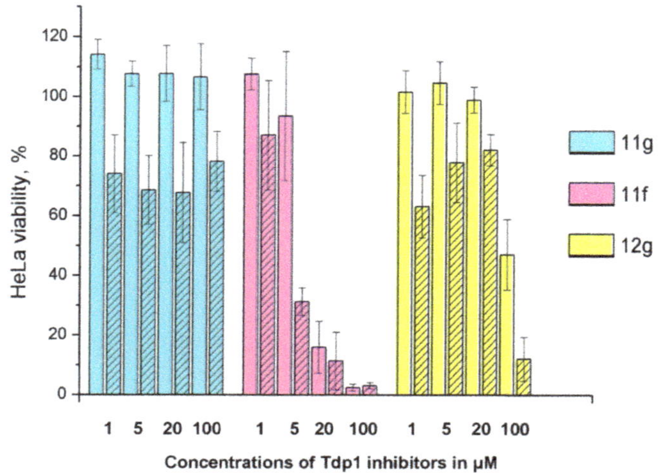

Figure 5. The influence of 2 μM topotecan on Tdp1 inhibitors' cytotoxicity. The unshaded histogram bars denote cell viability in the presence of a Tdp1 inhibitor. The hatched histogram bars indicate cell viability in the presence of a combination of a Tdp1 inhibitor with 2 μM of topotecan. The concentration of the Tdp1 inhibitor in μM is given under each pair of bars. Error bars show standard deviations.

11g is non-toxic in the concentration range used; 100% of living cells with a 100 μM maximum concentration. In the presence of topotecan, a significant decrease in cell survival is observed (~30%) at all concentrations. Compound **12g** has a low toxicity potential of 95 μM (CC_{50}). Topotecan weakly, but significantly reduces this value to 62 μM. Compound **11f** is inherently toxic, and 90% of the cells die at 20 μM. At lower concentrations, the effect of topotecan is significant. CC_{50} value for **11f** decreases three fold, from 11 to 3.7 μM, in the presence of topotecan. For other ligands (**11h**, **12c**, **d**, **f**, **h**), the effect of topotecan was negligible or unreliable.

Non-toxic concentrations of the berberine derivatives (5 μM) were then tested at different concentrations of topotecan. The most toxic compound **11f** caused 20% cell death at this concentration; the rest of the compounds were not toxic. In general, our Tdp1 inhibitors doubled the cytotoxic potential of topotecan as can be seen in Figure 6 and Table 3.

Table 3. The influence of the Tdp1 inhibitors at 5 μM on the cytotoxic potential of topotecan (Tpc).

Compounds	CC_{50}, μM–5 μM Tdp1 inhibitor	CC_{50}, μM–20 μM Tdp1 inhibitor
Tpc	6.8 ± 1.1	
Tpc + **11g**	3.5 ± 0.6	2.3 ± 0.5
Tpc + **12g**	2.9 ± 0.4	3.3 ± 1.1
Tpc + **11f**	1.7 ± 0.3	not determined

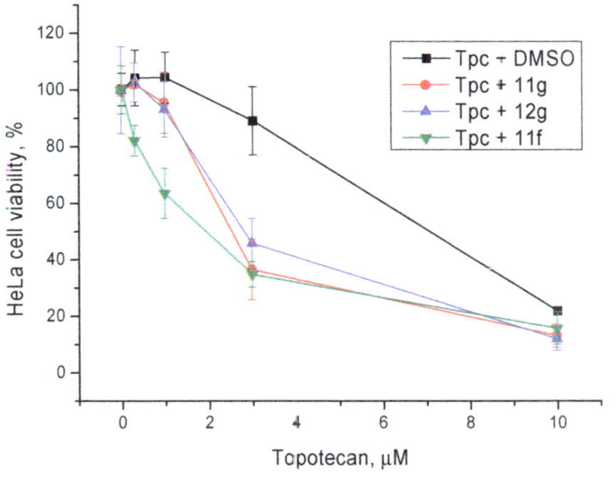

Figure 6. The influence of the Tdp1 inhibitors at 5 μM on topotecan cytotoxicity. Error bars show standard deviations.

For comparison, the concentration of the inhibitors was increased to 20 μM. Compound **11f** was not used due to its high toxicity. Again, compounds **12g** and **11g** had a significant effect, $p < 0.05$, confirmed by the Mann–Whitney U-test (Figure 7 and Table 3). It is interesting to note that the sensitizing effect of Tdp1 inhibitors was practically the same at 5 and 20 μM.

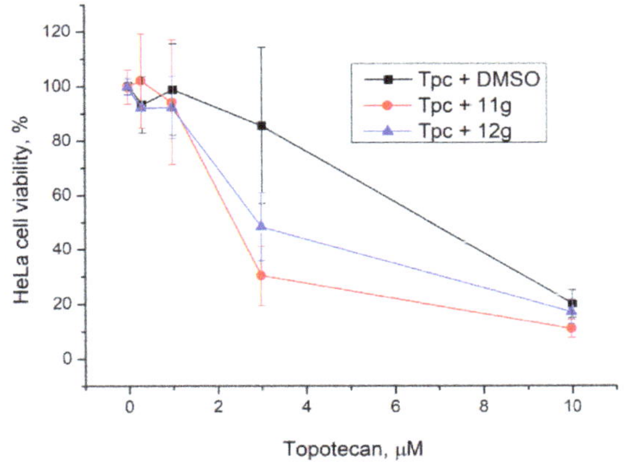

Figure 7. Influence of Tdp1 inhibitors at 20 μM on topotecan cytotoxicity. Error bars show standard deviations.

Derivatives **11g** and **12g** can be considered to be the lead compounds since they are, unlike **11f**, non-toxic (**11g**) or moderately toxic (**12g**) and have a pronounced sensitizing effect on topotecan.

2.6. Chemical Space

The calculated molecular descriptors MW (molecular weight), log *P* (water-octanol partition coefficient), HD (hydrogen bond donors), HA (hydrogen bond acceptors), PSA (polar surface area) and RB (rotatable bonds)) are given in Table S3. The MW of the ligands lies between 325.4 and 684.4 g mol^{-1} and falls into drug-like and Known Drug Space (KDS) regions. Log *P* spans from 1.6 to 5.4, i.e., over the three defined volumes in chemical space, with most in lead-like chemical space and only one ligand **12h** in KDS. The HD, HA, RB and PSA values are within the lead- and drug-like definitions (for the definition of lead-like, drug-like and Known Drug Space regions see ref. [38] and Table S4).

The Known Drug Indexes (KDIs) for the ligands were calculated to gauge the balance of the molecular descriptors (MW, log P, HD, HA, PSA and RB). This method is based on the analysis of drugs in clinical use, i.e., the statistical distribution of each descriptor is fitted to a Gaussian function and normalized to 1 resulting in a weighted index. Both the summation of the indexes (KDI_{2a}—Equation (1)) and multiplication (KDI_{2b}—Equation (2)) methods were used [39]. The numerical results are given in Table S3 in the supplementary information.

$$KDI_{2a} = I_{MW} + I_{\log P} + I_{HD} + I_{HA} + I_{RB} + I_{PSA} \quad (1)$$

$$KDI_{2b} = I_{MW} \times I_{\log P} \times I_{HD} \times I_{HA} \times I_{RB} \times I_{PSA} \quad (2)$$

The KDI_{2a} values range from 4.35 to 5.70 with a theoretical maximum of 6 and the average of 4.08 for known drugs. KDI_{2b} range from 0.06 to 0.73, with a theoretical maximum of 1 and with KDS average of 0.18. The berberine ligands can be considered reasonably well balanced in terms of their molecular descriptors and therefore biocompatible. The most active compound **12f**, has a KDI_{2a} value of 4.58 and KDI_{2b} of 0.12; the relatively low KDI_{2b} value can be explained by its high MW, which is compensated by the favorable values for the other descriptors resulting in a good KDI_{2a} value. The KDI_{2b} index is sensitive to any outliers since multiplication of small numbers leads to small numbers.

3. Materials and Methods

3.1. Chemistry

Berberine chloride was purchased from Tokyo Chemical Industry Co., Ltd (Tokyo, Japan). Methanesulfochloride and ethanesulfochloride were purchased from Acros Organics (Belgium), 1-propanesulfochloride and 1-buthanesulfochloride were purchased from Alfa Aesar (Karlsruhe, Germany). 48% aqueous HBr solution was purchased from Acros Organics (the Netherlands). All solvents used in the reactions were purified and dried. Bromine, sodium borohydride and triethylamine were purchased from Sigma-Aldrich. Column chromatography was carried out on neutral alumina LL40/250. All reactions were monitored by TLC analysis using Merck Aluminium oxide 60 F_{254} plastic sheets (Darmstadt, Germany), eluent CH_2Cl_2-MeOH.

The ^1H and ^{13}C NMR spectra were recorded on a Bruker AM-400 spectrometer (400.13 and 100.61 MHz) for 5–10% solutions of the compounds in $CDCl_3$ or DMSO-d_6 using the signal of the solvent $CDCl_3$ as the standard (δ 7.24 for 1H and δ 76.90 for 13C). The IR spectra were measured on a Vector 22 FTIR spectrometer in KBr pellets. High-resolution electrospray ionization (HRESI) mass spectra were carried out using a time-of-flight high-resolution mass spectrometer micrOTOF-Q (Bruker Daltonics, Germany) with an Agilent 1200 liquid chromatograph (Agilent Technologies, USA/Germany). Positive ion scanning in the range m/z = 100–3000. Drying gas (nitrogen) flow rate of 4 L/min; temperature—190 °C; sprayer pressure—1.0 bar.

The spectroscopic and analytical measurements were carried out at the Multi-access Chemical Service Center of the Siberian Branch of the Russian Academy of Sciences.

3.1.1. General Synthesis of Compounds **11a–g** and **12a–g**

Desired compounds **11a–d**, **f**, **g** and **12a–g** were synthesized according to the general procedure described previously [27], compound **11e** was also synthesized according to the procedure described previously [30].

3.1.2. General Procedure for the Synthesis of Sulfonates **10a–d**

2.5 mmol of triethylamine was added to the suspension of 1 mmol of berberrubine hydrobromide in 25 mL of methylene chloride. 1.5 mmol of sulfochloride in 3 mL of methylene chloride was added under stirring at room temperature. After 5 hours the precipitate was filtered, washed with methylene chloride and the pure product **10a–d** was isolated.

3.1.3. 10-Methoxy-9-methanesulfonyloxy-2,3-methylenedioxyprotoberberine chloride (**10a**)

Yield: 7%

Spectra NMR ^{1}H (400 MHz, DMSO-d$_6$, δ, ppm, *J*/Hz): 3.22 (2H, *t*, *J* = 6.07 Hz, H-5), 3.73 (3H, s, SCH_3), 4.12 (3H, s, OCH_3), 5.00 (2H, *t*, *J* = 5.72 Hz, H-6), 6.19 (2H, s, OCH_2O), 7.11 (1H, s, H-4), 7.82 (1H, s, H-1), 8.27 (1H, *d*, *J* = 9.06 Hz, H-12), 8.34 (1H, *d*, *J* = 9.06 Hz, H-11), 9.08 (1H, s, H-13), 9.72 (1H, s, H-8). Spectra NMR ^{13}C (100 MHz, DMSO-d$_6$, δ, ppm): 26.18 (C-5), 40.02 (SO_2CH_3), 55.70 (C-6), 57.44 (OCH_3), 102.23 (OCH_2O), 105.60 (C-1), 108.49 (C-4), 120.20 (C-13b), 120.86 (C-13), 121.79 (C-8a), 126.47 (C-12), 128.03 (C-11), 131.05 (C-4a), 131.44 (C-12a), 133.31 (C-13a), 138.59 (C-9), 144.04 (C-8), 147.77 (C-2), 150.17 (C-3), 151.89 (C-10). Spectra IR (cm^{-1}): 804.21, 925.71, 1033.71, 1097.35, 1224.63, 1263.20, 1365.42, 1506.20, 1610.34, 1623.84, 3039.40, 3320.97, 3635.33. MS (ESI): m/z (M^+) calcd for $C_{20}H_{18}NO_6S^+$, 400.085 found: 400.081.

3.1.4. 9-Ethanesulfonyloxy-10-methoxy-2,3-methylenedioxyprotoberberine chloride (**10b**)

Yield: 25%

Spectra NMR ^{1}H (400 MHz, DMSO-d$_6$, δ, ppm, *J*/Hz): 1.52 (3H, *t*, *J* = 7.32 Hz, CH_2CH_3), 3.22 (2H, *t*, *J* = 6.05 Hz, H-5), 3.89 (2H, q, J_1 = 3.50 Hz, J_2 = 10.96 Hz, CH_2CH_3), 4.12 (3H, s, OCH_3), 5.01 (2H, t, *J* = 5.91 Hz, H-6), 6.18 (2H, s, OCH_2O), 7.11 (1H, s, H-4), 7.82 (1H, s, H–1), 8.27 (1H, *d*, *J* = 9.48 Hz, H-12), 8.34 (1H, *d*, *J* = 9.48 Hz, H-11), 9.10 (1H, s, H-13), 9.64 (1H, s, H–8). Spectra NMR ^{13}C (100 MHz, DMSO-d$_6$, δ, ppm): 8.18 (CH_2CH_3), 26.19 (C-5), 47.00 (SCH_2), 55.83 (C–6), 57.44 (OCH_3), 102.20 (OCH_2O), 105.61 (C-1), 108.45 (C–4), 120.18 (C-13b), 120.90 (C-13), 121.98 (C-8a), 126.42 (C-12), 128.02 (C-11), 131.02 (C-4a), 131.27 (C-12a), 133.36 (C-13a), 138.63 (C-9), 143.91 (C-8), 147.75 (C-2), 150.17 (C-3), 151.71 (C-10). Spectra IR (cm^{-1}): 806.14, 923.78, 1033.71, 1097.35, 1164.85 1222.70, 1278.2063, 1363.49, 1506.20, 1608.42, 1621.92, 3023.98, 3423.19, 3629.54. MS (ESI): *m*/*z* (M^+) calcd for $C_{21}H_{20}NO_6S^+$, 414.101 found: 414.096.

3.1.5. 10-Methoxy-2,3-methylenedioxy-9-((propane-1-sulfonyl)oxy)protoberberine chloride (**10c**)

Yield: 34%

Spectra NMR ^{1}H (400 MHz, DMSO-d$_6$, δ, ppm, J/Hz): 1.11 (3H, t, *J* = 7.45 Hz, CH_2CH_3), 1.94–2.03 (2H, m, CH_2CH_3), 3.22 (2H, *t*, *J* = 5.72 Hz, H-5), 3.87 (2H, t, *J* = 7.58 Hz, SCH_2CH_2), 4.12 (3H, s, OCH_3), 5.01 (2H, *t*, *J* = 6.02 Hz, H-6), 6.19 (2H, s, OCH_2O), 7.11 (1H, s, H-4), 7.82 (1H, s, H-1), 8.27 (1H, *d*, *J* = 9.33 Hz, H-12), 8.34 (1H, d, J = 9.33 Hz, H-11), 9.10 (1H, s, H-13), 9.64 (1H, s, H-8). Spectra NMR ^{13}C (100 MHz, DMSO-d$_6$, δ, ppm): 12.49 (CH_2CH_3), 17.17 ($CH_2CH_2CH_3$), 26.16 (C-5), 53.27 (SCH_2), 55.80 (C-6), 57.46 (OCH_3), 102.18 (OCH_2O), 105.60 (C-1), 108.42 (C-4), 120.14 (C-13b), 120.86 (C-13), 121.93 (C-8a), 126.37 (C-12), 127.98 (C-11), 130.96 (C-4a), 131.23 (C-12a), 133.33 (C-13a), 138.55 (C-9), 143.91 (C-8), 147.70 (C-2), 150.12 (C-3), 151.69 (C-10). Spectra IR (cm^{-1}): 317.71, 925.71, 1037.56, 1095.42, 1162.92, 1222.70, 1263.20, 1363.49, 1506.20, 1610.34, 1621.92, 3023.98, 3427.04. MS (ESI): *m*/*z* (M^+) calcd for $C_{22}H_{22}NO_6S^+$, 428.116 found: 428.121.

3.1.6. 9-(Butane-1-sulfonyl)oxy-10-methoxy-2,3-methylenedioxy protoberberine chloride (**10d**)

Yield: 50%

Spectra NMR ^{1}H (400 MHz, DMSO-d$_6$, δ, ppm, *J*/Hz): 0.97 (3H, t, *J* = 7.34 Hz, CH$_2$*CH$_3$*), 1.48–1.58 (2H, m, *CH$_2$*CH$_3$), 1.89–1.97 (2H, m, SCH$_2$*CH$_2$*), 3.22 (2H, t, *J* = 5.96 Hz, H-5), 3.91 (2H, t, *J* = 7.68 Hz, SCH$_2$), 4.11 (3H, s, OCH$_3$), 5.02 (2H, t, *J* = 6.04 Hz), 6.18 (2H, c, OCH$_2$O), 7.10 (1H, s, H-4), 7.82 (1H, s, H-1), 8.27 (1H, *d*, *J* = 9.21 Hz, H-12), 8.34 (1H, *d*, *J* = 9.21 Hz, H-11), 9.13 (1H, s, H-13), 9.67 (1H, s, H-8). Spectra NMR ^{13}C (100 MHz, DMSO-d$_6$, δ, ppm): 13.44 (*CH$_3$*CH$_2$), 20.74 (*CH$_3$*CH$_2$), 25.29 (SCH$_2$*CH$_2$*), 26.19 (C-5), 51.75 (SCH$_2$), 55.80 (C-6), 57.46 (OCH$_3$), 102.20 (OCH$_2$O), 105.62 (C-1), 108.45 (C-4), 120.19 (C-13b), 120.91 (C-13), 121.98 (C-8a), 126.42 (C-12), 128.01 (C-11), 131.02 (C-4a), 131.30 (C-12a), 133.36 (C-13a), 138.62 (C-9), 143.97 (C-8), 147.75 (C-2), 150.16 (C-3), 151.72 (C-10). Spectra IR (cm^{-1}): 811.92, 925.71, 1035.63, 1099.28, 1222.70, 1265.13, 1365.42, 1504.27, 1610.34, 1619.99, 2960.33, 3425.11. MS (ESI): *m*/*z* (M$^+$) calcd for $C_{23}H_{24}NO_6S^+$, 442.132 found: 442.133.

3.2. *Biology*

3.2.1. Detection of Tdp1 Activity

The methodology has been reported in our previous work [30] and consists of fluorescence intensity measurement in a reaction of quencher removal from a fluorophore quencher-coupled DNA oligonucleotide catalyzed by Tdp1. The reaction was carried out at different concentrations of inhibitors (the control samples contained 1% of DMSO, Sigma, St. Louis, MO, USA). The reaction mixtures contained Tdp1 buffer (50 mM Tris-HCl pH 8.0, 50 mM NaCl, and 7 mM β-mercaptoethanol), 50 nM biosensor, and an inhibitor being tested. Purified Tdp1 (1.5 nM) triggered the reaction. The biosensor (5′-[FAM] AAC GTC AGGGTC TTC C [BHQ]-3′) was synthesized in the Laboratory of Biomedical Chemistry at the Institute of Chemical Biology and Fundamental Medicine (Novosibirsk, Russia).

The reactions were incubated on a POLARstar OPTIMA fluorimeter (BMG LABTECH, GmbH, Ortenberg, Germany) to measure fluorescence every 55 s (ex. 485/em. 520 nm) during the linear phase (here, data from minute 0 to minute 8). The values of IC_{50} were determined using a six-point concentration response curve in minimum three independent experiments and were calculated using MARS Data Analysis 2.0 (BMG LABTECH, GmbH, Ortenberg, Germany).

3.2.2. Cytotoxicity Assays

Cytotoxicity of the compounds to HeLa (human cervical cancer) cell line was examined using the EZ4U Cell Proliferation and Cytotoxicity Assay (Biomedica, Vienna, Austria), according to the manufacturer's protocols. The cells were grown in Iscove's modified Dulbecco's medium (IMDM) with 40 μg/mL gentamicin, 50 IU/mL penicillin, 50 μg/mL streptomycin (MP Biomedicals, Santa Ana, CA, USA), and 10% of fetal bovine serum (Biolot, St. Petersburg, Russia) in a 5% CO_2 atmosphere. After formation of a 30–50%-monolayer, the tested compounds were added to the medium. The volume of the added reagents was 1/100 of the total volume of the culture medium, and the amount of DMSO (Sigma, St. Louis, MO, USA) was 1% of the final volume. Control cells were grown in the presence of 1% DMSO. The cell culture was monitored for 3 days. To assess the influence of the inhibitors on the cytotoxic effect of topotecan (ACTAVIS GROUP PTC ehf., Bucharest, Romania), 50% cytotoxic concentrations of topotecan and of each inhibitor were determined to attain a defined single-agent effect. Then, minimum two independent tests were performed with each inhibitor in combination with topotecan. When using a combination of drugs, Tdp1 inhibitors were first added, then topotecan was added immediately (within 10–15 min).

3.3. *Molecular Modeling*

The compounds were docked against the crystal structure of TDP1 Tdp1 (PDB ID: 6DIE, resolution 1.78 Å) [32] which was obtained from the Protein Data Bank (PDB) [40,41]. The Scigress version

FJ 2.6 program [42] was used to prepare the crystal structure for docking, i.e., the hydrogen atoms were added, the co-crystallized ligand benzene-1,2,4-tricarboxylic acid was removed as well as crystallographic water molecules except HOH 814, 821 and 1078. The waters were set on toggle—bound or displaced by the ligand during docking—and spin—automatic optimization of the orientation of the hydrogen atoms. The Scigress software suite was also used to build the inhibitors and the MM2 [43] force field was used to optimize the structures. Furthermore, Scigress was used for the 10 ps MD runs at 1000 K; the MM2 force filed was used, 5 Å radius was defined around the Ligand and allowed to be flexible whereas the rest of the protein structure was held ridged (locked). First the binding pocket with the ligand was structurally optimized followed by the MD run. The docking center was defined as the position of a carbon on the ring of the co-crystallized benzene-1, 2, 4-tricarboxylic acid ($x = -6.052$, $y = -14.428$, $z = 33.998$) with 10 Å radius. Fifty docking runs were allowed for each ligand with default search efficiency (100%). The basic amino acids lysine and arginine were defined as protonated. Furthermore, aspartic and glutamic acids were assumed deprotonated. The GoldScore (GS) [44] and ChemScore (CS) [45,46] ChemPLP (Piecewise Linear Potential) [47] and ASP (Astex Statistical Potential) [48] scoring functions were implemented to predict the binding modes and relative energies of the ligands using the GOLD v5.4.1 software suite.

The QikProp 3.2 [49] software package was used to calculate the molecular descriptors of the molecules. The reliability of it QikProp established for the calculated descriptors. [50] The Known Drug Indexes (KDI) were calculated from the molecular descriptors as described by Eurtivong and Reynisson [40]. Five of the compounds (**9** and **10a–d**) carry a positive charge and QikProp does not compute charged molecules. The MW, Log P, HD and HA were derived using the Scigress version FJ 2.6 program [42] software, PSA and RB are not available in this software suite.

4. Conclusions

The berberine and tetrahydroberberine sulfonates and their brominated analogues were tested to evaluate their Tdp1 inhibitory activity. To our knowledge, this is the first report that documents berberine-based compounds as inhibitors of this enzyme. The IC_{50} values are in the 0.53 to 4 μM range. Of the alkyl sulfonates, only derivatives with bromine substitution on site 12 of tetrahydroberberrubine are active. While both tetrahydroberberrubine and 12-bromtetrahydroberberrubine derivatives containing polyfluoroaromatic substituents all have inhibitory activity with IC_{50} values of ~1 μM. According to the inhibitory activity, toxicity data and ability to sensitize topotecan against the HeLa cancer cell line the two most promising derivatives were identified—**11g** and **12g**. These results indicate that sulfonates of tetrahydroberberine have potential to be developed as new agents for anticancer therapy due to their inhibitory activity and lack of toxicity.

Supplementary Materials: Supplementary Materials can be found at http://www.mdpi.com/1422-0067/21/19/7162/s1.

Author Contributions: Chemistry investigation, E.D.G., I.V.N., R.A.B. and O.A.L.; in vitro investigation, A.L.Z., A.A.C., E.S.I., N.S.D., E.M.M. and R.O.A.; modeling, J.R.; methodology, N.F.S. and O.I.L.; project administration, K.P.V.; supervision, K.P.V.; writing—original draft, E.D.G., A.L.Z. and O.A.L.; writing—review and editing, K.P.V., J.R., N.F.S. and O.LL. All authors have read and agreed to the published version of the manuscript.

Funding: This study was funded by the Russian Science Foundation grant № 19-13-00040.

Acknowledgments: Authors would like to acknowledge the Multi-Access Chemical Research Center SB RAS for spectral and analytical measurements.

Conflicts of Interest: The authors declare no conflict of interest. The funders had no role in the design of the study; in the collection, analyses, or interpretation of data; in the writing of the manuscript, or in the decision to publish the results.

Abbreviations

Tdp1	Tyrosyl-DNA phosphodiesterase 1
PARP	Poly (ADP-ribose) polymerase

References

1. Zakharenko, A.L.; Lebedeva, N.A.; Lavrik, O.I. DNA Repair Enzymes as Promising Targets in Oncotherapy. *Russ. J. Org. Chem.* **2018**, *44*, 1–18. [CrossRef]
2. Ferri, A.; Stagni, V.; Barilà, D. Targeting the DNA Damage Response to Overcome Cancer Drug Resistance in Glioblastoma. *Int. J. Mol. Sci.* **2020**, *21*, 4910. [CrossRef]
3. Jiang, X.; Li, W.; Li, X.; Bai, H.; Zhang, Z. Current status and future prospects of PARP inhibitor clinical trials in ovarian cancer. *Cancer Manag. Res.* **2019**, *1*, 4371–4390. [CrossRef] [PubMed]
4. Comeaux, E.Q.; Waardenburg, R.C. Tyrosyl-DNA phosphodiesterase I resolves both naturally and chemically induced DNA adducts and its potential as a therapeutic target. *Drug Metab. Rev.* **2014**, *46*, 494–507. [CrossRef] [PubMed]
5. Pommier, Y.; Leo, E.; Zhang, H.; Marchand, C. DNA topoisomerases and their poisoning by anticancer and antibacterial drugs. *Chem. Biol.* **2010**, *17*, 421–433. [CrossRef] [PubMed]
6. Kciuk, M.; Marciniakand, B.; Kontek, R. Irinotecan—Still an Important Player in Cancer Chemotherapy: A Comprehensive Overview. *Int. J. Mol. Sci.* **2020**, *21*, 4919. [CrossRef]
7. Pommier, Y.; Huang, S.-y.N.; Gao, R.; Das, B.B.; Murai, J.; Marchand, C. Tyrosyl-DNA-phosphodiesterases (TDP1 and TDP2). *DNA Repair* **2014**, *19*, 114–129. [CrossRef] [PubMed]
8. Laev, S.S.; Salakhutdinov, N.F.; Lavrik, O.I. Tyrosyl-DNA phosphodiesterase inhibitors: Progress and potential. *Bioorg. Med. Chem.* **2016**, *24*, 5017–5027. [CrossRef] [PubMed]
9. Huang, S.N.; Pommier, Y.; Marchand, C. Tyrosyl-DNA Phosphodiesterase 1 (Tdp1) inhibitors. *Expert Opin. Ther. Pat.* **2011**, *21*, 1285–1292. [CrossRef] [PubMed]
10. Salomatina, O.V.; Popadyuk, I.I.; Zakharenko, A.L.; Zakharova, O.D.; Fadeev, D.S.; Komarova, N.I.; Reynisson, J.; Arabshahi, H.I.; Chand, R.; Volcho, K.P.; et al. Novel Semisynthetic Derivatives of Bile Acids as Effective Tyrosyl-DNA Phosphodiesterase 1 Inhibitors. *Molecules* **2018**, *23*, 679. [CrossRef]
11. Xiao, L.-G.; Zhang, Y.; Zhang, H.-L.; Li, D.; Gu, Q.; Tang, G.-H.; Yu, Q.; An, L.-K. Spiroconyone A, a new phytosterol with a spiro [5,6] ring system from Conyza japonica. *Org. Biomol. Chem.* **2020**, *18*, 5130–5136. [CrossRef] [PubMed]
12. Zakharenko, A.; Luzina, O.; Koval, O.; Nilov, D.; Gushchina, I.; Dyrkheeva, N.; Švedas, V.; Salakhutdinov, N.; Lavrik, O. Tyrosyl-DNA Phosphodiesterase 1 Inhibitors: Usnic Acid Enamines Enhance the Cytotoxic Effect of Camptothecin. *J. Nat. Prod.* **2016**, *79*, 2961–2967. [CrossRef] [PubMed]
13. Filimonov, A.S.; Chepanova, A.A.; Luzina, O.A.; Zakharenko, A.L.; Zakharova, O.D.; Ilina, E.S.; Dyrkheeva, N.S.; Kuprushkin, M.S.; Kolotaev, A.V.; Khachatryan, D.S.; et al. New Hydrazinothiazole Derivatives of Usnic Acid as Potent Tdp1 Inhibitors. *Molecules* **2019**, *24*, 3711. [CrossRef] [PubMed]
14. Zakharenko, A.L.; Luzina, O.A.; Sokolov, D.N.; Zakharova, O.D.; Rakhmanova, M.E.; Chepanova, A.A.; Dyrkheeva, N.S.; Lavrik, O.I.; Salakhutdinov, N.F. Usnic acid derivatives are effective inhibitors of tyrosyl-DNA phosphodiesterase 1. *Russ. J. Bioorg Chem.* **2017**, *43*, 84–90. [CrossRef]
15. Khomenko, T.; Zakharenko, A.; Odarchenko, T.; Arabshahi, H.J.; Sannikova, V.; Zakharova, O.; Korchagina, D.; Reynisson, J.; Volcho, K.; Salakhutdinov, N.; et al. New inhibitors of tyrosyl-DNA phosphodiesterase I (Tdp 1) combining 7-hydroxycoumarin and monoterpenoid moieties. *Bioorg. Med. Chem.* **2016**, *24*, 5573–5581. [CrossRef]
16. Ponomarev, K.Y.; Suslov, E.V.; Zakharenko, A.L.; Zakharova, O.D.; Rogachev, A.D.; Korchagina, D.V.; Zafar, A.; Reynisson, J.; Nefedov, A.A.; Volcho, K.P.; et al. Aminoadamantanes containing monoterpene-derived fragments as potent tyrosyl-DNA phosphodiesterase 1 inhibitors. *Bioorg. Chem.* **2018**, *76*, 392–399. [CrossRef]
17. Chepanova, A.A.; Li-Zhulanov, N.S.; Sukhikh, A.S.; Zafar, A.; Reynisson, J.; Zakharenko, A.L.; Zakharova, O.D.; Korchagina, D.V.; Volcho, K.P.; Salakhutdinov, N.F.; et al. Effective Inhibitors of Tyrosy l-DNA Phosphodiesterase 1 Based on Monoterpenoids as Potential Agents for Antitumor Therapy. *Russ. J. Bioorg. Chem.* **2019**, *45*, 647–655. [CrossRef]
18. Il'ina, I.V.; Dyrkheeva, N.S.; Zakharenko, A.L.; Sidorenko, A.Y.; Li-Zhulanov, N.S.; Korchagina, D.V.; Chand, R.; Ayine-Tora, D.M.; Chepanova, A.A.; Zakharova, O.D.; et al. Design, Synthesis, and Biological Investigation of Novel Classes of 3-Carene-Derived Potent Inhibitors of TDP1. *Molecules* **2020**, *25*, 3496. [CrossRef]

19. Zhang, X.-R.; Wang, H.-W.; Tang, W.-L.; Zhang, Y.; Yang, H.; Hu, D.-X.; Ravji, A.; Marchand, C.; Kiselev, E.; Ofori-Atta, K.; et al. Discovery, Synthesis, and Evaluation of Oxynitidine Derivatives as Dual Inhibitors of DNA Topoisomerase IB (TOP1) and Tyrosyl-DNA Phosphodiesterase 1 (TDP1), and Potential Antitumor Agents. *J. Med. Chem.* **2018**, *61*, 9908–9930. [CrossRef]
20. Zakharenko, A.L.; Luzina, O.A.; Sokolov, D.N.; Kaledin, V.I.; Nikolin, V.P.; Popova, N.A.; Patel, J.; Zakharova, O.D.; Chepanova, A.A.; Zafar, A.; et al. Novel tyrosyl-DNA phosphodiesterase 1 inhibitors enhance the therapeutic impact of topotecan on in vivo tumor models. *Eur. J. Med. Chem.* **2019**, *161*, 581–593. [CrossRef]
21. Khomenko, T.M.; Zakharenko, A.L.; Chepanova, A.A.; Ilina, E.S.; Zakharova, O.D.; Kaledin, V.I.; Nikolin, V.P.; Popova, N.A.; Korchagina, D.V.; Reynisson, J.; et al. Promising New Inhibitors of Tyrosyl-DNA Phosphodiesterase I (Tdp 1) Combining 4-Arylcoumarin and Monoterpenoid Moieties as Components of Complex Antitumor Therapy. *Int. J. Mol. Sci.* **2020**, *21*, 126. [CrossRef] [PubMed]
22. Pang, B.; Zhao, L.-H.; Zhou, Q.; Zhao, T.-Y.; Wang, H.; Gu, C.-J.; Tong, X.-L. Application of Berberine on Treating Type 2 Diabetes Mellitus. *Int. J. Endocrinol.* **2015**, *2015*, 905749:1–905749:13. [CrossRef] [PubMed]
23. Chu, M.; Zhang, M.-B.; Liu, Y.-C.; Kang, J.-R.; Chu, Z.-Y.; Yin, K.-L.; Ding, L.-Y.; Ding, R.; Xiao, R.-X.; Yin, Y.-N.; et al. Role of Berberine in the Treatment of Methicillin-Resistant *Staphylococcus aureus* Infections. *Sci. Rep.* **2016**, *6*, 24748:1–24748:9. [CrossRef] [PubMed]
24. Yu, H.H.; Kim, K.J.; Cha, J.D.; Kim, H.K.; Lee, Y.E.; Choi, N.Y.; You, Y.O. Antimicrobial activity of berberine alone and in combination with ampicillin or oxacillin against methicillin-resistant *Staphylococcus aureus*. *J. Med. Food.* **2005**, *8*, 454–461. [CrossRef]
25. Peng, L.; Kang, S.; Yin, Z.; Jia, R.; Song, X.; Li, L.; Li, Z.; Zou, Y.; Liang, X.; Li, L.; et al. Antibacterial activity and mechanism of berberine against Streptococcus agalactiae. *Int. J. Clin. Exp. Pathol.* **2015**, *8*, 5217–5223.
26. Tan, W.; Li, Y.; Chen, M.; Wang, Y. Berberine hydrochloride: Anticancer activity and nanoparticulate delivery system. *Int. J. Nanomed.* **2011**, *6*, 1773–1777. [CrossRef]
27. Nechepurenko, I.V.; Shirokova, E.D.; Khvostov, M.V.; Frolova, T.S.; Sinitsyna, O.I.; Maksimov, A.M.; Bredikhin, R.A.; Komarova, N.I.; Fadeev, D.S.; Luzina, O.A.; et al. Synthesis, hypolipidemic and antifungal activity of tetrahydroberberrubine sulfonates. *Russ. Chem. Bull. Int. Ed.* **2019**, *68*, 1052–1060. [CrossRef]
28. Liu, H.; Wang, J.; Zhang, R.; Cairns, N.; Liu, J. Compounds, Compositions and Methods for Reducing Lipid Levels. International Patent Application No. PCT/US2008/067762, 20 June 2008.
29. Iwasa, K.; Kamigauchi, M. 13-hydroxylation of tetrahydroberberine in cell suspension cultures of some Corydalis species. *Phytochemistry* **1996**, *41*, 1511–1515. [CrossRef]
30. Nechepurenko, I.V.; Boyarskikh, U.A.; Khvostov, M.V.; Baev, D.S.; Komarova, N.I.; Filipenko, M.L.; Tolstikova, T.G.; Salakhutdinov, N.F. Hypolipidemic Berberine Derivatives with a Reduced Aromatic Ring C. *Chem. Nat. Comp.* **2015**, *51*, 916–922. [CrossRef]
31. Zakharenko, A.L.; Khomenko, T.M.; Zhukova, S.V.; Koval, O.A.; Zakharova, O.D.; Anarbaev, R.O.; Lebedeva, N.A.; Korchagina, D.V.; Komarova, N.I.; Vasiliev, V.G.; et al. Synthesis and biological evaluation of novel tyrosyl-DNA phosphodiesterase 1 inhibitors with a benzopentathiepine moiety. *Bioorg. Med. Chem.* **2015**, *23*, 2044–2052. [CrossRef]
32. Lountos, G.T.; Zhao, X.Z.; Kiselev, E.; Tropea, J.E.; Needle, D.; Pommier, Y.; Burke, T.R.; Waugh, D.S. Identification of a Ligand Binding Hot Spot and Structural Motifs Replicating Aspects of Tyrosyl-DNA Phosphodiesterase I (TDP1) Phosphoryl Recognition by Crystallographic Fragment Cocktail Screening. *Nucleic Acids Res.* **2019**, *47*, 10134–10150. [CrossRef] [PubMed]
33. Martino, E.; Della Volpe, S.; Terribile, E.; Benetti, E.; Sakaj, M.; Centamore, A.; Sala, A.; Collina, S. The long story of camptothecin: From traditional medicine to drugs. *Bioorg. Med. Chem. Lett.* **2017**, *27*, 701–707. [CrossRef]
34. Bjornsti, M.A.; Kaufmann, S.H. Topoisomerases and cancer chemotherapy: Recent advances and unanswered questions. *F1000Research* **2019**, *8*, 1704. [CrossRef] [PubMed]
35. Bailly, C. Irinotecan: 25 years of cancer treatment. *Pharmacol. Res.* **2019**, *148*, 104398:1–104398:11. [CrossRef] [PubMed]
36. Perego, P.; Cossa, G.; Tinelli, S.; Corna, E.; Carenini, N.; Gatti, L.; De Cesare, M.; Ciusani, E.; Zunino, F.; Luison, E.; et al. Role of tyrosyl-DNA phosphodiesterase 1 and inter-players in regulation of tumor cell sensitivity to topoisomerase I inhibition. *Biochem. Pharmacol.* **2012**, *83*, 27–36. [CrossRef] [PubMed]

37. Brettrager, E.J.; van Waardenburg, R.C.A.M. Targeting Tyrosyl-DNA phosphodiesterase I to enhance toxicity of phosphodiester linked DNA-adducts. *Cancer Drug Resist.* **2019**, *2*, 1153–1163. [CrossRef] [PubMed]
38. Zhu, F.; Logan, G.; Reynisson, J. Wine Compounds as a Source for HTS Screening Collections. A Feasibility Study. *Mol. Inf.* **2012**, *31*, 847–855. [CrossRef]
39. Eurtivong, C.; Reynisson, J. The Development of a Weighted Index to Optimise Compound Libraries for High Throughput Screening. *Mol. Inf.* **2018**, *37*, 1800068:1–1800068:10. [CrossRef] [PubMed]
40. Berman, H.M.; Westbrook, J.; Feng, Z.; Gilliland, G.; Bhat, T.N.; Weissig, H.; Shindyalov, I.N.; Bourne, P.E. The Protein Data Bank. *Nucleic Acids Res.* **2000**, *28*, 235–242. [CrossRef]
41. Berman, H.; Henrick, K.; Nakamura, H. Announcing the Worldwide Protein Data Bank. *Nat. Struct. Biol.* **2003**, *10*, 980. [CrossRef]
42. Scigress Ultra V. F.J 2.6. (EU 3.1.7) Fujitsu Limited 2008–2016.
43. Allinger, N.L. Conformational Analysis. 130. MM2. A Hydrocarbon Force Field Utilizing V1 and V2 Torsional Terms. *J. Am. Chem. Soc.* **1977**, *99*, 8127–8134. [CrossRef]
44. Jones, G.; Willet, P.; Glen, R.C.; Leach, A.R.; Taylor, R. Development and Validation of a Genetic Algorithm for Flexible Docking. *J. Mol. Biol.* **1997**, *267*, 727–748. [CrossRef] [PubMed]
45. Eldridge, M.D.; Murray, C.; Auton, T.R.; Paolini, G.V.; Mee, P.M. Empirical Scoring Functions: I. The Development of a Fast Empirical Scoring Function to Estimate the Binding Affinity of Ligands in Receptor Complexes. *J. Comput. Aided Mol. Des.* **1997**, *11*, 425–445. [CrossRef]
46. Verdonk, M.L.; Cole, J.C.; Hartshorn, M.J.; Murray, C.W.; Taylor, R.D. Improved Protein-Ligand Docking using GOLD. *Proteins* **2003**, *52*, 609–623. [CrossRef] [PubMed]
47. Korb, O.; Stutzle, T.; Exner, T.E. Empirical Scoring Functions for Advanced Protein−Ligand Docking with PLANTS. *J. Chem. Inf. Model.* **2009**, *49*, 84–96. [CrossRef]
48. Mooij, W.T.M.; Verdonk, M.L. General and Targeted Statistical Potentials for Protein–ligand Interactions. *Proteins* **2005**, *61*, 272–287. [CrossRef]
49. *QikProp*; Version 3.2; Schrödinger: New York, NY, USA, 2009.
50. Ioakimidis, L.; Thoukydidis, L.; Naeem, S.; Mirza, A.; Reynisson, J. Benchmarking the Reliability of QikProp. Correlation between Experimental and Predicted Values. *QSAR Comb. Sci.* **2008**, *27*, 445–456. [CrossRef]

International Journal of
Molecular Sciences

MDPI

Article

Functional Role of N-Terminal Extension of Human AP Endonuclease 1 In Coordination of Base Excision DNA Repair via Protein–Protein Interactions

Nina Moor [1], Inna Vasil'eva [1] and Olga Lavrik [1,2,*]

1 Institute of Chemical Biology and Fundamental Medicine, Siberian Branch of the Russian Academy of Sciences, 630090 Novosibirsk, Russia; moor@niboch.nsc.ru (N.M.); iva@niboch.nsc.ru (I.V.)
2 Novosibirsk State University, 630090 Novosibirsk, Russia
* Correspondence: lavrik@niboch.nsc.ru

Received: 12 February 2020; Accepted: 27 April 2020; Published: 28 April 2020

Abstract: Human apurinic/apyrimidinic endonuclease 1 (APE1) has multiple functions in base excision DNA repair (BER) and other cellular processes. Its eukaryote-specific N-terminal extension plays diverse regulatory roles in interaction with different partners. Here, we explored its involvement in interaction with canonical BER proteins. Using fluorescence based-techniques, we compared binding affinities of the full-length and N-terminally truncated forms of APE1 (APE1NΔ35 and APE1NΔ61) for functionally and structurally different DNA polymerase β (Polβ), X-ray repair cross-complementing protein 1 (XRCC1), and poly(adenosine diphosphate (ADP)-ribose) polymerase 1 (PARP1), in the absence and presence of model DNA intermediates. Influence of the N-terminal truncation on binding the AP site-containing DNA was additionally explored. These data suggest that the interaction domain for proteins is basically formed by the conserved catalytic core of APE1. The N-terminal extension being capable of dynamically interacting with the protein and DNA partners is mostly responsible for DNA-dependent modulation of protein–protein interactions. Polβ, XRCC1, and PARP1 were shown to more efficiently regulate the endonuclease activity of the full-length protein than that of APE1NΔ61, further suggesting contribution of the N-terminal extension to BER coordination. Our results advance the understanding of functional roles of eukaryote-specific protein extensions in highly coordinated BER processes.

Keywords: APE1; protein–protein interactions; base excision repair; multifunctional disordered protein; fluorescence techniques

1. Introduction

Apurinic/apyrimidinic endonuclease 1 (APE1) is an essential protein in mammals with multiple functions in base excision DNA repair (BER), regulation of gene expression, RNA metabolism, and other specific cellular processes [1–3]. BER is the primary mechanism for correcting apurinic/apyrimidinic (AP) sites (created through N-glycosidic bond cleavage), modified bases, and single-strand breaks (SSBs) [4,5]. The major enzymatic function of APE1 in BER is the incision of AP sites, one of the most abundant types of oxidative DNA damage [6]. Additional activities of APE1, 3′-diesterase, and 3′-5′-exonuclease contribute to removing terminal blocking groups in BER DNA intermediates and to proofreading DNA mismatches introduced by DNA polymerase β (Polβ) [1,3]. APE1 is also capable of incising DNA at certain base lesions (nucleotide incision repair) and RNA at abasic site and specific regions, as well as of processing abasic and oxidized ribonucleotides embedded in the DNA [1–3,7,8]. A redox activity of APE1 is responsible for the regulation of DNA-binding activities of different transcription factors [1,2]. The DNA repair and redox activities of APE1 can play a collaborative role in different processes [9,10]. Interactions with multiple protein partners that modulate diverse enzymatic activities of APE1 and mediate its regulatory function in transcription are highly interconnected and dynamically regulated through various post-translational modifications of APE1 [1,2,11–13]. Dysregulation of multifunctional activities of APE1 is associated with various human pathologies, making APE1 a potential therapeutic target [1,5,12,14].

The human APE1 is composed of two structural domains and a disordered N-terminal region [15], formed mostly by the sequence highly conserved in mammals (Figure 1). The catalytically active core responsible for the enzyme activities in BER is created by the universally conserved sequence 62–318 [16]. The region required for the redox activity of APE1 is in the N-terminal domain [1]. The unique disordered region comprises the nuclear localization signal and multiple post-translational modification sites. This region, enriched in lysine residues, contributes to APE1 interaction with various DNA/RNA structures and to acetylation-mediated modulation of the enzyme DNA repair activity in vitro and in vivo [17–21]. Ubiquitination and acetylation of specific Lys residues of the N-terminal region modulate the expression level and different functions of APE1 in vivo via proteolytic degradation or limited N-terminal proteolysis, respectively [22,23]. The conserved catalytic core and the N-terminal extension are both required for the APE1 function at telomeric DNA substrates and their protective protein complexes [24,25]. The N-terminal extension of APE1 contributes to a different extent to the interaction with proteins involved in RNA processing and ribosome biogenesis [26]; its interaction with nucleophosmin (NPM1) is very unstable to be detected in the absence of the remaining C-terminal portion [20]. The first 35 residues are critical for the physical interaction of APE1 with X-ray repair cross-complementing protein 1 (XRCC1) functioning as a scaffold protein of BER [27]. However, involvement of the N-terminal extension of APE1 in the interaction with other canonical BER proteins still remains unexplored.

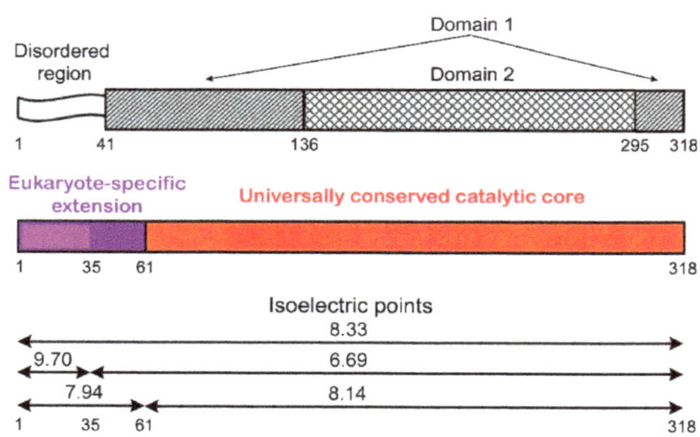

Figure 1. Structural organization of human apurinic/apyrimidinic endonuclease 1 (APE1). Protein is composed of two structural domains and disordered N-terminal region invisible in crystal structures [15]. Sequence 62–318 responsible for AP endonuclease activity is conserved in pro- and eukaryotes; first 35 residues of N-terminal eukaryote-specific extension (residues 1–61) are highly conserved in mammals [16]. Isoelectric points (calculated using ExPASy proteomics server) of full-length protein, its N-terminally truncated forms (APE1NΔ35, APE1NΔ61), the entire extension, and its mammalian-specific fragment are presented.

We demonstrated the usefulness of fluorescence-based techniques for the detection and quantification of physical interactions between various BER proteins, and for the detection of modulation of protein–protein interactions by DNA intermediates [28]. In the present study, we used the same approaches to explore the role of the N-terminal extension of APE1 in coordination of mammalian BER process via protein–protein interactions. The relative binding affinities of the full-length and two N-terminally truncated forms of APE1 (APE1NΔ35 and APE1NΔ61) for three functionally and structurally different BER proteins, downstream enzyme Polβ, scaffold XRCC1 protein, and poly(ADP-ribose) polymerase 1 (PARP1) responsible for detecting SSBs and initiating their repair, were determined in the absence and presence of model DNA intermediates of BER, and the DNA-induced rearrangements of the protein–protein complexes were examined. Additionally, the influence of the N-terminal truncation on the functional cooperation between APE1 and Polβ in the absence and presence of accessory XRCC1 and PARP1 proteins was explored.

2. Results

2.1. Contribution of N-terminal Extension of Human APE1 to the Interaction with Canonical BER Proteins

To explore involvement of the N-terminal extension of human APE1 in the interaction with other BER proteins, and to quantify the contribution of the entire eukaryote-specific region and its disordered mammalian-conserved fragment to the binding affinities, fluorescence-titration experiments were performed using fluorescein-labelled full-length APE1 and its N-terminally truncated forms (APE1NΔ35 and APE1NΔ61). The proteins N-terminally labelled with 5(6)-carboxyfluorescein (FAM) were prepared as previously described [28] and are detailed in Supplementary Materials. The change in the fluorescence intensity of a FAM-labelled protein (FAM-APE1, FAM-APE1NΔ35, or FAM-APE1NΔ61) was monitored in the presence of unlabeled protein partners added at increasing concentrations (Figure 2).

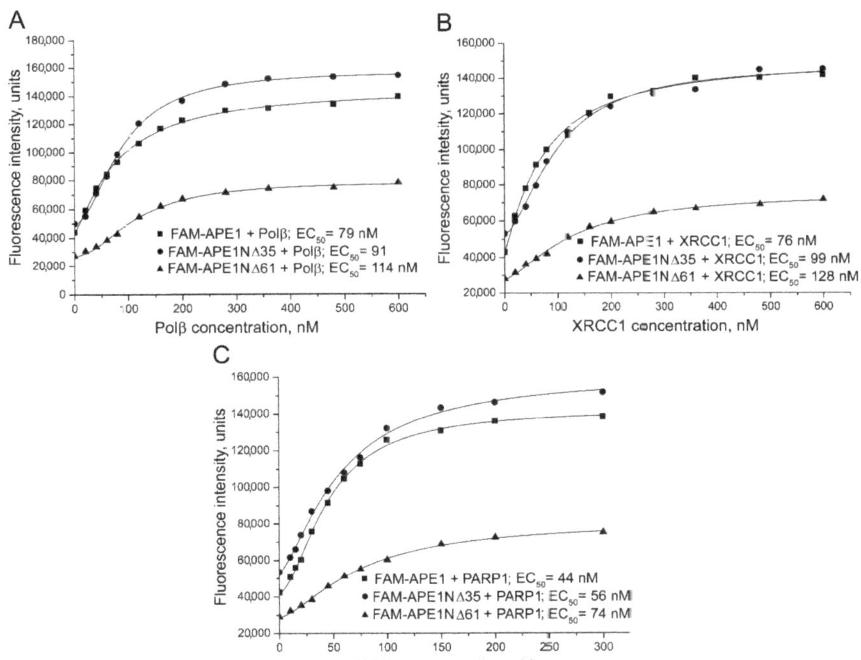

Figure 2. Fluorescence titration of 5(6)-carboxyfluorescein (FAM)-labelled APE1, APE1NΔ35, and APE1NΔ61 with (**A**) DNA polymerase β (Polβ) (**B**) X-ray repair cross-complementing (XRCC1) protein 1, and (**C**) poly(ADP-ribose) polymerase 1 (PARP1). FAM-labelled protein (40 nM) excited at 482 nm in absence or presence of increasing concentrations of protein partner, and relative fluorescence intensities monitored at 530 nm. Curves show best fits (R^2 values met or exceeded 0.93) of four-parameter equation; EC_{50} values derived from respective curves are presented. Data shown are representative of at least three independent experiments.

The fluorescence intensity of the FAM-labelled APE1 and its truncated forms increased in the presence of Polβ, XRCC1, or PARP1, indicating that the local environment of the fluorophore changed upon protein–protein association. Apparent equilibrium dissociation constants of the complexes determined by nonlinear regression analyses as effective concentrations (EC_{50} values) of the protein partners at the half-maximal increase in fluorescence intensity [29] are presented in Table 1. Removal of the first 35 amino acid residues slightly decreased the affinity of APE1 for all three proteins (a 1.2–1.3-fold increase in EC_{50}). Removal of the entire eukaryote-specific extension (residues 1–61) produced more significant effects: The binding affinity of APE1 for Polβ decreased 1.4-fold, and the affinity for PARP1 and XRCC1 decreased 1.7-fold.

Table 1. Binding parameters of protein–protein interactions determined for full-length and truncated forms of APE1 by fluorescence-based approaches.

Labelled Protein [a]	Protein Partner	EC_{50} [b], nM	Effect on Affinity [c]	E [d]
FAM-APE1	Polβ	84 ± 7		0.16 ± 0.02
FAM-APE1NΔ35	Polβ	97 ± 8 *	1.2	0.16 ± 0.02
FAM-APE1NΔ61	Polβ	120 ± 11 **	1.4	0.22 ± 0.02 *
FAM-APE1	XRCC1	76 ± 8		0.14 ± 0.01
FAM-APE1NΔ35	XRCC1	100 ± 10 **	1.3	0.11 ± 0.01 *
FAM-APE1NΔ61	XRCC1	130 ± 12 ***	1.7	0.09 ± 0.01 **
FAM-APE1	PARP1	45 ± 5		0.04 ± 0.01
FAM-APE1NΔ35	PARP1	57 ± 5 **	1.3	0.13 ± 0.02 **
FAM-APE1NΔ61	PARP1	75 ± 7 ***	1.7	0.10 ± 0.01 **

[a] Titration experiments were performed at the constant 40 nM concentration of FAM-labelled protein. Underlined data for FAM-APE1 (previously published [28]) reproduced to compare binding parameters of full-length and truncated forms. [b] Parameters derived from titration curves by fitting to four-parameter equation, where EC_{50} is half-maximal effective concentration of the protein partner, at which $F - F_0 = (F_\infty - F_0)/2$, where F, the fluorescence intensity of a solution containing the FAM-labelled protein and the binding partner at a given concentration (C); F_0, the fluorescence of a solution of the labelled protein alone; F_∞, the fluorescence of the labelled protein saturated with the binding partner. Values are the mean (± SD) of at least three independent experiments. Values determined for complexes formed by truncated forms of APE1 with each protein partner statistically different from those of the full-length protein: $p < 0.05$ (*), $p < 0.01$ (**), $p < 0.001$ (***); *t*-test, $n = 3$–4. [c] Effect of N-terminal truncation of APE1 on protein–protein affinity determined as the ratio of EC_{50} values for complexes formed by the truncated APE1 and by the full-length APE1 with each binding partner. [d] Fluorescence resonance energy transfer (FRET) efficiency calculated from the fractional decrease of fluorescence intensity, $E = 1 - F_{da}/F_d$, where F_{da} and F_d are fluorescence intensities of donor-labelled protein measured in the presence of the acceptor-labelled or the unlabeled protein partner, respectively. Values are the mean (± SD) of three independent experiments. Values determined for truncated forms of APE1, which were statistically different from those of the full-length protein in the respective complexes, are marked $p < 0.05$ (*), $p < 0.01$ (**).

To further explore influence of the N-terminal truncation on physical contacts between APE1 and BER proteins, we performed fluorescence resonance energy transfer (FRET) experiments. FRET is characterized by the efficiency of energy transfer (ET) from a fluorescent donor to an acceptor, which depends on the extent of the spectral overlap between donor emission and acceptor absorption, on the distance between the two fluorophores and on their relative orientation [30]. We chose fluorescein and tetramethylrhodamine as the donor–acceptor pair previously used to characterize interactions between BER proteins [28]. FAM-labelled APE1, APE1NΔ35, or APE1NΔ61 was titrated with Polβ, unlabeled or labelled with 5(6)-carboxytetramethylrhodamine (TMR) (Figure 3). The fluorescence intensity of the FAM-labelled protein increased less in the presence of TMR-Polβ than in the presence of Polβ, indicating participation of the donor- and acceptor-labelled proteins in FRET. FRET efficiencies determined for the FAM-APE1NΔ35-TMR-Polβ and FAM-APE1-TMR-Polβ pairs were identical (Table 1). FRET efficiency determined for the FAM-APE1NΔ61-TMR-Polβ pair was appreciably (6%) higher, indicating that, in this complex, the distance between two fluorophore probes and/or their relative orientation are more favorable for FRET. Analogous FRET measurements were performed for the other complexes. FRET signals detected for the XRCC1 complexes with the truncated forms of APE1 were lower than for the respective complex with the full-length APE1 (Table 1). On the other hand, efficiencies of FRET for the TMR-PARP1 pairs with FAM-APE1N35 and FAM-APE1NΔ61 were higher than for the respective pair with FAM-APE1. The differences in detected FRET signals evidently reflected changes in the localization of the FAM-labelled N-terminus within complexes formed by the full-length and N-terminally truncated forms of APE1.

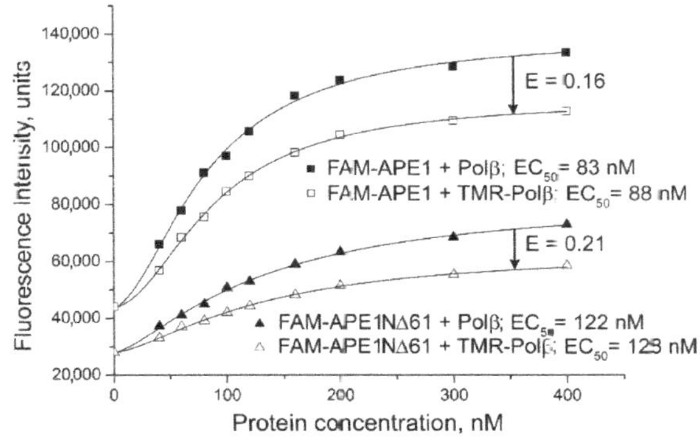

Figure 3. Characterization of Polβ interaction with APE1 and its truncated form by FRET. The FAM-labelled protein (40 nM FAM-APE1 or FAM-APE1NΔ61) excited at 432 nm in the absence or presence of increasing concentrations of unlabeled or TMR-labelled Polβ. FRET efficiency calculated from the fractional decrease (shown by arrow) of fluorescence intensity of the donor due to the presence of acceptor $E = 1 - F_{da}/F_d$, where F_{da} and F_d are fluorescence intensities measured in the presence of identical subsaturating concentrations of TMR-Polβ or Polβ (open or filled symbols, respectively), and determined at saturation by fitting to four-parameter equation. EC_{50} and E values determined for respective complexes are shown.

Taken together, these results indicate that the N-terminal extension of human APE1 is involved in the interactions of APE1 with other BER proteins. Small effects produced by the N-terminal truncation on the binding affinity constants suggest that the interaction domain for proteins is basically formed by the conserved catalytic core of APE1, and the N-terminal extension mediates dynamic interactions.

2.2. Involvement of N-Terminal Extension of APE in DNA-Dependent Modulation of Protein–Protein Interactions

Using fluorescence-based approaches, we showed in a previous study that DNA intermediates of BER induce rearrangements of various protein–protein complexes and modulate the strength of the interaction [28]. To explore the possible involvement of the N-terminal region of APE1 in this modulation, fluorescence titration and FRET experiments with the full-length and truncated forms of APE1 were performed in the absence and presence of model DNA ligands (shown in Figure S1). A double-stranded DNA with a synthetic abasic site (a tetrahydrofuran residue, F), AP-DNA, is an initial BER substrate of APE1. A 1-nucleotide gapped DNA with a 5′-F group at the margin of the gap models a product of the APE1-catalyzed incision (AP-DNA inc) and a stable analog of the Polβ substrate not processed by the 5′-deoxyribose phosphate lyase activity. A 1-nucleotide-gapped DNA (gap-DNA) and AP-DNA inc are substrates of both APE1 (in the 3′-5′-exonuclease reaction) and Polβ (in the DNA repair synthesis). APE1 and Polβ bind with the highest affinity AP-DNA and gap-DNA, respectively; their affinities for AP-DNA inc are relatively high and comparable, and those for the nonsubstrate double-stranded DNA are very low [31]. XRCC1 prefers 1-nt-gapped and nicked oligonucleotide duplexes, while there is no significant difference in the strength of PARP1 interaction with different duplex models of damaged DNA due to preferential binding to blunt ends [32–36]. Functional interactions between APE1, Polβ, XRCC1, and PARP1 on various DNA intermediates, contributing to regulation of BER, were demonstrated in previous studies [27,31,37–42] and are further detailed in Section 2.3.

Binding of the truncated forms of APE1 with each protein partner in the absence and presence of a given DNA intermediate was explored in parallel with the full-length APE1 (Figure 4). From these experiments, the effects produced by the model DNA on the quantitative characteristics of protein–protein interactions, apparent binding affinity constant and FRET efficiency, were determined (Table 2). AP-DNA inc produced the highest effect on the binding affinity of FAM-APE1 for Polβ (two-fold decrease), while its effects detected for the respective complexes formed by FAM-APE1NΔ35 and FAM-APE1NΔ61 were not statistically significant. The presence of intact AP-DNA revealed no significant effect on the binding affinity of both the full-length and truncated forms of APE1 for Polβ. Gap-DNA produced a small effect (1.3-fold decrease detected as statistically significant) for the FAM-APE1 complex. Changes in the efficiency of FRET detected for complexes of the truncated forms of APE1 with TMR-Polβ in the presence of the various DNA intermediates (a 7–9% decrease) were indicative of the DNA-induced rearrangement of the complexes. The different DNAs produced significantly different changes in the FRET signals only in the case of the full-length protein complex (a 5% increase vs. a 6–11% decrease). These combined results suggest that conformational changes in the APE1–Polβ complex caused by DNA binding involve both the conserved catalytic core of APE1 and the N-terminal extension, with the last being important to control the strength of interaction between proteins in the complex with damaged DNA depending on its type.

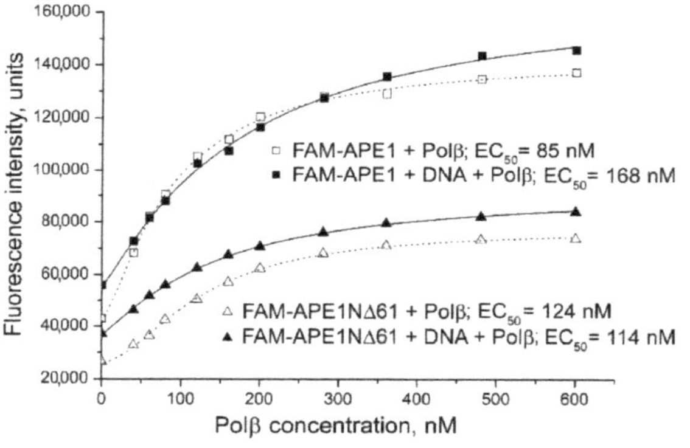

Figure 4. Influence of DNA intermediate of base excision DNA repair (BER) on the interaction of APE1 and its truncated form with Polβ. Fluorescence titration of FAM-labelled APE1 or APE1NΔ61 (40 nM) with Polβ was performed in the absence (open symbols) or presence (filled symbols) of incised AP-DNA (400 nM); higher values of F_0 measured in DNA presence are indicative of protein–DNA binding. Curves show best fits of four-parameter equation with R^2 values exceeding 0.98; EC_{50} values derived from respective curves are presented.

Both the structure and stability of the FAM-APE1 complex with XRCC1 were detected as described previously to be modulated in the presence of gap-DNA [28]: The binding affinity and FRET efficiency increased 1.5-fold and by 8%, respectively (Table 2). However, the parameters of FAM-APE1 interaction with XRCC1 were not affected by the presence of intact AP-DNA. Here, we revealed that interaction of FAM-APE1NΔ35 and FAM-APE1NΔ61 with XRCC1 was appreciably influenced by both gap-DNA and AP-DNA. For each of the truncated forms, the effects produced by the different DNAs on the strength of interaction with XRCC1 and on the FRET signals were very similar. Clearly, integrity of the N-terminal portion of APE1 is important to control the mode of APE1–XRCC1 interaction during the processing of distinct DNA intermediates.

We compared the influence of AP-DNA and gap-DNA on the interaction of the full-length and truncated forms of APE1 with PARP1 (Table 2). Significant effects on the interaction of FAM-APE1 with PARP1 were detected only in the presence of AP-DNA: Binding affinity decreased 1.7 fold and FRET efficiency increased by 5% (Table 2). Both AP-DNA and gap-DNA produced similar positive effects (a 1.4-fold increase) on the strength of interaction between FAM-APE1NΔ61 and PARP1. In the case of FAM-APE1NΔ35, no statistically significant effects of DNA intermediates on the protein interaction with PARP1 were detected. Interestingly, EC_{50} values determined for the complexes of FAM-APE1 and FAM-APE1NΔ61 with PARP1 were identical in the presence of gap-DNA, but the affinity of FAM-APE1 for PARP1 in the presence of AP-DNA was practically identical with that of FAM-APE1NΔ61 in the absence of DNA (76 nM vs. 75 nM). Evidently, the N-terminal extension is responsible for the destabilizing effect of the preferred binding substrate of APE1 on APE1–PARP1 interaction.

Table 2. Effects of BER DNA intermediates on protein–protein interactions.

Labelled Protein [a]	DNA [a]	Protein Partner	EC_{50} [b], nM	Effect on Affinity [c]	Effect on FRET Efficiency [d]
FAM-APE1	AP-DNA	Polβ	92 ± 7	1.1	+0.05 *
FAM-APE1	AP-DNA inc	Polβ	170 ± 13 ***	2.0	−0.06 **
FAM-APE1	gap-DNA	Polβ	110 ± 8 *	1.3	−0.11 **
FAM-APE1NΔ35	AP-DNA	Polβ	91 ± 7	0.94	−0.08 **
FAM-APE1NΔ35	AP-DNA inc	Polβ	110 ± 10	1.1	−0.0 8**
FAM-APE1NΔ61	AP-DNA	Polβ	110 ± 10	0.92	−0.08 **
FAM-APE1NΔ61	AP-DNA inc	Polβ	130 ± 11	1.1	−0.09 **
FAM-APE1NΔ61	gap-DNA	Polβ	140 ± 13	1.2	−0.07 *
FAM-APE1	AP-DNA	XRCC1	78 ± 7	1.0	−0.01
FAM-APE1	gap-DNA	XRCC1	51 ± 4 **	0.67	+0.08 **
FAM-APE1NΔ35	AP-DNA	XRCC1	68 ± 6 **	0.68	−0.03 *
FAM-APE1NΔ35	gap-DNA	XRCC1	65 ± 6 **	0.65	−0.03 *
FAM-APE1NΔ61	AP-DNA	XRCC1	99 ± 8 *	0.75	−0.06 **
FAM-APE1NΔ61	gap-DNA	XRCC1	97 ± 8 *	0.75	−0.05 **
FAM-APE1	AP-DNA	PARP1	76 ± 6 **	1.7	+0.05 **
FAM-APE1	gap-DNA	PARP1	54 ± 4	1.2	+0.02
FAM-APE1NΔ35	AP-DNA	PARP1	49 ± 4	0.86	−0.03
FAM-APE1NΔ35	gap-DNA	PARP1	48 ± 4	0.84	+0.03
FAM-APE1NΔ61	AP-DNA	PARP1	55 ± 5 *	0.73	−0.02
FAM-APE1NΔ61	gap-DNA	PARP1	54 ± 4 *	0.72	+0.04 *

[a] Titration experiments performed at the constant concentrations of FAM-labelled protein (40 nM) and DNA (400 nM). Underlined data for FAM-APE1 (previously published [28]) reproduced to compare binding parameters of full-length and truncated forms. [b] Parameters derived from titration curves by fitting to four-parameter equation. Values are the mean (± SD) of at least three independent experiments. Values determined for each protein pair in the presence of DNA, which were statistically different from the respective value in the absence of DNA, are marked $p < 0.05$ (*), $p < 0.01$ (**), $p < 0.001$ (***). [c] Effect of DNA on protein–protein affinity determined as the ratio of EC_{50} values in the presence or absence of DNA. [d] Increase (+) or decrease (−) in FRET efficiency between FAM- and TMR (5(6)-carboxytetramethylrhodamine)-labelled proteins in the presence of DNA. Statistically significant changes in E values produced by DNA are marked $p < 0.05$ (*), $p < 0.01$ (**).

2.3. *Influence of N-Terminal Truncation of APE1 on Functional Cooperation between BER Proteins*

Numerous studies showed that the enzymatic activities of APE1 on various DNA intermediates of BER are modulated by protein partners: The AP-endonuclease activity by Polβ and PARP1, and the 3′-5′-exonuclease activity by Polβ, PARP1, and XRCC1 [31,38–42]. The nucleotidyltransferase activity of Polβ is stimulated by APE1 and XRCC1, and inhibited by PARP1 [37–39,41,42]. In light of our results described above, we explored the possible involvement of the N-terminal extension of APE1 in the functional coupling of BER proteins to each other. First, we examined the influence of Polβ, XRCC1, and PARP1 on the AP endonuclease activity of APE1NΔ61 in comparison with the full-length enzyme (Figure 5). The activity of APE1 and APE1NΔ61 was enhanced by XRCC1 and inhibited by PARP1 in a concentration-dependent manner (Figure 5A). The effects observed for the full-length APE1 exceeded those for the truncated form at all tested concentrations of XRCC1 and PARP1. Stimulation of the AP endonuclease activity by Polβ shown previously by others [39] was detectable only for the full-length APE1 (Figure 5B). The stimulating effects produced by XRCC1 and Polβ present separately or together were fully suppressed by the addition of PARP1. The differences in effects detected for APE1 and APE1NΔ61 in the presence of all binary combinations of the protein partners were statistically insignificant. At the same time, the inhibiting effect on the activity of APE1 detected in the presence of the ternary protein combination was higher than that for APE1NΔ61. Thus, the N-terminal extension of APE1 being not essential for the major catalytic activity of APE1 may potentially contribute to the regulation of this activity by the BER protein partners.

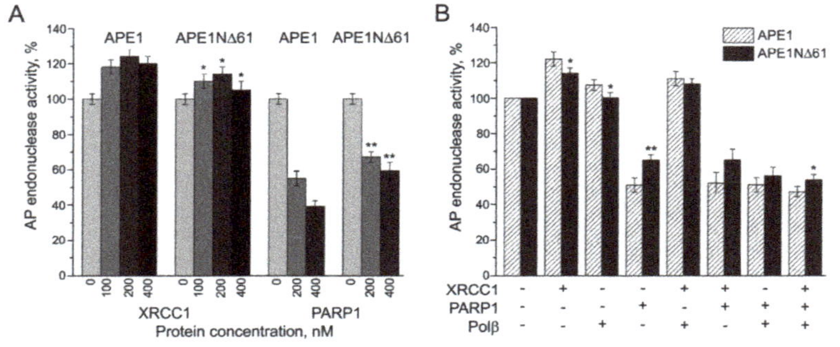

Figure 5. Comparison of effects produced by Polβ, XRCC1, and PARP1 on the AP endonuclease activity of APE1 and APE1NΔ61. Activity in incision of AP-DNA was measured in the absence or presence of Polβ, XRCC1, and PARP1, added separately at varied concentrations (indicated on the X axis, (**A**) or in different combinations at a constant concentration of each (200 nM) (**B**). In each set of experiments, activities were determined as initial rates of AP-DNA incision normalized to that of APE1/APE1NΔ61 in the absence of other proteins (taken as 100%). One representative experiment of the activity measurements is shown in Figure S2. Data are the mean (± SD) of three independent measurements. Effects detected for APE1NΔ61, which were statistically different from those for APE1, are marked $p < 0.05$ (*), $p < 0.01$ (**).

We next compared APE1 and APE1NΔ61 in modulating the nucleotidyltransferase activity of Polβ, using gap-DNA as the substrate (Figure 6). The concentration-dependent effects on the yields of products of gap-filling and strand-displacement DNA synthesis produced by the full-length and truncated forms of APE1 were very similar. No appreciable difference between the two APE1 forms was detected when their influence on the Polβ-catalyzed DNA synthesis in the presence of XRCC1, a strong binding partner of Polβ [28], was compared. Data suggested that the N-terminal extension of APE1 is not essential for the stimulation of Polβ-catalyzed DNA repair synthesis.

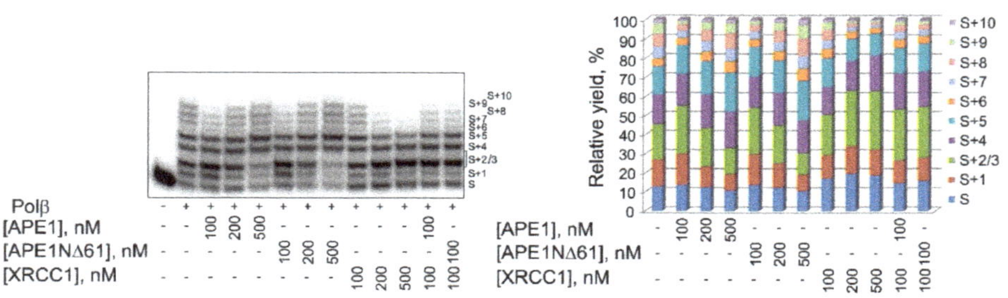

Figure 6. Comparison of effects produced by APE1 and APE1NΔ61 on Polβ activity. Polβ activity in DNA synthesis was measured on gap-DNA in the absence or presence of APE1, APE1NΔ61, and XRCC1, added separately at varied concentrations (100–500 nM) or together at a constant concentration (100 nM APE1/APE1NΔ61 and 100 nM XRCC1). Histograms present relative amounts of substrate (S) and products of gap-filling (S+1) and strand-displacement synthesis (S+2 to S+10). Reaction was performed as described in Materials and Methods. Data are representative of three independent experiments with very similar results.

2.4. Contribution of N-Terminal Extension of APE1 to AP-DNA Binding

Participation of the N-terminal extension of human APE1 in binding the intact and incised AP site-containing DNA was demonstrated by chemical footprinting assay [17]. Recently, oligomerization of APE1 on undamaged DNA and its dependence on the presence of the N-terminal extension was revealed by electron microscopy analysis [43]. However, there are no quantitative data on the relative contribution of the N-terminal extension and the conserved catalytic core to the protein affinity for the AP-DNA substrate. Here, we compared the full-length and truncated forms of APE1 in their modes and strength of AP-DNA binding using electrophoretic-mobility-shift assay (EMSA). Three types of complexes with different electrophoretic mobilities were detected for APE1 (Figure 7), while only two of them, designated as Complexes 1 and 3, could be visualized for APE1NΔ35 and APE1NΔ61. Fast migrating Complex 1 obviously represents a monomeric protein–DNA complex that is stabilized by interactions between AP-DNA and the conserved catalytic portion of APE1. Slow migrating Complex 3 most likely results from oligomerization of the monomeric complex via protein–protein interactions as previously proposed [43]. Complex 2 with intermediate mobility, formed exclusively by the full-length protein, is obviously stabilized by interactions between DNA and the N-terminal extension of APE1; the positively charged N-terminus may additionally neutralize the negative charge of DNA, thereby decreasing complex mobility. Another possibility is the dimerization of APE1 on the DNA, which is highly unstable in the absence of the N-terminus to be detected by the nonequilibrium EMSA technique. The affinity of APE1 for DNA estimated from the protein-concentration dependence of the total amount of bound DNA is 1.8-fold higher as compared to those of APE1NΔ35 and APE1NΔ61 (Table 3). At concentrations of APE1/APE1NΔ35/APE1NΔ61 around the EC_{50} value, the only detectable complex was Complex 1, indicating a lower stability of Complexes 2 and 3. Oligomerization was visualized at lower concentrations of APE1 as compared to those of APE1NΔ35 and APE1NΔ61. Our data provide quantitative evidence that the N-terminal extension of APE1 contributes to the additional stabilization of the complex with AP-DNA, and may control the mode of protein–DNA association (mono- and/or oligomerization) depending on protein concentration.

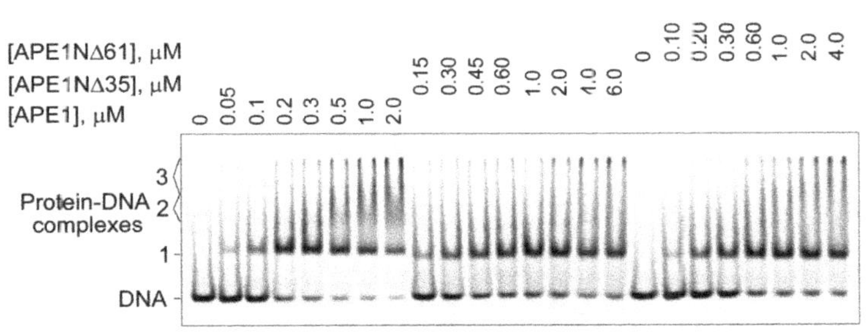

Figure 7. Comparison of APE1 and its N-terminally truncated forms in binding AP site-containing DNA. FAM-labelled AP-DNA was incubated with increasing concentrations of APE1/APE1NΔ35/APE1NΔ61 in the absence of metal ions. After incubation, protein–DNA complexes were separated from free DNA by native gel electrophoresis as described in Materials and Methods. Data are representative of three independent experiments.

Table 3. Parameters of AP-DNA binding by full-length and truncated forms of APE1.

Protein	EC_{50} [a], nM	Maximal Extent of Binding [a]
APE1	0.12 ± 0.02	0.96 ± 0.03
APE1NΔ35	0.20 ± 0.03	0.77 ± 0.03
APE1NΔ61	0.21 ± 0.03	0.84 ± 0.03

[a] Parameters derived from electrophoretic-mobility-shift-assay (EMSA) data as described in Materials and Methods; EC_{50} is the effective protein concentration, at which the extent of DNA binding is half of the maximal extent. Values are the mean (± SD) of at least three independent experiments.

3. Discussion

BER is an exceptionally efficient process evolved by mammalian cells to correct the most abundant DNA lesions. The BER process can proceed along one of different subpathways that involve distinct enzymes and accessory proteins/cofactors (Figure S3). The efficient repair of damaged DNA via the multistep process of each of the subpathways requires the coordinated action of enzymes catalyzing the sequential individual reactions [2–4]. Coordination is facilitated by multiple protein–protein interactions, reviewed previously [13,44]. Physical interaction and functional interplay between two major BER enzymes, APE1 and Polβ, were shown by various approaches [13,31,38,41,45]. XRCC1 functioning as a nonenzymatic scaffold protein directly interacts with multiple enzymatic components of BER, using all structural domains and flexible linkers [13,46]. PARP1, responsible for assembling DNA repair complexes via automodification at sites of DNA damage, forms direct and PAR-mediated contacts with various BER proteins, and modulates the catalytic activities of BER enzymes [4,13,35,40,41,47,48]. Physical and functional interaction between APE1 and PARP1 detected by various in vitro techniques as well as ADP-ribosylation of APE1 catalyzed by PARP1 via the unusual mechanism controlled by the structure of damaged DNA provide evidence of functional assistance between the proteins during DNA repair [28,41,45,49–53].

Here, we explored the contribution of the N-terminal extension of human APE1 to the interaction with Polβ, XRCC1, and PARP1, using fluorescence-based quantitative techniques. The eukaryote-specific extension of APE1 (residues 1–61) is dispensable for endonuclease activity [16], but essential for DNA binding and acetylation-mediated modulation of the enzyme DNA repair activity in vitro and in vivo [17–19,21]. Our data, obtained under true equilibrium conditions, clearly showed that the removal of the entire N-terminal extension (APE1NΔ61) or its mammalian-conserved fragment (APE1NΔ35) decreased the affinity of APE1 for all three proteins to small but measurable extents (Table 1). The small effects produced by N-terminal truncation (less than two-fold) indicate that the C-terminal catalytic core of human APE1, highly conserved through evolution, mainly contributes to the interaction with proteins, and the N-terminal extension forms an additional low-affinity binding site. Our results contradict previous data that demonstrated that the first 35 residues are absolutely required for APE1–XRCC1 interaction assayed by far-Western blotting [27]. This discrepancy could have resulted from disadvantages of the far-Western blotting assay: A significantly lower isoelectric point of APE1NΔ35 (6.69 vs. 8.33 for full-length APE1, Figure 1) might reduce efficiency of protein electrotransfer to nitrocellulose membrane [54]. The considerable part of the eukaryote-specific extension of APE1 is disordered and invisible in the crystal structure of the full-length protein [55]. Numerous studies showed that disordered protein regions are involved in dynamic interactions, contributing little to overall binding energy, but being essential for regulatory functions [56,57].

Disordered terminal tails characterized by clustering of positively charged residues are present predominantly in mammalian DNA-binding proteins [56]. They were proposed to have similar functions in the initial scanning of DNA by BER proteins via transient electrostatic binding, which is followed by specific binding of the lesion in the active site [56]. The N-terminus of human APE1 was shown to stabilize interaction with nonspecific RNA/DNA structures [18,20] and to increase the extent of APE1 oligomerization on undamaged long DNA [43]. Its involvement in binding a specific AP site-containing DNA substrate was demonstrated by protein footprinting with a lysine-reactive probe [16]. In our EMSA binding experiments, we detected a metastable, unique to the full-length APE1, complex with AP-DNA, which accumulated along with the most stable complex formed independently of N-terminal truncation (Figure 7). Compared to APE1, the truncated APE1NΔ35 and APE1NΔ61 forms were shown to form less stable monomeric and oligomeric complexes. These results provide the first quantitative evidence that the N-terminal extension forms an additional low-affinity DNA binding site in the whole protein and contributes to the stabilization of various forms of the APE1–DNA complex.

Previously, we demonstrated that the strength of APE1 interaction with Polβ, XRCC1, and PARP1, and the structure of APE1–protein complexes are modulated by model BER DNA intermediates to different extents depending on the type of damaged DNA [28]. This finding motivated us to explore the involvement of eukaryote-specific extension in this modulation, particularly in light of its capacity to dynamically bind both the DNA and the protein partners. The distinct effects produced by various DNA intermediates on protein–protein affinity binding constants of the full-length APE1 turned out to not be distinguishable for the N-terminally truncated forms (Table 2). These results indicate that the N-terminal extension of human APE1 plays a primary role in the DNA-dependent modulation of the strength of APE1 interaction with both the downstream enzyme and the accessory proteins of BER. This role is obviously related to the coordination of enzymatic functions during BER that is governed by DNA-binding specificity and protein–protein interactions. The interaction between APE1 and Polβ is stronger in the presence of intact AP-DNA than in the complex mimicking a step after the APE1-catalyzed incision, suggesting that the incised DNA intermediate is more effectively passed to Polβ immediately during the incision step. The conformationally flexible N-terminus of APE1, being highly adaptable to both the protein and DNA structure, may differently modulate direct and DNA-mediated protein–protein interactions in the ternary complex with the specific DNA depending on the relative affinities of the protein partners for the DNA. Indeed, changes in FRET signals reflecting structural DNA-induced rearrangements of the APE1–Polβ complex are quite dissimilar for different DNA intermediates. Footprinting experiments showed that intact and incised AP-DNAs protect the N-terminus of APE1 in accordance with the different stabilities of the complexes, but their protective action was undetectable in the ternary complexes with Polβ [17], suggesting higher flexibility of the N-terminal extension in the APE1–DNA–Polβ complexes than in the respective APE1–DNA complexes. The product release is the rate-limiting step of the APE1-catalyzed AP site incision, and this limitation can be overcome by the stimulatory action of Polβ [39]. Involvement of Polβ and PARP1 in the dynamics of APE1 function in vivo was recently shown [58]. Here, we showed that the major function of APE1 is also activated by XRCC1 (Figure 5), previously detected to stimulate the 3′-diesterase activities of APE1 [27]. A comparison of the effects produced by Polβ, XRCC1, and PARP1 on the activity of APE1 and APE1NΔ61 revealed more efficient regulation by all proteins (present separately) for the full-length enzyme. These results provide further evidence that the N-terminal extension of APE1 is involved in interactions with canonical BER proteins promoting the coordination of the process.

APE1 enhances the 5′-deoxyribose phosphate(dRp)-lyase and nucleotidyltransferase activities of Polβ [31,41]. Our comparison of APE1 and APE1NΔ61 in modulating the catalytic activity of Polβ in DNA synthesis, tested on the canonical substrate of short-patch BER, revealed no dependence of the stimulatory function of APE1 on the integrity of its N-terminal portion (Figure 6). No appreciable difference between the two APE1 forms was detected when their effects were compared in the presence of XRCC1, capable of forming the ternary complex with these enzymes [28]. The question of whether the N-terminal extension of APE1 contributes to the functional coupling with Polβ upon the processing of other DNA intermediates, more efficient substrates of APE1 in the 3′-5′-exonuclease reaction [59], remains open for future research.

APE1 stimulates the catalytic activities of different DNA glycosylases, upstream BER enzymes, via accelerating product release [43,60–64]. The coordination mechanism via protein–protein interactions was proposed for TDG (thymine DNA glycosylase) and MYH (MutY homolog DNA glycosylase) [60,62]; binding sites for MYH do not involve the N-terminal extension of APE1 [62]. Other studies suggested no physical interaction between APE1 and DNA glycosylases. The importance of the N-terminal extension of APE1 for the stimulatory function is explained by its participation in protein oligomerization along the DNA, proposed to promote dissociation of DNA glycosylases from the complex with product [43,64]. Recently we detected and characterized new, never previously predicted [65,66], activity of APE1 in binding poly(ADP-ribose) (PAR) [53]. Both the N-terminal extension and the conserved catalytic core of APE1 were found to be involved in PAR binding, with the first contributing for the most part to the interaction with small linear polymers. We propose that the interaction of APE1 with PAR, found to be less efficient as compared to that of XRCC1, may contribute to assembling DNA repair complexes during the PARP1-dependent processing of SSBs. Present and previous studies provide evidence that the N-terminal extension of APE1 may perform multiple, diverse functions in the coordination of BER.

4. Materials and Methods

4.1. Protein Expression and Purification

Recombinant human APE1, its N-terminally truncated forms (APE1NΔ35 and APE1NΔ61), and rat Polβ were produced by expression in *Escherichia coli* BL21(DE3)pLysS (BL21 E. coli strain carrying the lambda DE3 lysogen and the pLysS plasmid, which expresses T7 lysozyme) and purified as described previously [67–69]. Human PARP1 and human protein XRCC1 were produced by expression in *E. coli* Rosetta (DE3) and purified as described [70,71]. APE1 and Polβ expression vectors were kindly provided by S.H. Wilson (National Institute of Health, North Carolina, USA). The plasmid constructs used to express APE1NΔ35 and APE1NΔ61 were kindly provided by A.A. Ishchenko (Groupe Réparation de l'ADN, UMR 8126 CNRS, Univ Paris-Sud, Institut Gustave Roussy, France). The XRCC1 expression vector was a generous gift from J.P. Radicella (UMR217 CNRS/CEA, France). The plasmid construct used to express PARP1 was kindly provided by M. Satoh (Laval University, Quebec, Canada). The purified proteins were dialyzed against a solution containing 50 mM Tris-HCl, pH 8.0, 100 mM NaCl, 5 mM DTT (dithiotheitol), and 40% glycerol, and stored at −30 °C.

4.2. Fluorescence Studies of Protein–Protein Interactions

Fluorescent labelling of proteins on the terminal amino group using succinimidyl esters of 5(6)-carboxyfluorescein (FAM-SE) and 5(6)-carboxytetramethylrhodamine (TMR-SE) was performed as described [28]. The stoichiometry of protein labelling did not exceed 1 mole of dye per mole of protein, and FAM/TMR-labelled enzymes retained their specific activities (as detailed in Supplementary Materials, Table S1). Binding of proteins to each other was examined by fluorescence titration experiments. Fluorescence intensities of solutions of the FAM-labelled protein (APE1, APE1NΔ35, or APE1NΔ61) at a fixed concentration were measured in the absence and presence of varied concentrations of the partner (Polβ/PARP1/XRCC1), in a binding buffer containing 50 mM HEPES (4-(2-hydroxyethyl)-1-piperazineethanesulfonic acid), pH 8.0, 100 mM NaCl, and 4 mM DTT. Fluorescence intensity of the samples was measured in Corning 384-well black polypropylene assay plates using a CLARIOstar multifunctional microplate reader (BMG LABTECH GmbH, Germany); fluorescence probes were excited at 482 nm (482-16 filter), and the relative fluorescence intensities were detected at the emission maximum (530 nm, 530-40 filter). All measurements were carried out in duplicate for each specific condition, and the average values of fluorescence (with mean deviations not exceeding 2%) were taken for data analysis performed using MARS (Multivariate Adaptive Regression Splines) Data Analysis Software (BMG LABTECH GmbH, Germany). Data were plotted (F vs. C) and fitted by four-parameter logistic equation

$$F = F_0 + (F_\infty - F_0)/[1 + (EC_{50}/C)^n] \quad (1)$$

where F is the fluorescence intensity of a solution containing the FAM-labelled protein and the binding partner at a given concentration (C), F_0 is the fluorescence of a solution of the labelled protein alone, F_∞ is the fluorescence of the labelled protein saturated with the partner, EC_{50} is the concentration of the binding partner at which $F - F_0 = (F_\infty - F_0)/2$, and n is the Hill coefficient denoting the slope of the nonlinear curve.

To detect protein–protein interactions by the FRET approach, the fluorescence intensity of the FAM-labelled protein (donor probe) was measured in the absence and presence of varied concentrations of the TMR-labelled protein (acceptor probe). Measurements were performed in two series: (1) Donor probe + unlabeled partner, and (2) donor probe + acceptor probe (FRET pair). FRET efficiency (E) was calculated from the fractional decrease of the donor fluorescence (F_d) due to the presence of the acceptor (F_{da})

$$E = 1 - F_{da}/F_d. \quad (2)$$

In protein–protein binding experiments performed in the presence of model DNAs (prepared as described in Supplementary Materials), the FAM-labelled protein (40 nM) was premixed with the desired DNA, and fluorescence intensity was taken as a starting F_0 value. Conditions for the formation of ternary complexes were optimized by varying DNA concentration from 120 to 600 nM. Titration experiments were performed in binding buffer supplemented with 10 mM EDTA (ethylenediaminetetraacetic acid) (to suppress the AP-endonuclease and the 3′-5′-exonuclease activities of APE1 and its truncated forms).

4.3. AP Endonuclease Activity Assay

The endonuclease activity of APE1 and APE1NΔ61 was assayed in reaction mixtures (25–50 μL) containing 50 mM Tris-HCl, pH 8.0, 50 mM NaCl, 10 mM $MgCl_2$, 1 mM DTT, 0.1 mg/mL BSA (bovine serum albumin), and 100 nM 5′-^{32}P-labelled AP-DNA substrate (32 base-pair oligonucleotide with a synthetic AP site, prepared as described in Supplementary Materials). The reaction mixtures were preassembled on ice; when indicated, they were supplemented with Polβ, XRCC1, and PARP1 (at concentrations specified in the figure legends). Reaction was initiated by adding APE1 (APE1NΔ61) to a final concentration of 0.5 nM; the reaction mixtures were incubated at 37 °C for 0.5–4 min and terminated by the addition of denaturing PAGE sample buffer and heating for 2 min at 90 °C. The reaction products were separated by electrophoresis in 20% denaturing polyacrylamide gels. The gels were imaged on a Typhoon FLA 9500, and the amounts of DNA substrate and product were quantified using Quantity One Basic software.

4.4. DNA Synthesis by Polβ

The activity of Polβ in DNA synthesis was assayed in reaction mixtures (10 μL) containing 50 mM Tris-HCl, pH 8.0, 50 mM NaCl, 6 mM $MgCl_2$, 50 nM 5′-^{32}P-labelled gap-DNA substrate (32 base-pair oligonucleotide with a one-nucleotide gap prepared as described in Supplementary Materials), and 50 nM Polβ. Reaction mixtures were preassembled on ice; when indicated, they were supplemented with APE1, APE1NΔ61, and XRCC1 (at concentrations specified in the figure legends). The reaction was initiated by adding a mixture of dATP (2′-deoxyadenosine 5′-triphosphate), dGTP (2′-deoxyguanosine 5′-triphosphate), dTTP (2′-deoxythymidine 5′-triphosphate) and dCTP (2′-deoxycytidine 5′-triphosphate) to a final concentration of 10 μM for each nucleotide. Reaction mixtures were incubated at 37 °C for 30 min, and terminated by the addition of denaturing PAGE sample buffer and heating for 2 min at 90 °C. The reaction products were separated by electrophoresis in 20% denaturing polyacrylamide gels. The gels were imaged on a Typhoon FLA 9500, and the amounts of DNA substrate and products were quantified using Quantity One Basic software.

4.5. Electrophoretic-Mobility-Shift Assay

The affinity of APE1, APE1NΔ35, and APE1NΔ61 for AP-DNA was measured by electrophoretic-mobility-shift assay (EMSA). A desired protein (added at increasing from 0.05 to 2–6 μM concentrations) was incubated with FAM-labelled AP-DNA (50 nM) in a 10 μL mixture containing 50 mM Tris-HCl, pH 8.0, 100 mM NaCl, and 1 mM DTT at 4 °C for 30 min. After the addition of glycerol and bromophenol blue (to a final concentration of 5% and 0.1%, respectively), the incubation mixtures were electrophoresed at 4 °C on 5% nondenaturing PAG (polyacrylamide gel) in a 30 mM Tris-Borate-EDTA buffer. Gels were imaged on a Typhoon FLA 9500. Free and complexed DNA bands were quantified using Quantity One Basic software. To subtract free DNA co-migrated with the protein-bound DNA (due to partial overlapping of their migration zones as a result of DNA smearing) the portion of free DNA in the protein-containing samples (determined from radioactivity distribution between migration zones assigned to the free DNA and its complexes with proteins) was normalized to the portion of free DNA in the control sample without the protein. Data were fitted to a Hill equation: $\theta = \theta_\infty/[1 + (EC_{50}/C)^n]$, where θ is the portion of bound DNA (calculated as the complex amount divided by total DNA amount) at a given concentration (C) of the protein, θ_∞ is the maximal extent of DNA binding (i.e., the portion of DNA bound at saturating protein concentration), EC_{50} is the protein concentration at which $\theta = \theta_\infty/2$, and n is the Hill coefficient.

Supplementary Materials: Supplementary materials can be found at http://www.mdpi.com/1422-0067/21/9/3122/s1. Figure S1. Structures of DNA ligands used in the study; Figure S2. Influence of Polβ, XRCC1, and PARP1 on the AP endonuclease activity of APE1 and APE1NΔ61. Figure S3. Scheme illustrating stages of BER subpathways.

Author Contributions: Conceptualization, N.M.; methodology, N.M. and I.V.; validation, N.M. and I.V.; investigation, N.M. and I.V.; resources, N.M. and I.V.; writing—original draft preparation, N.M.; writing—review and editing, N.M., I.V., and O.L.; project administration, N.M. and O.L.; funding acquisition, O.L. All authors have read and agreed to the published version of the manuscript.

Funding: This research was funded by the Russian Science Foundation, grant number 19-14-00107 to O.L.

Acknowledgments: The authors are thankful to S.H. Wilson, A.A. Ishchenko, J.P. Radicella, and M. Satoh for providing us with recombinant plasmids coding for APE1, Polβ, XRCC1, and PARP1, and to A.-L. Haenni (Jacques Monod Institute, France) for careful reading of the manuscript and useful comments.

Conflicts of Interest: The authors declare no conflict of interest.

Abbreviations

APE1	apurinic/apyrimidinic endonuclease 1
BER	base excision repair
AP site	apurinic/apyrimidinic site
SSB	single-strand break
Polβ	DNA polymerase β
XRCC1	X-ray repair cross-complementing protein 1
PARP1	poly(ADP-ribose) polymerase 1
APE1NΔ35/APE1NΔ61	APE1 without 35/61 N-terminal amino acid residues
FAM	5(6)-carboxyfluorescein
TMR	5(6)-carboxytetramethylrhodamine
FRET	Fluorescence resonance energy transfer
AP-DNA/gap-DNA	DNA with a synthetic AP site/1-nucleotide gap
FAM-SE/TMR-SE	succinimidyl ester of 5(6)-carboxyfluorescein/5(6)-carboxytetramethylrhodamine

References

1. Li, M.; Wilson, D.M., 3rd. Human apurinic/apyrimidinic endonuclease 1. *Antioxid. Redox. Signal.* **2014**, *20*, 678–707. [CrossRef] [PubMed]
2. Antoniali, G.; Malfatti, M.C.; Tell, G. Unveiling the non-repair face of the Base Excision Repair pathway in RNA processing: A missing link between DNA repair and gene expression? *DNA Repair* **2017**, *56*, 65–74. [CrossRef] [PubMed]
3. Whitaker, A.M.; Freudenthal, B.D. APE1: A skilled nucleic acid surgeon. *DNA Repair (Amst)* **2018**, *71*, 93–100. [CrossRef] [PubMed]
4. Abbotts, R.; Wilson, D.M., 3rd. Coordination of DNA single strand break repair. *Free Radic. Biol. Med.* **2017**, *107*, 228–244. [CrossRef] [PubMed]
5. Whitaker, A.M.; Schaich, M.A.; Smith, M.R.; Flynn, T.S.; Freudenthal, B.D. Base excision repair of oxidative DNA damage: from mechanism to disease. *Front. Biosci. (Landmark Ed)* **2017**, *22*, 1493–1522. [CrossRef] [PubMed]
6. De Bont, R.; van Larebeke, N. Endogeneous DNA damage in humans: a review of quantitative data. *Mutagenesis* **2004**, *19*, 169–185. [CrossRef]
7. Malfatti, M.C.; Balachander, S.; Antoniali, G.; Koh, K.D.; Saint-Pierre, C.; Gasparutto, D.; Chon, H.; Crouch, R.J.; Storici, F.; Tell, G. Abasic and oxidized ribonucleotides embedded in DNA are processed by human APE1 and not by RNase H2. *Nucleic Acids Res.* **2017**, *45*, 11193–11212. [CrossRef]

8. Antoniali, G.; Serra, F.; Lirussi, L.; Tanaka, M.; D'Ambrosio, C.; Zhang, S.; Radovic, S.; Dalla, E.; Ciani, Y.; Scaloni, A.; et al. Mammalian APE1 controls miRNA processing and its interactome is linked to cancer RNA metabolism. *Nat. Commun.* **2017**, *8*, 797. [CrossRef]
9. Li, M.; Dai, N.; Wang, D.; Zhong, Z. Distinct APE1 activities affect the regulation of VEGF transcription under hypoxic conditions. *Comput. Struct. Biotechnol. J.* **2019**, *17*, 324–332. [CrossRef]
10. Frossi, B.; Antoniali, G.; Yu, K.; Akhtar, N.; Kaplan, M.H.; Kelley, M.R.; Tell, G.; Pucillo, C.E.M. Endonuclease and redox activities of human apurinic/apyrimidinic endonuclease 1 have distinctive and essential functions in IgA class switch recombination. *J. Biol. Chem.* **2019**, *294*, 5198–5207. [CrossRef]
11. Limpose, K.L.; Corbett, A.H. Doetsch, P.W. BERing the burden of damage: pathway crosstalk and posttranslational modification of base excision repair proteins regulate DNA damage management. *DNA Repair (Amst)* **2017**, *56*, 51–64. [CrossRef] [PubMed]
12. Thakur, S.; Dhiman, M.; Tell, G.; Mantha, A.K. A review on protein-protein interaction network of APE1/Ref-1 and its associated biological functions. *Cell Biochem. Funct.* **2015**, *33*, 101–112. [CrossRef] [PubMed]
13. Moor, N.A.; Lavrik, O.I. Protein-protein interactions in DNA base excision repair. *Biochemistry (Mosc)* **2018**, *83*, 411–422. [CrossRef] [PubMed]
14. Malfatti, M.C.; Antoniali, G.; Codrich, M.; Burra, S.; Mangiapane, G.; Dalla, E.; Tell, G. New perspectives in cancer biology from a study of canonical and non-canonical functions of base excision repair proteins with a focus on early steps. *Mutagenesis* **2020**, (in press). [CrossRef] [PubMed]
15. Gorman, M.A.; Morera, S.; Rothwell, D.G.; de La Fortelle, E.; Mol, C.D.; Tainer, J.A.; Hickson, I.D.; Freemont, P.S. The crystal structure of the human DNA repair endonuclease HAP1 suggests the recognition of extra-helical deoxyribose at DNA abasic sites. *EMBO J.* **1997**, *16*, 6548–6558. [CrossRef] [PubMed]
16. Izumi, T.; Mitra, S. Deletion analysis of human AP-endonuclease: minimum sequence required for the endonuclease activity. *Carcinogenesis* **1998**, *19*, 525–527. [CrossRef]
17. Yu, E.; Gaucher, S.P.; Hadi, M.Z. Probing conformational changes in Ape1 during the progression of base excision repair. *Biochemistry* **2010**, *49*, 3786–3796. [CrossRef]
18. Fantini, D.; Vascotto, C.; Marasco, D.; D'Ambrosio, C.; Romanello, M.; Vitagliano, L.; Pedone, C.; Poletto, M.; Cesaratto, L.; Quadrifoglio, F.; et al. Critical lysine residues within the overlooked N-terminal domain of human APE1 regulate its biological functions. *Nucleic Acids Res.* **2010**, *38*, 8239–8256. [CrossRef]
19. Lirussi, L.; Antoniali, G.; Vascotto, C.; D'Ambrosio, C.; Poletto, M.; Romanello, M.; Marasco, D.; Leone, M.; Quadrifoglio, F.; Bhakat, K.K.; et al. Nucleolar accumulation of APE1 depends on charged lysine residues that undergo acetylation upon genotoxic stress and modulate its BER activity in cells. *Mol. Biol. Cell* **2012**, *23*, 4079–4096. [CrossRef]
20. Poletto, M.; Vascotto, C.; Scognamiglio, P.L.; Lirussi, L.; Marasco, D.; Tell, G. Role of the unstructured N-terminal domain of the hAPE1 (human apurinic/apyrimidinic endonuclease 1) in the modulation of its interaction with nucleic acids and NPM1 (nucleophosmin). *Biochem. J.* **2013**, *452*, 545–557. [CrossRef]
21. Roychoudhury, S.; Nath, S.; Song, H.; Hegde, M.L.; Bellot, L.J.; Mantha, A.K.; Sengupta, S.; Ray, S.; Natarajan, A.; Bhakat, K.K. Human apurinic/apyrimidinic endonuclease (APE1) is acetylated at DNA damage sites in chromatin, and acetylation modulates its DNA repair activity. *Mol. Cell. Biol.* **2017**, *37*, e00401-16. [CrossRef]
22. Madlener, S.; Ströbel, T.; Vose, S.; Saydam, O.; Price, B.D.; Demple, B.; Saydam, N. Essential role for mammalian apurinic/apyrimidinic (AP) endonuclease Ape1/Ref-1 in telomere maintenance. *Proc. Natl Acad. Sci. USA* **2013**, *110*, 17844–17849. [CrossRef]
23. Burra, S.; Marasco, D.; Malfatti, M.C.; Antoniali, G.; Virgilio, A.; Esposito, V.; Demple, B.; Galeone, A.; Tell, G. Human AP-endonuclease (Ape1) activity on telomeric G4 structures is modulated by acetylatable lysine residues in the N-terminal sequence. *DNA Repair (Amst)* **2019**, *73*, 129–143. [CrossRef]
24. Bhakat, K.K.; Sengupta, S.; Adeniyi, V.F.; Roychoudhury, S.; Nath, S.; Bellot, L.J.; Feng, D.; Mantha, A.K.; Sinha, M.; Qiu, S.; et al. Regulation of limited N-terminal proteolysis of APE1 in tumor via acetylation and its role in cell proliferation. *Oncotarget* **2016**, *7*, 22590–22604. [CrossRef]
25. Scott, T.L.; Wicker, C.A.; Suganya, R.; Dhar, B.; Pittman, T.; Horbinski, C.; Izumi, T. Polyubiquitination of apurinic/apyrimidinic endonuclease 1 by Parkin. *Mol. Carcinog.* **2017**, *56*, 325–336. [CrossRef]
26. Vascotto, C.; Fantini, D.; Romanello, M.; Cesaratto, L.; Deganuto, M.; Leonardi, A.; Radicella, J.P.; Kelley, M.R.; D'Ambrosio, C.; Scaloni, A.; et al. APE1/Ref-1 interacts with NPM1 within nucleoli and plays a role in the rRNA quality control process. *Mol. Cell. Biol.* **2009**, *29*, 1834–1854. [CrossRef]

27. Vidal, A.E.; Boiteux, S.; Hickson, I.D.; Radicella, J.P. XRCC1 coordinates the initial and late stages of DNA abasic site repair through protein-protein interactions. *EMBO J.* **2001**, *20*, 6530–6539. [CrossRef]
28. Moor, N.A.; Vasil'eva, I.A.; Anarbaev, R.O.; Antson, A.A.; Lavrik, O.I. Quantitative characterization of protein-protein complexes involved in base excision DNA repair. *Nucleic Acids Res.* **2015**, *43*, 6009–6022. [CrossRef]
29. Lalonde, S.; Ehrhardt, D.W.; Loqué, D.; Chen, J.; Rhee, S.Y.; Frommer, W.B. Molecular and cellular approaches for the detection of protein-protein interactions: latest techniques and current limitations. *Plant J.* **2008**, *53*, 610–635. [CrossRef]
30. Wojtuszewski, K.; Harvey, J.J.; Han, M.K.; Knutson, J.R. Fluorescence detection of proximity. In *Protein Interactions; Biophysical Approaches for the Study of Complex Reversible Systems*; Schuck, P., Atassi, M.Z., Eds.; Springer: New York, USA, 2007; pp. 367–396.
31. Liu, Y.; Prasad, R.; Beard, W.A.; Kedar, P.S.; Hou, E.W.; Shock, D.D.; Wilson, S.H. Coordination of steps in single-nucleotide base excision repair mediated by apurinic/apyrimidinic endonuclease 1 and DNA polymerase β. *J. Biol. Chem.* **2007**, *282*, 13532–13541. [CrossRef]
32. D'Silva, I.; Pelletier, J.D.; Lagueux, D.; D'Amours, J.; Chaudhry, M.A.; Weinfeld, M.; Lees-Miller, S.P.; Poirier, G.G. Relative affinities of poly(ADP-ribose) polymerase and DNA-dependent protein kinase for DNA strand interruptions. *Biochim. Biophys. Acta* **1999**, *1430*, 119–126. [CrossRef]
33. Mani, R.S.; Karimi-Busheri, F.; Fanta, M.; Caldecott, K.W.; Cass, C.E.; Weinfeld, M. Biophysical characterization of human XRCC1 and its binding to damaged and undamaged DNA. *Biochemistry* **2004**, *43*, 16505–16514. [CrossRef] [PubMed]
34. Nazarkina, Z.K.; Khodyreva, S.N.; Marsin, S.; Lavrik, O.I.; Radicella, J.P. XRCC1 interactions with base excision repair DNA intermediates. *DNA Repair (Amst)* **2007**, *6*, 54–64. [CrossRef]
35. Khodyreva, S.N.; Prasad, R.; Ilina, E.S.; Sukhanova, M.V.; Kutuzov, M.M.; Liu, Y.; Hou, E.W.; Wilson, S.H.; Lavrik, O.I. Apurinic/apyrimidinic (AP) site recognition by the 5'-dRP/AP lyase in poly(ADP-ribose) polymerase-1 (PARP-1). *Proc. Natl Acad. Sci. USA* **2010**, *107*, 22090–22095. [CrossRef] [PubMed]
36. Lilyestrom, W.; van der Woerd, N.M.; Clark, J.; Luger, K. Structural and biophysical studies of human PARP-1 in complex with damaged DNA. *J. Mol. Biol.* **2010**, *395*, 983–994. [CrossRef] [PubMed]
37. Kubota, Y.; Nash, R.A.; Klungland, A.; Schär, P.; Barnes, D.E.; Lindahl, T. Reconstitution of DNA base excision-repair with purified human proteins: interaction between DNA polymerase β and the XRCC1 protein. *EMBO J.* **1996**, *15*, 6662–6670. [CrossRef]
38. Bennett, R.A.; Wilson, D.M., 3rd; Wong, D.; Demple, B. Interaction of human apurinic endonuclease and DNA polymerase β in the base excision repair pathway. *Proc. Natl Acad. Sci. USA* **1997**, *94*, 7166–7169. [CrossRef]
39. Masuda, Y.; Bennett, R.A.; Demple, B. Dynamics of the interaction of human apurinic endonuclease (Ape1) with its substrate and product. *J. Biol. Chem.* **1998**, *273*, 30352–30359. [CrossRef]
40. Lavrik, O.I.; Prasad, R.; Sobol, R.W.; Horton, J.K.; Ackerman, E.J.; Wilson, S.H. Photoaffinity labeling of mouse fibroblast enzymes by a base excision repair intermediate. Evidence for the role of poly(ADP-ribose) polymerase-1 in DNA repair. *J. Biol. Chem.* **2001**, *276*, 25541–25548. [CrossRef]
41. Sukhanova, M.V.; Khodyreva, S.N.; Lebedeva, N.A.; Prasad, R.; Wilson, S.H.; Lavrik, O.I. Human base excision repair enzymes apurinic/apyrimidinic endonuclease1 (APE1), DNA polymerase β and poly(ADP-ribose) polymerase 1: interplay between strand-displacement DNA synthesis and proofreading exonuclease activity. *Nucleic Acids Res.* **2005**, *33*, 1222–1229. [CrossRef]
42. Dyrkheeva, N.S.; Khodyreva, S.N.; Lavrik, O.I. Interaction of APE1 and other repair proteins with DNA duplexes imitating intermediates of DNA repair and replication. *Biochemistry (Mosc)* **2008**, *73*, 261–272. [CrossRef] [PubMed]
43. Kladova, O.A.; Bazlekowa-Karaban, M.; Baconnais, S.; Piétrement, O.; Ishchenko, A.A.; Matkarimov, B.T.; Iakovlev, D.A.; Vasenko, A.; Fedorova, O.S.; Le Cam, E.; et al. The role of the N-terminal domain of human apurinic/apyrimidinic endonuclease 1, APE1, in DNA glycosylase stimulation. *DNA Repair (Amst)* **2018**, *64*, 10–25. [CrossRef] [PubMed]
44. Moor, N.; Lavrik, O. Coordination of DNA Base Excision Repair by Protein-Protein Interactions. In *DNA Repair—An Update*; Mognato, M., Ed.; IntechOpen Ltd.: London, UK, 2019; Chapter 2; pp. 1–22. [CrossRef]
45. Vasil'eva, I.A.; Anarbaev, R.O.; Moor, N.A.; Lavrik, O.I. Dynamic light scattering study of base excision DNA repair proteins and their complexes. *Biochim. Biophys. Acta Proteins Proteom.* **2019**, *1867*, 297–305. [CrossRef]

46. London, R.E. The structural basis of XRCC1-mediated DNA repair. *DNA Repair (Amst)* **2015**, *30*, 90–103. [CrossRef]
47. Shall, S.; de Murcia, G. Poly(ADP-ribose) polymerase-1: what have we learned from the deficient mouse model? *Mutat. Res.* **2000**, *460*, 1–15. [CrossRef]
48. Dantzer, F.; de La Rubia, G.; Ménissier-De Murcia, J.; Hostomsky, Z.; de Murcia, G.; Schreiber, V. Base excision repair is impaired in mammalian cells lacking poly(ADP-ribose) polymerase-1. *Biochemistry* **2000**, *39*, 7559–7569. [CrossRef]
49. Cistulli, C.; Lavrik, O.I.; Prasad, R.; Hou, E.; Wilson, S.H. AP endonuclease and poly(ADP-ribose) polymerase-1 interact with the same base excision repair intermediate. *DNA repair (Amst)* **2004**, *3*, 581–591. [CrossRef]
50. Kutuzov, M.M.; Ilina, E.S.; Sukhanova, M.V.; Pyshnaya, I.A.; Pyshnyi, D.V.; Lavrik, O.I.; Khodyreva, S.N. Interaction of poly(ADP-ribose) polymerase 1 with apurinic/apyrimidinic sites within clustered DNA damage. *Biochemistry (Mosc)* **2011**, *76*, 147–156. [CrossRef]
51. Prasad, R.; Dyrkheeva, N.; Williams, J.; Wilson, S.H. Mammalian base excision repair: functional partnership between PARP-1 and APE1 in AP-site repair. *PLoS ONE* **2015**, *10*, e0124269. [CrossRef]
52. Liu, L.; Kong, M.; Gassman, N.R.; Freudenthal, B.D.; Prasad, R.; Zhen, S.; Watkins, S.C.; Wilson, S.H.; Van Houten, B. PARP1 changes from three-dimensional DNA damage searching to one-dimensional diffusion after auto-PARylation or in the presence of APE1. *Nucleic Acids Res.* **2017**, *45*, 12834–12847. [CrossRef]
53. Moor, N.A.; Vasil'eva, I.A.; Kuznetsov, N.A.; Lavrik, O.I. Human apurinic/apyrimidinic endonuclease 1 is modified in vitro by poly(ADP-ribose) polymerase 1 under control of the structure of damaged DNA. *Biochimie* **2020**, *168*, 144–155. [CrossRef]
54. Kurien, B.T.; Scofield, R.H. Western blotting: an introduction. *Methods Mol. Biol.* **2015**, *1312*, 17–30. [CrossRef]
55. Mol, C.D.; Izumi, T.; Mitra, S.; Tainer, J.A. DNA-bound structures and mutants reveal abasic DNA binding by APE1 and DNA repair coordination. *Nature* **2000**, *403*, 451–456. [CrossRef]
56. Hedge, M.L.; Hazra, T.K.; Mitra, S. Functions of disordered regions in mammalian early base excision repair proteins. *Cell. Mol. Life Sci.* **2010**, *67*, 3573–3587. [CrossRef]
57. Wright, P.; Dyson, H.J. Intrinsically disordered proteins in cellular signaling and regulation. *Nat. Rev. Mol. Cell Biol.* **2015**, *16*, 18–29. [CrossRef]
58. Janoshazi, A.K.; Horton, J.K.; Zhao, M.L.; Prasad, R.; Scappini, E.L.; Tucker, C.J.; Wilson, S.H. Shining light on the response to repair intermediates in DNA of living cells. *DNA Repair (Amst)* **2020**, *85*, 102749. [CrossRef]
59. Dyrkheeva, N.S.; Khodyreva, S.N.; Sukhanova, M.V.; Safronov, I.V.; Dezhurov, S.V.; Lavrik, O.I. 3′-5′ exonuclease activity of human apurinic/apyrimidinic endonuclease 1 towards DNAs containing dNMP and their modified analogs at the 3′ end of single strand DNA break. *Biochemistry (Mosc)* **2006**, *71*, 200–210. [CrossRef]
60. Waters, T.R.; Gallinari, P.; Jiricny, J.; Swann, P.F. Human thymine DNA glycosylase binds to apurinic sites in DNA but is displaced by human apurinic endonuclease 1. *J. Biol. Chem.* **1999**, *274*, 67–74. [CrossRef]
61. Baldwin, M.R.; O'Brien, P.J. Defining the functional footprint for recognition and repair of deaminated DNA. *Nucleic Acids Res.* **2012**, *40*, 11638–11647. [CrossRef]
62. Luncsford, P.J.; Manvilla, B.A.; Patterson, D.N.; Malik, S.S.; Jin, J.; Hwang, B.J.; Gunther, R.; Kalvakolanu, S.; Lipinski, L.J.; Yuan, W.; et al. Coordination of MYH DNA glycosylase and APE1 endonuclease activities via physical interactions. *DNA Repair (Amst)* **2013**, *12*, 1043–1052. [CrossRef]
63. Esadze, A.; Rodriguez, G.; Cravens, S.L.; Stivers, J.T. AP-Endonuclease 1 accelerates turnover of human 8-oxoguanine DNA glycosylase by preventing retrograde binding to the abasic-site product. *Biochemistry* **2017**, *56*, 1974–1986. [CrossRef]
64. Popov, A.V.; Grin, I.R.; Dvornikova, A.P.; Matkarimov, B.T.; Groisman, R.; Saparbaev, M.; Zharkov, D.O. Reading targeted DNA damage in the active demethylation pathway: Role of accessory domains of eukaryotic AP endonucleases and thymine-DNA glycosylases. *J. Mol. Biol.* **2020**, (in press). [CrossRef]
65. Gagné, J.P.; Isabelle, M.; Lo, K.S.; Bourassa, S.; Hendzel, M.J.; Dawson, V.L.; Dawson, T.M.; Poirier, G.G. Proteome-wide identification of poly(ADP-ribose) binding proteins and poly(ADP-ribose)-associated protein complexes. *Nucleic Acids Res.* **2008**, *36*, 6959–6976. [CrossRef]
66. Teloni, F.; Altmeyer, M. Readers of poly(ADP-ribose): designed to be fit for purpose. *Nucleic Acids Res.* **2016**, *44*, 993–1006. [CrossRef]

67. Strauss, P.R.; Beard, W.A.; Patterson, T.A.; Wilson, S.H. Substrate binding by human apurinic/apyrimidinic endonuclease indicates a Briggs-Haldane mechanism. *J. Biol. Chem.* **1997**, *272*, 1302–1307. [CrossRef] [PubMed]
68. Kumar, A.; Widen, S.G.; Williams, K.R.; Kedar, P.; Karpel, R.L.; Wilson, S.H. Studies of the domain structure of mammalian DNA polymerase β. Identification of a discrete template binding domain. *J. Biol. Chem.* **1990**, *265*, 2124–2131. [PubMed]
69. Daviet, S.; Couvé-Privat, S.; Gros, L.; Shinozuka, K.; Ide, H.; Saparbaev, M.; Ishchenko, A.A. Major oxidative products of cytosine are substrates for the nucleotide incision repair pathway. *DNA Repair (Amst)* **2007**, *6*, 8–18. [CrossRef] [PubMed]
70. Sukhanova, M.V.; Khodyreva, S.N.; Lavrik, O.I. Poly(ADP-ribose) polymerase-1 inhibits strand-displacement synthesis of DNA catalyzed by DNA polymerase β. *Biochemistry (Mosc)* **2004**, *69*, 558–568. [CrossRef]
71. Belousova, E.A.; Vasil'eva, I.A.; Moor, N.A.; Zatsepin, T.S.; Oretskaya, T.S.; Lavrik, O.I. Clustered DNA lesions containing 5-formyluracil and AP site: repair via the BER system. *PLoS One* **2013**, *8*, e68576. [CrossRef]

International Journal of
Molecular Sciences

MDPI

Article

Molecular Mechanisms of PARP-1 Inhibitor 7-Methylguanine

Dmitry Nilov [1,*], Natalya Maluchenko [2], Tatyana Kurgina [3,4], Sergey Pushkarev [5], Alexandra Lys [2], Mikhail Kutuzov [3], Nadezhda Gerasimova [2], Alexey Feofanov [2,6], Vytas Švedas [1,5], Olga Lavrik [3,4] and Vasily M. Studitsky [2,7,*]

1 Lomonosov Moscow State University, Belozersky Institute of Physicochemical Biology, Lenin Hills 1, bldg. 40, 119991 Moscow, Russia; vytas@belozersky.msu.ru
2 Lomonosov Moscow State University, Biology Faculty, Lenin Hills 1, bldg. 12, 119992 Moscow, Russia; mal_nat@mail.ru (N.M.); lys-alex-bio-msu@yandex.ru (A.L.); shordome@gmail.com (N.G.); avfeofanov@yandex.ru (A.F.)
3 Institute of Chemical Biology and Fundamental Medicine, Siberian Branch of the Russian Academy of Sciences, Lavrentiev avenue 8, 630090 Novosibirsk, Russia; t.a.kurgina@gmail.com (T.K.); kutuzov.mm@mail.ru (M.K.); lavrik@niboch.nsc.ru (O.L.)
4 Novosibirsk State University, Pirogov str. 2, 630090 Novosibirsk, Russia
5 Lomonosov Moscow State University, Faculty of Bioengineering and Bioinformatics, Lenin Hills 1, bldg. 73, 119991 Moscow, Russia; spush.bio@gmail.com
6 Shemyakin-Ovchinnikov Institute of Bioorganic Chemistry, Russian Academy of Sciences, Miklukho-Maklaya str. 16/10, 117997 Moscow, Russia
7 Fox Chase Cancer Center, Cottman Avenue 333, Philadelphia, PA 19111-2497, USA
* Correspondence: nilov@belozersky.msu.ru (D.N.); vasily.studitsky@fccc.edu (V.M.S.)

Received: 28 January 2020; Accepted: 17 March 2020; Published: 20 March 2020

Abstract: 7-Methylguanine (7-MG), a natural compound that inhibits DNA repair enzyme poly(ADP-ribose) polymerase 1 (PARP-1), can be considered as a potential anticancer drug candidate. Here we describe a study of 7-MG inhibition mechanism using molecular dynamics, fluorescence anisotropy and single-particle Förster resonance energy transfer (spFRET) microscopy approaches to elucidate intermolecular interactions between 7-MG, PARP-1 and nucleosomal DNA. It is shown that 7-MG competes with substrate NAD^+ and its binding in the PARP-1 active site is mediated by hydrogen bonds and nonpolar interactions with the Gly863, Ala898, Ser904, and Tyr907 residues. 7-MG promotes formation of the PARP-1–nucleosome complexes and suppresses DNA-dependent PARP-1 automodification. This results in nonproductive trapping of PARP-1 on nucleosomes and likely prevents the removal of genotoxic DNA lesions.

Keywords: 7-methylguanine; poly(ADP-ribose) polymerase 1; inhibitor; nucleosome; trapping; docking; molecular dynamics; fluorescence anisotropy; spFRET microscopy

1. Introduction

Recently, various chemotherapy regimens including inhibitors of DNA repair enzyme poly(ADP-ribose) polymerase 1 (PARP-1), which selectively target BRCA-deficient tumors, have been extensively evaluated [1–4]. The underlying concept is that the cell death can be induced by simultaneous inhibition/inactivation of two key DNA repair molecules, PARP-1 and BRCA (breast cancer susceptibility protein), involved in different pathways of removing DNA lesions. PARP-1 binds to DNA breaks and synthesizes a signal polymer poly(ADP-ribose) (PAR) from NAD^+ molecules to activate the excision repair proteins [5–8]. Upon PARP-1 inhibition, the number of double-strand DNA breaks is increased; however, they can still be removed using homologous recombination (i.e., through an alternative mechanism involving BRCA protein). If BRCA is deficient, genome instability reaches the critical level, resulting in cancer cell death [9–12]. A first-in-class PARP-1 inhibitor, a phthalazine derivative olaparib, was approved by FDA in 2014 and positioned as an innovative drug for the treatment of ovarian cancer in people with hereditary BRCA mutations [1,13,14]. Success of olaparib has inspired further studies of PARP inhibition, but also revealed serious side effects (in particular, hematological toxicity) that accompany the use of synthetic inhibitors [15–19]. The toxicity is likely related to the nonselective interaction with numerous NAD^+-binding proteins (including other PARP family members, such as PARP-2) as well as to nonspecific effects on the organism characteristic for the synthetic molecules.

One way to reduce adverse effects of chemotherapy might be the use of natural PARP inhibitors instead of synthetic compounds. Although strong suppression of PARP seems to be inherently toxic due to an important role played by these proteins in the organism, attempts are continuing to find the proper balance between efficacy of natural inhibitors and their toxicity. For example, PARP inhibitors were identified among caffeine metabolites [20,21] and 2,5-diketopiperazines from chicken essence (a food supplement in Asian countries) [22]. Recently we have shown that a natural nitrogenous base, 7-methylguanine (7-MG), inhibits PARP-1 in vitro and accelerates apoptotic death of BRCA-deficient breast cancer cells induced by cisplatin and doxorubicin [23]. 7-MG has an attractive predicted profile of pharmacokinetics and toxicity and exerts no significant adverse effects on the organism in preliminary in vivo tests [24]. It contains a lactam group built in an aromatic scaffold (a common structural feature of effective PARP inhibitors [25,26]) and is expected to form substrate-specific interactions with the Gly863 and Tyr907 residues in the PARP-1 active site.

In this article, we present the results of a study of the 7-MG inhibition mechanism, which includes molecular modeling of the 7-MG binding to PARP-1, kinetics analysis of its ability to suppress the PARP-1-catalysed synthesis of PAR, and microscopy analysis of interaction between 7-MG, PARP-1 and nucleosomal DNA.

2. Results

2.1. MD Modeling

Molecular dynamics (MD) modeling of the complex of multidomain human PARP-1 with DNA and 7-MG (Figure S1) has been performed for the first time. 7-MG was docked into its putative binding region (binding site of the NAD^+ nicotinamide group), and the obtained complex was subjected to MD simulation in explicit solvent. The system contained 703 amino acid residues, 52 nucleotides, 2 Zn^{2+} ions, 45 Na^+ ions, and 73,302 water molecules (232,862 atoms). Analysis of the 20-ns equilibrium simulation trajectory revealed the following important intermolecular interactions. The lactam group and 2-amino group of 7-MG form hydrogen bonds with the Gly863 residue (Figure 1 and Table 1). 7-Methyl group forms a hydrophobic contact with the Ala898 side chain, and purine rings stack with the Tyr907 side chain. An additional hydrogen bond is formed between the 7-MG lactam group and the Ser904 side chain. This interaction is characterized by an increased mean distance between atoms (Table 1) because Ser904 periodically forms an alternative hydrogen bond with the Trp861 backbone and thus can be observed in two possible conformations (Figure S2a,b). However, the hydrogen bond of Ser904 with 7-MG is more prevalent, with the occupancy of about 60% (Figure S2c). As demonstrated in Figures S3 and S4, the position of 7-MG and its hydrogen bonds with protein were stable during the simulation. Noticeably, we did not observe significant changes in the PARP-1 multidomain organization upon 7-MG binding (compared with the PARP-1–DNA crystal structure); thus the inhibitor apparently does not affect the interaction between PARP-1 and DNA in the case of double-stranded oligonucleotides.

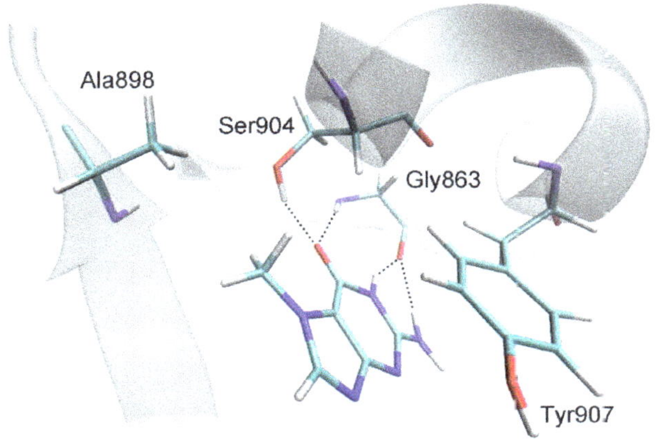

Figure 1. Interactions of 7-MG molecule in the PARP-1 active site revealed by MD simulation: hydrogen bonds with Gly863 and Ser904, π-stacking of purine rings with Tyr907, and hydrophobic contact between the 7-MG methyl group and Ala898.

Table 1. Distances and angles describing 7-MG position in the PARP-1 active site determined by the 20-ns MD simulation. Mean values are presented with standard deviations.

Distance, Å	
7-MG:CO:O ··· Gly863:H	2.0 ± 0.2
7-MG:CO:O ··· Ser904:OG:HG	2.5 ± 0.7
7-MG:NH:H ··· Gly863:O	1.9 ± 0.1
7-MG:NH_2:H ··· Gly863:O	2.4 ± 0.3
7-MG:CH_3:C ··· Ala898:CB	4.0 ± 0.4
C(7-MG fused rings) ··· C(Tyr907 benzene ring) [1]	3.6 ± 0.2
Angle, deg	
7-MG:CO:O ··· Gly863:H ··· Gly863:N	162 ± 11
7-MG:CO:O ··· Ser904:OG:HG ··· Ser904:OG	134 ± 34
7-MG:NH:N ··· 7-MG:NH:H ··· Gly863:O	153 ± 11
7-MG:NH_2:N ··· 7-MG:NH_2:H ··· Gly863:O	138 ± 10

[1] Distance between the geometric center of 7-MG fused rings and the center of the Tyr907 benzene ring.

2.2. Fluorescence Anisotropy Analysis

PARP-1 inhibition by 7-MG was studied using analysis of the fluorescence anisotropy of PARP-1 complexes with a labeled double-stranded oligonucleotide. Upon PARP-1 binding to DNA, the anisotropy level is increased due to a decrease in the fluorophore mobility. The addition of the NAD^+ substrate enables synthesis of negatively charged PAR and automodification of the enzyme, leading to the dissociation of the PARP-1–DNA complexes accompanied by a decrease in anisotropy. The observed dissociation rate is proportional to the PARP-1-catalyzed reaction rate [27].

We have found that 7-MG does not affect the oligonucleotide structure by itself and does not significantly interfere with the binding of PARP-1 to DNA (Figure S5), but inhibits the dissociation of the PARP-1–DNA complex (due to suppression of PARP-1 catalytic activity) Figure 2 shows a typical plot of the reaction rate as a function of 7-MG concentration. The absolute IC_{50} [28] was found to be 162 ± 4 μM at 100 μM NAD^+ concentration. To determine the type of inhibition, we calculated apparent K_M and V_{max} values at various 7-MG concentrations (Figure 3). The 7-MG addition altered only the K_M^{app} leaving the V_{max} value the same, which displays the competitive enzyme inhibition; the corresponding K_i value was found to be 61 ± 9 μM.

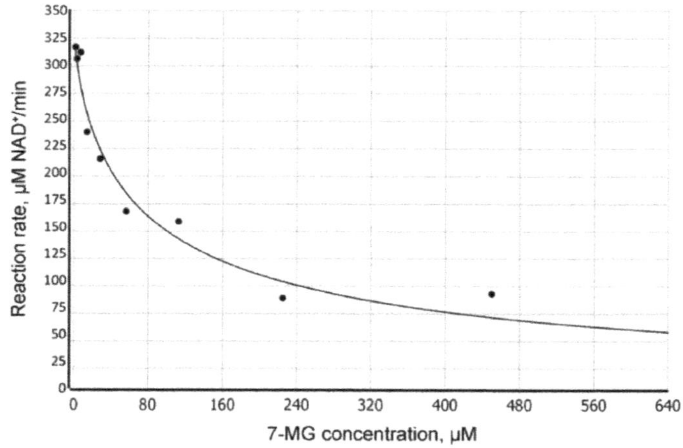

Figure 2. Dependence of the PARP-1-catalyzed reaction rate on the concentration of 7-MG inhibitor determined by fluorescence anisotropy (100 µM NAD^+ concentration).

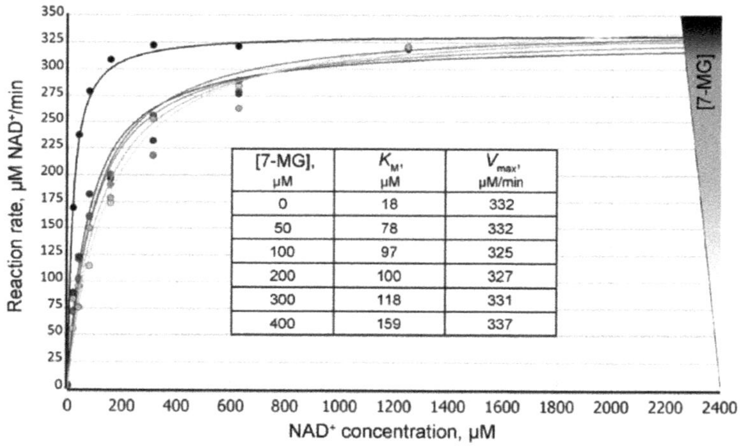

Figure 3. Dependence of the PARP-1-catalyzed reaction rate on the NAD^+ concentration at different concentrations of 7-MG added to the reaction mixture. Insert: calculated $K_M{}^{app}$ values increase with increasing 7-MG concentrations, thus demonstrating the competitive inhibition mechanism.

2.3. *spFRET Analysis*

To model the inhibitor activity in the chromatin environment, we have studied effects of 7-MG on the PARP-1 complexes with fluorescently labeled nucleosomes P147 and P167 using single-particle Förster resonance energy transfer (spFRET) microscopy. Cy3 and Cy5 labels were introduced in the neighboring DNA gyres and served to probe DNA conformation near the entrance of DNA into the nucleosome by measuring FRET efficiency between the labels.

spFRET microscopy revealed the presence of two subpopulations for both P147 and P167 nucleosomes in solution, which differed in FRET efficiency (presented as the proximity ratio E_{PR} [29] in Figure 4). These subpopulations were observed in the calculated E_{PR} profiles of nucleosomes (i.e., frequency distributions of nucleosomes by E_{PR}) as two peaks. In agreement with our previous work [30], the major peaks ($E_{PR} \approx 0.79$ for P147 and $E_{PR} \approx 0.52$ for P167) can be assigned to nucleosome subpopulations with tightly wrapped nucleosomal DNA, while the minor peaks ($E_{PR} \approx 0.03$ for P147 and P167) can be related to nucleosomes with partially unwrapped DNA and/or free DNA. Differences in the E_{PR} profiles of P147 and P167 nucleosomes are related with the presence of the 20-bp linker DNA arm in the P167 that seems to affect both wrapping of DNA on the surface of the histone octamer and DNA "breathing" (temporal spontaneous partial DNA unwrapping) near the entrance of DNA into nucleosome.

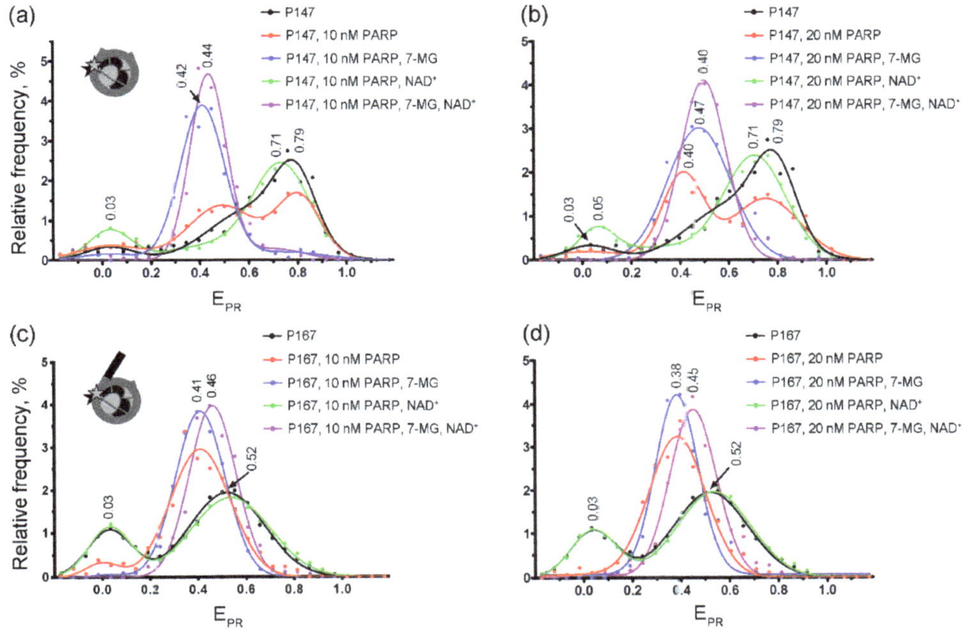

Figure 4. spFRET studies of 7-MG effects at the nucleosome level. Typical frequency distributions of P147 (**a**,**b**) and P167 (**c**,**d**) nucleosomes by E_{PR} in different mixtures are shown. Conditions: 1 nM nucleosomes, 10 nM or 20 nM PARP-1, 100 μM NAD^+, 450 μM 7-MG. Inserts: schemes of nucleosome structure and positions of fluorescent labels (asterisks).

The formation of the complexes between nucleosomes (P147 or P167) and PARP-1 resulted in the appearance of a new peak characterized by $E_{PR} \approx 0.38$–0.41 in the E_{PR} profiles (Figure 4). Relative intensity of this peak indicates that only a part of P147 nucleosomes formed complexes at the studied PARP-1 concentrations (10–20 nM), and this subpopulation was increased after the increase in the PARP-1 concentration. In contrast, nearly all P167 nucleosomes formed the complexes even at 10 nM concentration of PARP-1, indicating higher affinity of PARP-1 to nucleosomes with a linker DNA arm. The PARP-1-induced shift of the main E_{PR} peak to lower values is related to structural changes in nucleosomal DNA, which are accompanied by an increase in the distance between neighboring DNA gyres in the region of Cy3 and Cy5 label position, i.e., at the H2A–H2B interface (+13 bp position) and the H4–H2B interface (+91 bp position) [30]. Although 7-MG alone did not affect the nucleosome structure (Figure S6), its addition to P147 together with PARP-1 resulted in a considerable increase of the fraction of nucleosomes with the E_{PR} distribution having maximum at 0.42–0.47 (Figure 4a,b), thus indicating that (i) 7-MG is involved in the formation of complexes of nucleosomes with PARP-1, (ii) 7-MG only weakly disturbs the structure of nucleosomal DNA in the PARP-1–nucleosome complexes, (iii) 7-MG facilitates PARP-1 binding to nucleosomes. Similarly, spFRET analysis demonstrated the ability of 7-MG to promote PARP-1 binding to P167 nucleosome; however, the effect was less pronounced than in the case of P147 nucleosome (Figure 4c,d).

Incubation of PARP-1–nucleosome complexes with NAD^+ resulted in the changes of the E_{PR} profiles characterized by disappearance of the peak at $E_{PR} \approx 0.38$–0.41 (a signature of nucleosome-PARP-1 complexes) and appearance of the peak in the region of $E_{PR} \approx 0.52$ (for P167) or $E_{PR} \approx 0.71$ (for P147), which was characteristic for intact nucleosomes (Figure 4). The data indicate that, as expected, PARP-1 automodification resulted in its dissociation from nucleosomal DNA and almost complete recovery of the intact nucleosome structure. The presence of 7-MG has blocked the described changes in the E_{PR} profiles of enzyme–nucleosome complexes induced by addition of NAD^+: only a small shift of the peak at $E_{PR} \approx 0.38$–0.47 was observed without appearance of the subpopulation of free nucleosomes (no peaks with a maximum at $E_{PR} \approx 0.52$ for P167 or $E_{PR} \approx 0.71$–0.79 for P147). This effect of 7-MG is likely related to inhibition of PARP-1 enzymatic activity that prevents the automodification of PARP-1 and its dissociation from the nucleosomes.

3. Discussion

In our earlier study, we have modeled a complex of PARP-1 with 7-MG [23]; however, the preliminary data were obtained using an isolated catalytic domain of chicken PARP-1 and might be incomplete. Here we report the MD model of human multidomain PARP-1 complex with DNA fragment and 7-MG that most accurately describes interactions between the inhibitor and PARP-1 bound to a DNA double-strand break. In the refined model, 7-MG formed polar and hydrophobic interactions, in particular hydrogen bonds with Gly863, similar to the NAD^+ substrate and known effective PARP-1 inhibitors [15,25,26]. One important difference from chicken PARP-1 model was the formation of an additional hydrogen bond between 7-MG and Ser904 which apparently contributes to the binding of the inhibitor with human PARP-1. Revealed interactions of 7-MG in the PARP-1 active site and the stability of the enzyme–inhibitor complex during MD simulation suggest that 7-MG occupies the binding site of the NAD^+ nicotinamide group. Kinetics analysis of the ability of 7-MG to suppress PAR synthesis was performed using a recently developed fluorescent method for the real-time measurement of PARP-1 activity [27]; it corroborated conclusions of MD modeling that 7-MG is a competitive PARP-1 inhibitor.

To confirm the inhibitory properties at a more complex level, 7-MG effects have been investigated using spFRET microscopy, an advanced technique allowing analysis of structurally different subpopulations of nucleosomes in heterogeneous samples [31–33]. Mononucleosomes used in our study represented a convenient model of DNA double-strand break in the chromatin environment. Although 7-MG exerted no significant effect on the PARP-1 interaction with isolated double-stranded oligonucleotides (as shown using MD modeling and fluorescence anisotropy experiments), spFRET data showed that it promotes PARP-1 binding to nucleosomes. It seems that the inhibitor stabilizes the interaction between PARP-1 and nucleosomal DNA. We propose that the relative orientation of PARP-1 domains undergoes changes to fit the nucleosome structure, and bound 7-MG stabilizes the new PARP-1 conformation. The formed enzyme–nucleosome complexes are nonproductive because of the concomitant 7-MG inhibition of PARP-1 catalytic activity. PARP-1 molecules fail to regulate their dissociation from DNA via PAR synthesis and get trapped on DNA. The trapped PARP-1 complexes are considered to be even more deleterious for cancer cells than unrepaired DNA strand breaks, because PARP-1 protein tightly bound to DNA interferes with transcription, replication, and DNA repair [34–36]. Noticeably, the most effective PARP-1 inhibitors, including olaparib, display the strong ability to trap PARP [37,38].

In conclusion, the molecular mechanisms of a promising anticancer compound, 7-MG, can be outlined as follows. (1) 7-MG forms substrate-specific interactions in the PARP-1 active site and inhibits the synthesis of PAR, a signal polymer that induces the reorganization of chromatin structure and recruits DNA repair proteins to eliminate the damage. (2) 7-MG inhibits the dissociation of PARP-1 from the DNA damage site in the context of nucleosome and likely prevents further steps in DNA repair, as well as DNA replication and transcription, inducing cancer cell death. Despite the fact that 7-MG is a weaker inhibitor compared to some synthetic PARP-1 inhibitors, we believe that this natural compound has more favorable profile of pharmacokinetics and toxicity, and therefore can be considered as a promising new component of chemotherapy.

4. Materials and Methods

4.1. Molecular Modeling

The molecular model of human PARP-1 bound to DNA was constructed on the basis of the 4dqy crystal structure (chains A, B, C, M, and N) [39,40]. The coordinates of the missing loop 576–583 in the WGR domain were transferred from the 2cr9 structure. The coordinates of the loop 645–661 between the WGR and catalytic domains were predicted with the Modeller 9.20 program [41]. Next, the protein and DNA structure was optimized with AmberTools 15 and Amber 14 [42,43], according to the following protocol. Hydrogen atoms were added to the structure considering ionization of amino acid residues, and then it was solvated by 12 Å-thick layer of TIP3P water; sodium ions were added to neutralize the negative net charge. The two-stage energy minimization was then performed to relax the solvated system. At the first stage (2500 steps of the steepest descent algorithm + 2500 steps of the conjugate gradient algorithm), the DNA coordinates of the protein and DNA were kept fixed by the positional restraints of 2 kcal/(mol·Å^2) on heavy atoms. The second minimization stage (5000 steepest descent steps + 5000 conjugate gradient steps) was carried out without restraints.

To obtain the starting model of the PARP-1–DNA–7-MG complex, water molecules and ions were removed from the optimized PARP-1–DNA complex and then it was subjected to molecular docking with Lead Finder 1.1.16 [44,45]. The 7-MG molecule was docked into the active site of the PARP-1 catalytic domain using the genetic algorithm. The obtained ternary complex was re-solvated and re-optimized using the protocol described above, and subsequently studied through molecular dynamics simulation. The system was heated up from 0 to 300 K with positional restraints of 1 kcal/(mol·Å^2) on the protein, DNA, and 7-MG atoms (250 ps, constant volume) and equilibrated at 300 K (500 ps, constant pressure). Lastly, a 20 ns trajectory of equilibrium simulation was calculated and analyzed. Control data for energy minimization and MD simulation are provided in Table S1. The *ff14SB* force field [46] was used to describe the protein and DNA with molecular mechanics, and recently developed parameters [23] were used to describe the 7-MG molecule. VMD 1.9.2 was used for the visualization of structures [47].

4.2. Fluorescence Anisotropy Assay

PARP-1 protein was obtained from the insect cells using baculovirus expression system. A suspension of Hi5 cells in serum-free medium (2·10^6 cells/mL) was infected with baculovirus (10 pfu/mL) containing cDNA of PARP-1, a kind gift of V. Schreiber (Strasbourg, France). Insect cells were then collected by centrifugation during 10 min at 1000× g. The purification of PARP-1 was performed according to the earlier described protocol [48]. The fluorescein-labeled DNA duplex used in kinetics experiments is provided in Table S2.

Real-time measurements of the PARP-1 activity were based on recently developed fluorescent method [27]. The reaction mixture contained a buffer (50 mM Tris-HCl, pH 8.0, 50 mM NaCl, 1 mM DTT, 5 mM $MgCl_2$), 200 nM PARP-1, 100 nM DNA, and 7-MG (0–450 μM). The reaction was started by adding NAD^+ (100 μM). Fluorescence anisotropy measurements were performed at 25 °C using the CLARIOstar multifunctional microplate reader (BMG LABTECH, Ortenberg, Germany). The fluorescent probes were excited at 495 nm, and the fluorescence intensity was detected at 520 nm. To determine the IC_{50} value, the experiment was done in triplicate. To determine the type of enzyme inhibition, reaction mixtures containing the buffer, 7-MG (0–400 μM) and NAD^+ (0–2500 μM) were used, and the reaction was started by adding a mixture of PARP-1 (200 nM) and DNA (100 nM).

Anisotropy was calculated using formula:

$$A = \frac{(I_{\parallel} - I_{\perp})}{(I_{\parallel} + 2I_{\perp})} \quad (1)$$

where $I_{\parallel}$—fluorescence intensity parallel to the plane-polarized exited light, $I_{\perp}$—fluorescence intensity perpendicular to the light.

4.3. spFRET Microscopy

Mononucleosomes were assembled using 147 or 167 bp DNA fragments containing strong nucleosome positioning sequence 603 (147 bp-long) [49,50]. Nucleosomes P147 contained only the 603 sequence, while P167 included an additional 20-bp linker. Cy3 and Cy5 fluorophore labels (Biotech Industry Ltd., Moscow, Russia) were introduced at +13 and +91 bp (from the beginning of the 603 sequence) to enable spFRET-microscopy detection of changes in the nucleosome structure. Label positions were selected in the neighboring DNA gyres based on the nucleosome crystal structure [51,52]. Sequences of DNA template and fluorescently labeled oligonucleotide primers are shown in Table S3. Nucleosomes were assembled with donor chromatin (lacking H1 histone) from chicken erythrocytes [53,54] using salt dialysis against buffers (10 mM Tris pH 7.5, 0.2 mM EDTA, 0.1% NP-40, 5 mM 2-mercaptoethanol) with decreasing concentrations of NaCl (1, 0.5, 0.2, 0.01 M). The nucleosome assembly was controlled by a native polyacrylamide gel (4%) electrophoresis. Fluorescently-labeled nucleosomes (1 nM) were incubated with PARP-1 for 20 min in a buffer containing 50 mM Tris-HCl pH 8.0, 40 mM NaCl, 1mM DTT at 25 °C in siliconized tubes. To activate PARP-1, NAD^+ was added to final concentration of 100 μM and the mixture was incubated for 15 min. In experiments with 7-MG, PARP-1 was preincubated with 450 μM of the inhibitor for 15 min and mixed with the nucleosomes.

Fluorescence of single nucleosomes and their complexes was measured during their free diffusion through the focus of a laser beam (wavelength of 514.5 nm) with the LSM 710 ConfoCor 3 confocal microscope (Zeiss, Oberkochen, Germany) as described elsewhere [32]. To characterize the distance between Cy3 (donor) and Cy5 (acceptor) dyes in neighboring gyres, a proximity ratio E_{PR} was calculated for each nucleosome:

$$E_{PR} = \frac{(I_{Aa} - \alpha \times I_{Dd})}{[I_{Aa} + (1-\alpha) \times I_{Dd}]} \quad (2)$$

where I_{Aa} indicates the fluorescent intensity of the Cy5 label (Cy5 channel), I_{Dd} indicates intensity of the Cy3 label (Cy3 channel) and α is a coefficient of spectral cross-talk calculated as:

$$\alpha = \frac{I_{Da}}{I_{Dd}} \quad (3)$$

where I_{Da} is a fluorescent intensity of the Cy3 in the Cy5 channel. Data collected from at least 2000 individual particles have been used to plot a relative frequency distribution of nucleosomes by E_{PR}.

Supplementary Materials: Supplementary materials can be found at http://www.mdpi.com/1422-0067/21/6/2159/s1.

Author Contributions: Conceptualization, D.N.; investigation, N.M., T.K., S.P., A.L. and M.K.; methodology, A.F., T.K. and D.N.; writing—original draft preparation, project administration, and funding acquisition, D.N. and N.G.; supervision and writing—review and editing, V.Š., O.L. and V.M.S. All authors have read and agreed to the published version of the manuscript.

Funding: Fluorescence anisotropy studies were funded by the Russian Science Foundation, grant number 19-74-10072; spFRET analysis and molecular modeling were funded by the Russian Foundation for Basic Research, grant number 17-00-00163 (17-00-00132, 17-00-00097).

Acknowledgments: The research was carried out using the equipment of the shared research facilities of HPC computing resources at Lomonosov Moscow State University.

Conflicts of Interest: The authors declare no conflict of interest.

Abbreviations

PARP-1	poly(ADP-ribose)polymerase 1
PAR	poly(ADP-ribose)
7-MG	7-methylguanine
MD	molecular dynamics
spFRET	Single-particle Förster resonance energy transfer

References

1. Frampton, J.E. Olaparib: A review of its use as maintenance therapy in patients with ovarian cancer. *BioDrugs* **2015**, *29*, 143–150. [CrossRef]
2. Mittica, G.; Ghisoni, E.; Giannone, G.; Genta, S.; Aglietta, M.; Sapino, A.; Valabrega, G. PARP inhibitors in ovarian cancer. *Recent Pat. Anticancer Drug Discov.* **2018**, *13*, 392–410. [CrossRef]
3. Zimmer, A.S.; Gillard, M.; Lipkowitz, S.; Lee, J.M. Update on PARP inhibitors in breast cancer. *Curr. Treat. Options Oncol.* **2018**, *19*, 21. [CrossRef]
4. Keung, M.Y.T.; Wu, Y.; Vadgama, J.V. PARP inhibitors as a therapeutic agent for homologous recombination deficiency in breast cancers. *J. Clin. Med.* **2019**, *8*, 435. [CrossRef]
5. Alemasova, E.E.; Lavrik, O.I. Poly(ADP-ribosyl)ation by PARP1: Reaction mechanism and regulatory proteins. *Nucleic Acids Res.* **2019**, *47*, 3811–3827. [CrossRef]
6. Drenichev, M.S.; Mikhailov, S.N. Poly(ADP-ribose)–a unique natural polymer structural features, biological role and approaches to the chemical synthesis. *Nucleosides Nucleotides Nucleic Acids* **2015**, *34*, 258–276. [CrossRef]
7. Hassler, M.; Ladurner, A.G. Towards a structural understanding of PARP1 activation and related signalling ADP-ribosyl-transferases. *Curr. Opin. Struct. Biol.* **2012**, *22*, 721–729. [CrossRef]
8. Ray Chaudhuri, A.; Nussenzweig, A. The multifaceted roles of PARP1 in DNA repair and chromatin remodeling. *Nat. Rev. Mol. Cell Biol.* **2017**, *18*, 610–621. [CrossRef]
9. Bryant, H.E.; Schultz, N.; Thomas, H.D.; Parker, K.M.; Flower, D.; Lopez, E.; Kyle, S.; Meuth, M.; Curtin, N.J.; Helleday, T. Specific killing of BRCA2-deficient tumours with inhibitors of poly(ADP-ribose) polymerase. *Nature* **2005**, *434*, 913–917. [CrossRef]
10. Martin, S.A.; Lord, C.J.; Ashworth, A. DNA repair deficiency as a therapeutic target in cancer. *Curr. Opin. Genet. Dev.* **2008**, *18*, 80–86. [CrossRef]
11. Ferraris, D.V. Evolution of poly(ADP-ribose) polymerase-1 (PARP-1) inhibitors. From concept to clinic. *J. Med. Chem.* **2010**, *53*, 4561–4584. [CrossRef] [PubMed]
12. Lord, C.J.; Tutt, A.N.; Ashworth, A. Synthetic lethality and cancer therapy: Lessons learned from the development of PARP inhibitors. *Annu. Rev. Med.* **2015**, *66*, 455–470. [CrossRef] [PubMed]
13. Heo, Y.A.; Dhillon, S. Olaparib tablet: A review in ovarian cancer maintenance therapy. *Target Oncol.* **2018**, *13*, 801–808. [CrossRef] [PubMed]
14. Caulfield, S.E.; Davis, C.C.; Byers, K.F. Olaparib: A novel therapy for metastatic breast cancer in patients with a BRCA1/2 mutation. *J. Adv. Pract. Oncol.* **2019**, *10*, 167–174. [PubMed]
15. Jain, P.G.; Patel, B.D. Medicinal chemistry approaches of poly ADP-ribose polymerase 1 (PARP1) inhibitors as anticancer agents-a recent update. *Eur. J. Med. Chem.* **2019**, *165*, 198–215. [CrossRef]
16. Malyuchenko, N.V.; Kotova, E.Y.; Kulaeva, O.I.; Kirpichnikov, M.P.; Studitskiy, V.M. PARP1 inhibitors: Antitumor drug design. *Acta Naturae* **2015**, *7*, 27–37. [CrossRef]
17. Sonnenblick, A.; de Azambuja, E.; Azim, H.A., Jr.; Piccart, M. An update on PARP inhibitors–moving to the adjuvant setting. *Nat. Rev. Clin. Oncol.* **2015**, *12*, 27–41. [CrossRef]
18. Wang, Y.Q.; Wang, P.Y.; Wang, Y.T.; Yang, G.F.; Zhang, A.; Miao, Z.H. An update on poly(ADP-ribose)polymerase-1 (PARP-1) inhibitors: Opportunities and challenges in cancer therapy. *J. Med. Chem.* **2016**, *59*, 9575–9598. [CrossRef]
19. Ohmoto, A.; Yachida, S. Current status of poly(ADP-ribose) polymerase inhibitors and future directions. *Onco Targets Ther.* **2017**, *10*, 5195–5208. [CrossRef]
20. Geraets, L.; Moonen, H.J.; Wouters, E.F.; Bast, A.; Hageman, G.J. Caffeine metabolites are inhibitors of the nuclear enzyme poly(ADP-ribose)polymerase-1 at physiological concentrations. *Biochem. Pharmacol.* **2006**, *72*, 902–910. [CrossRef]

21. Geraets, L.; Haegens, A.; Weseler, A.R.; Brauers, K.; Vernooy, J.H.; Wouters, E.F.; Bast, A.; Hageman, G.J. Inhibition of acute pulmonary and systemic inflammation by 1,7-dimethylxanthine. *Eur. J. Pharmacol.* **2010**, *629*, 132–139. [CrossRef] [PubMed]
22. Nilov, D.K.; Yashina, K.I.; Gushchina, I.V.; Zakharenko, A.L.; Sukhanova, M.V.; Lavrik, O.I.; Švedas, V.K. 2,5-Diketopiperazines: A new class of poly(ADP-ribose)polymerase inhibitors. *Biochemistry (Mosc.)* **2018**, *83*, 152–158. [CrossRef]
23. Nilov, D.K.; Tararov, V.I.; Kulikov, A.V.; Zakharenko, A.L.; Gushchina, I.V.; Mikhailov, S.N.; Lavrik, O.I.; Švedas, V.K. Inhibition of poly(ADP-ribose) polymerase by nucleic acid metabolite 7-methylguanine. *Acta Naturae* **2016**, *8*, 108–115. [CrossRef] [PubMed]
24. Nilov, D.; Kirsanov, K.; Antoshina, E.; Maluchenko, N.; Feofanov, A.; Kurgina, T.; Zakharenko, A.; Khodyreva, S.; Gerasimova, N.; Studitsky, V.; et al. 7-Methylguanine: A natural DNA repair inhibitor and a promising anticancer compound. *FEBS Open Bio* **2018**, *8*. [CrossRef]
25. Ruf, A.; de Murcia, G.; Schulz, G.E. Inhibitor and NAD^+ binding to poly(ADP-ribose) polymerase as derived from crystal structures and homology modeling. *Biochemistry* **1998**, *37*, 3893–3900. [CrossRef] [PubMed]
26. Jagtap, P.; Szabó, C. Poly(ADP-ribose) polymerase and the therapeutic effects of its inhibitors. *Nat. Rev. Drug Discov.* **2005**, *4*, 421–440. [CrossRef]
27. Kurgina, T.A.; Anarbaev, R.O.; Sukhanova, M.V.; Lavrik, O.I. A rapid fluorescent method for the real-time measurement of poly(ADP-ribose) polymerase 1 activity. *Anal. Biochem.* **2018**, *545*, 91–97. [CrossRef]
28. Sebaugh, J.L. Guidelines for accurate EC50/IC50 estimation. *Pharm. Stat.* **2011**, *10*, 128–134. [CrossRef]
29. Valieva, M.E.; Armeev, G.A.; Kudryashova, K.S.; Gerasimova, N.S.; Shaytan, A.K.; Kulaeva, O.I.; McCullough, L.L.; Formosa, T.; Georgiev, P.G.; Kirpichnikov, M.P.; et al. Large-scale ATP-independent nucleosome unfolding by a histone chaperone. *Nat. Struct. Mol. Biol.* **2016**, *23*, 1111–1116. [CrossRef]
30. Sultanov, D.C.; Gerasimova, N.S.; Kudryashova, K.S.; Maluchenko, N.V.; Kotova, E.Y.; Langelier, M.F.; Pascal, J.M.; Kirpichnikov, M.P.; Feofanov, A.V.; Studitsky, V.M. Unfolding of core nucleosomes by PARP-1 revealed by spFRET microscopy. *AIMS Genet.* **2017**, *4*, 21–31. [CrossRef]
31. Buning, R.; van Noort, J. Single-pair FRET experiments on nucleosome conformational dynamics. *Biochimie* **2010**, *92*, 1729–1740. [CrossRef] [PubMed]
32. Kudryashova, K.S.; Chertkov, O.V.; Nikitin, D.V.; Pestov, N.A.; Kulaeva, O.I.; Efremenko, A.V.; Solonin, A.S.; Kirpichnikov, M.P.; Studitsky, V.M.; Feofanov, A.V. Preparation of mononucleosomal templates for analysis of transcription with RNA polymerase using spFRET. *Methods Mol. Biol.* **2015**, *1288*, 395–412. [PubMed]
33. Lyubitelev, A.V.; Kudryashova, K.S.; Mikhaylova, M.S.; Malyuchenko, N.V.; Chertkov, O.V.; Studitsky, V.M.; Feofanov, A.V.; Kirpichnikov, M.P. Change in linker DNA conformation upon histone H1.5 binding to nucleosome: Fluorescent microscopy of single complexes. *Moscow Univ. Biol. Sci. Bull.* **2015**, *71*, 108–113. [CrossRef]
34. Murai, J.; Huang, S.Y.; Das, B.B.; Renaud, A.; Zhang, Y.; Doroshow, J.H.; Ji, J.; Takeda, S.; Pommier, Y. Trapping of PARP1 and PARP2 by clinical PARP inhibitors. *Cancer Res.* **2012**, *72*, 5588–5599. [CrossRef] [PubMed]
35. Shen, Y.; Aoyagi-Scharber, M.; Wang, B. Trapping poly(ADP-ribose) polymerase. *J. Pharmacol. Exp. Ther.* **2015**, *353*, 446–457. [CrossRef]
36. Pommier, Y.; O'Connor, M.J.; de Bono, J. Laying a trap to kill cancer cells: PARP inhibitors and their mechanisms of action. *Sci. Transl. Med.* **2016**, *8*, 362ps17. [CrossRef]
37. Lord, C.J.; Ashworth, A. PARP inhibitors: Synthetic lethality in the clinic. *Science* **2017**, *355*, 1152–1158. [CrossRef]
38. Gerasimova, N.S.; Valieva, M.E.; Sultanov, D.C.; Kudryashova, K.S.; Maluchenko, N.V.; Kotova, E.S.; Kirpichnikov, M.P.; Studitsky, V.M.; Feofanov, A.V. Complexes of nucleosomal nanoparticles with proteins: spFRET microscopy study of olaparib and PARP-1 binding to core nucleosomes. In *Microscopy and Imaging Science: Practical Approaches to Applied Research and Education*; Méndez-Vilas, A., Ed.; Formatex Research Center: Badajoz, Spain, 2017; pp. 55–61.
39. Langelier, M.F.; Planck, J.L.; Roy, S.; Pascal, J.M. Structural basis for DNA damage-dependent poly(ADP-ribosyl)ation by human PARP-1. *Science* **2012**, *336*, 728–732. [CrossRef]
40. Langelier, M.F.; Eisemann, T.; Riccio, A.A.; Pascal, J.M. PARP family enzymes: Regulation and catalysis of the poly(ADP-ribose) posttranslational modification. *Curr. Opin. Struct. Biol.* **2018**, *53*, 187–198. [CrossRef]
41. Sali, A.; Blundell, T.L. Comparative protein modelling by satisfaction of spatial restraints. *J. Mol. Biol.* **1993**, *234*, 779–815. [CrossRef]

42. Case, D.A.; Berryman, J.T.; Betz, R.M.; Cerutti, D.S.; Cheatham, T.E., 3rd; Darden, T.A.; Duke, R.E.; Giese, T.J.; Gohlke, H.; Goetz, A.W.; et al. *AMBER 2015*; University of California: San Francisco, CA, USA, 2015; pp. 1–883.
43. Salomon-Ferrer, R.; Case, D.A.; Walker, R.C. An overview of the Amber biomolecular simulation package. *WIREs Comput. Mol. Sci.* **2013**, *3*, 198–210. [CrossRef]
44. Stroganov, O.V.; Novikov, F.N.; Stroylov, V.S.; Kulkov, V.; Chilov, G.G. Lead finder: An approach to improve accuracy of protein-ligand docking, binding energy estimation, and virtual screening. *J. Chem. Inf. Model.* **2008**, *48*, 2371–2385. [CrossRef]
45. Zakharenko, A.L.; Sukhanova, M.V.; Khodyreva, S.N.; Novikov, F.N.; Stroylov, V.S.; Nilov, D.K.; Chilov, G.G.; Švedas, V.K.; Lavrik, O.I. Improved procedure of the search for poly(ADP-Ribose) polymerase-1 potential inhibitors with the use of the molecular docking approach. *Mol. Biol. (Mosc.)* **2011**, *45*, 517–521. [CrossRef]
46. Maier, J.A.; Martinez, C.; Kasavajhala, K.; Wickstrom, L.; Hauser, K.E.; Simmerling, C. ff14SB: Improving the accuracy of protein side chain and backbone parameters from ff99SB. *J. Chem. Theory Comput.* **2015**, *11*, 3696–3713. [CrossRef]
47. Humphrey, W.; Dalke, A.; Schulten, K. VMD: Visual molecular dynamics. *J. Mol. Graph.* **1996**, *14*, 33–38. [CrossRef]
48. Amé, J.C.; Kalisch, T.; Dantzer, F.; Schreiber, V. Purification of recombinant poly(ADP-ribose) polymerases. *Methods Mol. Biol.* **2011**, *780*, 135–152.
49. Kulaeva, O.I.; Gaykalova, D.A.; Pestov, N.A.; Golovastov, V.V.; Vassylyev, D.G.; Artsimovitch, I.; Studitsky, V.M. Mechanism of chromatin remodeling and recovery during passage of RNA polymerase II. *Nat. Struct. Mol. Biol.* **2009**, *16*, 1272–1278. [CrossRef]
50. Gaykalova, D.A.; Kulaeva, O.I.; Volokh, O.; Shaytan, A.K.; Hsieh, F.K.; Kirpichnikov, M.P.; Sokolova, O.S.; Studitsky, V.M. Structural analysis of nucleosomal barrier to transcription. *Proc. Natl. Acad. Sci. USA* **2015**, *112*, E5787–E5795. [CrossRef]
51. Vasudevan, D.; Chua, E.Y.D.; Davey, C.A. Crystal structures of nucleosome core particles containing the '601' strong positioning sequence. *J. Mol. Biol.* **2010**, *403*, 1–10. [CrossRef]
52. Morozov, A.V.; Fortney, K.; Gaykalova, D.A.; Studitsky, V.M.; Widom, J.; Siggia, E.D. Using DNA mechanics to predict in vitro nucleosome positions and formation energies. *Nucleic Acids Res.* **2009**, *37*, 4707–4722. [CrossRef]
53. Ausio, J.; Dong, F.; van Holde, K.E. Use of selectively trypsinized nucleosome core particles to analyze the role of the histone "tails" in the stabilization of the nucleosome. *J. Mol. Biol.* **1989**, *206*, 451–463. [CrossRef]
54. Kornberg, R.D.; LaPointe, J.W.; Lorch, Y. Preparation of nucleosomes and chromatin. *Methods Enzymol.* **1989**, *170*, 3–14. [PubMed]

International Journal of
Molecular Sciences

MDPI

Article

Design, Synthesis and Molecular Modeling Study of Conjugates of ADP and Morpholino Nucleosides as A Novel Class of Inhibitors of PARP-1, PARP-2 and PARP-3

Yuliya V. Sherstyuk [1], Nikita V. Ivanisenko [2], Alexandra L. Zakharenko [1], Maria V. Sukhanova [1], Roman Y. Peshkov [3], Ilia V. Eltsov [3], Mikhail M. Kutuzov [1], Tatiana A. Kurgina [1], Ekaterina A. Belousova [1], Vladimir A. Ivanisenko [2], Olga I. Lavrik [1], Vladimir N. Silnikov [1] and Tatyana V. Abramova [1,*]

1 Institute of Chemical Biology and Fundamental Medicine SB RAS, Lavrent'ev Ave, 8, 630090 Novosibirsk, Russia; yuliya.tarasenko2012@gmail.com (Y.V.S); sashaz@niboch.nsc.ru (A.L.Z.); mary@niboch.nsc.ru (M.V.S.); kutuzov.mm@mail.ru (M.M.K.); t.a.kurgina@gmail.com (T.A.K.); rina@niboch.nsc.ru (E.A.B.); lavrik@niboch.nsc.ru (O.I.L.); silnik@niboch.nsc.ru (V.N.S.)

2 Federal Research Centre, Institute of Cytology and Genetics SB RAS, Lavrent'ev Ave, 10, 630090 Novosibirsk, Russia; n.ivanisenko@gmail.com (N.V.I.); salix@bionet.nsc.ru (V.A.I.)

3 Novosibirsk State University, Pirogova St., 2, 630090 Novosibirsk, Russia; peshkov@nioch.nsc.ru (R.Y.P.); eiv@fen.nsu.ru (I.V.E.)

* Correspondence: abramova@niboch.nsc.ru

Received: 29 November 2019; Accepted: 24 December 2019; Published: 27 December 2019

Abstract: We report on the design, synthesis and molecular modeling study of conjugates of adenosine diphosphate (ADP) and morpholino nucleosides as potential selective inhibitors of poly(ADP-ribose)polymerases-1, 2 and 3. Sixteen dinucleoside pyrophosphates containing natural heterocyclic bases as well as 5-haloganeted pyrimidines, and mimicking a main substrate of these enzymes, nicotinamide adenine dinucleotide (NAD+)-molecule, have been synthesized in a high yield. Morpholino nucleosides have been tethered to the β-phosphate of ADP via a phosphoester or phosphoramide bond. Screening of the inhibiting properties of these derivatives on the autopoly(ADP-ribosyl)ation of PARP-1 and PARP-2 has shown that the effect depends upon the type of nucleobase as well as on the linkage between ADP and morpholino nucleoside. The 5-iodination of uracil and the introduction of the P–N bond in NAD+-mimetics have shown to increase inhibition properties. Structural modeling suggested that the P–N bond can stabilize the pyrophosphate group in active conformation due to the formation of an intramolecular hydrogen bond. The most active NAD+ analog against PARP-1 contained 5-iodouracil 2′-aminomethylmorpholino nucleoside with IC50 126 ± 6 μM, while in the case of PARP-2 it was adenine 2′-aminomethylmorpholino nucleoside (IC50 63 ± 10 μM). In silico analysis revealed that thymine and uracil-based NAD+ analogs were recognized as the NAD+-analog that targets the nicotinamide binding site. On the contrary, the adenine 2′-aminomethylmorpholino nucleoside-based NAD+ analogs were predicted to identify as PAR-analogs that target the acceptor binding site of PARP-2, representing a novel molecular mechanism for selective PARP inhibition. This discovery opens a new avenue for the rational design of PARP-1/2 specific inhibitors.

Keywords: morpholino nucleosides; molecular modeling; NAD+ analogs; DNA repair; PARP

1. Introduction

Poly(ADP-ribose)polymerases (PARP; EC 2.4.2.30) belong to a family of eukaryotic proteins with diverse cellular functions mainly related to DNA repair, maintenance of genomic stability and cell death. The PARP family members [1] perform the catalytic activity of transferring ADP-ribose from the beta-nicotinamide adenine dinucleotide molecule (NAD+) to the acceptors; however, only three of them PARP-1, PARP-2 and PARP-3 possess DNA-dependent (ADP-ribose)transferase activity [1,2]. PARP-1 and PARP-2 catalyze the synthesis of a long stretch of poly(ADP-ribose polymers, whereas PARP-3 is a mono(ADP-ribose)transferase [3]. Acceptors of ADP-ribose, which is synthesized by PARP-1–3, can be both proteins and DNA [4–6]. The catalytic domain (CAT) of PARP-1–3 consists of two subdomains, a helical domain (HD) and the (ADP-ribosyl)transferase (ART) domain. This ART domain contains the active site, and is highly conserved in all members of the PARP family [7]. HD is an autoinhibitory domain to the PARP catalytic activity, and is conserved among DNA damage-dependent PARPs including PARP-1, PARP-2 and PARP-3 [7]. Due to a high homology between the catalytic domains of PARP-1 and PARP-2, small molecule inhibitors of PARP-1 usually possess the inhibitory affect to PARP-2 [8].

PARP-3 has two distinct functions: (a) participation in double strand break (DSB) repair pathway(s) and (b) the regulation of mitotic progression [9]. PARP-3 inhibitors sensitize breast cancer cells to vinorelbine, a vinca alkaloid used for the treatment of metastatic breast cancer [10]. Thus, there is a need for selective inhibitors of PARP-3, both for probing the functions of PARP-3 as an enzyme and evaluating its potential as a therapeutic target.

PARPs have been used as pharmacological targets for the treatment of different types of tumors with defects in DNA repair using synthetic lethality. PARP inhibition is considered a promising cancer treatment strategy, and a number of PARP inhibitors are currently undergoing clinical trials; for a review, see [11]. Four PARPi, olaparib, rucaparib, niraparib and talazoparib, have already been approved in the U.S. or Europe, mainly for the treatment of BRCA-deficient cancer [12–14].

Nicotinamide adenine dinucleotide (Figure 1) plays a key role in such vital processes, as maintaining the integrity of the genome, energy supply, cell death and others [15]. NAD+-metabolizing enzymes are considered by researchers as targets for the treatment of a variety of human diseases, including cancer, multiple sclerosis, neurodegeneration and Huntington's disease [16]. Nicotinamide, which is one of the structure elements of NAD+ and one of the major products of ADP-ribozylation, inhibits PARP-1 [17]. A large number of existed so far enzyme inhibitors, including some approved by the United States Food and Drug Administration (FDA), have been developed on the basis of nicotinamide [18]. Binding features of nicotinamide analogs include the formation of hydrogens bonds between the lactam or the carboxamide group of the inhibitor with the Gly863 of backbone chain and Ser904 of the side chain, as well as pi-stacking interactions with Tyr907 [19–21]. In spite of a great amount of developed analogs of nicotinamide for PAPR-1–3 inhibition, there is an increasing interest in creating novel ones, as evidenced by the growing number of publications in scientific journals [22–24]. However, a pharmaceutical usage of existing PARPi is still limited for a number of reasons: the presence of side effects, the appearance of tumor resistance to PARPi and the different efficacy in combination with traditional chemotherapy [14,25]. Therefore, the development of new structural classes of compounds with improved properties, such as selectivity to different types of PARP enzymes, increased inhibitor efficacy, lessened toxicity and higher bioavailability, remains an urgent task.

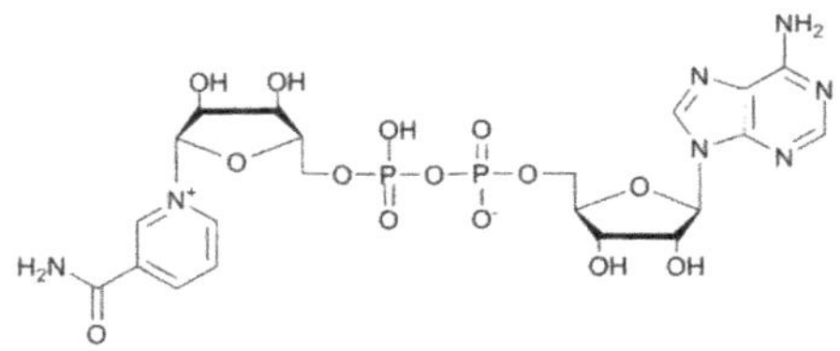

Figure 1. The structure of the beta-oxidized nicotinamide adenine dinucleotide (NAD+)-molecule.

Natural and modified nucleosides and their phosphorylated derivatives are indispensable tools in searching new anticancer, antiviral and antibacterial drugs [26,27]. However, a small number of studies on nucleoside derivatives as PARPi have been reported in the literature, although some of them exhibit moderate PARP-1 inhibition activity [28]. It has been shown that some thymidine derivatives modified at the 5- and/or 5′-position inhibited PARP-1 more efficiently than 3-aminobenzamide, which is the first generation of the PARP-1 inhibitors and the closest structural analog of nicotinamide [29]. Among phosphorylated adenosines (AMP, ADP, ATP, 3′,5′-cycloAMP, 3′,5′-diphosphoadenosine) a 3′,5′-diphosphoadenosine inhibits PARP-1 most effectively [30]. Ueda's group show that 5-substituted derivatives of uridine and uracil exhibited an inhibitory activity against PARP-1 too [31]. Additionally, a number of small molecules containing uracil derivatives could be seen as a class of potent PARP-1 inhibitors [32]. There are known PARP-1 inhibitors with a high activity constructed on the base of isoindolinone derivatives and their conjugates with adenosine joined by various aliphatic spacers [33]. This fact may be explained by the dual binding of both parts of the inhibitor, nicotinamide- and adenosine-mimicking. According to the literature data, compounds binding simultaneously into the nicotinamide and adenosine binding sites of the active center of PARP family enzymes could be more selective inhibitors [34].

The NAD+ molecule may be modified at adenine or nicotinamide heterocyclic bases, ribose moieties or the pyrophosphate chain. The investigation of the applicability of fluorescent NAD+ analogs modified at different positions of the adenine moiety has been performed [35]. Recently, fluorescent NAD+ derivatives have been used for real-time cellular imaging of protein poly(ADP-ribosyl)ation [36]. Clickable NAD+ analogs modified at C8 of adenine are promising tools to illuminate the ADP-ribosylated proteome and investigate the molecular mechanisms carried out by individual PARPs upon different cellular signals [37]. Moreover, the well-known non-hydrolyzable NAD+ analog of benzamide adenine dinucleotide (BAD) has been successfully applied to establish the mechanism of PARP-1 interaction with DNA substrate [38]. All these studies demonstrate the power of modified NAD+ derivatives as potential tools for goals of molecular and cellular biology.

Previously, some naturally occurring dinucleoside 5′,5′-pyrophosphates were investigated as potential PARP-1 inhibitors [39]. Diadenosine 5′,5′-tetraphosphate was shown to inhibit the ADP-ribosylation of histone H1 by PARP from the bovine thymus [40]. However, there are published significantly fewer modern investigations devoted to the study of dinucleoside pyrophosphates or NAD+ derivatives as inhibitors of enzymes using NAD+ as a substrate and not as a cofactor. These were found to inhibit NAD kinase [41], bacterial DNA ligases [42] and CD38 NAD glycohydrolase [43]. Recently, we have developed a number of NAD+ mimetics presented by conjugates of ADP and nicotinamide riboside analogs [44,45].

It was shown that some of those provided an inhibitory effect to the autopoly(ADP-ribosyl)ation of PARP-1; IC_{50} of the most effective conjugate consisting of ADP and morpholino-glycine thymine nucleoside [46] was 41.5 ± 3.5 μM [45]. The combination of data from in silico models of PARP-1 with the NAD+-molecule [47] and the crystal structure of the CAT-ΔHD domain of PARP-1 with the non-hydrolyzable NAD+ analog of BAD [38] suggests the requirement of conducting a rational searching for the details of mechanisms of action and specificity of new NAD+ analogs, being the conjugates of ADP and morpholino nucleosides.

Morpholino nucleosides are widely used for the synthesis of oligonucleotide mimetics [48]. However, there are only few examples of the utilization of morpholino nucleoside monomers or their phosphorylated derivatives for the goal of molecular biology and biochemistry [49,50]. In the present study we describe the synthesis, inhibition activity and the molecular modeling of the novel conjugates combining from ADP and morpholino nucleosides (morpholino nucleoside adenine dinucleotides, MorXppA), where morpholino nucleosides mimic the nicotinamide riboside fragment of the NAD+ molecule (Figure 2).

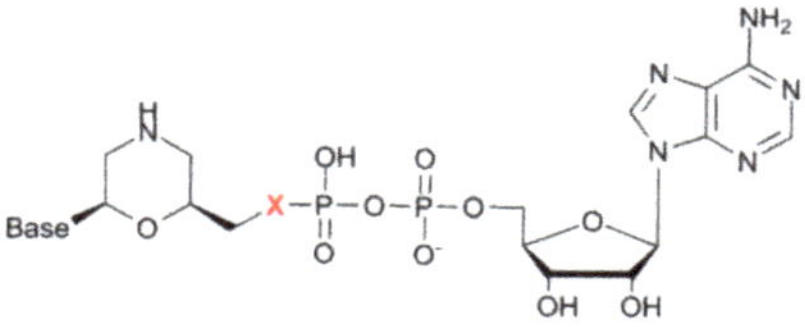

Figure 2. Morpholino nucleoside adenosine dinucleotides (MorXppA).

2. Results and Discussion

2.1. Chemistry

2.1.1. Synthesis of ADP Conjugates Containing Phosphoester (P–O) Bond

Recently we have proposed a versatile method for the synthesis of ADP conjugates functionalized at the terminal phosphate [44,45]. In that study, we developed an effective protocol for the coupling of two monoester phosphate derivatives under the action of the redox coupling pair triphenylphosphine/2,2′-dipyridyldisulfide ($Ph_3P/(PyS)_2$) in the presence of 1-methylimidazole (MeIm). We applied the same strategy for the synthesis of the sought-for pyrophosphates MorXppA **4**, where X represents an oxygen atom (Figure 2, Scheme 1).

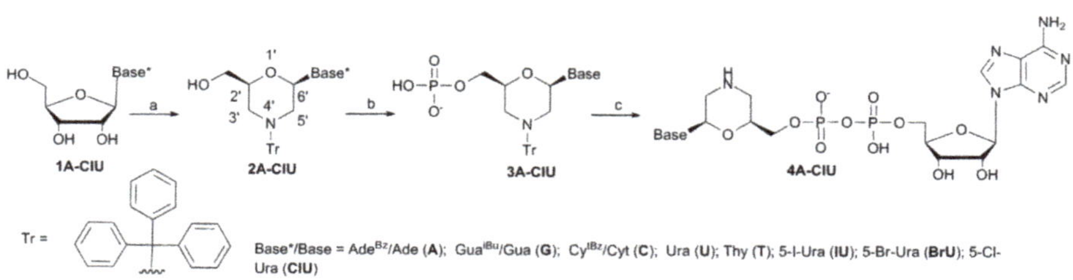

Scheme 1. Synthesis of the conjugates of adenosine diphosphate (ADP) with 2′-hydroxymethylmorpholino nucleosides. Reagents and conditions: (a) $NaIO_4$, $EtOH/H_2O$, 15 min; $(NH_4)_2B_4O_7{\cdot}4H_2O$, Et_3N, 1.5 h; $NaBH_3CN$, 40 min; trifluoroacetic acid (TFA), pH 3–4, 1 h; TrCl, Et_3N, dimethylformamide (DMF), 3 h; yield 60%–70%; (b) $POCl_3$, Py, –15 °C, 15 min; 1 M triethylammonium bicarbonate (TEAB), yield 75%–90%; (c) $Ph_3P/(PyS)_2$, MeIm, 1,3-dimethyl-2-imidazolidinone (DMI); n-Bu_3NH^+ salt of AMP; conc. aq. NH_3 for compounds **4A,G,C**; 80% aq. AcOH (v/v); yield 70%–80%.

At the first step, we provided preparation of *N*-Tr-protected morpholino nucleosides. The standard protocol for nucleosides with natural heterocyclic bases was Summerton's procedure [51]. Recently we have published a detailed protocol for the synthesis of 5-iodopyrimidine morpholino nucleosides based on this method [52]. Continuing our research, we have succeeded in obtaining *N*-Tr-protected morpholino nucleosides **2** from acyl-*N*-base-protected (when necessary) ribonucleosides **1** without chromatographic purification in an overall yield of 60%–70%.

To obtain monophosphate **3U**, we treated protected morpholino nucleoside **2U** with 2 eq of $POCl_3$ in dry pyridine (Py) under cooling in ice bath according to [53]. Unfortunately, the product **3U** was obtained in a yield of less than 50% after reverse phase chromatography (RPC). We varied the excess of $POCl_3$ (2–4 eq), the temperature (0 –−15 °C) and reaction time (5–30 min) and found optimal conditions for the monophosphorylation of nucleosides **2** (Scheme 1). The monophoshates **3A,G,C,IU** were obtained without chromatographic purification in a yield of over 90%; and monophoshates **3U,T,BrU,ClU** in a yield of 75%–80% after RPC.

To prepare ADP conjugates **4**, we activated a phosphate group in derivatives **3** by $Ph_3P/(PyS)_2$ in a presence of MeIm in dry 1,3-dimethyl-2-imidazolidinone (DMI) at room temperature for 15–20 min according to [44]. Further coupling of phosphoro(*N*-methyl)imidazolidate of morpholino nucleoside with AMP was carried out in situ in dry DMI for 1 h. After chromatographic purification and deblocation procedures, we obtained target conjugates **4** in an overall yield of 70%–80% (Scheme 1).

2.1.2. Synthesis of ADP Conjugates Containing Phosphoramide (P–N) Bond

For the synthesis of these conjugates we prepared 2′-aminomethylmorpholino nucleosides **7** (Scheme 2).

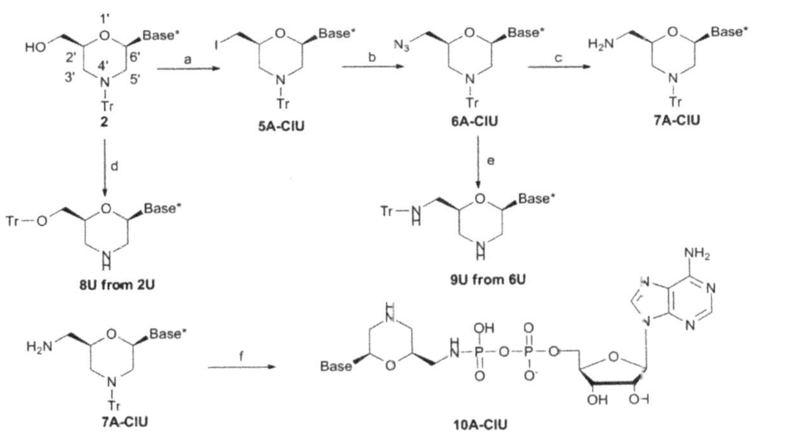

Base*/Base = Ade^{Bz}/Ade (**A**); Gua^{iBu}/Gua (**G**); Cy^{tBz}/Cyt (**C**); Ura (**U**); Thy (**T**); 5-I-Ura (**IU**); 5-Br-Ura (**BrU**); 5-Cl-Ura (**ClU**)

Scheme 2. Synthesis of the conjugates of ADP with 2′-aminomethylmorpholino nucleosides. Reagents and conditions: (a) Ph_3P, Im, DCE, I_2, 0 °C → rt, 5 h; (b) NaN_3, DMF, 12 h; (c) H_2, 10% Pd/C, MeOH; (d) Ph_3P, CBr_4, DMI; (e) Ph_3P (2 eq), Py; conc. aq. NH_3; (f) ADP n-Bu_3N salt, Ph_3P, $(PyS)_2$, MeIm, DMI; conc. aq. NH_3 for compounds **10A,G,C**; 80% aq. AcOH (*v*/*v*).

An azido group is a convenient source for amino functions. In the literature, several methods for obtaining 5′-azidoribonucleosides containing natural heterocyclic bases are known. Methods are based on conversion of the hydroxyl group to halogen [54–59], sulfonate [60–62] or phosphate ester group [63] under anhydrous conditions. The further reaction with lithium or sodium azide leads to the formation of 5′-azidoribosyl derivatives in an overall yield of 14%–92% Similar approaches were proposed for the synthesis of 2′-aminomethylmorpholino nucleosides containing natural heterocyclic bases. One of them is based on the one-pot synthesis of 2′-azidomethylmorpholino nucleosides under the treatment of compounds **2A,G,C,U,T** with $Ph_3P/CBrCl_3/NaN_3$ in dry dimethylformamide (DMF). After reduction of the N_3-group and a full deblocation procedure, 2′-aminomethylmorpholino nucleosides were obtained in an overall yield of 39%–64% [46,64].

Another procedure is based on obtaining 2′-sulfonylmethylmorpholino nucleosides by the treatment of compounds **2A,G,C,T** with mesyl chloride in dry pyridine and a further substitution of mesyl with an azide anion. After the reduction of the N_3-group 2′-aminomethylmorpholino nucleosides were obtained in an overall yield of 34%–45% [65].

A literature search for the synthesis of 5′-azidonucleosides containing 5-halosubstituted pyrimidines provided only a few reports. Most of suggested methods are based on the ribosylation of modified heterocyclic bases with 5′-azidosugar derivatives [66–68]. In 1976, Prusoff et al. reported a synthesis of the 5′-azido-5-halouridine by substitution of the sulfonate group of 5′-tosyl-5-halouridine with the azide one in a yield of 33%–73% [69]. The attempts to reduce the N_3-group of 5′-azido-5-chlorouridine under the action of H_2/PtO_2 resulted in the dehalogenation of 5-chlorouridine. 5′-Amino-5-halouridine was obtained by the halogenation of 5′-aminouridine with $Hg(OAc)_2$ and *N*-bromosuccinimide (NBS) or I_2. Later, in a search for obtaining 5′-aminoarabinonucleosides containing 5-bromo- or 5-iodouracil, Prusoff et al. found that the halogenation of 5′-aminoarabinouridine by NBS or *N*-iodosuccinimide (NIS) did not lead to the formation of the target compounds [70]. In addition, attempts to replace a sulfonate group of 5′-tosyl-5-iodoarabinouridine with the azido one under the action of lithium azide failed. The only successful approach was the halogenation of 5′-azidoarabinouridine by NBS or NIS in a yield of 94% (for the Br-derivative) and 52% (for the I-derivative), and then the reduction of this azido function in the presence of Ph_3P in Py in a yield of 56% (for the Br-derivative) and 41% (for the I-derivative). The authors noted that the reduction of the azido group by $NaBH_4$ or H_2/Pd/C led to dehalogenation of the 5-bromo- or 5-iodouracil derivatives.

We did not find any data concerning the synthesis of 2′-aminomethylmorpholino nucleosides **7IU, BrU, ClU** containing 5-halosubstituited uracil or any modified heterocyclic base in the literature. Thus, at first, we tried to synthesize the azido derivative **6BrU** following the one-pot procedure using $Ph_3P/CBrCl_3/NaN_3$ in dry DMF according to the published method [46]. However, our attempts failed. Under using DMI instead of DMF and changing the order of the reagent's addition [58], we obtained azido derivative **6BrU** in a yield of 24% (Scheme 2). Since the one-pot method did not provide a satisfactory yield of the target compound, we applied a "step-by-step" approach and tested the method on the compound **2U**. To our surprise, after treatment of the compound **2U** with $Ph_3P/CBrCl_3$ or Ph_3P/CBr_4 in dry MeCN or DMI we observed a migration of the Tr-group from the 4′-*N* position to 2′-OH-methyl one giving *O*-tritylated morpholino nucleoside **8U**. Then, we tried to convert the hydroxyl group of morpholino nucleosides **2** into an iodine function and applied the I_2/Ph_3P/Im reagent [71]. The reaction of 2′-hydroxymethylmorpholino nucleosides **2** with this reagent in a variety of solvents (DMI, *N*-methyl-2-pyrrolidone, DCE, DCM, MeCN, THF) leads to a quantitative conversion of the compounds **2** into the compounds **5** in 5 h according to HPLC analysis. Further treatment of the reaction mixture with aq. sol. of $NaHSO_3$ and $NaHCO_3$ and chromatographic purification results in iodine derivatives **5** in a yield of 85%–93%. When the reaction mixture was treated with aq. sol. of $NaHCO_3$ only a partial detritylation of the target products were observed. Reaction of the hydroxyl group substitution with iodine also proceeds when pyridine was used instead of imidazole in the reagent mixture I_2/Ph_3P/Im [55,71,72]. However, when the reaction of **2T** with I_2/Ph_3P/Py was carried out in DCE, we observed the formation of two products **5T** and **8T** in a ratio of 1:2.

The reaction of the compounds **5** with sodium azide in DMF afforded morpholino nucleosides **6** in a quantitative yield. We tried to reduce the azido to an amino group by Pd/C-catalyzed hydrogenation and by Ph_3P/Py treatment. However, in a case of treatment of the compound **6U** with Ph_3P in dry Py we observed the reduction of the azido group and the migration of the trityl moiety to the 2′-aminomethyl position of the morpholine ring resulted in the compound **9U**. The successful reduction of azido derivatives **6** was carried out by Pd/C-catalyzed hydrogenation. In this way, 2′-aminomethylmorpholino nucleosides **7** were obtained in a nearly quantitative yield by "step-by-step" approach. It should be noted that under reducing the 2′-azidmethylmorpholino nucleosides containing 5-halo-substituited uracil **6IU, BrU, ClU** by both Pd/C-catalyzed hydrogenation and Ph_3P/pyridine, we did not observe any dehalogenation of the heterocyclic base [52].

The synthesis of ADP conjugates **10** containing the phosphoramide (P–N) bond was carried out by $Ph_3P/(PyS)_2$/MeIm activation of the β-phosphate of ADP and further coupling with the

2′-aminomethylmorpholino nucleosides **7** (Scheme 2). Previously, the approach consisting in the $Ph_3P/(PyS)_2$/MeIm activation of terminal phosphate in oligonucleotides to obtain phosphoroamide derivatives was widely used [73]. After chromatographic purification and a deblocation procedure ADP conjugates **10** were obtained in an overall yield of 70%–75%.

2.2. PARP-1, PARP-2 and PARP-3 Inhibition

2.2.1. Inhibition of PARP-1 and PARP-2 by ADP Conjugates

The effects of 16 compounds MorXppA **4** and **10** on the autopoly(ADP-ribosyl)ation of PARP-1 and PARP-2 were studied. The inhibitory properties of NAD+ analogs are summarized in Table 1. As it can be seen from the data presented in Table 1, phosphodiester dinucleotide pyrophosphates **4** containing morpholino nucleosides conjugated to ADP through the P–O bond have a little effect on the activity of PARP-1 and PARP-2 (residual enzyme activity in the presence of 1 mM compounds is 40%–90%). Among compounds **4** an inhibition efficiency of PARP-1 increases depending on the heterocyclic base of morpholino nucleoside in the order Cyt < Ura < Ade < 5-Cl-Ura < Gua < Thy < 5-Br-Ura < 5-I-Ura. In a case of PARP-2, this order has been changed, (Cyt, Ura) < 5-Cl-Ura < Gua < (5-Br-Ura, Thy) < Ade < < 5-I-Ura for conjugates **4.** The only compound that has a significant effect on the activity of both enzymes is the 5-iodouracil derivative **4IU**.

Table 1. Residual PARP-1 and PARP-2 activity (%) in the presence of 1 mM compounds **4** and **10** or IC_{50} (μ).

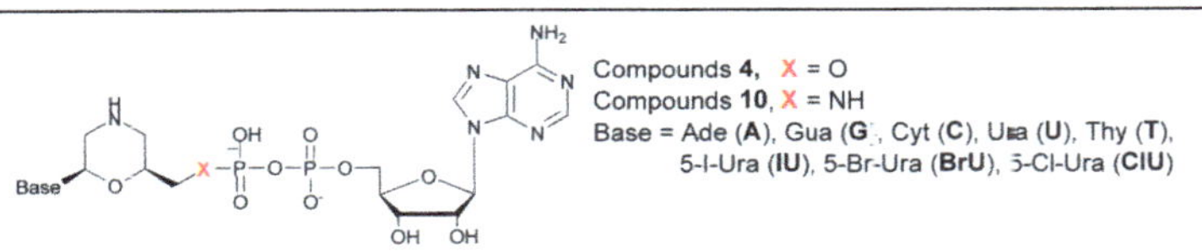

Morpholino nucleoside adenosine dinucleotides (MorXppA)

Compound	PARP-1	PARP-2	Compound	PARP-1	PARP-2
4A	80% ± 14%	421 ± 6 μM	**10A**	353 ± 4 μM	63 ±10 μM
4G	45.5% ± 3.5%	34% ± 18%	**10G**	74% ± 21%	224 ± 24 μM
4C	96.0% ± 2.8%	66.0% ± 8.5%	**10C**	87.5% ± 7.8%	34.5% ± 0.7%
4U	93.0% ± 9.9%	67% ± 4%	**10U**	89% ± 26%	33% ± 5%
4T	40% ± 20%	474 ± 14 μM	**10T**	220 ± 50 μM	136 ± 2 μM
4IU	255 ± 5 μM	160 ± 10 μM	**10IU**	126 ± 6 μM	110 ± 4 μM
4BrU	36.0% ± 8.5%	474 ± 71 μM	**10BrU**	60% ± 27%	254 ± 100 μM
4ClU	57.5% ± 7.8%	46.8% ± 9.5%	**10ClU**	46% ± 5%	47% ± 3%

Analogs of these compounds with the P–N bond **10** generally have a more pronounced effect on the activity of PARP-1 and PARP-2. The residual activity of the enzymes in 1 mM concentration of compounds is from 10% to 90%. The 5-iodouracil derivative **10IU** is also active, with the IC_{50} value for PARP-1 being equal to 126 μM. Close to **10IU** in effectiveness of inhibition of both enzymes is the thymidine derivative **10T** with IC_{50} value for PARP-1 220 μM. Activity of conjugates **10** with the P–N bond against PARP-1 changes as Ura < Cyt < Gua < 5-Br-Ura < 5-Cl-Ura < Ade < <Thy < <5-I-Ura depending on the heterocyclic base. In a case of PARP-2, this order has been changed, 5-Cl-Ura < (Cyt, Ura) < (5-Br-Ura, Gua) < Thy,5-I-Ura < <Ade for conjugates **10**.

The most active compound in the series **4** against both enzymes is the 5-I-Ura-containing analog **4IU** with IC_{50} 255 ± 5 μM for PARP-1 and 160 ± 10 μM for PARP-2. However, compounds **10** containing a P–N bond generally have a more pronounced effect on the activity of both enzymes.

It is interesting that the most active NAD+ mimetic among the **10A,G,C,U,T,IU** series is different for PARP-1 and PARP-2 in contrast to **4** series: for PARP-1 it is compound **10IU** with IC_{50} 126 ± 6 μM, for PARP-2 it is compound **10A** with IC_{50} 63 ± 10 μM. The influence of the 2′-aminomethyl group of morpholino nucleosides is more pronounced in the inhibition of PARP-2 than of PARP-1 (see adenine- and thymine-containing compounds **4A** and **10A**, and **4T** and **10T**). On the other hand, compounds **4IU** and **10IU** inhibit PARP-1 almost equally independently of the P–O or P–N bond presence.

2.2.2. Inhibition of PARP-1 and PARP-2 by Natural and Modified Nucleosides

Morpholino nucleosides and ribonucleosides were used to study the contribution of the morpholine cycle to enzyme inhibition. The inhibitory properties of these compounds are shown in Table 2. Natural ribonucleosides were weakly active as inhibitors except **rThd** (IC_{50} values 277 μfor PARP-1 and 374 μfor PARP-2), while iodine- and bromine-containing derivatives inhibited both enzymes. The 5-I-uridine was five times more effective than 5-Br-uridine for both enzymes. The ratios of IC_{50} values for PARP-1:PARP-2 are also approximately 1:1. The 5-Cl-uridine had a little effect on the activity of PARP-1, while PARP-2 was inhibited with an IC_{50} value of 196 μM. Basically, the replacement of ribose by the morpholine cycle leads to an increase in the inhibitory ability of the compounds.

Table 2. Residual PARP-1 and PARP-2 activity (%) or IC_{50}, μ, in the presence of 1 mM morpholino nucleosides and ribonucleosides.

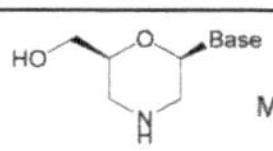

Morpholino nucleosides **11**

Base = Ade (**A**), Gua (**G**), Cyt (**C**), Ura (**U**), Thy (**T**), 5-I-Ura (**IU**), 5-Br-Ura (**BrU**), 5-Cl-Ura (**ClU**)

			Ribonucleosides		
Compound	**PARP-1**	**PARP-2**	**Compound**	**PARP-1**	**PARP-2**
11A	82% ± 6%	90 % ± 11%	**Ado**	77% ± 21%	86% ± 1%
11G	139% ± 21 μ	64% ± 25%	**Guo**	60% ± 13%	46% ± 25%
11C	200% ± 20 μ	58% ± 6%	**Cyd**	81% ± 16%	85% ± 4%
11U	91% ± 24%	97% ± 8%	**Urd**	7% 4 ± 20%	96% ± 10%
11T	148 ± 53 μ	114 ± 32 μ	**rThd**	277 ± 107 μ	330 ± 61 μ
11IU	53 ± 12 μ	85 ± 9 μ	**5-I-Urd**	49 ± 9 μ	26.5 ± 3.5 μ
11BrU	61 ± 16 μ	210 ± 48 μ	**5-Br-Urd**	244 ± 83 μ	190 ± 16 μ
11ClU	233 ± 25 μ	378 μ	**5-Cl-Urd**	82% ±17 %	196 ± 25 μ

Among the morpholino nucleosides, Ade- and Ura-containing derivatives (**11A** and **11U**, respectively, Table 2) were the least active against both enzymes. For PARP-1, the inhibition activity of nucleosides increases depending on the heterocyclic base of the morpholino nucleoside in the order Ade, Ura < (Cyt, 5-Cl-Ura) < (Gua, Thy) < (Ura, 5-Br-Ura); for PARP-2 the order is (Ade, Ura) < Cyt < Gua < 5-Cl-Ura < 5-Br-Ura < Thy < 5-I-Ura.

Thus, morpholino nucleosides proved to be more effective inhibitors of both enzymes than dinucleotides. Regardless of the other part of the molecule, i.e., nucleobase, linker type, the presence of the adenosine part of the molecule, the most effective inhibitors contained 5-iodouracil. The 5-iodouracil derivatives are characterized by the ratio of IC_{50} values for PARP-1 and PARP-2 as 1:1. Adenine dinucleotide **10A** with the P–N bond stands apart inhibiting the activity of PARP-2 but not PARP-1, with the efficacy comparable to 5-iodoururacil containing derivatives. Adenine containing dinucleotide **4A** with the P–O bond is a little active against both enzymes, whereas adenine mononucleosides are completely inactive.

2.2.3. Determination of the Type of Inhibition

We conducted kinetic studies of the inhibition of PARP-1 by the most effective inhibitors — NAD+ mimetics. For studies we selected compounds **4IU, 10IU, 11IU**. For comparison, kinetic parameters of the autopoly(ADP-ribosyl)ation in the presence of compound **10A**, which more effectively inhibits PARP-2, were determined. The 3-aminobenzamide (**3-AB**) is a well known PARP inhibitor [74,75] and has been taken for comparison.

The type of inhibition was determined based on at what stage of the reaction the inhibitor binds to the enzyme (before or after the substrate binding) and where it binds (at the substrate binding site or elsewhere). There are four types of inhibition: competitive, whereby an inhibitor binding occurs at the substrate-binding site; noncompetitive, where an inhibitor binds at the allosteric center; mixed with inhibitor binding at both the active and allosteric center; or uncompetitive under an inhibitor binding to the enzyme-substrate complex. Determination of the type of inhibition is necessary to confirm the computer simulation data and to plan further modifications of the inhibitor structure. The type of inhibition can be determined from the plots of the reaction parameters, such as maximum rates, Vmax and the apparent Michaelis constant, Km, values, versus inhibitor concentration [76]. So, under conditions of competitive inhibition, the Vmax value does not change, and the Km value does grow. Noncompetitive inhibition leads to a decrease of Vmax and retention of the Km values. Mixed inhibition is a superposition of the first two, and is characterized by a decrease in the reaction rate and an increase in Km values. In the case of uncompetitive inhibition, both values increase.

To determine the type of inhibition of the studied compounds, the activity of PARP-1 was measured at presence of three different inhibitor concentrations at nine NAD+ substrate concentrations by the fluorescence anisotropy technique [77]. Based on the obtained data, dependence graphs of the initial reaction rate on the substrate concentration were constructed, and reaction parameters such as maximum reaction rate (Vmax) and apparent Michaelis constant (Km) values were found (Tables S3 and S4, Supporting). To determine the type of inhibition, we plotted the dependence graph of Vmax and Km values on the inhibitor concentration using OriginPro 8.6.0 software. (Figure 3). As can be seen from Table S3 and Figure 3, the Km values have been grown, and the maximum reaction rate remains unchanged; that is, compounds **10IU** and **11IU** are competitive inhibitors for PARP-1 catalyzed reaction. Therefore, the results of modeling of the binding type of these compounds in the enzyme's active center are legitimate. Compound **4IU** may be characterized as a mixed inhibitor, although its influence on Km and Vmax values is weak. Compound **10A** turned out to be a mixed inhibitor because the Km values grow and the Vmax values decreased. These results indicate that compounds containing adenine and iodouracil differ in their binding mode.

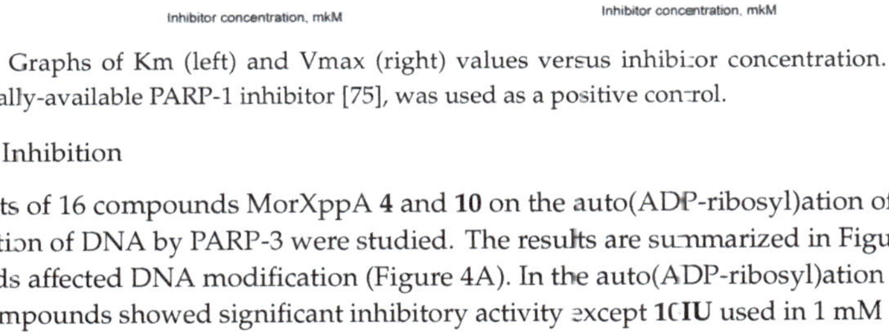

Figure 3. Graphs of Km (left) and Vmax (right) values versus inhibitor concentration. **3-AB**, a commercially-available PARP-1 inhibitor [75], was used as a positive control.

2.2.4. PARP-3 Inhibition

The effects of 16 compounds MorXppA **4** and **10** on the auto(ADP-ribosyl)ation of PARP-3 and ADP-ribosylation of DNA by PARP-3 were studied. The results are summarized in Figure 4. None of the compounds affected DNA modification (Figure 4A). In the auto(ADP-ribosyl)ation reaction, also none of the compounds showed significant inhibitory activity except **10IU** used in 1 mM concentration (Figure 4B).

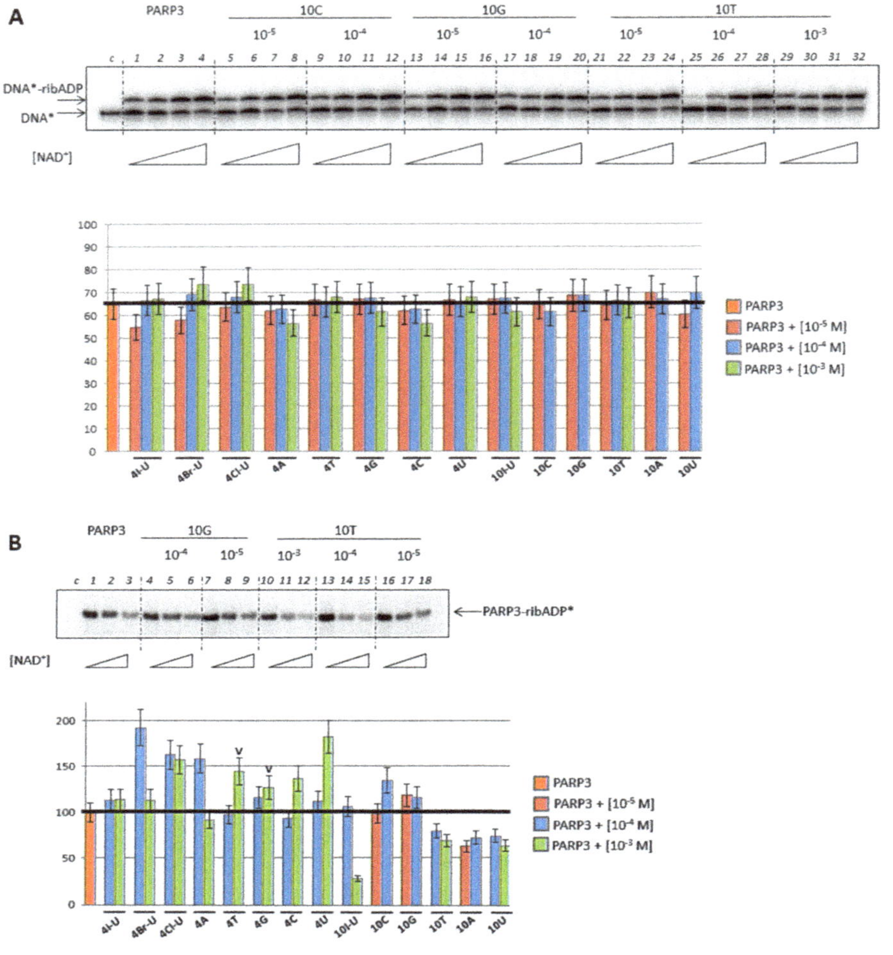

Figure 4. Basic activity of PARP-3 in DNA (**A**) and protein ADP-ribosylation (**B**) and the influence of the inhibitors on these reactions. (**A**) Activity of PARP-3 on [^{32}P]-labeled one-window gapped DNA substrate in the absence (lanes 1–4) or presence of inhibitors in the different concentrations (lanes 5–32) on the upper panel. The reactions were performed using increasing concentration of NAD+. Lane c corresponds to initial electrophoretic mobility of the DNA substrate. The chart on the bottom panel is reflected of the reaction yield of the ADP-ribosylated DNA in percentage terms. (**B**) Activity of PARP-3 on gap1 DNA substrate in the absence (lanes 1–3) or presence of inhibitors in the different concentrations (lanes 4–18) on the upper panel. The reactions were performed using increasing concentration of NAD+ in the presence of [^{32}P]-labeled NAD+. Lane c corresponds to reaction mixture without PARP-3. The chart on the bottom panel is reflected of the reaction yield of the ADP-ribosylated PARP-3 in the presence of inhibitor normalized on the yield of the autoribosylation of PARP-3 in percentage terms. The ticks on the chart mark the bars with the 0.5 mM of inhibitor in the experiment.

2.3. Structural Studies

2.3.1. Binding Modes of Guanine and 2,4-dioxopyrimidine Containing NAD– Analogs to PARP-1 and PARP-2

Based on observed activity of 2,4-dihydroxypyrimidine nucleosides (including thymine and uracil derivatives) against PARP-1/2, one can assume that they can form canonical hydrogen bonds with an active site of PARP-1 and PARP-2, similar to those observed at the binding with natural substrate NAD+ (Figure 5). We carried out a molecular docking study to assess the efficiency of morpholino and ribonucleosides binding with the donor binding site of NAD+ (Figure 6A).

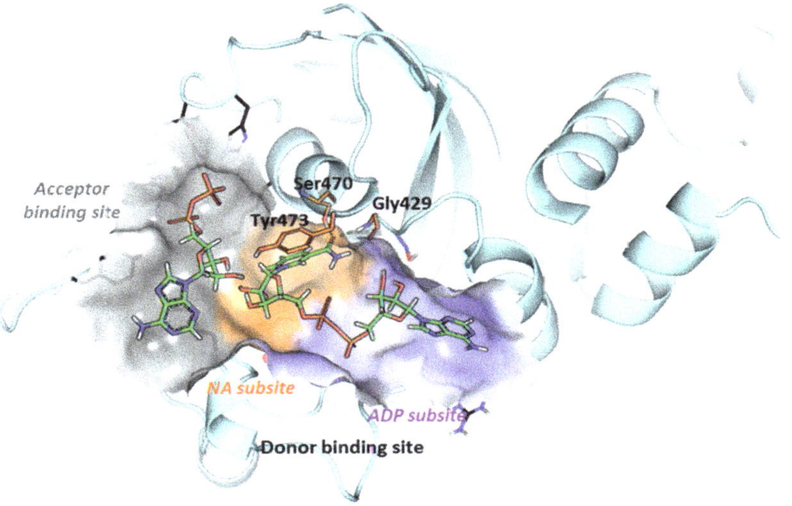

Figure 5. Structural model of the PARP-2 catalytic domain in complex with NAD+ at the donor binding site and ADP fragment at the acceptor binding site. The molecular surface illustrates ADP binding subsites of acceptor substrate (gray color), nicotinamide riboside fragment of donor NAD+ substrate (orange color) and the ADP fragment of donor NAD+ substrate (blue color). HD domain is not shown for simplicity. Substrates are shown in green color.

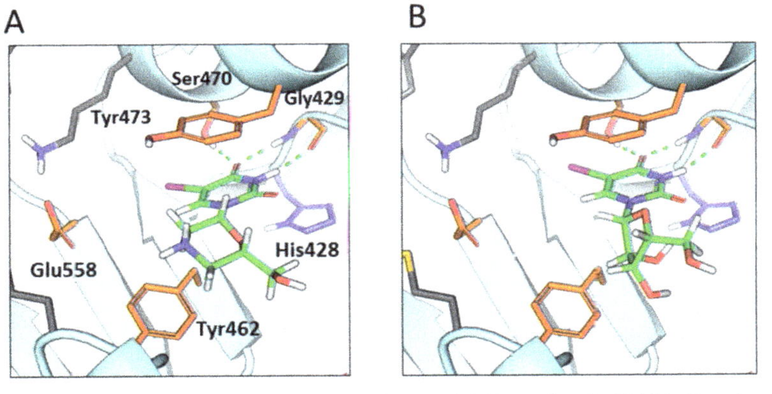

Figure 6. Predicted binding poses of **11IU** (**A**) and **5-I-Urd** (**B**) bound to the NA binding site of the PARP-2 catalytic domain. Hydrogen bonds are depicted as green dashed lines. Small molecules are shown in green color.

It includes the formation of hydrogen bonds with Gly863/Ser904 of PARP-1 or Gly429/Ser470 of PARP-2 and pi-stacking interaction with Tyr907 and Tyr473 of PARP-1 and PARP-2 correspondingly. The 5-halogen derivatives **11IU** (XP score −8.25 and −9.857 for PARP-1 and PARP-2 respectively (Figure 6A)) and **5-I-Urd** (XP score −8.328 and −8.902 for PARP-1 and PARP-2 respectively (Figure 6B)) were predicted to be among the top ranking binding scores due to enhanced hydrophobic interactions of iodine in the nicotinamide binding pocket, which is in agreement with experimental data.

Besides uracil-containing derivatives only guanine-containing compounds **Guo** (XP score −8.387 and −9.582 for PARP-1 and PARP-2 respectively) and **11G** (XP score −9.615 and −9.207 for PARP-1 and PARP-2 respectively) were able to form canonical hydrogen bonds in the nicotinamide binding subpocket of PARP-1/2, and have shown relatively high scoring function values. At the same time the presence of unsatisfied hydrogen bond acceptor groups of guanine residue located in the hydrophobic region of the nicotinamide binding subpocket likely leads to decreased binding affinity, observed in vitro. Lack of unsatisfied hydrogen bond groups in the hydrophobic nicotine amide subpocket is necessary according to analysis of all known co-crystallized PARP-1/2 inhibitors. Lack of this penalty likely leads to an overestimation of the scoring function value in case of PARP-1/2.

11A, **Ado**, **11C** and **Cyd** were unable to form canonical hydrogen bonds in the nicotine amide binding pocket with PARP1/2, and in agreement with experimental data have shown lowest scoring function values with dG > −7.7.

We found out that morpholino nucleoside derivatives in general have stronger inhibition activity to PARP-1/2 in comparison with corresponding ribonucleosides (Table 2). We could not correlate these observations with XP scoring functional values, but according to molecular docking predictions, the ribose fragment of ribonucleosides has a different localization in the PARP-1/2 catalytic site in comparison with the corresponding ribose fragment of the NAD+ substrate (Figure 5, Figure 6B). Presumably, it leads to reduced hydrophobic interactions in the subpocket formed by Tyr473, while the morpholine fragment occupies the binding site volume more efficiently (Figure 6A).

Overall, the conducted analysis shows that uracil-containing compounds, including thymine, can be used for the development of efficient PARP-1/2 inhibitors.

On the next step, we carried out a prediction of the binding modes of uracil-based NAD+ analogs synthesized in the current work, including **10T, 10IU, 4T** and **4IU**. We could not obtain the binding pose of the compound mimicking the natural substrate NAD+ binding pose using a molecular docking tool, expecting that uracil-containing derivatives would be bound to the nicotinamide binding subpocket, and adenosine would be bound to the adenine subpocket simultaneously (Figure 7A). One of the reasons could be due to the low propensity of the compound to fit the donor binding site or to the high number of conformational degrees of freedom of compounds. Therefore, we applied an alternative approach by combining the results of **5-I-Urd** and **11IU** molecular docking results and predicted the localization of the ADP fragment of the donor NAD+ nucleoside fragment based on the crystal structure of PARP-1 bound to BAD [38]. The predicted fragments were manually linked to each other, and molecular mechanics minimization using Schrödinger software with constraints on the heavy atoms of the PARP-1/2 and AMP fragment of compounds was carried out. By applying this protocol, we succeeded to obtain the molecular pose of compound **10IU** bounded to nicotinamide and ADP binding sites simultaneously in a similar way as the NAD+ substrate (Figure 7B). The visual inspection of the obtained binding poses shows the close proximity of the negatively-charged phosphate group and carbonyl oxygen of the pyrimidine group, potentially leading to a reduction of binding affinity. Another feature is the distant localization of the positively-charged group of the morpholine fragment from negatively-charged side chain of Glu558 of PARP-2. This side chain is involved in hydrogen bond formation with the ribose group of NAD+, while lack of this interaction can potentially lead to reduction of the binding affinity of **10IU.** The obtained results can explain why addition of the ADP group to **11IU** and **5-I-Urd** does not enhance its activity, and it also provides the basis for further compound optimization.

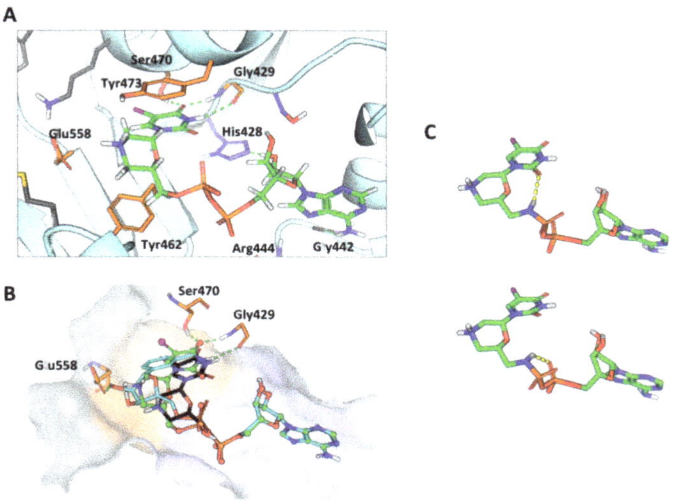

Figure 7. Predicted binding mode of **10IU** with the donor binding site of the PARP-2 catalytic domain. (**A**) The close-view of the donor binding site. (**B**) Structural alignment of the binding poses of **10IU** (green color) with NAD+ (cyan color) and **5-I-Urd** (black color). (**C**) Representative conformations of **10IU** from the molecular dynamics trajectory, the conformation with the intramolecular hydrogen bond with the non-bridging α-phosphate oxygen of ADP (bottom) and the conformation with the hydrogen bond with carbonyl oxygen of the heterocyclic base of the modified nucleoside (top) are shown. The molecular surface of the binding site is shown. Hydrogen bonds are depicted as green dashed lines.

As can be seen from Table 1, a substitution of P–O to P–N bond leads to an enhanced activity of the most tested NAD+ analogs. These effects can be explained by the formation of an intramolecular hydrogen bond that stabilizes pyrophosphate group conformation. To further study this phenomenon we carried out 100 ns molecular dynamics simulation of the complex of PARP-1 with the 5-I-uracil-containing ADP conjugate **10IU** with positional restraints on the heavy atoms of the adenine-ribose fragment of the compound and the CA-atoms of the protein. In this obtained MD trajectory, two types of hydrogen bonds of the P–NH group were observed: the hydrogen bond with the non-bridging α-phosphate oxygen of ADP (Figure 7C) and the hydrogen bond with the carbonyl oxygen of heterocyclic base of the modified nucleoside (Figure 7C). It has to be noted that in the first case the stabilization is independent from a type of heterocyclic base of morpholino nucleosides and is expected to be observed in all studied NAD+ analogs that is in agreement with in vitro data.

2.3.2. Micromolar Inhibition of PARP-2 Activity by Targeting Acceptor Substrate Binding Site

As mentioned above, the dependence of compounds **4** efficacy on the heterocyclic base looks like Cyt < Ura < Ade < 5-Cl-Ura < Gua < Thy < 5-Br-Ura < 5-I-Ura for PARP-1 and (Cyt, Ura) < 5-Cl-Ura < Gua < (5-Br-Ura, Thy) < Ade < < 5-I-Ura for PARP-2. The main discrepancy between PARP-1 and PARP-2 series is the position of the Ade-containing conjugate **4A**: this compound does not influence on the activity of PARP-1, but weakly inhibits PARP-2. For P–N-containing compounds **10**, these series turn into Ura < Cyt < Gua < Ade << Thy << 5-I-Ura for PARP-1 and (Cyt, Ura) < Gua < Thy < 5-I-Ura << Ade for PARP-2. We noticed that these series were almost the same for PARP-1, but they had significant difference for PARP-2 in the position of the Ade-containing inhibitor. Namely, compound **10A** weakly inhibits PARP-1 (similar to its P–O analog **4A**), but proves to be the most effective for PARP-2. These results are counterintuitive due to the low activity of **10A** and **Ado** compounds predicted in silico and observed in vitro. Indeed, **10A** and **Ado** cannot be efficiently bound by the nicotinamide subpocket according to the molecular docking predictions. In particular, this is due to a lack of pharmacophores necessary for the formation of critical hydrogen bonds with the NA site, including lactam or carboxamide groups [20].

Therefore, first we assumed that the adenine-MorXpp fragment of **10A** can compete for binding with the ADP fragment of NAD+. It has to be noted that NAD+ can be bound to PARP-1/2 at least at two binding sites of PARP-1/2, including donor (NAD+) and acceptor (PAR) binding ones (Figure 5). On the first step we evaluated the efficiency of adenine-MorXpp binding to the donor ADP binding site using molecular docking simulation. Molecular docking predictions show that adenine-MorXpp is not expected to bind more efficiently than Ado-5′-pp or NAD+ in the donor binding site, based both on bindings score values and visual inspection of high-ranked binding poses. Visual inspection indicates that the morpholine group does not form hydrogen bonds with residues His862 and Ser864 of PARP-1 or His428 and Ser430 of PARP-2, while the hydroxyl group of ribose in Ado-5′-pp does. The conducted analysis strongly suggests that **10A** cannot bind to donor NA and ADP binding sites.

On the next step we evaluated the hypothesis whether **10A** and **4A** can compete for binding with the acceptor substrate. We found that the adenine-MorXpp fragment of these compounds can efficiently mimic the interactions of the ADP fragment of the acceptor substrate. The latter was predicted based on the crystal structure of PARP from *Gallus gallus* (red junglefowl, PDB identifier 1A26; [78]). Moreover, the NH^+ moiety of the morpholine ring of the **10A** compound can form a salt bridge with Glu988 or Glu558 residues of PARP-1 or PARP-2, respectively. In this case, the morpholine ring can mimic an interaction of the 2′–OH group of adenosine and expected to further enhance binding affinity in comparison with natural acceptor substrate (Figure 8A). Visual inspection of binding poses shows that PO→PN substitution leads to the formation of an additional intramolecular hydrogen bond with phosphate oxygen leading to an increased stability of interaction, which is supported by an enhanced docking score and enhanced in vitro activity. In accordance with the analyzed activity of the **10IU** compound we suggest that PO→PN substitution can be used as a general strategy to stabilize the active conformation of the compounds containing diphosphate groups.

It has to be noted that the binding mode of **10A** with the acceptor binding site of PARP-2 is characterized by the high solvent exposure of the compound and multiple polar interactions, which include at least 10 hydrogen bonds. The hydrophilic nature of the stabilizing interactions with the PARP-1/2 of the compound can explain the moderate activity of the **10A** compound, and provides further strategies of compound optimization.

The observed selectivity of **10A** to PARP-2 can be explained by the variable region of PARP-1/2 in proximity to the acceptor binding site. In particular, loops of PARP-1 (978–986) and PARP-2 (544–556) have a distinct conformation and amino acid composition (Figure 8A–C), while according to the structural model, Asn555 of PARP-2 is involved in the formation of a hydrogen bond with phosphate oxygen of **10A** and replaced by Leu985 in the case of PARP-1 (Figure 8B,C). Additionally, PARP-1 lacks stabilizing interactions with Tyr552 due to shortening of the corresponding loop. This is supported by the lower XP binding score for PARP-1 that was −7.975 and −10.574 in case of PARP-2.

Furthermore, lack of activity of **10A** against PARP-3 supports the before suggested mechanism of action. Indeed, PARP-3 structure was reported to be different from PARP-1/2 and characterized as mono(ADP-ribose) transferase [79,80]. In accordance with this data, molecular docking predicted that **10A** does not have the same mode of binding to PARP-3 as to PARP-1/2 nor other MorXppA compounds. Analysis of the binding site revealed that this could be due to steric hindrance caused by Arg408 and Lys421 residues, as well as the hindrance of the hydrogen bond acceptor and donor groups of Gln512 and Arg408, correspondingly, by small molecule (Figure 8D).

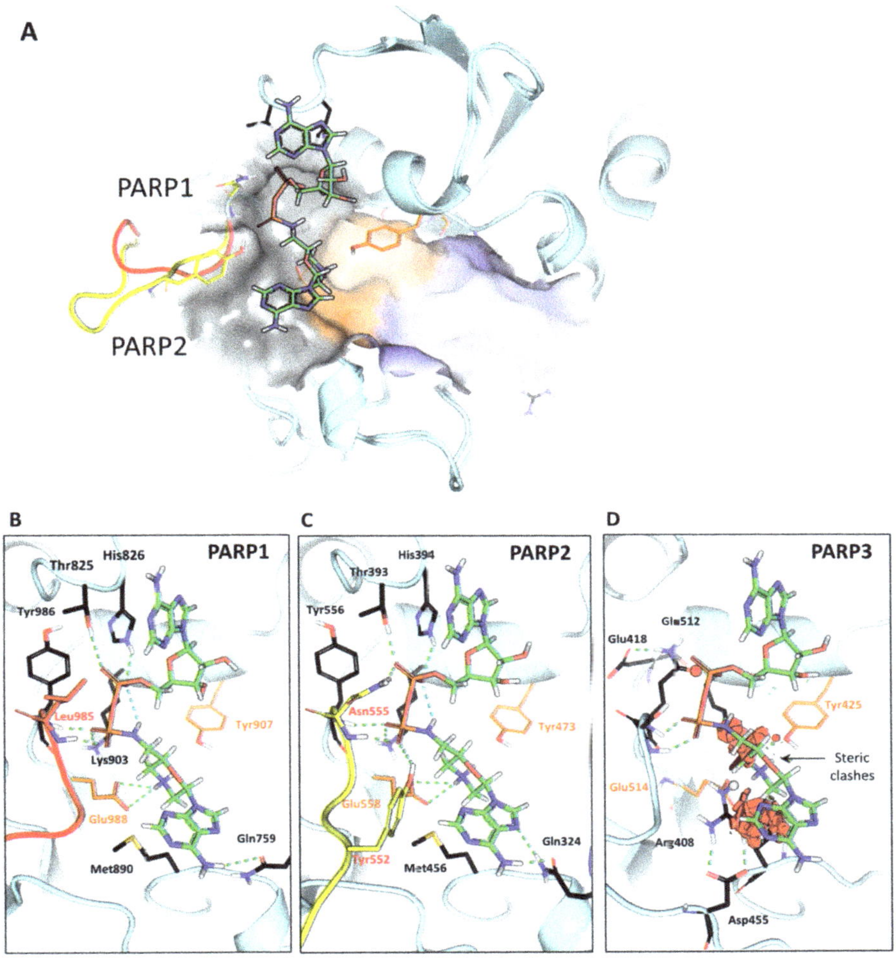

Figure 8. Predicted binding pose of **10A** with the acceptor binding site of the PARP-1/2 catalytic domain. (**A**) The structural alignment of PARP-1 and PARP-2. Variable loops are indicated in red and yellow colors for PARP-1 and PARP-2, respectively. Detailed view of **10A** interaction with PARP-1 (**B**) and PARP-2 (**C**) is shown. (**D**) PARP-3 acceptor binding site with superimposed binding pose of **10A** from PARP-2/**10A** complex. Steric clashes are shown with red disks. Unsatisfied hydrogen bond donor and acceptor atoms of the PARP-3 acceptor binding site hindered by ligand are shown as spheres. Hydrogen bonds are depicted as dashed lines. HD domain is not shown for simplicity.

On the whole, structural analysis of the binding modes of **10A** together with experimental data suggests that this compound acts by a different mechanism than known PARP-1/2 potent inhibitors and binds to the acceptor substrate binding site, but not to the donor NAD+ binding site. Additionally, the presence of variable regions in proximity of the acceptor binding site may lead to enhanced PARP-2 specificity of **10A**.

3. Materials and Methods

3.1. Chemistry

General information, NMR, mass and IR spectra can be found in Supplementary Materials.

3.1.1. General Procedure for the Synthesis of Morpholino Nucleosides **2**

Ribonucleosides **1** (4 mmol) were suspended in EtOH (80 mL). A warm solution of $NaIO_4$ (4.2 mmol, 0.9 g) in water (4 mL) was added to the suspension under vigorous stirring. In 15 min, $(NH_4)_2B_4O_7{\cdot}4H_2O$ (4.8 mmol, 1.26 g) was added. The pH of the reaction mixture was maintained between 8.5 and 9.0 by adding triethylamine (TEA, total amount 0.6–1.0 mL). After 1.5 h of stirring, the precipitate was separated by filtration and washed with EtOH (2 × 4 mL). $NaCNBH_3$ (5.2 mmol, 0.326 g) was added to the combined filtrates and the stirring was continued. In 1 h, trifluoroacetic acid (TFA) was added till pH 3.0–4.0. In 2 h, solvents were evaporated and the residue was dried by coevaporation with MeCN (3×10 mL) and toluene (3×10 mL), suspended in DMF (16 mL) and Et_3N (12 mmol, 1.7 mL). Triphenylmethyl chloride (TrCl) (3.5 mmol, 0.975 g) was added to the suspension, and the stirring was performed overnight. The reaction was quenched by the addition of MeOH (5 mL); the solution was evaporated up to half of a volume and poured into the water (300 mL). The precipitate was separated by filtration and dried. After that, the precipitate was dissolved in CH_2Cl_2 (10 mL), and the target morpholino nucleoside was precipitated by the petroleum ether (250 mL). The suspension was cooled (−20 °C, 2 h) and the precipitate was separated by filtration, washed by petroleum ether, and dried under vacuum.

3.1.2. General Procedure for the Synthesis of Morpholino Nucleosides **11**

Morpholino nucleosides **2A,G,C** (0.1 mmol) were dissolved in 1.5 mL of EtOH, conc. aq. ammonia (3 mL) was added to each solution and the reaction mixtures were stirred at room temperature. After the deblocation of heterocyclic bases was completed (1–2 days), the reaction mixtures were evaporated. Aq. 80% AcOH (1.5 mL) was added to the residues and to the morpholino nucleosides **2U,T,IU,BrU,ClU** (0.1 mmol). In 1.5 h, all reaction mixtures were diluted with water (5 mL) and washed by CH_2Cl_2 (5 × 5 mL). Aqueous layers were evaporated, coevaporated with toluene (3 × 5 mL) and again with water (3 × 5 mL). After drying in vacuum, morpholino nucleosides 11 were obtained in a yield of 90% as partial acetate salts.

3.1.3. General Procedure for the Phosphorylation of Morpholino Nucleosides **2**

Morpholino nucleosides **2** (0.1 mmol) were dissolved in dry Py (1 mL). The solution was cooled in an ice bath (−15 °C). $POCl_3$ (0.4 mmol, 0.037 mL) was added to the solution under stirring. In 20 min, 1 M TEAB (2 mL) was added to the cold reaction mixture. The reaction mixture was stirred for 10 min at room temperature. A) Compounds **2A,G,C,IU**: the reaction mixture was distributed between CH_2Cl_2 (10 mL) and water (10 mL). An aqueous layer was washed with CH_2Cl_2 (10 mL). Combined organic layers were dried (Na_2SO_4), filtrated and evaporated. CH_2Cl_2 (1 mL) was added to the residue, and the target products **3A,G,C,IU** were precipitated by petroleum ether (12 mL). The suspension was cooled (−20 °C, 2 h), the precipitate was separated by centrifugation, washed by petroleum ether and dried under vacuum. B) Compounds **2U,T,BrU,ClU**: the reaction mixture was distributed between CH_2Cl_2 (10 mL) and water (10 mL). An aqueous layer was evaporated, and the target monophosphates **3U,T,BrU,ClU** were purified by RPC in a gradient of MeCN (0%–75 %) in water. Appropriate fractions were combined and evaporated. The residue was dried under vacuum. Morpholino nucleotides **3A-ClU** were obtained as TEA salts

3.1.4. General Procedure for the Synthesis of Conjugates **4**

Morpholino nucleotides **3A-ClU** (TEA salt, 0.05 mmol) and Ph_3P (0.15 mmol, 0.039 g) were dissolved in 1,3-dimethyl-2-imidazolidinone (DMI, 0.5 mL). *N*-Methylimidazole (MeIm, 0.6 mmol, 0.05 mL) and $(PyS)_2$ (0.15 mmol, 0.03 g) were added to the solution, and the reaction mixture was stirred for 15 min. Adenosine 5′-phosphate (n-Bu_3N salt, 0.2 mmol in 0.5 mL of DMI) was added to the reaction mixture and stirring was continued for 1 h. Conc. aq. NH_3 (2 mL) was added to the reaction mixture in case of the synthesis of conjugates **4A,G,C**, and after the deblocation of heterocyclic bases was completed (24 h at room temperature) the reaction mixture was evaporated up to its initial volume (1 mL). Then the reaction mixture was poured into Et_2O (20 mL) and cooled (−20 °C, 2 h). The precipitate was separated by centrifugation, washed by Et_2O and dried. *N*-Tr-protected conjugates **4A-ClU** were purified by RPC in a linear gradient of MeCN in water in the presence of 0.1 M NH_4HCO_3. Appropriate fractions were collected and evaporated. Traces of a buffer were removed by repeating the evaporation from aq. EtOH. To remove the Tr protective group, the residue was dissolved in 80% aq. AcOH (1 mL). In 1 h, the conjugates **4A-ClU** were precipitated by 4% $NaClO_4$ in acetone (10 mL), cooled (−20 °C, 2 h), and centrifuged. The precipitate was washed with acetone and dried.

3.1.5. General Procedure for the Synthesis of 2′-aminomethylmorpholino Nucleosides **7**

Morpholino nucleosides **2A-ClU** (0.7 mmol), Ph_3P (1.4 mmol, 0.367 g) and imidazole (1.4 mmol, 0.095 g) were dissolved in dichloroethane (DCE) (2.8 mL). The solution was cooled in an ice bath (−5 °C). A solution of I_2 (0.9 mmol, 0.23 g) in DCE (0.4 mL) was added under vigorous stirring. The cooling was removed and the reaction mixture was stirred for 5 h at r.t. The reaction mixture was diluted with CH_2Cl_2 (20 mL) and washed with aq. 5 M $NaHSO_3$ (20 mL), sat. aq. $NaHCO_3$ (20 mL) and water (2×20 mL). The organic layer was dried (Na_2SO_4), filtered and evaporated. 2′-Iodomethylmorpholino nucleosides **5A-ClU** were purified by silica gel chromatography in a gradient of acetone in CH_2Cl_2 (0%–20%). Appropriate fractions were combined and evaporated. The residue was dissolved in CH_2Cl_2 (1 mL) and precipitated with petroleum ether (15 mL). The suspension was cooled (−20 °C, 2 h), the precipitate was separated by centrifugation, washed with petroleum ether and dried under vacuum. Morpholino nucleosides **5A-ClU** (0.5 mmol) were dissolved in DMF (2 mL), then NaN_3 (2.5 mmol, 0.163 g) was added. The suspension was stirred overnight at r.t. and then was diluted with CH_2Cl_2 (20 mL) and washed with water (4×10 mL). The organic layer was dried with Na_2SO_4 and filtered. The evaporated residue was dissolved in CH_2Cl_2 (1 mL) and precipitated by petroleum ether (15 mL). The suspension was cooled (−20 °C, 2 h), the precipitate was separated by centrifugation, washed by petroleum ether and dried under vacuum. 10% Pd/C (0.01 mmol, 10 mg) was added to a solution of 2′-azidomethylmorpholino nucleosides **6A-ClU** (0.1 mmol) in MeOH (1 mL). The mixture was stirred for 48 h under H_2. The reaction mixture was filtered and evaporated. The residue was dissolved in CH_2Cl_2 (0.1 mL) and precipitated by petroleum ether (1.5 mL). The suspension was cooled (−20 °C, 2 h), the precipitate was separated by centrifugation, washed by petroleum ether and dried under vacuum.

3.1.6. General Procedure for the Synthesis of Conjugates **10**

ADP (n-Bu_3N salt, 0.02 mmol) was dissolved in dimethyl sulfoxide (DMSO) (0.2 mL) and activated as described above for morpholino nucleotides **3A-ClU**. In 1 h, a solution of 2′-aminomethylmorpholino nucleosides **7A-ClU** (0.06 mmol) in DMI (0.2 mL) and n-Bu_3N (0.06 mmol, 0.014 mL) were added to the reaction mixture. In 3 h, the reaction was completed. Further treatment of the reaction mixture, deblocation and purification of target conjugates **10A-ClU** were carried out as described for conjugates **4A-ClU**.

3.2. Biology

3.2.1. Materials

The oligonucleotides were purchased from Biosset (Novosibirsk, Russia). The reagents, solvents and the basic components of buffers were purchased from Sigma (St. Louis, MO, USA), Acros organics (Waltham, MA, USA) or Promega (Madison, WI, USA). γ–[^{32}P]-ATP and α–[^{32}P]-ATP (with specific activities of 5000 and 3000 Ci/mmol, respectively) were from the Laboratory of Biotechnology (Institute of Chemical Biology and Fundamental Medicine, Novosibirsk, Russia). T4 polynucleotide kinase was from Biosan (Novosibirsk, Russia). SUMO fusion expression vector pETHSUL (GenBank: EF205333.1) and pSUPER vector coding for the catalytic domain of the S. cerevisiae SUMO hydrolase dtUD1 were kindly provided by Dr. P. Loll (Drexel University, Philadelphia, USA). His6 and Strep-II double-tagged dtUD1 hydrolase was expressed and purified as described in ref. [81]. Human recombinant poly(ADP-ribose)-polymerase 1 (EC 2.4.2.30) was prepared as described previously [82]. Murine recombinant poly(ADP-ribose)-polymerase 2 (EC 2.4.2.30) was purified as described previously [83]. Plasmid encoding human PARP-3 was kindly provided by Dr. A. Ishchenko (Gustave Roussy, Université Paris-Saclay, France). Poly(ADP-ribose)-polymerase 3 (EC 2.4.2.30) was purified according to [6].

3.2.2. Oligonucleotide Substrates

The oligodeoxyribonucleotides were 5′-[^{32}P]-phosphorylated with T4 polynucleotide kinase as described. Unreacted γ–[^{32}P]-ATP was removed using MicroSpinTM G-25 column (Amersham Pharmacia Biotech, Little Chalfont, UK, and GE Healthcare, Chicago, IL, USA). The 5′-[^{32}P]-phosphorylated oligonucleotides were precipitated by 4% $LiClO_4$ in acetone and dissolved in water. Complementary oligodeoxynucleotides were annealed as described [6]. The following oligonucleotide sequences were used for the construction of DNA duplexes: whole strands 5′-(d)GGC TTC ATC GTT GTC TCA GAC CTG GTG GAT ACC G-3′; upstream oligonucleotides 5′-(d)CGG TAT CCA CCA GGT CTG-3′ with or without 5′-phosphate groups; downstream oligonucleotide 5′-p-(d)GAC AAC GAT GAA GCC-3′.

3.2.3. Synthesis of [^{32}P]-NAD+

The radioactive NAD+ was synthesized from α–[^{32}P]-ATP. The reaction mixture containing 1 mM ATP, 10 MBq of α–[^{32}P]-ATP, 20 mM MgCl2, 2 mM beta-nicotinamide mononucleotide, and 5 mg/mL nicotinamide nucleotide adenylyltransferase 1 in 25 mM Tris-HCl, pH 7.5, was incubated at 37 °C for 60 min, then stopped by heating to 80 °C for 5 min. The denatured protein was removed by centrifugation.

3.2.4. PARP-1 and PARP-2 Enzyme Assay

The reaction of autopoly(ADP-ribosyl)ation was carried out as follows: for PARP-1 50 mM tris-HCl, pH 8.0, 20 mM $MgCl_2$, 150 mM NaCl, and 7 mM β-mercaptoethanol, activated DNA 2 OE/mL, 0.3 mM [^{32}P] NAD+ at 37 °C. The reaction was initiated by adding PARP-1 to 200 nM and the reaction mixtures were incubated for 1.5 min. For PARP-2: 50 mM tris-HCl, pH 8.0, 3 mM spermin, 150 mM NaCl, and 7 mM β-mercaptoethanol, activated DNA 2 OE/mL, 0.6 mM [^{32}P] NAD+ at 37 °C. The reaction was initiated by adding PARP-2 to 800 nM and the reaction mixtures were incubated for 5 min. The reaction was stopped by placing 10 µL aliquots onto Whatman 1 paper filters soaked with 5 % TCA. The filters were washed with 5% TCA for four times and dried in air after the removal of TCA with 90% ethanol. The incorporation of radioactivity into the product was calculated with a Tri-Carb 2800 scintillation counter (Perkin Elmer, Waltham, MA, USA) according to the Cherenkov method or using Typhoon FLA 9500 scanner (GE Healthcare, Chicago, IL, USA). To determine the IC_{50} value of inhibitors (concentration of the compound required reducing the enzyme activity by 50%), we studied the activity of the enzyme at different concentrations of inhibitors. Measurements were done in at least two independent experiments. IC_{50} values were calculated using the Origin Pro 8.0 software by nonlinear regression analysis.

The determination of the kinetic parameters of inhibition Km and Vmax was carried out as described [76]. In all experiments, the points on the experimental curves represent the average of a minimum three independent experiments. Standard deviation did not exceed 10%.

3.2.5. Inhibitors Screening, ADP-ribosylation of DNA by PARP-3

The reactions were performed in the presence of 2 mM $MgCl_2$ as a cofactor in HDB buffer solution containing HEPES-KOH, pH 8.6, 0.25 mg/mL BSA and 0.5 mM DTT. The reaction mixtures (final volume 10 μL) contained 0.02 μM [^{32}P]-DNA substrate, 0.1 μM PARP-3, 5, 10, 100 or 500 μM NAD+ in the absence or presence of various concentrations of the inhibitor, which are indicated on the figures. The reaction was carried out for 20 min at 37 °C. The mixtures were separated in a standard 20% denaturing acrylamide gel. The gels were dried and subjected to autoradiography and/or phosphorimaging for quantitation using the Typhoon imaging system (GE Healthcare Life Sciences, Chicago, IL, USA). Quantitative processing was carried out using OriginPro7.5, Microcal Software (Origin Systems, Houston, TX, USA). In all experiments, the points on the experimental curves represent the average of minimum three independent experiments. Standard deviation did not exceed 10%.

3.2.6. Inhibitors Screening, ADP-ribosylation of PARP-3

The reactions were performed in the presence of 2 mM $MgCl_2$ as a cofactor in HDB buffer. The reaction mixtures (final volume 10 μL) contained 0.1 μM DNA substrate, 0.5 μM PARP-3, 5 μM [^{32}P]-NAD+ and 0, 35 or 95 μM NAD+ in the absence or presence of various concentrations of the inhibitor, which are indicated on the figures. The reaction mixtures were incubated for 20 min at 37 °C then terminated by adding the 5x gel loading Laemmli sample buffer and heating at 85 °C for 10 min. The mixtures were resolved on a 10% SDS-PAG as described in [84] The gels were dried and subjected to autoradiography and/or phosphorimaging for quantitation using the Typhoon imaging system from GE Healthcare Life Sciences. Quantitative processing was carried out using OriginPro7.5, Microcal Software, USA. In all experiments, the points on the experimental curves represent the average of a minimum three independent experiments. Standard deviation did not exceed 10%.

3.3. *Structural Studies*

Reference crystal structures of PARP-1 and PARP-2 were taken from the PDB database with identifiers 4ZZZ [85] and 3KJD [86], correspondingly. Localization of the ADP fragment of acceptor NAD+ was predicted by a structural alignment of the reference PARP-1/2 crystal structures with the crystal structure of PARP from *Gallus gallus* in complex with NAD− (PDB identifier 1A26, [78]). The localization of donor NAD+ was predicted by structural alignment with catalytic domain of PARP-1 (PDB identifier 6BHV) [38] in complex with BAD, structural analog of NAD+. In this case C1 carbon of benzamide was manually replaced by nitrogen. Structural alignment was carried out using Maestro (Schrodinger Inc.). Obtained complexes were minimized with Protein Preparation Wizard (Schrodinger Inc.). Preliminary the side chains of amino acid residues containing steric hindrance with BAD were removed and re-added using Prime Module (Schrödinger Inc.).

3D structures of small molecules under study were prepared using LigPrep module (Schrödinger Inc.). Molecular docking was carried out using Glide in the extra-precision mode (XP Score). Small molecule binding pose with the best XP score were selected [87]. Two grids were calculated for the molecular docking study. First grid was calculated based on the center of mass of the acceptor ADP fragment of NAD+ and the second was calculated based on the center mass of the donor NAD+ ligand. Molecular dynamics simulation was carried out using the pmemd module of Amber12 software [88]. The RESP charges were calculated using the ANTECHAMBER module based on ab initio calculations of electrostatic potentials using theHF/6-31G* level of theory using Gaussian 03 [89,90]. Amberff99SB and GAFF force fields were used for the simulation of molecular dynamics of protein and small molecules, respectively. Implicit solvent generalized Born (GB) Onufriev was used for MD simulation with mbondi2 radii set [91].

Molecular dynamics protocol included 1000 steps of steepest descent minimization, 100 ps heating from 0K to 300K, and 100 ns production molecular dynamics simulation. Langevin thermostat with a 5 ps^{-1} collision rate was used [92]; 1 kcal/mole restraints on protein CA atoms and heavy atoms of the adenosine fragment of NAD+ analogs were applied at all steps of MD simulation.

4. Conclusions

Morpholino nucleosides are readily available derivatives of natural ribonucleosides that found a broad application in the synthesis of morpholino oligonucleotides as nucleic acid mimetics. These are important and widely used molecular tools in molecular biology and are now even in clinic for an antisense-based treatment. However, a high potential of morpholino nucleoside monomers as terminators in gene sequencing, polymerase inhibitors (in phosphorylated form), or as a part of other low molecular weight biological regulators is not investigated in its entirety. In this regard, it is essential that the morpholine ring is a well-known and important pharmacophore widely used in numerous medicines [93]. The present study allowed for a broader involvement of morpholino nucleosides into a pool of potentially biologically active compounds. We proposed a novel class of PARP-1 and PARP-2 inhibitors consisting of various types of morpholino nucleosides and ADP [45].

The highest inhibition activity against PARP-1 was found for ADP conjugates containing 5-iodouracil morpholino nucleosides. These compounds were predicted to target the NA binding site by a common mechanism to known for potential PARP-1 inhibitors.

Surprisingly, in relation to PARP-2 the most active compound with IC_{50}~50 μM was dinucleotide-containing adenine morpholino nucleoside. The activity of 2′-aminomethylmorpholino nucleoside of adenine could be achieved by targeting to the acceptor substrate binding site by mimicking interactions of the PAR substrate. In particular, it has been shown that the amine group of the morpholine ring can mimic an interaction of the 2′–OH group of adenine riboside with Glu558. The replacement of P–O to P–N bond leads to the enhanced stabilization of pyrophosphate conformation due to the formation of an intramolecular hydrogen bond, and followed to a subsequently higher inhibition activity. Stabilizing interactions with variable regions lead to enhanced selectivity in relation to PARP-2. Strikingly, these predictions are validated by a fluorescence anisotropy assay that suggests the mixed mode of inhibitory activity by targeting the active and allosteric centers simultaneously. This is in contrast to 5-iodouracil morpholino nucleosides as well as the **3-AB** compound, a well known PARP inhibitor targeting nicotinamide binding site.

Lack of activity of **10A** against PARP-3 that has a different amino acid composition of the acceptor binding site as well as substrates selectivity further confirms the predicted mechanism of action. It has to be noted that only 5-iodouracil morpholino nucleosides **10IU** demonstrated low activity against PARP-3 at mM concentration. These data are in accordance with generally observed up to two orders lower activity of known PARP-1/2 inhibitors against PARP-3, targeting the donor nicotinamide binding pocket [94].

Interestingly, it has been shown that occupancy of the NAD+-binding site with benzamide adenine dinucleotide (BAD) locks PARP-1 on a DNA break [38]. It makes a design of new NAD+ analogs a promising strategy to create a wide spectrum of the compounds for future clinical usage under PARP inhibitor therapy. Conducted analysis suggests that compounds based on 2′-aminomethylmorpholino nucleosides target PARP-1/2 by a novel molecular mechanism distinct from a known class of potent inhibitors directed on the NA donor subsite, but not on the acceptor substrate binding site. It opens new horizons for the development of additional classes of PARP-1/2 selective inhibitors.

Taking into account our enzyme assay and molecular modeling calculations, a new strategy for conformation stabilization and enhancement of the activity and selectivity of a novel class of NAD+ analogs may be proposed.

Supplementary Materials: Supplementary materials can be found at http://www.mdpi.com/1422-0067/21/1/214/s1.

Author Contributions: T.V.A, O.I.L. and V.N.S. have conceptualized the research. Y.V.S and T.V.A performed the synthesis. N.V.I. and V.A.I. carried out molecular modeling study. R.Y.P. and I.V.E. performed NMR analysis. A.L.Z., M.V.S., M.M.K., T.A.K. and E.A.B. carried out an enzyme purification and enzyme inhibition assays. T.V.A., A.L.Z., N.V.I., Y.V.S. and E.A.B. wrote the manuscript. All authors have read and agreed to the published version of the manuscript.

Funding: The work was partially supported by the Russian Foundation for Basic Research (RFBR, Grant No. 18-04-00352: the design of the study, the collection, analyses and interpretation of data, the writing of the manuscript, the decision to publish the results, compound synthesis and molecular modeling; and Grant No. 17-00-00097: the collection, analyses and interpretation of data, PARP-1 and PARP-2 assay); the Russian Science Foundation (RSF, Grant No. 17-74-20075: the writing of the manuscript, the decision to publish the results PARP-3 purification and assay); and by Russian State funded budget projects of ICBFM SB RAS No AAAA-A17-117020210022-4: the collection, analyses and interpretation of data, the writing of the manuscript, the decision to publish the results.

Conflicts of Interest: The authors declare no conflict of interest.

References

1. Amé, J.C.; Spenlehauer, C.; de Murcia, G. The PARP superfamily. *Bioessays* **2004**, *26*, 882–893. [CrossRef] [PubMed]
2. Gupte, R.; Liu, Z.; Kraus, W.L. PARPs and ADP-ribosylation: Recent advances linking molecular functions to biological outcomes. *Genes Dev.* **2017**, *31*, 101–126. [CrossRef] [PubMed]
3. Vyas, S.; Matic, I.; Uchima, L.; Rood, J.; Zaja, R.; Hay, R.T.; Ahel, I.; Chang, P. Family-wide analysis of poly(ADP-ribose) polymerase activity. *Nat. Commun.* **2014**, *5*, 4426. [CrossRef] [PubMed]
4. Lüscher, B.; Bütepage, M.; Eckei, L.; Krieg, S.; Verheugd, P.; Shilton, B.H. ADP-Ribosylation, a multifaceted posttranslational modification involved in the control of cell physiology in health and disease. *Chem Rev.* **2018**, *118*, 1092–1136. [CrossRef] [PubMed]
5. Talhaoui, I.; Lebedeva, N.A.; Zarkovic, G.; Saint-Pierre, C.; Kutuzov, M.M.; Sukhanova, M.V.; Matkarimov, B.T.; Gasparutto, D.; Saparbaev, M.K.; Lavrik, O.I.; et al. Poly(ADP-ribose) polymerases covalently modify strand break termini in DNA fragments in vitro. *Nucleic Acids Res.* **2016**, *44*, 9279–9295. [PubMed]
6. Belousova, E.A.; Ishchenko, A.A.; Lavrik, O.I. Dna is a new target of Parp3. *Sci Rep.* **2018**, *8*, 4176. [CrossRef] [PubMed]
7. Hottiger, M.O.; Hassa, P.O.; Lüscher, B.; Schüler, H.; Koch-Nolte, F. Toward a unified nomenclature for mammalian ADP-ribosyltransferases. *Trends Biochem. Sci.* **2010**, *35*, 208–219. [CrossRef]
8. Steffen, J.D.; Brody, J.R.; Armen, R.S.; Pascal, J.M. Structural implications for selective targeting of PARPs. *Front. Oncol.* **2013**, *3*, 301. [CrossRef]
9. Boehler, C.; Dantzer, F. PARP-3, a DNA-dependent PARP with emerging roles in double-strand break repair and mitotic progression. *Cell Cycle* **2011**, *10*, 1023–1024. [CrossRef]
10. Sharif-Askari, B.; Amrein, L.; Aloyz, R.; Panasci, L. PARP3 inhibitors ME0328 and olaparib potentiate vinorelbine sensitization in breast cancer cell lines. *Breast Cancer Res. Treat.* **2018**, *172*, 23–32. [CrossRef]
11. Rajawat, J.; Shukla, N.; Mishra, D.P. Therapeutic Targeting of Poly(ADP-Ribose) Polymerase-1 (PARP1) in cancer: Current developments, therapeutic strategies, and future opportunities. *Med. Res. Rev.* **2017**, *37*, 1461–1491. [CrossRef] [PubMed]
12. Walsh, C. Targeted therapy for ovarian cancer: The rapidly evolving landscape of PARP inhibitor use. *Minerva Ginecol.* **2017**, *70*, 150–170. [PubMed]
13. Exman, P.; Barroso-Sousa, R.; Tolaney, S.M. Evidence to date: Talazoparib in the treatment of breast cancer. *Onco Targets Ther.* **2019**, *12*, 5177–5187. [CrossRef] [PubMed]
14. Ohmoto, A.; Yachida, S. Current status of poly(ADP-ribose) polymerase inhibitors and future directions. *Onco Targets Ther.* **2017**, *10*, 5195–5208. [CrossRef]
15. Katsyuba, E.; Auwerx, J. Modulating NAD+ metabolism, from bench to bedside. *EMBO J.* **2017**, *36*, 2670–2683. [CrossRef]

16. Khan, J.A.; Forouhar, F.; Tao, X.; Tong, L. Nicotinamide adenine dinucleotide metabolism as an attractive target for drug discovery. *Expert Opin. Ther. Targets.* **2007**, *11*, 695–705. [CrossRef]
17. Clark, J.B.; Ferris, G.M.; Pinder, S. Inhibition of nuclear NAD nucleosidase and poly ADP-ribose polymerase activity from rat liver by nicotinamide and 5′-methyl nicotinamide. *Biochim. Biophys. Acta.* **1971**, *238*, 82–85. [CrossRef]
18. Yuan, Z.; Chen, J.; Li, W.; Li, D.; Chen, C.; Gao, C.; Jiang, Y. PARP inhibitors as antitumor agents: A patent update (20132–015). *Expert. Opin. Ther. Pat.* **2017**, *27*, 363–382. [CrossRef]
19. Cepeda, V.; Fuertes, M.A.; Castilla, J.; Alonso, C.; Quevedo, C.; Soto, M.; Pérez, J.M. Poly(ADP-ribose) polymerase-1 (PARP-1) inhibitors in cancer chemotherapy. *Recent Pat. Anticancer Drug Discov.* **2006**, *1*, 39–53. [CrossRef]
20. Ferraris, D.V. Evolution of poly (ADP-ribose) polymerase-1 (PARP-1) inhibitors. From concept to clinic. *J Med Chem.* **2010**, *53*, 4561–4584. [CrossRef]
21. Penning, T.D. Small-molecule PARP modulators-current status and future therapeutic potential. *Curr. Opin. Drug Discov. Devel.* **2010**, *13*, 577–586. [PubMed]
22. Zhou, Q.; Ji, M.; Zhou, J.; Jin, J.; Xue, N.; Chen, J.; Xu, B.; Chen, X. Poly (ADP-ribose) polymerases inhibitor, Zj6413, as a potential therapeutic agent against breast cancer. *Biochem. Pharmacol.* **2016**, *107*, 29–40. [CrossRef] [PubMed]
23. Wang, B.; Chu, D.; Feng, Y.; Shen, Y.; Aoyagi-Scharber, M.; Post, L.E. Discovery and characterization of (8*S*,9*R*)-5-fluoro-8-(4-fluorophenyl)-9-(1-methyl-1*H*-1,2,4-triazol-5-yl)-2,7,8,9-tetrahydro-3*H*-pyrido[4,3,2-de]phthalazin-3-o (BMN 673, Talazoparib), a novel, highly potent, and orally efficacious poly(ADP-ribose) polymerase-1/2 inhibitor, as an anticancer agent. *J. Med. Chem.* **2016**, *59*, 335–357. [PubMed]
24. Li, H.; Hu, Y.; Wang, X.; He, G.; Xu, Y.; Zhu, Q. Novel tricyclic poly (ADP-ribose) polymerase-1/2 inhibitors with potent anticancer chemopotentiating activity: Design, synthesis and biological evaluation. *Bioorg. Med. Chem.* **2016**, *24*, 4731–4740. [CrossRef]
25. Oplustil O'Connor, L.; Rulten, S.L.; Cranston, A.N.; Odedra, R.; Brown, H.; Jaspers, J.E.; Jones, L.; Knights, C.; Evers, B.; Ting, A.; et al. The PARP inhibitor AZD2461 provides insights into the role of PARP3 inhibition for both synthetic lethality and tolerability with chemotherapy in preclinical models. *Cancer Res.* **2016**, *76*, 6084–6094. [CrossRef]
26. Jordheim, L.P.; Durantel, D.; Zoulim, F.; Dumontet, C. Advances in the development of nucleoside and nucleotide analogues for cancer and viral diseases. *Nat. Rev. Drug Discov.* **2013**, *12*, 447–464. [CrossRef]
27. Maffioli, S.I.; Zhang, Y.; Degen, D.; Carzaniga, T.; Del Gatto, G.; Serina, S.; Monciardini, P.; Mazzetti, C.; Guglierame, P.; Candiani, G.; et al. Antibacterial nucleoside-analog inhibitor of bacterial RNA polymerase. *Cell* **2017**, *169*, 1240–1248. [CrossRef]
28. Efremova, A.S.; Zakharenko, A.L.; Shram, S.I.; Kulikova, I.V.; Drenichev, M.S.; Sukhanova, M.V.; Khodyreva, S.N.; Myasoedov, N.F.; Lavrik, O.I.; Mikhailov, S.N. Disaccharide pyrimidine nucleosides and their derivatives: A novel group of cell-penetrating inhibitors of poly(ADP-ribose) polymerase 1. *Nucleosides, Nucleotides Nucleic Acids.* **2013**, *32*, 510–528. [CrossRef]
29. Pivazyan, A.D.; Birks, E.M.; Wood, T.G.; Lin, T.S.; Prusoff, W.H. Inhibition of poly(ADP-ribose)polymerase activity by nucleoside analogs of thymidine. *Biochem. Pharmacol.* **1992**, *44*, 947–953. [CrossRef]
30. Toledano, E.; Ogryzko, V.; Danchin, A.; Ladant, D.; Mechold, U. 3′-5′ Phosphoadenosine phosphate is an inhibitor of PARP-1 and a potential mediator of the lithium-dependent inhibition of PARP-1 in vivo. *Biochem. J.* **2012**, *443*, 485–490. [CrossRef]
31. Banasik, M.; Komura, H.; Shimoyama, M.; Ueda, K. Specific inhibitors of poly (ADP-ribose) synthetase and mono (ADP-ribosyl) transferase. *J. Biol. Chem.* **1992**, *267*, 1569–1575.
32. Steinhagen, H.; Gerisch, M.; Mittendorf, J.; Schlemmer, K.H.; Albrecht, B. Substituted uracil derivatives as potent inhibitors of poly (ADP-ribose) polymerase-1 (PARP-1). *Bioorg. Med. Chem. Lett.* **2002**, *12*, 3187–3190. [CrossRef]
33. Jagtap, P.G.; Southan, G.J.; Baloglu, E.; Ram, S.; Mabley, J.G.; Marton, A.; Salzman, A.; Szabo, C. The discovery and synthesis of novel adenosine substituted 2,3-dihydro-1H-isoindol-1-ones: Potent inhibitors of poly(ADP-ribose) polymerase-1 (PARP-1). *Bioorg. Med. Chem. Lett.* **2004**, *14*, 81–85. [CrossRef] [PubMed]
34. Ekblad, T.; Camaioni, E.; Schüler, H.; Macchiarulo, A. PARP inhibitors: Polypharmacology versus selective inhibition. *FEBS J.* **2013**, *280*, 3563–3575. [CrossRef] [PubMed]

35. Wallrodt, S.; Simpson, E.L.; Marx, A. Investigation of the action of poly(ADP-ribose)-synthesising enzymes on NAD+ analogues. *Beilstein J. Org. Chem.* **2017**, *13*, 495–501. [CrossRef] [PubMed]
36. Buntz, A.; Wallrodt, S.; Gwosch, E.; Schmalz, M.; Beneke, S.; Ferrando-May, E.; Marx, A.; Zumbusch, A. Real-time cellular imaging of protein poly(ADP-ribos)ylation. *Angew. Chem. Int. Ed. Engl.* **2016**, *55*, 11256–11260. [CrossRef] [PubMed]
37. Gibson, B.A.; Zhang, Y.; Jiang, H.; Hussey, K.M.; Shrimp, J.H.; Lin, H.; Schwede, F.; Yu, Y.; Kraus, W.L. Chemical genetic discovery of PARP targets reveals a role for PARP-1 in transcription elongation. *Science* **2016**, *353*, 45–50. [CrossRef]
38. Langelier, M.F.; Zandarashvili, L.; Aguiar, P.M.; Black, B.E.; Pascal, J.M. NAD+ analog reveals PARP-1 substrate-blocking mechanism and allosteric communication from catalytic center to DNA-binding domains. *Nat. Commun.* **2018**, *9*, 844. [CrossRef]
39. Banasik, M.; Stedeford, T.; Strosznajder, R.P. Natural inhibitors of poly(ADP-ribose) polymerase 1. *Mol. Neurobiol.* **2012**, *46*, 55–63. [CrossRef]
40. Tanaka, Y.; Matsunami, N.; Yoshihara, K. Inhibition of ADP-ribosylation of histone by diadenosine 5′, 5″ -P(1), P(4)-tetraphosphate. *Biochem. Biophys. Res. Commun.* **1981**, *99*, 837–843. [CrossRef]
41. Bonnac, L.; Chen, L.; Pathak, R.; Gao, G.; Ming, Q.; Bennett, E.; Felczak, K.; Kullberg, M.; Patterson, S.E.; Mazzola, F.; et al. Probing binding requirements of NAD kinase with modified substrate (NAD) analogues. *Bioorg. Med. Chem. Lett.* **2007**, *17*, 1512–1515. [CrossRef] [PubMed]
42. Pergolizzi, G.; Cominetti, M.M.D.; Butt, J.N.; Field, R.A.; Bowater, R.P.; Wagner, G.K. Base-modified NAD and AMP derivatives and their activity against bacterial DNA ligases. *Org. Biomol. Chem.* **2015**, *13*, 6380–6398. [CrossRef] [PubMed]
43. Wang, S.; Zhu, W.; Wang, X.; Li, J.; Zhang, K.; Zhang, L.; Zhao, Y.-J.; Lee, H.C.; Zhang, L. Design, synthesis and SAR studies of NAD analogues as potent inhibitors towards CD38 NADase. *Molecules* **2014**, *19*, 15754–15767. [CrossRef] [PubMed]
44. Sherstyuk, Y.V.; Zakharenko, A.L.; Kutuzov, M.M.; Chalova, P.V.; Sukhanova, M.V.; Lavrik, O.I.; Silnikov, V.N.; Abramova, T.V. A versatile strategy for the design and synthesis of novel ADP conjugates and their evaluation as potential poly(ADP-ribose) polymerase 1 inhibitors. *Mol. Divers.* **2017**, *21*, 101–113. [CrossRef]
45. Sherstyuk, Y.V.; Zakharenko, A.L.; Kutuzov, M.M.; Sukhanova, M.V.; Lavrik, O.I.; Silnikov, V.N.; Abramova, T.V. Synthesis of a series of NAD+ analogues, potential inhibitors of PARP 1, using ADP conjugates functionalized at the terminal phosphate group. *Russ. J. Bioorgan. Chem.* **2017**, *43*, 76–83. [CrossRef]
46. Abramova, T.V.; Belov, S.S.; Tarasenko, Y.V.; Silnikov, V.N. Solid-phase-supported synthesis of morpholino-glycine oligonucleotide mimics. *Beilstein J. Org. Chem.* **2014**, *10*, 1151–1158. [CrossRef]
47. Ivanisenko, N.V.; Zhechev, D.A.; Ivanisenko, V.A. Structural modeling of NAD+ binding modes to PARP-1. *Russ. J. Genetics: Applied Res.* **2017**, *7*, 574–579. [CrossRef]
48. Summerton, J.E. Invention and early history of morpholinos: From pipe dream to practical products. *Methods Mol. Biol.* **2017**, *1565*, 1–15.
49. Marciacq, F.; Sauvaigo, S.; Issartel, J.-P.; Mouret, J.-F.; Molko, D. Synthesis and enzymatic incorporation of morpholino thymidine-5-triphosphate in DNA fragments. *Tetrahedron Lett.* **1999**, *40*, 4673–4676. [CrossRef]
50. Tarasenko, Y.V.; Abramova, T.V.; Mamatuk, V.I.; Silnikov, V.N. Effective synthesis of fluorescently labeled morpholino nucleoside yriphosphate derivatives. *Nucleosides Nucleotides Nucleic Acids* **2016**, *35*, 32–42. [CrossRef]
51. Summerton, J.E.; Weller, D.D. Uncharged Morpholino-based Polymers Having Achiral Intersubunit Linkages. US Patent Application No. 5,034,506, 23 July 1991.
52. Vohtancev, I.P.; Sherstyuk, Y.V.; Silnikov, V.N.; Abramova, T.V. Effective synthesis of 5-iodo derivatives of pyrimidine morpholino nucleosides. *Org. Prep. Proced. Int.* **2018**, *50*, 332–342. [CrossRef]
53. Abramova, T.V.; Bakharev, P.A.; Vasilyeva, S.V.; Silnikov, V.N. Synthesis of morpholine nucleosides triphosphates. *Tetrahedron Lett.* **2004**, *45*, 4361–4364. [CrossRef]
54. Yamamoto, I.; Sekine, M.; Hata, T. One-step synthesis of 5′-azido-nucleosides. *J. Chem. Soc. Perkin T. 1* **1980**, 306–310. [CrossRef]
55. Mesmaeker, A.; Lesueur, C.; Bévièrre, M.-O.; Waldner, A.; Fritsch, V.; Wolf, R.M. Stark erhöhte affinität modifizierter oligonucleotide mit in ihrer konformation eingeschränkten furanose-ringen für komplementäre RNA-stränge. *Angew. Chem.* **1996**, *108*, 2960–2964. [CrossRef]

56. Robins, M.J.; Doboszewski, B.; Nilsson, B.L.; Peterson, M.A. Synthesis of amide-linked [(3′)CH_2CO-NH(5′)] nucleoside analogues of small oligonucleotides. *NucleosidesNucleotides Nucleic Acids* **2006**, *19*, 69–86. [CrossRef] [PubMed]
57. Dean, D.K. An improved synthesis of 5′-amino-5′-deoxyguanosine. *Synth. Commun.* **2002**, *32*, 1517–1521. [CrossRef]
58. Peterson, T.V.; Streamland, T.U.B.; Awad, A.M. A tractable and efficient one-pot synthesis of 5′-azido-5′-deoxyribonucleosides. *Molecules* **2014**, *19*, 2434–2444. [CrossRef]
59. Kotikam, V.; Rozners, E. Concurrent hydrogenation of three functional groups enables synthesis of C3′-homologated nucleoside amino acids. *Org. Lett.* **2017**, *19*, 4122–4125. [CrossRef]
60. Zhang, W.; Ntai, I.; Bolla, M.L.; Malcolmson, S.J.; Kahne, D.; Kelleher, N.L.; Walsh, C.T. Nine enzymes are required for assembly of the pacidamycin group of peptidyl nucleoside antibiotics. *J. Am. Chem. Soc.* **2011**, *133*, 5240–5243. [CrossRef]
61. Trunkfield, A.E.; Gurcha, S.S.; Besra, G.S.; Bugg, T.D.H. Inhibition of *Escherichia coli* glycosyltransferase MurG and *Mycobacterium tuberculosis* Gal transferase by uridine-linked transition state mimics. *Bioorg. Med. Chem.* **2010**, *18*, 2651–2663. [CrossRef]
62. Yeoh, K.K.; Butters, T.D.; Wilkinson, B.L.; Fairbanks, A.J. Probing replacement of pyrophosphate via click chemistry; synthesis of UDP-sugar analogues as potential glycosyl transferase inhibitors. *Carbohydr. Res.* **2009**, *344*, 586–591. [CrossRef] [PubMed]
63. Liu, F.; Austin, D.J. A General synthesis of 5′-azido-5′-deoxy-2′,3′-*O*-isopropylidene nucleosides. *J. Org. Chem.* **2001**, *66*, 8643–8645. [CrossRef] [PubMed]
64. Abramova, T.V.; Kassakin, M.F.; Lomzov, A.A.; Pyshnyi, D.V.; Silnikov, V.N. New oligonucleotide analogues based on morpholine subunits joined by oxalyl diamide tether. *Bioorg. Chem.* **2007**, *35*, 258–275. [CrossRef] [PubMed]
65. Sinha, S.; Pattanayak, S.; Paul, S.; Nandi, B. Morpholino-based Antisense Agent. US Patent Appliction No. 9,914,745, 13 March 2018.
66. Xavier, N.M.; Gonçalves-Pereira, R.; Jorda, R.; Řezníčková, E.; Kryštof, V.; Oliveira, M.C. Synthesis and antiproliferative evaluation of novel azido nucleosides and their phosphoramidate derivatives. *Pure Appl. Chem.* **2017**, *89*, 1267–1281. [CrossRef]
67. Al-Masoudi, N.A.; Pfleiderer, W. Synthesis and reactions of 1-(5-azido-5-deoxy-3-*O*-*p*-toluenesulfonyl-β-D-xylofuranosyl) derivatives of 5-alkyl- and 5-halo-pyrimidines. *Carbohydr. Res.* **1995**, *275*, 95–105. [CrossRef]
68. Ugarkar, B.G.; DaRe, J.M.; Kopcho, J.J.; Browne, C.E.; Schanzer, J.M.; Wiesner, J.B.; Erion, M.D. Adenosine kinase inhibitors. 1. Synthesis, enzyme inhibition, and antiseizure activity of 5-iodotubercidin analogues. *J. Med. Chem.* **2000**, *43*, 2883–2893. [CrossRef]
69. Lin, T.-S.; Neenan, J.P.; Cheng, Y.-C.; Prusoff, W.H. Synthesis and antiviral activity of 5- and 5′-substituted thymidine analogs. *J. Med. Chem.* **1976**, *19*, 495–498. [CrossRef]
70. Schinazi, R.F.; Chen, M.S.; Prusoff, W.H. Antiviral and antineoplastic activities of pyrimidine arabinosyl nucleosides and their 5′-amino derivatives. *J. Med. Chem.* **1979**, *22*, 1273–1277. [CrossRef]
71. Garegg, P.J.; Regberg, T.; Stawiński, J.; Strömberg, R.J. Nucleoside phosphonates: Part 7. Studies on the oxidation of nucleoside phosphonate esters. *Chem. Soc. Perkin Tras. 1* **1987**, 1269–1273. [CrossRef]
72. Huang, J.; McElroy, E.B.; Widlanski, T.S. Synthesis of sulfonate-linked DNA. *J. Org. Chem.* **1994**, *59*, 3520–3521. [CrossRef]
73. Grimm, G.N.; Boutorine, A.S.; Helene, C. Rapid routes of synthesis of oligonucleotide conjugates from non-protected oligonucleotides and ligands possessing different nucleophilic or electrophilic functional groups. *NucleosidesNucleotides Nucleic Acids* **2000**, *19*, 1943–1965. [CrossRef] [PubMed]
74. Rankin, P.W.; Jacobson, E.L.; Benjamin, R.C.; Moss, J.; Jacobson, M.K. Quantitative studies of inhibitors of polyADP-ribosylation in vitro and in vivo. *J. Biol. Chem.* **1989**, *264*, 4312–4317. [PubMed]
75. Purnell, M.R.; Whish, W.J. Novel inhibitors of poly(ADP-ribose) synthetase. *Biochem. J.* **1980**, *185*, 775–777. [CrossRef] [PubMed]
76. Marangoni, A.G. *Enzyme Kinetics. A Modern Approach*; John Wiley & Sons INC.: New York, NY, USA, 2003; p. 248.
77. Kurgina, T.A.; Anarbaev, R.O.; Sukhanova, M.V.; Lavrik, O.I. A rapid fluorescent method for the real-time measurement of poly(ADP-ribose) polymerase 1 activity. *Anal. Biochem.* **2018**, *545*, 91–97. [CrossRef]

78. Ruf, A.; Rolli, V.; de Murcia, G.; Schulz, G.E. The mechanism of the elongation and branching reaction of poly (ADP-ribose) polymerase as derived from crystal structures and mutagenesis. *J. Mol. Biol.* **1998**, *278*, 57–65. [CrossRef]
79. Lindgren, A.E.; Karlberg, T.; Thorsell, A.G.; Hesse, M.; Spjut, S.; Ekblad, T.; Andersson, C.D.; Pinto, A.F.; Weigelt, J.; Hottiger, M.O.; et al. PARP inhibitor with selectivity toward ADP-ribosyltransferase ARTD3/PARP3. *ACS Chem. Biol.* **2013**, *8*, 1698–1703. [CrossRef]
80. Vyas, S.; Chesarone-Cataldo, M.; Todorova, T.; Huang, Y.H.; Chang, P. A systematic analysis of the PARP protein family identifies new functions critical for cell physiology. *Nat. Commun.* **2013**, *4*, 2240. [CrossRef]
81. Weeks, S.D.; Drinker, M.; Loll, P.J. Ligation independent cloning vectors for expression of SUMO fusions. *Protein Expr. Purif.* **2007**, *53*, 40–50. [CrossRef]
82. Sukhanova, M.V.; Khodyreva, S.N.; Lavrik, O.I. Poly(ADP-ribose) polymerase-1 inhibits strand-displacement synthesis of DNA catalyzed by DNA polymerase beta. *Biochemistry* **2004**, *69*, 558–568.
83. Amé, J.C.; Rolli, V.; Schreiber, V.; Niedergang, C.; Apiou, F.; Decker, P.; Muller, S.; Höger, T.; Ménissier-de Murcia, J.; de Murcia, G. PARP-2, A novel mammalian DNA damage-dependent poly(ADP-ribose) polymerase. *J. Biol. Chem.* **1999**, *274*, 17860–17868. [CrossRef]
84. Laemmli, U.K. Cleavage of structural proteins during the assembly of the head of bacteriophage T4. *Nature* **1970**, *227*, 680–685. [CrossRef]
85. Papeo, G.; Posteri, H.; Borghi, D.; Busel, A.A.; Caprera, F.; Casale, E.; Ciomei, M.; Cirla, A.; Corti, E.; D'Anello, M.; et al. Discovery of 2-[1-(4,4-difluorocyclohexyl)piperidin-4-yl]-6-fluoro-3-oxo-2,3 -dihydro-1H-isoindole-4-carboxamide (NMS-P118): A potent, orally available, and highly selective PARP-1 inhibitor for cancer therapy. *J. Med Chem.* **2015**, *58*, 6875–6898. [CrossRef] [PubMed]
86. Karlberg, T.; Hammarstroöm, M.; Schuützt, P.; Svensson, L.; Schuüler, H. Crystal structure of the catalytic domain of human PARP2 in complex with PARP inhibitor ABT-888. *Biochemistry* **2010**, *49*, 1056–1058. [CrossRef] [PubMed]
87. Friesner, R.A.; Murphy, R.B.; Repasky, M.P.; Frye, L.L.; Greenwood, J.R.; Halgren, T.A.; Sanschagrin, P.C.; Mainz, D.T. Extra Precision Glide: Docking and scoring incorporating a model of hydrophobic enclosure for protein-ligand complexes. *J. Med. Chem.* **2006**, *49*, 6177–6196. [CrossRef]
88. Case, D.A.; Darden, T.A.; Cheatham, T.E., III; Simmerling, C.L.; Wang, J.; Duke, R.E.; Luo, R.; Walker, R.C.; Zhang, W.; Merz, K.M.; et al. *AMBER 12*; University of California: San Francisco, CA, USA, 2012; Available online: https://ambermd.org/ (accessed on 28 November 2019).
89. Wang, J.; Wang, W.; Kollman, P.A. Antechamber: An accessory software package for molecular mechanical calculations. *J. Am. Chem. Soc.* **2001**, *222*, U403.
90. Frisch, M.; Trucks, G.W.; Schlegel, H.; Scuseria, G.E.; Robb, M.A.; Cheeseman, J.R.; Montgomery, J.; Vreven, T.; Kudin, K.-N.; Burant, J.; et al. Gaussian 03, revision C. 02. 2004. Available online: https://gaussian.com (accessed on 28 November 2019).
91. Onufriev, A.; Bashford, D.; Case, D.A. Modification of the generalized Born model suitable for macromolecules. *J. Phys. Chem. B.* **2000**, *104*, 3712–3720. [CrossRef]
92. Pastor, R.W.; Brooks, B.R.; Szabo, A. An analysis of the accuracy of Langevin and molecular dynamics algorithms. *Mol. Phys.* **1988**, *65*, 1409–1419. [CrossRef]
93. Naim, M.J.; Alam, O.; Alam, M.J.; Alam, P.; Shrivastava, N. A review on pharmacological profile of morpholine derivatives. *Int. J. Pharmacol. Pharmaceut. Sci.* **2015**, *3*, 40–51.
94. Thorsell, A.G.; Ekblad, T.; Karlberg, T.; Loöw, M.; Pinto, A.F.; Trésaugues, L.; Moche, M.; Cohen, M.S.; Schuüler, H. Structural basis for potency and promiscuity in poly (ADP-ribose) polymerase (PARP) and tankyrase inhibitors. *J. Med. Chem.* **2016**, *60*, 1262–1271. [CrossRef]

International Journal of
Molecular Sciences

MDPI

Article

Promising New Inhibitors of Tyrosyl-DNA Phosphodiesterase I (Tdp 1) Combining 4-Arylcoumarin and Monoterpenoid Moieties as Components of Complex Antitumor Therapy

Tatyana M. Khomenko [1], Alexandra L. Zakharenko [2], Arina A. Chepanova [2], Ekaterina S. Ilina [2], Olga D. Zakharova [2], Vasily I. Kaledin [3], Valeriy P. Nikolin [3], Nelly A. Popova [3,4], Dina V. Korchagina [1], Jóhannes Reynisson [5], Raina Chand [5], Daniel M. Ayine-Tora [6], Jinal Patel [6], Ivanhoe K. H. Leung [6], Konstantin P. Volcho [1,4,*], Nariman F. Salakhutdinov [1,4] and Olga I. Lavrik [2,4,7]

1 N.N. Vorozhtsov Novosibirsk Institute of Organic Chemistry, 9 acad. Lavrentjev ave., 630090 Novosibirsk, Russia; chomenko@nioch.nsc.ru (T.M.K.); korchaga@nioch.nsc.ru (D.V.K.) anvar@nioch.nsc.ru (N.F.S.)

2 Novosibirsk Institute of Chemical Biology and Fundamental Medicine, 8, acad. Lavrentjev ave., 630090 Novosibirsk, Russia; sashaz@niboch.nsc.ru (A.L.Z.); arinachepanova@mail.ru (A.A.C.); katya.plekhanova@gmail.com (E.S.I.); garonna3@mail.ru (O.D.Z.); lavrik@niboch.nsc.ru (O.I.L.)

3 Institute of Cytology and Genetics, 10, acad. Lavrentjev Ave., 630090 Novosibirsk, Russian; kaledin@bionet.nsc.ru (V.I.K.); nikolin@bionet.nsc.ru (V.P.N.); nelly@bionet.nsc.ru (N.A.P.)

4 Novosibirsk State University, V. Zelman Institute for Medicine and Psychology and Department of Natural Sciences, 2, Pirogova str., 630090 Novosibirsk, Russia

5 School of Pharmacy and Bioengineering, Keele University, Hornbeam Building, Staffordshire ST5 5BG, UK; j.reynisson@keele.ac.uk

6 School of Chemical Sciences, The University of Auckland, Private Bag 92019, 1142 Auckland, New Zealand; rcha387@aucklanduni.ac.nz (R.C.); dayi479@aucklanduni.ac.nz (D.M.A.-T.); jpat649@aucklanduni.ac.nz (J.P.); i.leung@auckland.ac.nz (I.K.H.L.)

7 Department of Physical and Chemical Biology and Biotechnology, Altai State University, 61, Lenina Ave., 656049 Barnaul, Russia

* Correspondence: volcho@nioch.nsc.ru

Received: 3 December 2019; Accepted: 20 December 2019; Published: 23 December 2019

Abstract: Tyrosyl-DNA phosphodiesterase 1 (Tdp1) is an important DNA repair enzyme in humans, and a current and promising inhibition target for the development of new chemosensitizing agents due to its ability to remove DNA damage caused by topoisomerase 1 (Top1) poisons such as topotecan and irinotecan. Herein, we report our work on the synthesis and characterization of new Tdp1 inhibitors that combine the arylcoumarin (neoflavonoid) and monoterpenoid moieties. Our results showed that they are potent Tdp1 inhibitors with IC_{50} values in the submicromolar range. In vivo experiments with mice revealed that compound **3ba** (IC_{50} 0.62 μM) induced a significant increase in the antitumor effect of topotecan on the Krebs-2 ascites tumor model. Our results further strengthen the argument that Tdp1 is a druggable target with the potential to be developed into a clinically-potent adjunct therapy in conjunction with Top1 poisons.

Keywords: coumarin; neoflavone; DNA repair enzymes; Tdp1 inhibitor; cancer; tumor; topotecan; topoisomerase 1 inhibitors; molecular modeling; chemical space

1. Introduction

Natural and synthetic coumarins (2*H*-chromen-2-one) demonstrate diverse biological activities, and are often considered as a privileged scaffold [1–6]. In particular, a large number of coumarin derivatives with high antitumor activity have been found in recent years [7–18]. Natural derivatives of 7-hydroxycoumarin containing terpene fragments have also attracted the attention of the medicinal chemistry community [19–21]. The best-known compound of this structural type is auraptene (Figure 1), for which a variety of biological activities are known, including antitumor properties [22].

One of the current approaches to increase the efficacy of clinically-established antitumor therapy is the inhibition of DNA repair enzymes that counteract the effect of DNA-damaging chemotherapy agents [23–25]. One of these important enzymes is tyrosyl-DNA phosphodiesterase 1 (Tdp1) [26]. Tdp1 is involved in the repairing of damaged DNA, including the removal of lesions caused by topoisomerase 1 (Top1) inhibitors. Top1 inhibitors such as the camptothecin derivatives (CPTs), topotecan (tpc), and irinotecan are well-established antitumor agents [27] that are widely used [28]. Thus, Tdp1 reduces the impact of Top1 poisons, resulting in diminished DNA damage and reduced efficacy of this class of chemotherapeutic drugs.

Because Tdp1 repairs Top1/DNA cleavage complexes induced by CPTs, inhibitors of Tdp1 can enhance the sensitivity of cancer cells to CPT analogues [29]. Furthermore, increased Tdp1 expression counteracts the cytotoxicity of CPTs [30,31], and is frequently observed in cancers resistant to CPT therapy [31–33]. Convincing evidence exists from preclinical studies that the ratio of Tdp1/Top1 activity influences cellular sensitivity to Top1 inhibitors [34,35], and that the suppression of Tdp1 activity leads to an increase in the sensitivity of tumor cells to CPTs [32,36–39]. It is believed that targeted short-term treatment with a potent Tdp1 inhibitor will not lead to serious poisoning in normal cells. Indeed, it was shown that Tdp1-/- knockout mice were fertile and had a normal life expectancy, with no signs of premature aging [40]. Until now, no inhibitors of the Tdp1 enzyme have reached human clinical testing.

To date, many Tdp1 inhibitors have been identified. A major class of Tdp1 inhibitors comprises those based on natural products including usnic acid derivatives [41–45], coumarins [46], adamantanes [47–49], nucleoside analogs [50], dehydroabietylamine derivatives [51], chromenes [52], bile acids derivatives [53], and fungal products [54–56]. There are also early reports of Tdp1 inhibition based on diamidines [57], antibiotics [58,59], steroids [60], and other compounds [61]. Nevertheless, only a few Tdp1 inhibitors have been tested in cell- or cancer-models. Synergy with tpc has been demonstrated in vivo for just two Tdp1 inhibitors, both of which were derivatives of usnic acid, a natural product [43,44].

Previously, virtual screening of the InterBioScreen natural product library [62] and subsequent testing identified that 3-methoxybenzyl, a derivative of 7-hydroxycoumarin, annelated with the cyclohexane ring **1** (Figure 1) as a new structural type of Tdp1 inhibitors [46]. Further optimization of the inhibitor, including the replacement of the aromatic substituent in the phenolic group with bulky monoterpenoid substituents, allowed us to increase the potency by almost an order of magnitude, thereby reaching nanomolar activity [46]. Most importantly, the use of compound **2** in non-toxic doses significantly increased the cytotoxic activity of CPT in human cancer cells [46].

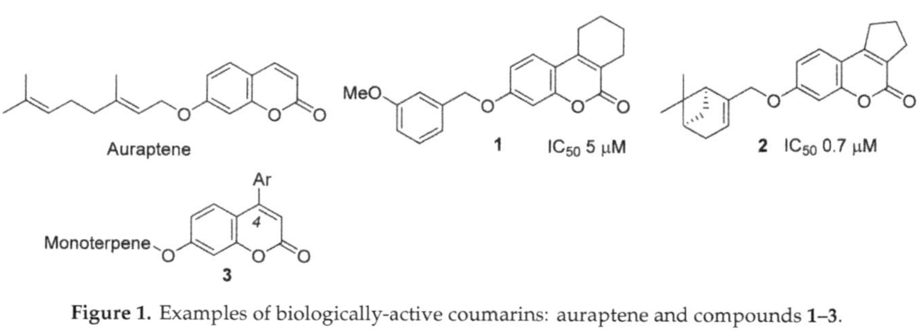

Figure 1. Examples of biologically-active coumarins: auraptene and compounds **1–3**.

Based on molecular modeling, it was predicted that the attachment of an aromatic substituent at the fourth position of coumarin would be promising for enhanced binding. Note that 4-arylcoumarins are often considered a separate group of natural products, called neoflavones. Natural and synthetic neoflavones have low toxicities and exhibit a broad spectrum of biological activity, in particular against tumors [7]. The aim of this work was to synthesize neoflavone derivatives of structural type 3 (Figure 1) by varying both aromatic and monoterpene substituents to determine their inhibitory activity against Tdp1, and to study the synergistic effect with tpc, a clinically-important Top1 inhibitor, in in vivo experiments. As a result, it was discovered that arylcoumarins containing monoterpenoid substituents are indeed potent Tdp1 inhibitors and, most importantly, are able to enhance the antitumor activity of tpc in animal models.

2. Results and Discussion

2.1. Chemistry

The main approach to producing 4-arylcoumarins unsubstituted at the hydroxy group is acid-catalyzed Pechmann condensation between resorcinol **4** and ester of β-keto-carboxylic acids **5** [63] (Scheme 1). Using this approach, we synthesized 7-hydroxy-4-arylcoumarins **6a–d** with yields of 63-81% by interaction of resorcinol **4** with esters **5a–d** (Scheme 1). Ester **5a** is commercially available, while compounds **5b–d** were obtained by the reaction of substituted acetophenones **5b–d** with diethyl carbonate in the presence of sodium hydride.

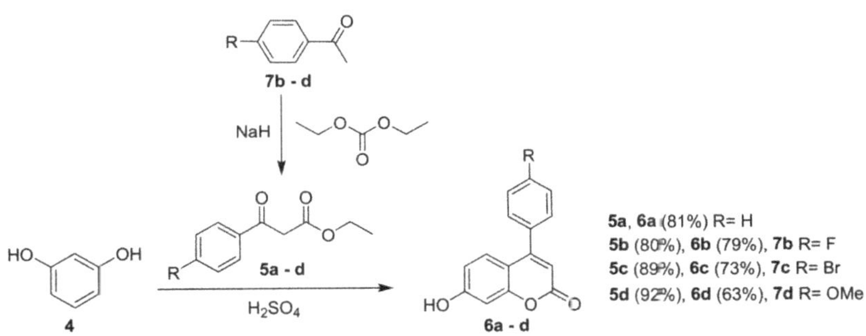

Scheme 1. Synthesis of 7-hydroxy-4-arylcoumarins **6a–d.**

Monoterpenoid bromides **8a–d** were obtained from the corresponding alcohols (geraniol, (−)-nopol, (−)- and (+)-myrtenols) by interaction with PBr_3 according to the procedure [46] (Scheme 2). Geraniol and (−)-nopol were purchased from commercial sources, while (−)- and (+)-myrtenols were synthesized from (−)-myrtenal and (+)-α-pinene in accordance with the methods [46]. The choice of monoterpenoids was based on the results we obtained previously, i.e., when both the absolute configuration of the pinane cycle and the length of the bridge played important roles [46]. A further consideration was the desire to compare the activity data obtained with bulky bicyclic substituents with corresponding data for products with acyclic monoterpene fragments. Note that coumarin-containing derivatives of geraniol can be considered as analogues of the natural coumarin auraptene, having the same monoterpenoid fragment (Figure 1).

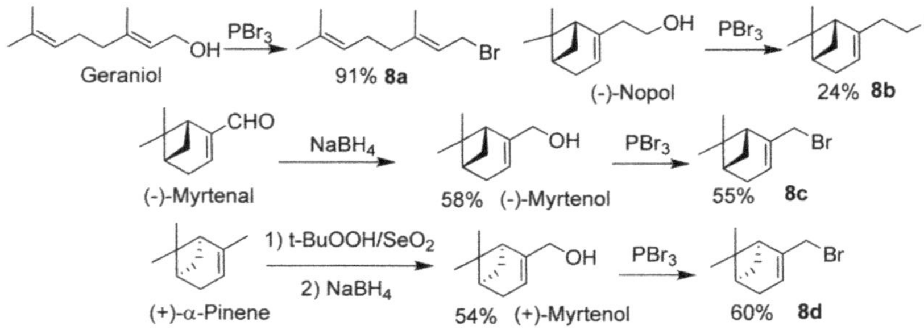

Scheme 2. Synthesis of monoterpenoid bromides **8a–d**.

The target monoterpenoid-arylcoumarin hybrids **3** were synthesized by the reaction of 7-hydroxycoumarins **6a–d** with monoterpenoid bromides **8a–d** using DBU in DMF (Scheme 3). To compare and identify the importance of the monoterpenoid fragment, compounds **10a, c, d** containing a benzyl substituent were also synthesized. The products were purified by recrystallization or column chromatography, and obtained with yields of 12–65%. In the case of the synthesis of nopol derivatives, a low conversion was observed, and double purification on SiO_2 was required, for example, for compounds **3ab** or **3cb** with yields less than 20%).

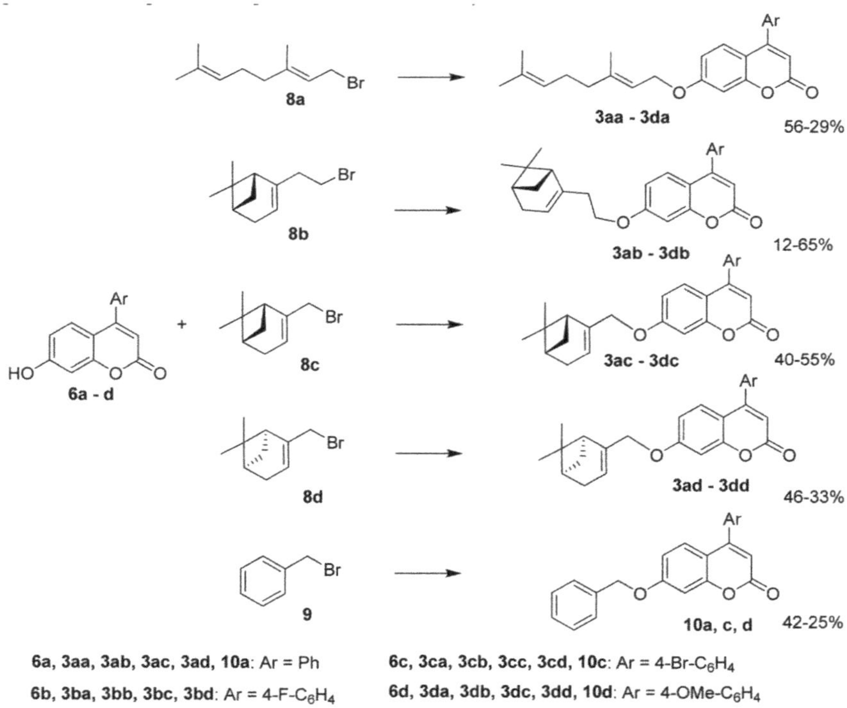

Scheme 3. Synthesis of monoterpenoid-arylcoumarin hybrids.

2.2. Biology

A previously designed [64], real-time, hexadecameric oligonucleotide biosensor with 5(6)-carboxyfluorescein (FAM) at the 5′ end and fluorophore quencher BHQ1 (Black Hole Quencher-1) at the 3′-end was used to determine the inhibitory properties of the new compounds.

The results of the Tdp1 assay for the arylcoumarin derivatives are shown in Figure 2 and Supplementary Table S1. All arylcoumarin derivatives containing a geraniol residue (**3aa–3da**) showed high inhibitory activity, with IC_{50} values in the submicromolar range; compound **3ac** with a bromine atom in the aromatic ring was slightly less active. Among the derivatives of nopol **3ab–3db**, only the fluorine derivative **3bb** showed a markedly lower activity; the remaining compounds were comparable in activity with the derivatives of geraniol. Since it was necessary to use column chromatography to obtain nopol-arylcoumarin hybrids, which complicates and increases the cost of the synthesis process, geraniol-containing inhibitors are more promising for further studies.

Almost all derivatives of (−)- and (+)-myrtenols (**3ac–3dc** and **3ad–3dd**, respectively) showed similar inhibitory activity with IC_{50} values in the 0.4-1.0 μM range, except compound **3cc**. Interestingly, compounds **10a, c, d** containing a benzyl substituent instead of monoterpenoid fragments were significantly less active than most of their monoterpenoid-containing analogues, with an IC_{50} in the micromolar range.

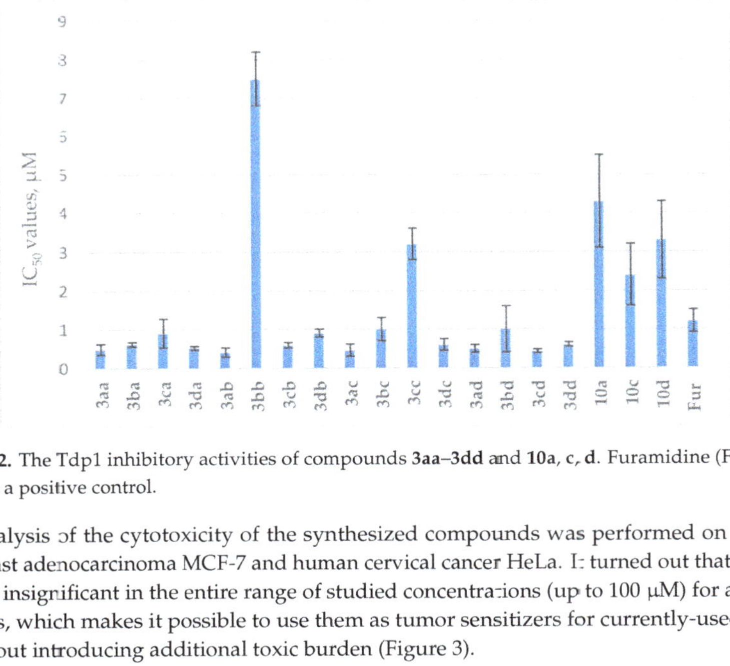

Figure 2. The Tdp1 inhibitory activities of compounds **3aa–3dd** and **10a**, c, **d**. Furamidine (Fur) was used as a positive control.

An analysis of the cytotoxicity of the synthesized compounds was performed on cell lines of human breast adenocarcinoma MCF-7 and human cervical cancer HeLa. It turned out that cytotoxicity is absent or insignificant in the entire range of studied concentrations (up to 100 μM) for all the tested compounds, which makes it possible to use them as tumor sensitizers for currently-used antitumor drugs without introducing additional toxic burden (Figure 3).

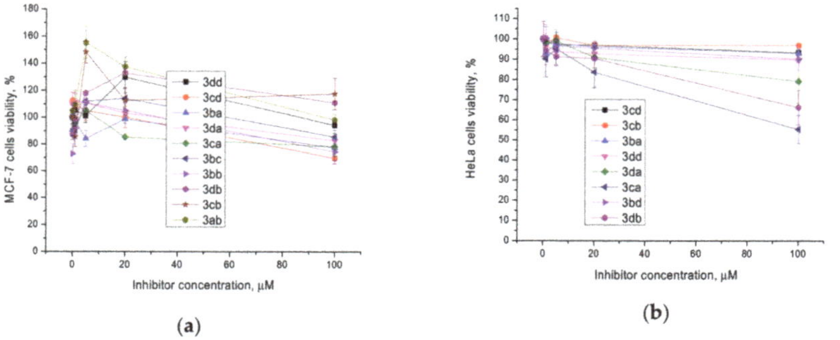

(a) (b)

Figure 3. The effect of compounds **3** on the survival of cells of the lines MCF-7 (**a**) and HeLa (**b**).

Since most monoterpene-arylcoumarin hybrids showed comparable inhibitory activity against Tdp1 (~0.5 μM) and no or limited cytotoxicity, we selected a candidate for subsequent studies based on the following considerations. Derivatives of nopol were the most complex compounds to synthesize and purify. Therefore, they were excluded from further consideration. Since we previously obtained contradictory results in in vivo experiments with the myrtenol derivative **2** (unpublished data), in this work, we decided to focus on geraniol derivatives for the in vivo studies. Among the three derivatives of geraniol that showed similar inhibitory activity (Figure 2), we selected compound **3ba** containing a fluorine atom in the para position of the aryl substituent, which can contribute to greater metabolic stability of the inhibitor [65]. In addition to the activity assay that we reported above, we wanted to confirm the interactions between compound **3ba** and Tdp1 before progressing to in vivo studies. Using an intrinsic tryptophan fluorescence quenching assay that we previously applied to study the binding interactions of Tdp1 and its inhibitors [43,45,48,49,52], we evaluated the binding of compound **3ba** to recombinant Tdp1. Clear quenching of the Tdp1 intrinsic fluorescence was observed upon the addition of **3ba** (Supplementary Figure S1). Titration experiments were then performed to determine the dissociation constant (K_D) of compound **3ba** with Tdp1. A K_D value of 63.0 ± 11 μM was obtained, indicating that it is a reasonable binder to the enzyme. This confirmed that the inhibition efficacy of compound **3ba** was due to binding to the enzyme, and gave us confidence to progress with this compound towards in vivo studies.

A study of the influence of **3ba** on the antitumor effect of tpc (topotecan) was performed using a murine Krebs-2 carcinoma model. An ascitic tumor model combines the advantages of in vitro and in vivo approaches in studying the cytotoxic effect of compounds, since ascitic cells grow in the context of the organism (in vivo), and the intraperitoneal administration of drug ensures its direct contact with tumor cells (in vitro). The experiments were performed using female C57BL/6 mice, which were injected intraperitoneally with 2×10^5 ascitic cells on day zero. The mice were divided into six groups of ten animals each. Control group 1 did not receive treatment; group 2 received tpc in a single dose of 0.5 mg/kg of body weight intraperitoneally after 2 days; group 3 received tpc as described above, and **3ba** at a dose of 80 mg/kg intraperitoneally; group 4 received tpc and **3ba** 40 mg/kg; group 5 received tpc and **3ba** 20 mg/kg; and finally, group 6 received **3ba** 80 mg/kg only.

The combined use of **3ba** at a maximum concentration of 80 mg/kg with tpc (group 3) led to a significant decrease in the weight of the ascitic tumor compared to the use of only tpc (Figure 4). The **3ba** dose of 40 mg/kg also caused a decrease in ascites weight, although the difference between groups 2 and 4 was not significant. The dose of **3ba** 20 mg/kg co-administered with tpc, as well as the use of **3ba** (80 mg/kg) in the absence of tpc, did not affect tumor growth.

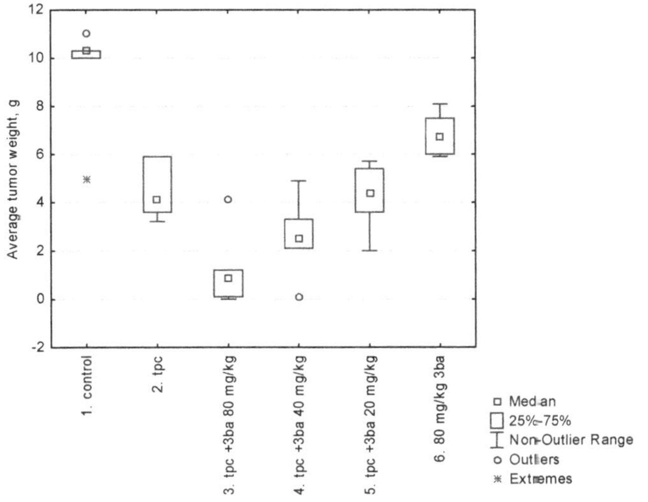

Figure 4. Box plot of the **3ba** influence on the antitumor effect of tpc against Krebs-2 carcinoma with intraperitoneal administration. $P_{1-2} = 0.002$; $P_{1-3} = 0.00013$; $P_{2-3} = 0.04$. The differences between group 2 and groups 4–6 are not significant.

The number of tumor cells in ascites (Figure 5) in the control group (1750 million per mouse) and in group 2, which received only tpc (950 mln), was significantly different, i.e., by half, p = 0.005. The size of ascites in group 3 (tpc + 80 mg/kg **3ba**) was very small; we managed to extract ascitic fluid to determine the number of cells in only one mouse; the number of tumor cells in this mouse was 250 mln. Differences in other groups are not significant.

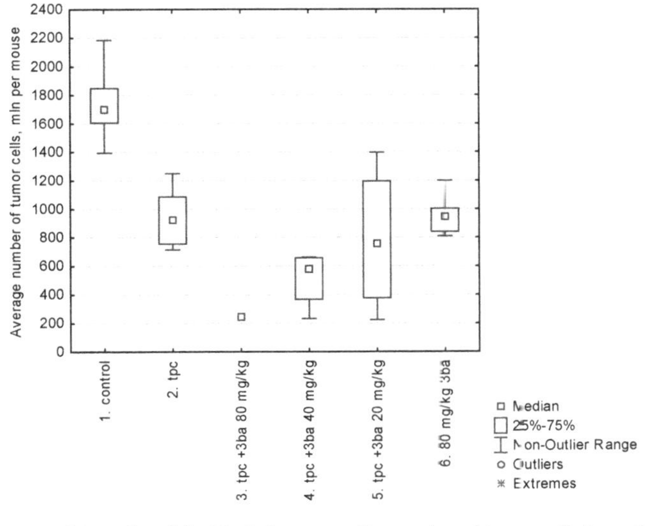

Figure 5. A box plot of the **3ba** influence on the number of tumor cells in ascites.

We then examined the effect of **3ba** in combination with tpc on the lifespan of mice. C57BL/6 mice were intraperitoneally inoculated with the ascites variant of Lewis carcinoma. Group 1: control without treatment; group 2: tpc 0.5 mg/kg intraperitoneally; group 3: tpc and **3ba** intraperitoneally 120 mg/kg; and group 4: **3ba** 120 mg/kg only. The results are given in Figure 6.

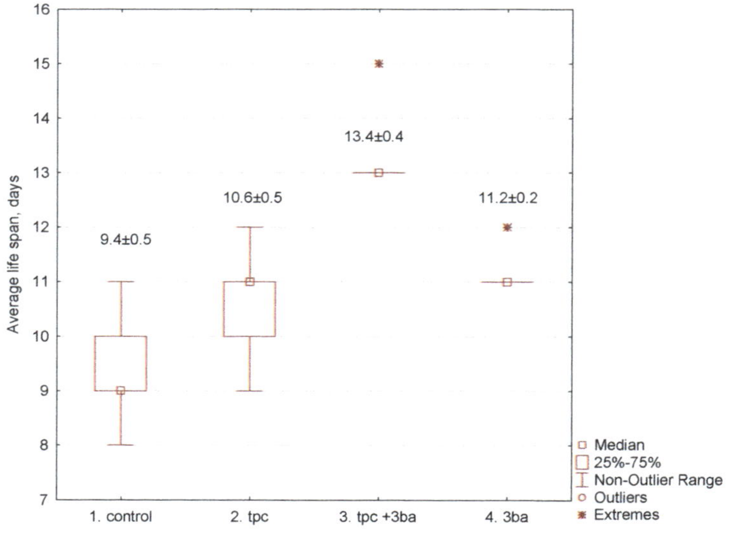

Figure 6. The influence of **3ba** in combination with tpc on the lifespan of mice. The numbers above the boxes indicate the average lifespan in the group.

When using a combination of tpc and **3ba**, a significant increase in lifespan was noted by 26% ($p = 0.0065$) compared with mice receiving only tpc, and by 42% compared with the control group ($p = 0.0002$). Monotherapy with tpc or **3ba** at selected doses prolonged the life of experimental animals unreliably, i.e., by 13–19%, > 0.05.

2.3. In Silico

2.3.1. Molecular Modeling

The 19 compounds were docked into the binding site of Tdp1 (PDB ID: 6DIE, resolution 1.78 Å) [66] with three water molecules (HOH 814, 821 and 1078). It has been shown that keeping these crystalline water molecules improves the prediction quality of the docking scaffold [45]. The modeling shows that all the ligands have a plausible binding mode and good scores with the four scoring functions used, i.e., Astex Statistical Potential (ASP) [67], improved Piecewise Linear Potential (ChemPLP) [68], ChemScore (CS) [69,70] and GoldScore (GS) [71]; the results are given in Table S2, Supplementary Information. Considering **3ba**, one of the most active compounds, the coumarin moiety occupies the hydrophilic binding region, which contains amino acids such as threonine and glutamic acid, whilst the alkene side chain occupies the hydrophobic region formed by isoleucine, leucine, and phenylalanine. The carbonyl on the benzopyrone group forms hydrogen bonds with the amine side chain groups of Lys495 and Asn516. The predicted binding mode of **3ba** is shown in Figure 7.

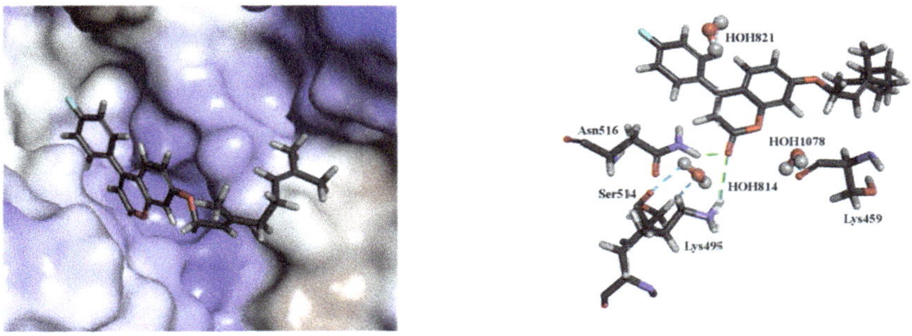

(a) (b)

Figure 7. The docked configuration of **3ba** in the binding site of Tdp1 as predicted using the ChemPLP scoring function. (**a**) The protein surface is rendered. The ligand occupies the binding pocket. Blue depicts a hydrophilic region with a partial positive charge on the surface; brown depicts hydrophobic region with a partial negative charge and grey shows neutral areas. (**b**) Hydrogen bonds are shown as green lines between the ligand and residues Lys495 and Asn516. The water molecules also form hydrogen bonds with Ser514 and Lys459.

2.3.2. Chemical Space

The calculated molecular descriptors (MW (molecular weight), log *P* (water-octanol partition coefficient), HD (hydrogen bond donors), HA (hydrogen bond acceptors), PSA (polar surface area), and RB (rotatable bonds)) are given in Table S3. The log *P* values range from 4.4 and 6.3, lying between the drug-like and Known Drug Space (KDS), while the HD and PSA values are within the lead-like space (for the definition of lead-like, drug-like, and KDS regions, see [72] and Table S4). The molecular weight of the ligands is between 326.4 and 453.4 g·mol^{-1}, falling in the drug-like chemical space. The main issue with these ligands is their relatively high lipophilicity, with Log *P* values reaching into the KDS.

The Known Drug Indexes (KDIs) of each ligand were calculated to gauge the balance of the molecular descriptor of the ligands (Table S5, Supplementary Information). This method is based on the statistical analysis of drugs in clinical use (KDS) and a weighted index for each of the six molecular descriptors used; both the summation (KDI_{2a}) and multiplication (KDI_{2b}) methods were used [73]. The KDI_{2a} values range from 4.7 to 5.3, with a theoretical maximum of 6 and an average of 4.08 for known drugs. KDI_{2b} ranges from 0.2 to 0.5, with a theoretical maximum of 1 and a KDS average of 0.18. This indicates that the majority of the ligands are well balanced. The most potent ligand **3ba** has KDI_{2a} of 4.90 and KDI_{2b} of 0.27, while drugs with high bioavailability (>50%) have average KDI_{2a} of 4.43 and KDI_{2b} of 0.21, which shows that **3ba** has a very good balance of physicochemical properties for bioavailability.

3. Materials and Methods

3.1. Chemistry Section

General Information. Reagents and solvents were purchased from commercial suppliers (Sigma-Aldrich, Acros) and used as received. GC-MS: *Agilent 7890A* gas chromatograph equipped with a quadrupole mass spectrometer *Agilent 5975C* as a detector; quartz column HP-5MS (copolymer 5%–diphenyl–95%–dimethylsiloxane) of length 30 m, internal diameter 0.25 mm and stationary phase film thickness 0.25 µm. Optical rotation: polAAr 3005 spectrometer. ^{1}H and ^{13}C NMR: *Bruker DRX-500* apparatus at 500.13 MHz (^{1}H) and 125.76 MHz (^{13}C) and *Bruker Avance—III 600* apparatus at 600.30 MHz (^{1}H) and 150.95 MHz (^{13}C), *J* in Hz; structure determinations by analyzing the ^{1}H NMR spectra, including ^{1}H–^{1}H double resonance spectra and ^{1}H–^{1}H 2D homonuclear correlation,

J-modulated ^{13}C NMR spectra (JMOD), and ^{13}C–^{1}H 2D heteronuclear correlation with one-bond (C–H COSY, ^{1}J(C,H) = 160 Hz, HSQC, ^{1}J(C,H) = 145 Hz) and long-range spin-spin coupling constants (COLOC, $^{2,3}J$(C,H) = 10 Hz, HMBC, $^{2,3}J$(C,H) = 7 Hz). HR-MS: DFS Thermo Scientific spectrometer in a full scan mode (15–500 m/z, 70 eV electron impact ionization, direct sample administration).

Spectral and analytical investigations were carried out at the Multi-Access Chemical Research Center of Siberian Branch of Russian Academy of Sciences. All product yields are given for pure compounds purified by recrystallization from ethanol or isolated by column chromatography (SiO_2; 60–200 μ; *Macherey-Nagel*). The purity of the target compounds was determined by GC-MS methods. All of the target compounds reported in this paper have purities of no less than 95%.

3.1.1. Synthesis of Compounds **5b–d**

General procedure. To a stirred mixture of sodium hydride (3 mol equiv), washed with hexane (3 × 15 mL), and diethyl carbonate (4 mol equiv) in 50 mL of tetrahydrofuran (THF), the corresponding substituted acetophenone (1 mol equiv) was added dropwise over 30 min. The reaction mixture was refluxed for 4 h, and then poured into ice water, acidified with 5 mL of glacial acetic acid, and extracted with EtOAc (3 × 100 mL). The combined organic phase was washed with saturated sodium bicarbonate, brine, and water, and then dried over anhydrous Na_2SO_4 and evaporated in vacuo. The crude products were purified by silica gel column chromatography eluting with dichloromethane to afford **5b–d**. The yields of **5b, 5c,** and **5d** were 80%, 89%, and 92%, respectively.

3.1.2. Synthesis of Compounds **6a–d**

Syntheses were carried out from resorcinol **4** and appropriate β-keto esters (**5a–d**) in accordance with [46]. Conc. H_2SO_4 (2 mL, 37.6 mmol) was added dropwise to cooled (0–5 °C) solution of resorcinol **4** (2.3 g, 21 mmol) and appropriate β-keto esters (**5a–d**) (21 mmol) in dry ethanol (5 mL) with vigorously stirring. The reaction mixture was stirred at room temperature for 15 min, and then heated at 60 °C until it congealed. It was then left overnight at room temperature. Finally, it was poured into ice water (50 mL). The resulting solid was filtered off and crystallized from ethanol–water. The yields of **6a, 6b, 6c,** and **6d** were 81%, 79%, 73%, and 63% respectively.

3.1.3. Synthesis of Compounds **8a–d**

(+)-Myrtenal was synthesized according to the procedure [46] by the oxidation of (+)-α-pinene using t-BuOOH/SeO_2 system with a 57% yield. (−)- The (+)-myrtenols were synthesized from the corresponding aldehydes via reduction to alcohols with $NaBH_4$, as described above. $NaBH_4$ (10.3 mmol) was added to a cooled (0–5 °C) solution of 10.3 mmol of the appropriate aldehyde in methanol (20 mL), and the reaction mixture was stirred for 3 h at room temperature. Then, 5% aqueous HCl was added to reach a pH of 4–5. The solvent was distilled off and the product was extracted using ether and dried with Na_2SO_4. The solvent was evaporated; the resulting alcohols (58% and 54% yields) were used in the synthesis without purification.

Bromides **8a–d** were synthesized from geraniol, (−)-nopol, and (−)- and (+)-myrtenols via the aforementioned reaction with PBr_3. PBr_3 (8.9 mmol) was added to a cooled (0–5 °C) solution of the corresponding monoterpenoid alcohols (26.7 mmol) in dry ether (30 mL), and the reaction mixture was stirred for 2 h at room temperature. Saturated aqueous $NaHCO_3$ was added, and the product was extracted with ether. The extracts were washed with brine, dried with Na_2SO_4, and evaporated. Compounds **8a**, **8c**, and **8d** (with yields of 91%, 55%, and 60%, respectively) were sufficiently pure and used for the next step without purification. The compound **8b** was purified by column chromatography on SiO_2, eluent–hexane (yield 24%).

3.1.4. Synthesis of Compounds **3aa–3da**, **3ab–3db**, **3ac–3dc**, **3ad–3dd**, and **10a**, **c**, **d**

Compounds **3aa–3da, 3ab–3db, 3ac–3dc, 3ad–3dd, and 10a, c, d** were synthesized from coumarins **6a–d** and the corresponding bromides **8a–d, 9** using DBU and DMF.

DBU (1.0 mmol) and corresponding bromide **8a–d, 9** (0.75 mmol) were added to compound **6a–d** (0.5 mmol) in dry DMF (5 mL) at room temperature under stirring. The reaction mixture was stirred at room temperature for 15 min and then heated at 60 °C for 5 h. H_2O (15 mL) was added and the product was extracted with ethyl acetate. The extracts were washed with brine, dried with Na_2SO_4, and evaporated. The products **3aa–3da, 3ab–3db, 3ac–3dc, 3ad–3dd, and 10a, c, d** were isolated in the individual form **a**) by recrystallization from ethanol; or **b**) by column chromatography on silica gel using eluent–hexane, a solution containing from 25 to 100% ethyl acetate in hexane, and ethanol.

(E)-7-(3,7-Dimethylocta-2,6-dienyloxy)-4-phenyl-2H-chromen-2-one ***3aa***. Yield 56%, method **a**. M.p. 54 °C. HRMS: 374.1879 [M]$^+$; calcd. 374.1877 $(C_{25}H_{26}O_3)^+$. ^{1}H NMR ($CDCl_3$,δ ppm, J, Hz): 1.58 (br.s, 3H, CH_3-24), 1.64 (m, 3H, all $J < 1.5$, CH_3-23), 1.74 (m, 3H, all $J < 1.5$, CH_3-25), 2.03-2.15 (m, 4H, 2H-19, 2H-20), 4.60 (d, 2H, $J_{16,17} = 6.6$, 2H-16), 5.06 (tm, 1H, $J_{21,20} = 6.8$, other $J < 1.5$, H-21), 5.45 (tm, 1H, $J_{17,16} = 6.6$, other $J < 1.5$, H-17), 6.18 (s, 1H, H-3), 6.77 (dd, 1H, $J_{7,6} = 8.9$, $J_{7,9} = 2.5$, H-7), 6.87 (d, 1H, $J_{9,7} = 2.5$, H-9), 7.34 (d, 1H, $J_{6,7} = 8.9$, H-6), 7.39-7.43 (m, 2H, H-11, H-15), 7.46-7.51 (m, 3H, H-12, H-13, H-14). ^{13}C NMR (δ ppm, $CDCl_3$): 155.85 (s, C-1), 161.12 (s, C-2), 111.61 (d, C-3), 155.70 (s, C-4), 112.28 (s, C-5), 127.75 (d, C-6), 112.81 (d, C-7), 161.99 (s, C-8), 101.75 (d, C-9), 135.52 (s, C-10), 128.23 (d, C-11, C-15), 128.66 (d, C-12, C-14), 129.40 (d, C-13), 65.38 (t, C-16), 118.34 (d, C-17), 142.19 (s, C-18), 39.37 (t, C-19), 26.11 (t, C-20), 123.49 (d, C-21), 131.78 (s, C-22), 25.50 (k, C-23), 17.56 (k, C-24), 16.63 (k, C-25).

(E)-7-(3,7-Dimethylocta-2,6-dienyloxy)-4-(4-fluorophenyl)-2H-chromen-2-one **3ba**. Yield 35%, method **a**. M.p. 72 °C. HRMS: 392.1778 [M]$^+$; calcd. 392.1782 $(C_{25}H_{25}FO_3)^+$. ^{1}H NMR ($CDCl_3$,δ ppm, J, Hz): 1.58 (s, 3H, CH_3-24), 1.64 (s, 3H, CH_3-23), 1.74 (s, 3H, CH_3-25), 2.02-2.16 (m, 4H, 2H-19, 2H-20), 4.60 (d, 2H, $J_{16,17} = 6.5$, 2H-16), 5.06 (tm, 1H, $J_{21,20} = 6.8$, other $J < 2$, H-21), 5.45 (tm, 1H, $J_{17,16} = 6.5$, other $J < 2$, H-17), 6.16 (s, 1H, H-3), 6.78 (dd, 1H, $J_{7,6} = 8.9$, $J_{7,9} = 2.5$, H-7), 6.87 (d, 1H, $J_{9,7} = 2.5$, H-9), 7.19 (dd, 2H, $J_{12,11} = J_{14,15} = 8.7$, $J_{12(14),F} = 8.7$, H-12, H-14), 7.30 (d, 1H, $J_{6,7} = 8.9$, H-6), 7.41 (dd, 2H, $J_{11,12} = J_{15,14} = 8.7$, $J_{11(15),F} = 5.3$, H-11, H-15). ^{13}C NMR (δ ppm, $CDCl_3$, $J_{C,F}$, Hz): 155.86 (s, C-1), 160.98 (s, C-2), 111.77 (d, C-3), 154.66 (s, C-4), 112.17 (s, C-5), 127.51 (d, C-6), 112.95 (d, C-7), 162.12 (d, C-8), 101.84 (d, C-9), 131.52 (s, $^4J = 3.4$, C-10), 130.18 (d, $^3J = 8.3$, C-11, C-15), 115.89 (d, $^2J = 21.4$, C-12, C-14), 163.35 (s, $^1J = 250.0$, C-13), 65.43 (t, C-16), 118.29 (d, C-17), 142.30 (s, C-18), 39.39 (t, C-19), 26.13 (t, C-20), 123.49 (d, C-21), 131.82 (s, C-22), 25.51 (k, C-23), 17.57 (k, C-24), 16.64 (k, C-25).

(E)-4-(4-Bromophenyl)-7-(3,7-dimethylocta-2,6-dienyloxy)-2H-chromen-2-one **3ca**. Yield 40%, method **a**. M.p. 80 °C. HRMS: 452.0979 [M]$^+$; calcd. 452.0982 $(C_{25}H_{25}BrO_3)^+$. ^{1}H NMR ($CDCl_3$,δ ppm, J, Hz): 1.58 (br.s, 3H, CH_3-24), 1.64 (m, 3H, all $J < 1.5$, CH_3-23), 1.74 (br.s, 3H, CH_3-25), 2.03-2.14 (m, 4H, 2H-19, 2H-20), 4.59 (d, 2H, $J_{16,17} = 6.6$, 2H-16), 5.06 (tm, 1H, $J_{21,20} = 6.7$, other $J \leq 1.5$, H-21), 5.44 (tm, 1H, $J_{17,16} = 6.6$, other $J < 1.5$, H-17), 6.16 (s, 1H, H-3), 6.78 (dd, 1H, $J_{7,6} = 8.9$, $J_{7,9} = 2.5$, H-7), 6.87 (d, 1H, $J_{9,7} = 2.5$, H-9), 7.28 (d, 1H, $J_{6,7} = 8.9$, H-6), 7.29 (br.d, 2H, $J_{11,12} = J_{15,14} = 8.5$, H-11, H-15), 7.63 (br.d, 2H, $J_{12,11} = J_{14,15} = 8.5$, H-12, H-14). ^{13}C NMR (δ ppm, $CDCl_3$): 155.82 (s, C-1), 160.92 (s, C-2), 111.65 (d, C-3), 154.52 (s, C-4), 111.83 (s, C-5), 127.43 (d, C-6), 112.99 (d, C-7), 162.13 (c, C-8), 101.78 (d, C-9), 134.31 (s, C-10), 129.84 (d, C-11, C-15), 131.98 (d, C-12, C-14), 123.87 (s, C-13), 65.39 (t, C-16), 118.16 (d, C-17), 142.38 (s, C-18), 39.37 (t, C-19), 26.08 (t, C-20), 123.45 (d, C-21), 131.84 (s, C-22), 25.54 (k, C-23), 17.59 (k, C-24), 16.65 (k, C-25).

(E)-7-(3,7-Dimethylocta-2,6-dienyloxy)-4-(4-methoxyphenyl)-2H-chromen-2-one **3da.** Yield 29%, method **b**. HRMS: 404.1980 [M]$^+$; calcd. 404.1982 $(C_{26}H_{28}O_4)^+$. ^{1}H NMR ($CDCl_3$,δ ppm, J, Hz): 1.58 (s, 3H, CH_3-24), 1.64 (m, 3H, all $J < 2.0$, CH_3-23), 1.74 (m, 3H, all $J < 1.5$, CH_3-25), 2.03-2.15 (m, 4H, 2H-19, 2H-20), 3.86 (s, 3H, CH_3-26), 4.59 (d, 2H, $J_{16,17} = 6.6$, 2H-16), 5.06 (tm, 1H, $J_{21,20} = 6.8$, other $J < 1.5$, H-21), 5.45 (tm, 1H, $J_{17,16} = 6.6$, other $J < 1.5$, H-17), 6.16 (s, 1H, H-3), 6.78 (dd, 1H, $J_{7,6} = 8.9$, $J_{7,9} = 2.5$, H-7), 6.86 (d, 1H, $J_{9,7} = 2.5$, H-9), 7.00 (br.d, 2H, $J_{12,11} = J_{14,15} = 8.7$, H-12, H-14), 7.37 (br.d, 2H, $J_{11,12} = J_{15,14} = 8.7$, H-11, H-15), 7.41 (d, 1H, $J_{6,7} = 8.9$, H-6). ^{13}C NMR (δ ppm, $CDCl_3$): 155.85 (s, C-1), 161.34 (s, C-2), 111.08 (d, C-3), 155.40 (s, C-4), 112.39 (s, C-5), 127.78 (d, C-6), 112.74 (d, C-7), 161.87 (s, C-8), 101.70 (d, C-9), 127.74 (s, C-10), 129.73 (d, C-11, C-15), 114.12 (d, C-12, C-14), 160.60 (s, C-13), 65.34 (t, C-16), 118.31 (d, C-17), 142.21 (s, C-18), 39.38 (t, C-19), 26.10 (t, C-20), 123.48 (d, C-21), 131.80 (s, C-22), 25.52 (k, C-23), 17.57 (k, C-24), 16.63 (k, C-25), 55.28 (k, C-26).

7-(2-((1R,5S)-6,6-Dimethylbicyclo[3 1.1]hept-2-en-2-yl)ethoxy)-4-phenyl-2H-chromen-2-one **3ab**. Yield 12%, method **b**. HRMS: 386.1872 [M]$^+$; calcd. 386.1877 ($C_{26}H_{26}O_3$)$^+$. $[\alpha]^{22}_{589} = -20.0$ (c = 0.53, $CHCl_3$). ^{1}H NMR ($CDCl_3$,δ ppm, J, Hz): 0.81 (c, 3H, CH_3-26), 1.16 (d, 1H, $J_{24a,24s}$ = 8.6, H-24a), 1.26 (s, 3H, CH_3-25), 2.05-2.11 (m, 2H, H-21, H-23), 2.20 (dm, 1H, $J_{20,20'}$ = 17.6, other J < 3.5, H-20), 2.26 (dm, 1H, $J_{20',20}$ = 17.6, other J < 3.5, H-20′), 2.37 (ddd, 1H, $J_{24s,24a}$ = 8.6, $J_{24s,21}$ = $J_{24s,23}$ = 5.6, H-24s), 2.40-2.52 (m, 2H, 2H-17), 3.98-4.06 (m, 2H, 2H-16), 5.32-5.36 (m, 1H, H-19), 6.18 (s, 1H, H-3), 6.74 (dd, 1H, $J_{7,6}$ = 8.9, $J_{7,9}$ = 2.5, H-7), 6.85 (d, 1H, $J_{9,7}$ = 2.5, H-9), 7.34 (d, 1H, $J_{6,7}$ = 8.9, H-6), 7.39-7.44 (m, 2H, H-11, H-15), 7.46-7.51 (m, 3H, H-12, H-13, H-14). ^{13}C NMR (δ ppm, $CDCl_3$): 155.91 (s, C-1), 161.11 (s, C-2), 111.63 (d, C-3), 155.71 (s, C-4), 112.30 (s, C-5), 127.80 (d, C-6), 112.63 (d, C-7), 162.07 (s, C-8), 101.50 (d, C-9), 135.54 (s, C-10), 128.25 (d, C-11, C-15), 128.68 (d, C-12, C-14), 129.42 (d, C-13), 66.90 (t, C-16), 36.07 (t, C-17), 143.89 (s, C-18), 118.99 (d, C-19), 31.27 (t, C-20), 40.64 (d, C-21), 37.97 (c, C-22), 45.79 (d, C-23), 31.54 (t, C-24), 26.18 (k, C-25), 21.08 (k, C-26).

7-(2-((1R,5S)-6,6-Dimethylbicyclo[3.1.1]hept-2-en-2-yl)ethoxy)-4-(4-fluorophenyl)-2H-chromen-2-one **3bb**. Yield 37%, method **b**. HRMS: 404.1774 [M]$^+$; calcd. 404.1782 ($C_{26}H_{25}FO_3$)$^+$. $[\alpha]^{22}_{589} = -17.4$ (c = 0.78, $CHCl_3$). ^{1}H NMR ($CDCl_3$,δ ppm, J, Hz): 0.81 (s, 3H, CH_3-26), 1.16 (d, 1H, $J_{24a,24s}$ = 8.6, H-24a), 1.26 (s, 3H, CH_3-25), 2.06-2.10 (m, 2H, H-21, H-23), 2.20 (dm, 1H, $J_{20,20'}$ = 17.7, H-20), 2.27 (dm, 1H, $J_{20',20}$ = 17.7, H-20′), 2.36 (ddd, 1H, $J_{24s,24a}$ = 8.6, $J_{24s,21}$ = $J_{24s,23}$ = 5.6, H-24s), 2.40-2.51 (m, 2H, 2H-17), 4.02 (t, 2H, $J_{16,17}$ = 7.0, 2H-16), 5.32-5.36 (m, 1H, H-19), 6.16 (s, 1H, H-3), 6.75 (dd, 1H, $J_{7,6}$ = 8.9, $J_{7,9}$ = 2.5, H-7), 6.84 (d, 1H, $J_{9,7}$ = 2.5, H-9), 7.16-7.21 (m, 2H, $J_{12,11}$ = $J_{14,15}$ = 8.7, $J_{12(14),F}$ =8.7, H-12, H-14), 7.30 (d, 1H, $J_{6,7}$ = 8.9, H-6), 7.38-7.43 (m, 2H, $J_{11,12}$ = $J_{15,14}$ = 8.7, $J_{11(15),F}$ = 5.3, H-11, H-15). ^{13}C NMR (δ ppm, $CDCl_3$, $J_{C,F}$, Hz): 155.90 (s, C-1), 160.94 (s, C-2), 111.76 (d, C-3), 154.64 (s, C-4), 112.16 (s, C-5), 127.54 (d, C-6), 112.74 (d, C-7), 162.18 (s, C-8), 101.57 (d, C-9), 131.51 (s, ^4J = 3.5, C-10), 130.18 (d, ^3J = 8.3, C-11, C-15), 115.88 (d, ^2J = 21.8, C-12, C-14), 66.93 (t, C-16), 36.05 (t, C-17), 143.86 (s, C-18), 119.02 (d, C-19), 31.27 (t, C-20), 40.64 (d, C-21), 37.97 (s, C-22), 45.79 (d, C-23), 31.54 (t, C-24), 26.17 (k, C-25), 21.08 (k, C-26).

4-(4-Bromophenyl)-7-(2-((1R,5S)-6,6-dimethylbicyclo[3.1.1]hept-2-en-2-yl)ethoxy)-2H-chromen-2-one **3cb**. Yield 15%, method **b**. $[\alpha]^{26.6}_{589} = -17.9$ (c = 0.58, $CHCl_3$). HRMS: 464.0988 [M]$^+$; calcd. 450.0982 ($C_{26}H_{25}BrO_3$)$^+$. ^{1}H NMR ($CDCl_3$,δ ppm, J, Hz): 0.80 (s, 3H, CH_3-26), 1.15 (d, 1H, $J_{24a,24s}$ = 8.6, H-24a), 1.25 (s, 3H, CH_3-25), 2.06-2.09 (m, 2H, H-21, H-23), 2.19 (dm, 1H, $J_{20,20'}$ = 17.6, H-20), 2.26 (dm, 1H, $J_{20',20}$ = 17.6, H-20′), 2.36 (ddd, 1H, $J_{24s,24a}$ = 8.6, $J_{24s,21}$ = $J_{24s,23}$ = 5.6, H-24s), 2.40-2.51 (m, 2H, 2H-17), 3.98-4.04 (m, 2H, 2H-16), 5.33-5.36 (m, 1H, H-19), 6.16 (s, 1H, H-3), 6.75 (dd, 1H, $J_{7,6}$ = 8.9, $J_{7,9}$ = 2.5, H-7), 6.84 (d, 1H, $J_{9,7}$ = 2.5, H-9), 7.28 (d, 1H, $J_{6,7}$ = 8.9, H-6), 7.28-7.31 (m, 2H, $J_{11,12}$ = $J_{15,14}$ = 8.4, H-11, H-15), 7.61-7.64 (m, 2H, $J_{12,11}$ = $J_{14,15}$ = 8.4, H-12, H-14). ^{13}C NMR (δ ppm, $CDCl_3$): 155.86 (s, C-1), 160.87 (s, C-2), 111.65 (d, C-3), 154.49 (s, C-4), 111.81 (s, C-5), 127.44 (d, C-6), 112.79 (d, C-7), 162.19 (s, C-8), 101.50 (d, C-9), 134.30 (s, C-10), 129.84 (d, C-11, C-15), 131.97 (d, C-12, C-14), 123.86 (s, C-13), 66.87 (t, C-16), 36.01 (t, C-17), 143.78 (s, C-18), 119.03 (d, C-19), 31.23 (t, C-20), 40.54 (d, C-21), 37.95 (s, C-22), 45.65 (d, C-23), 31.51 (t, C-24), 26.14 (k, C-25), 21.08 (k, C-26).

7-(2-((1R,5S)-6,6-Dimethylbicyclo[3.1.1]hept-2-en-2-yl)ethoxy)-4-(4-methoxyphenyl)-2H-chromen-2-one **3db**. Yield 65%, method **b**. HRMS: 415.1906 [M-H]$^+$; calcd. 415.1904 ($C_{27}H_{27}O_4$)$^+$. $[\alpha]^{22}_{589} = -15.7$ (c = 0.75, $CHCl_3$). ^{1}H NMR ($CDCl_3$,δ ppm, J, Hz): 0.81 (s, 3H, CH_3-27), 1.16 (d, 1H, $J_{25a,25s}$ = 8.6, H-25a), 1.26 (s, 3H, CH_3-26), 2.05-2.11 (m, 2H, H-21, H-24), 2.19 (dm, 1H, ^2J = 17.6, H-21), 2.26 (dm, 1H, ^2J = 17.6, H-21′), 2.36 (ddd, 1H, $J_{25s,25a}$ = 8.6, $J_{25s,22}$ = $J_{25s,24}$ = 5.6, H-25s), 2.40-2.51 (m, 2H, 2H-18), 3.86 (s, 3H, CH_3-16), 4.02 (t, 2H, $J_{17,18}$ = 7.0, H-17), 5.32-5.36 (m, 1H, H-20), 6.15 (s, 1H, H-3), 6.75 (dd, 1H, $J_{7,6}$ = 8.9, $J_{7,9}$ = 2.5, H-7), 6.83 (d, 1H, $J_{9,7}$ = 2.5, H-9), 7.01 (br.d, 2H, $J_{12,11}$ = $J_{14,15}$ = 8.7, H-12, H-14), 7.37 (br.d, 2H, $J_{11,12}$ = $J_{15,14}$ = 8.7, H-11, H-15), 7.40 (d, 1H, $J_{6,7}$ = 8.9, H-6). ^{13}C NMR (δ ppm, $CDCl_3$): 155.93 (s, C-1), 161.28 (s, C-2), 111.12 (d, C-3), 155.38 (s, C-4), 112.43 (s, C-5), 127.81 (d, C-6), 112.54 (d, C-7), 161.97 (s, C-8), 101.49 (d, C-9), 127.79 (s, C-10), 129.74 (d, C-11, C-15), 114.16 (d, C-12, C-14), 160.64 (s, C-13), 55.30 (k, C-16), 66.88 (t, C-17), 36.07 (t, C-18), 143.90 (s, C-19), 118.97 (d, C-20), 31.27 (t, C-21), 40.64 (d, C-22), 37.97 (s, C-23), 45.79 (d, C-24), 31.54 (t, C-25), 26.17 (k, C-26), 21.08 (k, C-27).

7-(((1R,5S)-6,6-Dimethylbicyclo[3.1.1]hept-2-en-2-yl)methoxy)-4-phenyl-2H-chromen-2-one **3ac**. Yield 40%, method **a**. M.p. 106 °C. HRMS: 372.1717 $[M]^+$; calcd. 372.1720 $(C_{25}H_{24}O_3)^+$. $[\alpha]_{589}^{27.3} = -25.33$ (c = 1.02, EtOH). ^1H NMR ($CDCl_3$,δ ppm, J, Hz): 0.82 (s, 3H, CH_3-25), 1.18 (d, 1H, ^2J = 8.7, H-23a), 1.28 (s, 3H, CH_3-24), 2.08-2.14 (m, 1H, H-20), 2.20 (ddd, 1H, $J_{22,20} = J_{22,23s}$ = 5.6, $J_{22,18}$ = 1.4, H-22), 2.26 (dm, 1H, ^2J = 18.1, H-19), 2.33 (dm, 1H, ^2J = 18.1, H-19′), 2.40 (ddd, 1H, ^2J = 8.7, $J_{23s,20}$= $J_{23s,22}$ = 5.6, H-23s), 4.44 (dm, 1H, ^2J = 12.4, other $J \leq 2.0$, H-16), 4.47 (dm, 1H, ^2J = 12.4, other $J \leq 2.0$, H-16′), 5.61-5.64 (m, 1H, H-18), 6.19 (s, 1H, H-3), 6.77 (dd, 1H, $J_{7,6}$ = 8.9, $J_{7,9}$ = 2.5, H-7), 6.89 (d, 1H, $J_{9,7}$ = 2.5, H-9), 7.33 (d, 1H, $J_{6,7}$ = 8.9, H-6), 7.40-7.43 (m, 2H, H-11, H-15), 7.47-7.51 (m, 3H, H-12, H-13, H-14). ^{13}C NMR (δ ppm, $CDCl_3$): 155.81 (s, C-1), 161.24 (s, C-2), 111.65 (d, C-3), 155.76 (s, C-4), 112.32 (s, C-5), 127.68 (d, C-6), 112.94 (d, C-7), 162.14 (s, C-8), 102.00 (d, C-9), 135.55 (s, C-10), 128.27 (d, C-11, C-15), 128.68 (d, C-12, C-14), 129.42 (d, C-13), 71.12 (t, C-16), 142.92 (s, C-17), 121.34 (d, C-18), 31.18 (t, C-19), 40.71 (d, C-20), 38.00 (s, C-21), 43.11 (d, C-22), 31.40 (t, C-23), 26.02 (k, C-24), 20.97 (k, C-25).

7-(((1R,5S)-6,6-Dimethylbicyclo[3.1.1]hept-2-en-2-yl)methoxy)-4-(4-fluorophenyl)-2H-chromen-2-one **3bc**. Yield 53%, method **b**. $[\alpha]_{589}^{22.0} = -15.6$ (c = 0.68, $CHCl_3$). HRMS: 390.1624 $[M]^+$; calcd. 390.1626 $(C_{25}H_{23}FO_3)^+$. ^1H NMR ($CDCl_3$,δ ppm, J, Hz): 0.81 (s, 3H, CH_3-25), 1.17 (d, 1H, $J_{23a,23s}$ = 8.7, H-23a), 1.28 (s, 3H, CH_3-24), 2.11 (ddtd, 1H, $J_{20,22} = J_{20,23s}$ = 5.6, $J_{20,19}$ = 2.9, $J_{20,18}$ = 1.3, H-20), 2.20 (ddd, 1H, $J_{22,20} = J_{22,23s}$ = 5.6, $J_{22,18}$ = 1.4, H-22), 2.26 (dm, 1H, $J_{19,19'}$ = 18.0, other J < 3.5, H-19), 2.33 (dm, 1H, $J_{19',19}$ = 18.0, other J < 3.5, H-19′), 2.40 (ddd, 1H, $J_{23s,23a}$ = 8.7, $J_{23s,20}$= $J_{23s,22}$ = 5.6, H-23s), 4.41-4.49 (m, 2H, 2H-16), 5.61-5.64 (m, 1H, H-18), 6.16 (s, 1H, H-3), 6.78 (dd, 1H, $J_{7,6}$ = 8.9, $J_{7,9}$ = 2.5, H-7), 6.88 (d, 1H, $J_{9,7}$ = 2.5, H-9), 7.16-7.21 (m, 2H, $J_{12,11} = J_{14,15}$ = 8.8, $J_{12(14),F}$ = 8.6, H-12, H-14), 7.29 (d, 1H, $J_{6,7}$ = 8.9, H-6), 7.38-7.43 (m, 2H, $J_{11,12} = J_{15,14}$ = 8.8, $J_{11(15),F}$ = 5.3, H-11, H-15). ^{13}C NMR (δ ppm, $CDCl_3$, $J_{C,F}$, Hz): 155.80 (s, C-1), 161.02 (s, C-2), 111.78 (d, C-3), 154.66 (s, C-4), 112.18 (s, C-5), 127.42 (d, C-6), 113.03 (d, C-7), 162.23 (s, C-8), 102.07 (d, C-9), 131.52 (s, ^4J = 3.5, C-10), 130.19 (d, ^3J = 8.3, C-11, C-15), 115.88 (d, ^2J = 21.8, C-12, C-14), 71.14 (t, C-16), 142.86 (s, C-17), 121.39 (d, C-18), 31.18 (t, C-19), 40.69 (d, C-20), 38.00 (s, C-21), 43.09 (d, C-22), 31.39 (t, C-23), 26.00 (k, C-24), 20.97 (k, C-25).

4-(4-Bromophenyl)-7-(((1R,5S)-6,6-dimethylbicyclo[3.1.1]hept-2-en-2-yl)methoxy)-2H-chromen-2-one **3cc**. Yield 36%, method **a**. M.p. 130 °C. HRMS: 450.0826 $[M]^+$; calcd. 450.0825 $(C_{25}H_{23}O_3Br)^+$. $[\alpha]_{589}^{26.4}$ = −13.9 (c = 0.52, $CHCl_3$). ^1H NMR ($CDCl_3$,δ ppm, J, Hz): 0.80 (s, 3H, CH_3-25), 1.16 (d, 1H, $J_{23a,23s}$ = 8.7, H-23a), 1.28 (s, 3H, CH_3-24), 2.08-2.12 (m,1H, H-20), 2.19 (ddd, 1H, $J_{22,20} = J_{22,23s}$ = 5.6, $J_{22,18}$ = 1.3, H-22), 2.25 (dm, 1H, $J_{19,19'}$ = 18.0, other J < 3.5, H-19), 2.33 (dm, 1H, $J_{19',19}$ = 18.0, other J < 3.5, H-19′), 2.40 (ddd, 1H, $J_{23s,23a}$ = 8.7, $J_{23s,20}$= $J_{23s,22}$ = 5.6, H-23s), 4.43-4.48 (m, 2H, 2H-16), 5.60-5.63 (m, 1H, H-18), 6.16 (s, 1H, H-3), 6.78 (dd, 1H, $J_{7,6}$ = 8.9, $J_{7,9}$ = 2.5, H-7), 6.88 (d, 1H, $J_{9,7}$ = 2.5, H-9), 7.27 (d, 1H, $J_{6,7}$ = 8.9, H-6), 7.30 (d, 2H, $J_{11,12} = J_{15,14}$ = 8.5, H-11, H-15), 7.63 (d, 2H, $J_{12,11} = J_{14,15}$ = 8.5, H-12, H-14). ^{13}C NMR (δ ppm, $CDCl_3$): 155.76 (s, C-1), 160.97 (s, C-2), 111.67 (d, C-3), 154.53 (s, C-4), 111.85 (s, C-5), 127.33 (d, C-6), 113.10 (d, C-7), 162.25 (s, C-8), 102.02 (d, C-9), 134.32 (s, C-10), 129.85 (d, C-11, C-15), 131.98 (d, C-12, C-14), 123.87 (s, C-13), 71.12 (t, C-16), 142.77 (s, C-17), 121.43 (d, C-18), 31.15 (t, C-19), 40.60 (d, C-20), 37.99 (s, C-21), 43.00 (d, C-22), 31.37 (t, C-23), 25.98 (k, C-24), 20.96 (k, C-25).

7-(((1R,5S)-6,6-Dimethylbicyclo[3.1.1]hept-2-en-2-yl)methoxy)-4-(4-methoxyphenyl)-2H-chromen-2-one **3dc**. Yield 55%, method **b**. M.p. 124 °C. $[\alpha]_{589}^{30.6} = -17.3$ (c = 0.82, $CHCl_3$). HRMS: 402.1820 $[M]^+$; calcd. 402.1827 $(C_{26}H_{26}O_4)^+$. ^1H NMR ($CDCl_3$,δ ppm, J, Hz): 0.81 (s, 3H, CH_3-26), 1.17 (d, 1H, $J_{24a,24s}$ = 8.7, H-24a), 1.28 (s, 3H, CH_3-25), 2.10 (ddtd, 1H, $J_{21,23} = J_{21,24s}$ = 5.6, $J_{21,20}$ = 2 8, $J_{21,19}$ = 1.3, H-21), 2.20 (ddd, 1H, $J_{23,21} = J_{23,24s}$ = 5.6, $J_{23,19}$ = 1.4, H-23), 2.25 (dm, 1H, $J_{20,20'}$ = 18.0, other J < 3.0, H-20), 2.33 (dm, 1H, $J_{20',20}$ = 18.0, other J < 3.5, H-20′), 2.40 (ddd, 1H, $J_{24s,24a}$ = 8.7, $J_{24s,21}$= $J_{24s,23}$ = 5.6, H-24s), 3.85 (s, 3H, CH_3-16), 4.41-4.49 (m, 2H, 2H-17), 5.60-5.63 (m, 1H, H-19), 6.14 (s, 1H, H-3), 6.77 (dd, 1H, $J_{7,6}$ = 8.9, $J_{7,9}$ = 2.5, H-7), 6.86 (d, 1H, $J_{9,7}$ = 2.5, H-9), 7.00 (br.d, 2H, $J_{12,11} = J_{14,15}$ = 8.8, H-12, H-14), 7.36 (br.d, 2H, $J_{11,12} = J_{15,14}$ = 8.8, H-11, H-15), 7.39 (d, 1H, $J_{6,7}$ = 8.9, H-6). ^{13}C NMR (δ ppm, $CDCl_3$): 155.80 (s, C-1), 161.30 (s, C-2), 111.11 (d, C-3), 155.36 (s, C-4), 112.41 (s, C-5), 127.67 (d, C-6), 112.79 (d, C-7), 161.99 (s, C-8), 101.96 (d, C-9), 127.77 (s, C-10), 129.72 (d, C-11, C-15), 114 13 (d, C-12, C-14), 160.62 (s, C-13), 55.27 (k, C-16), 71.06 (t, C-17), 142.91 (s, C-18), 121.26 (d, C-19), 31.15 (t, C-20), 40.67 (d, C-21), 37.97 (s, C-22), 43.07 (d, C-23), 31.37 (t, C-24), 25.99 (k, C-25), 20.94 (k, C-26).

7-(((1S,5R)-6,6-Dimethylbicyclo[3.1.1]hept-2-en-2-yl)methoxy)-4-phenyl-2H-chromen-2-one **3ad**. Yield 46%, method **b**. $[\alpha]_{589}^{26.6}$= +23.33 (*c*=1.02, EtOH). HRMS: 372.1718 $[M]^+$; calcd. 372.1720 $(C_{25}H_{24}O_3)^+$. The ^1H and ^{13}C NMR spectra of **3ad** correspond to the spectra of the enantiomer **3ac**.

7-(((1S,5R)-6,6-Dimethylbicyclo[3.1.1]hept-2-en-2-yl)methoxy)-4-(4-fluorophenyl)-2H-chromen-2-one **3bd**. Yield 35%, method **a**. M.p. 129 °C. HRMS: 390.1628 $[M]^+$; calcd. 390.1626 $(C_{25}H_{23}FO_3)^+$. $[\alpha]_{589}^{22.0}$ = +22.3 ($CHCl_3$, *c* = 0.53). The ^1H and ^{13}C NMR spectra of **3bd** correspond to the spectra of the enantiomer **3bc**.

4-(4-Bromophenyl)-7-(((1S,5R)-6,6-dimethylbicyclo[3.1.1]hept-2-en-2-yl)methoxy)-2H-chromen-2-one ***3cd***. Yield 39%, method **a**. M.p. 138 °C. HRMS: 450.0825 $[M]^+$; calcd. 450.0820 $(C_{25}H_{23}BrO_3)^+$. $[\alpha]_{589}^{26.7}$ = +22.4 (*c* = 0.58, $CHCl_3$). The ^1H and ^{13}C NMR spectra of **3cd** correspond to the spectra of the enantiomer **3cc**.

7-(((1S,5R)-6,6-Dimethylbicyclo[3.1.1]hept-2-en-2-yl)methoxy)-4-(4-methoxyphenyl)-2H-chromen-2-one ***3dd***. Yield 33%, method **a**. M.p. 112 °C. HRMS: 402.1823 $[M]^+$; calcd. 402.1827 $(C_{26}H_{26}O_4)^+$. $[\alpha]_{589}^{22.0}$ = +20.8 (*c* = 0.72, $CHCl_3$). The ^1H and ^{13}C NMR spectra of **3dd** correspond to the spectra of the enantiomer **3dc**.

7-(Benzyloxy)-4-phenyl-2H-chromen-2-one **10a**. Yield 42%, method **b**. M.p. 92 °C. HRMS: 328.1093 $[M]^+$; calcd. 328.1094 $(C_{22}H_{16}O_3)^+$. ^1H NMR ($CDCl_3$,δ ppm, *J*, Hz): 5.13 (s, 2H, 2H-16), 6.19 (s, 1H, H-3), 6.85 (dd, 1H, $J_{7,6}$ = 8.9, $J_{7,9}$ = 2.5, H-7), 6.94 (d, 1H, $J_{9,7}$ = 2.5, H-9), 7.31-7.35 (m, 1H, H-20), 7.35–7.44 (m, 7H, H-6, H-11, H-15, H-18, H-19, H-21, H-22), 7.46-7.52 (m, 3H, H-12, H-13, H-14). ^{13}C NMR (δ ppm, $CDCl_3$): 155.80 (s, C-1), 161.00 (s, C-2), 111.86 (d, C-3), 155.62 (s, C-4), 112.62 (s, C-5), 127.88 (d, C-6), 112.80 (d, C-7), 161.71 (s, C-8), 102.09 (d, C-9), 135.44 (s, C-10), 128.23 (d, C-11, C-15), 128.68 (d, C-12, C-14), 129.44 (d, C-13), 70.39 (t, C-16), 135.67 (s, C-17), 127.35 (d, C-18, C-22), 128.62 (d, C-19, C-21), 128.23 (d, C-20).

7-(Benzyloxy)-4-(4-bromophenyl)-2H-chromen-2-one **10c**. Yield 25%, method **a**. M.p. 128 °C. HRMS: 406.0204 $[M]^+$; calcd. 406.0199 $(C_{22}H_{15}BrO_3)^+$. ^1H NMR ($CDCl_3$,δ ppm, *J*, Hz): 5.13 (s, 2H, 2H-16), 6.17 (s, 1H, H-3), 6.86 (dd, 1H, $J_{7,6}$ = 8.9, $J_{7,9}$ = 2.5, H-7), 6.94 (d, 1H, $J_{9,7}$ = 2.5, H-9), 7.29 (dm, 2H, $J_{11,12}$ = $J_{15,14}$ = 8.5, H-11, H-15), 7.31 (d, 1H, $J_{6,7}$ = 8.9, H-6), 7.32-7.35 (m, 1H, H-20), 7.37-7.43 (m, 4H, H-18, H-19, H-21, H-22), 7.63 (dm, 2H, $J_{12,11}$ = $J_{14,15}$ = 8.5, H-12, H-14). ^{13}C NMR (δ ppm, $CDCl_3$): 155.79 (s, C-1), 160.81 (s, C-2), 111.92 (d, C-3), 154.46 (s, C-4), 112.18 (s, C-5), 127.56 (d, C-6), 113.02 (d, C-7), 161.86 (s, C-8), 102.14 (d, C-9), 134.24 (s, C-10), 129.83 (d, C-11, C-15), 132.01 (d, C-12, C-14), 123.92 (s, C-13), 70.42 (t, C-16), 135.54 (s, C-17), 127.38 (d, C-18, C-22), 128.66 (d, C-19, C-21), 128.30 (d, C-20).

7-(Benzyloxy)-4-(4-methoxyphenyl)-2H-chromen-2-one **10d**. Yield 34%, method **b**. M.p. 143 °C. HRMS: 358.1198 $[M]^+$; calcd. 358.1200 $(C_{23}H_{18}O_4)^+$. ^1H NMR ($CDCl_3$,δ ppm, *J*, Hz): 3.86 (s, 3H, CH_3-16), 5.12 (s, 2H, 2H-17), 6.16 (s, 1H, H-3), 6.85 (dd, 1H, $J_{7,6}$ = 8.9, $J_{7,9}$ = 2.5, H-7), 6.93 (d, 1H, $J_{9,7}$ = 2.5, H-9), 7.01 (dm, 2H, $J_{12,11}$ = $J_{14,15}$ = 8.7, H-12, H-14), 7.30-7.35 (m, 1H, H-21), 7.35–7.45 (m, 7H, H-6, H-11, H-15, H-19, H-20, H-22, H-23). ^{13}C NMR (δ ppm, $CDCl_3$): 155.82 (s, C-1), 161.18 (s, C-2), 111.34 (d, C-3), 155.30 (s, C-4), 112.75 (s, C-5), 127.90 (d, C-6), 112.72 (d, C-7), 161.61 (s, C-8), 102.07 (d, C-9), 127.68 (s, C-10), 129.72 (d, C-11, C-15), 114.16 (d, C-12, C-14), 160.65 (s, C-13), 55.29 (k, C-16), 70.37 (t, C-17), 135.70 (s, C-18), 127.36 (d, C-19, C-23), 128.61 (d, C-20, C-22), 128.22 (d, C-21).

3.2. Biology Section

Real-Time Detection of Tdp1 Activity. The Tdp1 activity measurements were carried out as described [64]. Briefly, Tdp1-biosensor with a final concentration of 50 nM was incubated in a volume of 200 μL containing buffer (50 mM Tris-HCl pH8.0, 50 mM NaCl, 7 mM β-mercaptoethanol) supplemented with purified 1.3 nM Tdp1. The reactions were incubated in a POLARstar OPTIMA fluorimeter, BMG LABTECH, GmbH, to measure fluorescence every 1 min (Ex485/Em520 nm). Tdp1 inhibition was calculated by comparing the rate of increase in fluorescence in the presence of the compound to that of DMSO control wells. IC_{50} values were determined using a 6-point concentration response curve. The data were imported into the MARS Data Analysis 2.0 program (BMG LABTECH), and the slope during the linear phase (here data from 0 to 7 min) was calculated.

Cell Culture Assays. Tumor cells from human mammary adenocarcinoma cell line MCF-7 and cervical cancer cell line HeLa (~2000 cells per well) were incubated for 24 h at 37 °C in IMDM medium (5% CO_2), and then treated with the synthesized derivatives. After 72 h of cell incubation, the relative amount of alive cells was determined using standard colorimetric MTT test [74] or EZ4U Cell Proliferation and Cytotoxicity Assay (Biomedica, Austria), as per the manufacturer's protocols.

Binding Assay. Synthetic DNA encoding human Tdp1 (residues 149-608) was cloned into pET-28a (+) (GenScript), which was then transformed into *Escherichia coli* BL21 (DE3) for recombinant protein production. Protein production was induced with 1 mM IPTG at 28 °C with overnight incubation. Purification of Tdp1 was performed using affinity and size exclusion chromatography. Intrinsic protein fluorescence was measured using PerkinElmer EnSpire Multimode Reader. The Tdp1 concentration was 10 μM, and the compound concentrations were 25 μM, 50 μM, 75 μM, 100 μM, 150 μM, and 250 μM. The buffer was composed of 20 mM Tris and 250 mM NaCl, pH 8. The excitation wavelength was 280 nm and the intrinsic fluorescence was measured at 350 nm. Compound control was performed using the buffer and compound only. The total volume per well was 30 μL. Dissociation constants (K_D) were calculated using the following formula, that takes nonspecific binding into account.

$$I = \frac{Imax \times [L_T]}{K_D + [L_T]} + Ns[L_T]$$

In this formula, I indicates changes in fluorescence intensity from the titration, Imax indicates the maximum fluorescence intensity change, $[L_T]$ is the titration ligand concentration, and Ns is the non-specific term. Non-linear curve fitting was conducted using SigmaPlot 13.0 (Systat Software, San Jose, CA, USA). Experiments were conducted in triplicate and the errors shown are standard derivations.

Lab animals. Three to four-month-old male and female C57Bl/6 mice from the breeding colony of the Institute of Cytology and Genetics, SB RAS, were used in the study. The animals were kept on sawdust in plastic cages with 5–7 mice per cage, with free access to ground food ("Laboratorkorm", Moskow, Russia) and tap water. All experiments were performed in accordance with protocols approved by the Animal Care and Use Committee of the Institute of Cytology and Genetics. Also, all experimental procedures were performed in accordance with the Directive 2010/63/EU for animal experiments.

Tumor models. The experimental tumor used was Lewis Lung Carcinoma (LLC) and Krebs-2. The animals were treated with tpc and the Tdp1 inhibitor **3ba** two days after tumor transplantation. The tumor was transplanted into the muscles of the thigh by 0.2 million cells per mouse. Tpc (Sindan Pharma SRL, Romania) was administered intraperitoneally at a single dose of 0.5 mg/kg; Tdp1 inhibitor **3ba** was administrated intraperitoneally at a single dose of 20, 40, or 80 mg/kg (for Krebs-2), or 120 mg/kg (for LLC) in 15% dimethyl sulfoxide (DMSO)–10% Tween-80 suspension in water (0.2 mL of suspension per mouse) simultaneously with tpc. Control mice were injected with a DMSO-Tween-80 mixture into the stomach.

The antitumor effect was assessed by the size and weight of the solid tumors at 18 days after transplantation. For estimations of daily gain in volume, the tumor nodules were periodically measured with a caliper.

Statistical analysis. The experimenter measuring and calculating the primary animal data (tumor size, lifespan) was blinded. After unblinding, the animal data were statistically processed using oneway ANOVA. Post-hoc testing was completed using Turkey's Honestly Significant Difference (HSD). $p < 0.05$ was considered to be statistically significant. The statistical package STATISTICA version 12.5 was used for analysis. All results are expressed as mean ± SEM.

3.3. Modeling Section

Molecular modeling and chemical space. The compounds were docked against the crystal structure of Tdp1 (PDB ID: 6DIE, resolution 1.78 Å) [75], which was obtained from the Protein Data Bank (PDB) [76,77]. The Scigress version FJ 2.6 program [78] was used to prepare the crystal structure for docking, i.e., the hydrogen atoms were added, and the cocrystallized ligand benzene-1,2,4-tricarboxylic acid was removed, as well as crystallographic water molecules, except HOH 814, 821, and 1078. The Scigress software suite was also used to build the inhibitors, and the MM2 [77] force field was used to optimize the structures. The docking centre was defined as the position of a carbon on the ring of benzene-1, 2, 4-tricarboxylic acid ($x = -6.052$, $y = -14.428$, $z = 33.998$) with 10 Å radius. Fifty docking runs were allowed for each ligand with the default search efficiency (100%). The basic amino acids lysine and arginine were defined as protonated. Furthermore, aspartic and glutamic acids were assumed to be deprotonated. The GoldScore(GS) [71] and ChemScore (CS) [69,70], ChemPLP (Piecewise Linear Potential) [68], and ASP (AstexStatistical Potential) [67] scoring functions were implemented to validate the predicted binding modes and relative energies of the ligands using the GOLD v5.4.1 software suite (The Cambridge Crystallographic Data Centre, Cambridge, UK). The QikProp 3.2 [79] software package (Schrödinger, New York, USA) was used to calculate the molecular descriptors of the molecules; the reliability of this method has been established for the calculated descriptors [80].

4. Conclusions

Overall, we reported the synthesis and evaluation of novel Tdp1 inhibitors that combine the arylcoumarin and monoterpenoid moieties. Our results found that these compounds are good Tdp1 inhibitors with IC_{50} in the submicromolar or low submicromolar ranges. Compound **3ba** showed a significant increase in the antitumor effect of tpc on Krebs-2 ascites in an in vivo tumor model. In addition, these compounds presented the good physicochemical properties required for oral bioavailability, making them good candidates for further development. Thus, this type of arylcoumarin-monoterpenoid hybrids represents an excellent starting point for the further development of adjuvant therapy against cancer in combination with Top 1 poisons.

Supplementary Materials: Supplementary materials can be found at http://www.mdpi.com/1422-0067/21/1/126/s1.

Author Contributions: Chemistry investigation, T.M.K., D.V.K. and K.P.V.; In vitro investigation, A.L.Z.; A.A.C., E.S.I., O.D.Z., J.P., I.K.H.L.; In vivo investigation, V.I.K., V.P.N., N.A.P., Modeling, J.R., R.C., D.M.A.-T.; Methodology, N.F.S. and O.I.L.; Project administration, K.P.V.; Supervision, K.P.V.; Writing—original draft, A.L.Z. and T.M.K.; Writing—review & editing, K.P.V., J.R., I.K.H.L., N.F.S., O.I.L. All authors have read and agreed to the published version of the manuscript.

Funding: This study was funded by the Russian Science Foundation grant N° 19-13-00040. A.A. Chepanova is grateful to Russian State funded budget project of ICBFM SB RAS N° AAAA-A17-117020210022-4 for financial support for Tdp1 purification.

Acknowledgments: Authors would like to acknowledge the Multi-Access Chemical Research Center SB RAS for spectral and analytical measurements.

Conflicts of Interest: The authors declare no conflict of interest. The funders had no role in the design of the study; in the collection, analyses, or interpretation of data; in the writing of the manuscript, or in the decision to publish the results.

Abbreviations

Tdp1	Tyrosyl-DNA phosphodiesterase 1
Top1	topoisomerase 1
CPTs	camptothecin derivatives
tpc	topotecan

References

1. Stefanachi, A.; Leonetti, F.; Pisani, L.; Catto, M.; Carotti, A. Coumarin: A Natural, Privileged and Versatile Scaffold for Bioactive Compounds. *Molecules* **2018**, *23*, 250. [CrossRef] [PubMed]
2. Schneider, P.; Schneider, G. Privileged Structures Revisited. *Angew. Chem. Int. Ed.* **2017**, *56*, 7971–7974. [CrossRef] [PubMed]

3. Riveiro, M.E.; De Kimpe, N.; Moglioni, A.; Vázquez, R.; Monczor, F.; Shayo, C.; Davio, C. Coumarins: Old Compounds with Novel Promising Therapeutic Perspectives. *Curr. Med. Chem.* **2010**, *17*, 1325–1338. [CrossRef] [PubMed]
4. Ibrar, A.; Shehzadi, S.A.; Saeed, F.; Khan, I. Developing hybrid molecule therapeutics for diverse enzyme inhibitory action: Active role of coumarin-based structural leads in drug discovery. *Bioorg. Med. Chem.* **2018**, *26*, 3731–3762. [CrossRef]
5. Khomenko, T.M.; Zarubaev, V.V.; Orshanskaya, I.R.; Kadyrova, R.A.; Sannikova, V.A.; Korchagina, D.V.; Volcho, K.P.; Salakhutdinov, N.F. Anti-influenza activity of monoterpene-containing substituted coumarins. *Bioorg. Med. Chem. Lett.* **2017**, *27*, 2920–2925. [CrossRef]
6. Singh, H.; Singh, J.V.; Bhagat, K.; Gulati, H.K.; Sanduja, M.; Kumar, N.; Kinarivala, N.; Sharma, S. Rational approaches, design strategies, structure activity relationship and mechanistic insights for therapeutic coumarin hybrids. *Bioorg. Med. Chem.* **2019**, *27*, 3477–3510. [CrossRef]
7. Batran, R.Z.; Kassem, A.F.; Abbas, E.M.H.; Elseginy, S.A.; Mounier, M.M. Design, synthesis and molecular modeling of new 4-phenylcoumarin derivatives as tubulin polymerization inhibitors targeting MCF-7 breast cancer cells. *Bioorg. Med. Chem.* **2018**, *26*, 3474–3490. [CrossRef]
8. Dhawan, S.; Kerru, N.; Awolade, P.; Singh-Pillay, A.; Saha, S.T.; Kaur, M.; Jonnalagadda, S.B.; Singh, P. Synthesis, computational studies and antiproliferative activities of coumarin-tagged 1,3,4-oxadiazole conjugates against MDA-MB-231 and MCF-7 human breast cancer cells. *Bioorg. Med. Chem.* **2018**, *26*, 5612–5623. [CrossRef]
9. Yamaguchi, Y.; Nishizono, N.; Kobayashi, D.; Yoshimura, T.; Wada, K.; Oda, K. Evaluation of synthesized coumarin derivatives on aromatase inhibitory activity. *Bioorg. Med. Chem. Lett.* **2017**, *27*, 2645–2649. [CrossRef]
10. Lingaraju, G.S.; Balaji, K.S.; Jayarama, S.; Anil, S.M.; Kiran, K.R.; Sadashiva, M.P. Synthesis of new coumarin tethered isoxazolines as potential anticancer agents. *Bioorg. Med. Chem. Lett.* **2018**, *28*, 3606–3612. [CrossRef]
11. Wang, Y.; Liu, H.; Lu, P.; Mao, R.; Xue, X.; Fan, C.; She, J. Design, Synthesis, and In Vitro Evaluation of Novel 3,7-Disubstituted Coumarin Derivatives as Potent Anticancer Agents. *Chem. Biol. Drug Des.* **2015**, *86*, 637–647. [CrossRef] [PubMed]
12. Bisi, A.; Cappadone, C.; Rampa, A.; Farruggia, G.; Sargenti, A.; Belluti, F.; Di Martino, R.M.C.; Malucelli, E.; Meluzzi, A.; Iotti, S.; et al. Coumarin derivatives as potential antitumor agents: Growth inhibition, apoptosis induction and multidrug resistance reverting activity. *Eur. J. Med. Chem.* **2017**, *127*, 577–585. [CrossRef] [PubMed]
13. Ayati, A.; Bakhshaies, T.O.; Moghimi, S.; Esmaeili, R.; Majidzadeh-A, K.; Safavi, M.; Firoozpour, L.; Emami, S.; Foroumadi, A. Synthesis and biological evaluation of new coumarins bearing 2,4-diaminothiazole-5-carbonyl moiety. *Eur. J. Med. Chem.* **2018**, *155*, 483–491. [CrossRef] [PubMed]
14. Xu, Z.; Zhao, S.J.; Deng, J.-L.; Wang, Q.; Lv, Z.-S.; Fan, Y.-L. Design, Synthesis, and Evaluation of Tetraethylene Glycol Tethered Isatin–Coumarin Hybrids as Novel Anticancer Agents. *J. Heterocycl. Chem.* **2009**, *56*, 400–405. [CrossRef]
15. Nagarsenkar, A.; Guntuku, L.; Prajapti, S.K.; Guggilapu, S.D.; Sonar, R.; Vegi, G.M.N.; Babu, B.N. Umbelliferone–oxindole hybrids as novel apoptosis inducing agents. *New J. Chem.* **2017**, *41*, 12604–12610. [CrossRef]
16. Rao, Y.J.; Goud, E.Y.; Hemasri, Y.; Jain, N.; Gabriella, S. Synthesis and Antiproliferative Activity of 6,7-Aryl/Hetaryl Coumarins. *Russ. J. Gen. Chem.* **2016**, *86*, 184–189.
17. Zhang, L.; Xu, Z. Coumarin-containing hybrids and their anticancer activities. *Eur. J. Med. Chem.* **2019**, *181*, 111587. [CrossRef]
18. Goud, N.S.; Pooladanda, V.; Mahammad, G.S.; Jakkula, P.; Gatreddi, S.; Qureshi, I.A.; Alvala, R.; Godugu, C.; Alvala, M. Synthesis and biological evaluation of morpholines linked coumarin–triazole hybrids as anticancer agents. *Chem. Biol. Drug Des.* **2019**, *94*, 1919–1929. [CrossRef]
19. Fiorito, S.; Ianni, F.; Preziuso, F.; Epifano, F.; Scotti, L.; Bucciarelli, T.; Genovese, S. UHPLC-UV/Vis Quantitative Analysis of Hydroxylated and O-prenylated Coumarins in Pomegranate Seed Extracts. *Molecules* **2019**, *24*, 1963. [CrossRef]
20. Ito, C.; Matsui, T.; Tokuda, H.; Tan, H.T.W.; Itoigawa, M. Cancer chemopreventive constituents from Melicope lunu-ankenda. *Phytochem. Lett.* **2017**, *20*, 172–176. [CrossRef]

21. Sun, S.; Phrutivorapongkul, A.; Dibwe, D.F.; Balachandran, C.; Awale, S. Chemical Constituents of Thai Citrus hystrix and Their Antiausterity Activity against the PANC-1 Human Pancreatic Cancer Cell Line. *J. Nat. Prod.* **2018**, *81*, 1877–1883. [CrossRef] [PubMed]
22. Genovese, S.; Epifano, F. Auraptene: A Natural Biologically Active Compound with Multiple Targets. *Curr. Drug Targets* **2011**, *12*, 381–386. [CrossRef] [PubMed]
23. Jain, P.G.; Patel, B.D. Medicinal chemistry approaches of poly ADP-Ribose polymerase 1(PARP1) inhibitors as anticancer agents—A recent update. *Eur. J. Med. Chem.* **2019**, *165*, 198–215. [CrossRef] [PubMed]
24. Laev, S.; Salakhutdinov, N.; Lavrik, O. Tyrosyl–DNA phosphodiesterase inhibitors: Progress and potential. *Bioorg. Med. Chem.* **2016**, *24*, 5017–5027. [CrossRef]
25. Zakharenko, A.; Dyrkheeva, N.; Lavrik, O. Dual DNA topoisomerase 1 and tyrosyl-DNA phosphodiesterase 1 inhibition for improved anticancer activity. *Med. Res. Rev.* **2019**, *39*, 1427–1441. [CrossRef]
26. Kawale, A.S.; Povirk, L.F. Tyrosyl–DNA phosphodiesterases: Rescuing the genome from the risks of relaxation. *Nucleic Acids Res.* **2018**, *46*, 520–537. [CrossRef]
27. Ohmoto, A.; Yachida, S. Current status of poly(ADP-ribose) polymerase inhibitors and future directions. *Onco Targets Ther.* **2017**, *26*, 5195–5208. [CrossRef]
28. Bailly, C. Irinotecan: 25 years of cancer treatment. *Pharmacol. Res.* **2019**, *148*, 104398. [CrossRef]
29. Dean, R.A.; Fam, H.K.; An, J.; Choi, K.; Shimizu, Y.; Jones, S.J.; Boerkoel, C.F.; Interthal, H.; Pfeifer, T.A. Identification of a putative Tdp1 inhibitor (CD00509) by in vitro and cell-based assays. *J. Biomol. Screen.* **2014**, *19*, 1372–1382. [CrossRef]
30. Barthelmes, H.U.; Habermeyer, M.; Christensen, M.O.; Mielke, C.; Interthal, H.; Pouliot, J.J.; Boege, F.; Marko, D. TDP1 overexpression in human cells counteracts DNA damage mediated by topoisomerases I and II. *J. Biol. Chem.* **2004**, *279*, 55618–55625. [CrossRef]
31. Liu, C.; Zhou, S.; Begum, S.; Sidransky, D.; Westra, W.H.; Brock, M.; Califano, J.A. Increased expression and activity of repair genes TDP1 and XPF in non-small cell lung cancer. *Lung Cancer* **2007**, *55*, 303–311. [CrossRef] [PubMed]
32. Fam, H.K.; Walton, C.; Mitra, S.A.; Chowdhury, M.; Osborne, N.; Choi, K.; Sun, G.; Wong, P.C.; O'Sullivan, M.J.; Turashvili, G.; et al. TDP1 and PARP1 deficiency are cytotoxic to rhabdomyosarcoma cells. *Mol. Cancer Res.* **2013**, *11*, 1179–1192. [CrossRef] [PubMed]
33. Pappo, A.S.; Lyden, E.; Breitfeld, P.; Donaldson, S.S.; Wiener, E.; Parham, D.; Crews, K.R.; Houghton, P.; Meyer, W.H. Two consecutive phase II window trials of irinotecan alone or in combination with vincristine for the treatment of metastatic rhabdomyosarcoma: The Children's Oncology Group. *J. Clin. Oncol.* **2007**, *25*, 362–369. [CrossRef] [PubMed]
34. Meisenberg, C.; Ward, S.E.; Schmid, P.; El-Khamisy, S.F. TDP1/TOP1 Ratio as a Promising Indicator for the Response of Small Cell Lung Cancer to Topotecan. *J. Cancer Sci. Ther.* **2014**, *6*, 258–267. [CrossRef] [PubMed]
35. Wang, W.; Rodriguez-Silva, M.; Acanda de la Rocha, A.M.; Wolf, A.L.; Lai, Y.; Liu, Y.; Reinhold, W.C.; Pommier, Y.; Chambers, J.W.; Tse-Dinh, Y.C. Tyrosyl-DNA Phosphodiesterase 1 and Topoisomerase I Activities as Predictive Indicators for Glioblastoma Susceptibility to Genotoxic Agents. *Cancers* **2019**, *11*, 1416. [CrossRef] [PubMed]
36. Keil, A.; Frese-Schaper, M.; Steiner, S.K.; Korner, M.; Schmid, R.A.; Frese, S. The Topoisomerase I Inhibitor Irinotecan and the Tyrosyl-DNA Phosphodiesterase 1 Inhibitor Furamidine Synergistically Suppress Murine Lupus Nephritis. *Arthritis Rheumatol.* **2015**, *67*, 1858–1867. [CrossRef]
37. He, X.; van Waardenburg, R.C.; Babaoglu, K.; Price, A.C.; Nitiss, K.C.; Nitiss, J.L.; Bjornsti, M.A.; White, S.W. Mutation of a conserved active site residue converts tyrosyl-DNA phosphodiesterase I into a DNA topoisomerase I-dependent poison. *J. Mol. Biol.* **2007**, *372*, 1070–1081. [CrossRef]
38. Interthal, H.; Chen, H.J.; Kehl-Fie, T.E.; Zotzmann, J.; Leppard, J.B.; Champoux, J.J. SCAN1 mutant Tdp1 accumulates the enzyme—DNA intermediate and causes camptothecin hypersensitivity. *EMBO J.* **2005**, *24*, 2224–2233. [CrossRef]
39. Murai, J.; Huang, S.Y.; Das, B.B.; Dexheimer, T.S.; Takeda, S.; Pommier, Y. Tyrosyl-DNA phosphodiesterase 1 (TDP1) repairs DNA damage induced by topoisomerases I and II and base alkylation in vertebrate cells. *J. Biol. Chem.* **2012**, *287*, 12848–12857. [CrossRef]
40. Katyal, S.; El-Khamisy, S.F.; Russell, H.R.; Li, Y.; Ju, L.; Caldecott, K.W.; McKinnon, P.J. TDP1 facilitates chromosomal single-strand break repair in neurons and is neuroprotective in vivo. *EMBO J.* **2007**, *26*, 4720–4731. [CrossRef]

41. Zakharenko, A.; Luzina, O.; Koval, O.; Nilov, D.; Gushchina, I.; Dyrkheeva, N.; Švedas, V.; Salakhutdinov, N.; Lavrik, O. Tyrosyl-DNA phosphodiesterase 1 inhibitors: Usnic acid enamines enhance the cytotoxic effect of camptothecin. *J. Nat. Prod.* **2016**, *79*, 2961–2967. [CrossRef] [PubMed]
42. Zakharova, O.; Luzina, O.; Zakharenko, A.; Sokolov, D.; Filimonov, A.; Dyrkheeva, N.; Chepanova, A.; Ilina, E.; Ilyina, A.; Klabenkova, K.; et al. Synthesis and evaluation of aryliden- and hetarylidenfuranone derivatives of usnic acid as highly potent Tdp1 inhibitors. *Bioorg. Med. Chem.* **2018**, *26*, 4470–4480. [CrossRef] [PubMed]
43. Zakharenko, A.L.; Luzina, O.A.; Sokolov, D.N.; Kaledin, V.I.; Nikolin, V.P.; Popova, N.A.; Patel, J.; Zakharova, O.D.; Chepanova, A.A.; Zafar, A.; et al. Novel tyrosyl-DNA phosphodiesterase 1 inhibitors enhance the therapeutic impact of topotecan on in vivo tumor models. *Eur. J. Med. Chem.* **2019**, *161*, 581–593. [CrossRef] [PubMed]
44. Koldysheva, E.V.; Men'shchikova, A.P.; Lushnikova, E.L.; Popova, N.A.; Kaledin, V.I.; Nikolin, V.P.; Zakharenko, A.L.; Luzina, O.A.; Salakhutdinov, N.F.; Lavrik, O.I. Antimetastatic Activity of Combined Topotecan and Tyrosyl-DNA Phosphodiesterase-1 Inhibitor on Modeled Lewis Lung Carcinoma. *Bull. Exp. Biol. Med.* **2019**, *166*, 661–666. [CrossRef]
45. Filimonov, A.S.; Chepanova, A.A.; Luzina, O.A.; Zakharenko, A.L.; Zakharova, O.D.; Ilina, E.S.; Dyrkheeva, N.S.; Kuprushkin, M.S.; Kolotaev, A.V.; Khachatryan, D.S.; et al. New Hydrazinothiazole Derivatives of Usnic Acid as Potent Tdp1 Inhibitors. *Molecules* **2019**, *24*, 3711. [CrossRef]
46. Khomenko, T.; Zakharenko, A.; Odarchenko, T.; Arabshahi, H.J.; Sannikova, V.; Zakharova, O.; Korchagina, D.; Reynisson, J.; Volcho, K.; Salakhutdinov, N.; et al. New inhibitors of tyrosyl-DNA phosphodiesterase I (Tdp 1) combining 7-hydroxycoumarin and monoterpenoid moieties. *Bioorg. Med. Chem.* **2016**, *24*, 5573–5581. [CrossRef]
47. Ponomarev, K.Y.; Suslov, E.V.; Zakharenko, A.L.; Zakharova, O.D.; Rogachev, A.D.; Korchagina, D.V.; Zafar, A.; Reynisson, J.; Nefedov, A.A.; Volcho, K.P.; et al. Aminoadamantanes containing monoterpene-derived fragments as potent tyrosyl-DNA phosphodiesterase 1 inhibitors. *Bioorg. Chem.* **2018**, *76*, 392–399. [CrossRef]
48. Mozhaitsev, E.S.; Zakharenko, A.L.; Suslov, E.V.; Korchagina, D.V.; Zakharova, O.D.; Vasil'eva, I.A.; Chepanova, A.A.; Black, E.; Patel, J.; Chand, R.; et al. Novel Inhibitors of DNA Repair Enzyme TDP1 Combining Monoterpenoid and Adamantane Fragments. *Anticancer Agents Med. Chem.* **2019**, *19*, 463–472. [CrossRef]
49. Chepanova, A.A.; Mozhaitsev, E.S.; Munkuev, A.A.; Suslov, E.V.; Korchagina, D.V.; Zakharova, O.D.; Zakharenko, A.L.; Patel, J.; Ayine-Tora, D.M.; Reynisson, J.; et al. The Development of Tyrosyl-DNA Phosphodiesterase 1 Inhibitors. Combination of Monoterpene and Adamantine Moieties via Amide or Thioamide Bridges. *Appl. Sci.* **2019**, *9*, 2767. [CrossRef]
50. Komarova, A.O.; Drenichev, M.S.; Dyrkheeva, N.S.; Kulikova, I.V.; Oslovsky, V.E.; Zakharova, O.D.; Zakharenko, A.L.; Mikhailov, S.N.; Lavrik, O.I. Novel group of tyrosyl-DNA-phosphodiesterase 1 inhibitors based on disaccharide nucleosides as drug prototypes for anti-cancer therapy. *J. Enzym. Inhib. Med. Chem.* **2018**, *33*, 1415–1429. [CrossRef]
51. Kovaleva, K.; Oleshko, O.; Mamontova, E.; Yarovaya, O.; Zakharova, O.; Zakharenko, A.; Kononova, A.; Dyrkheeva, N.; Cheresiz, S.; Pokrovsky, A.; et al. Dehydroabietylamine Ureas and Thioureas as Tyrosyl-DNA Phosphodiesterase 1 Inhibitors That Enhance the Antitumor Effect of Temozolomide on Glioblastoma Cells. *J. Nat. Prod.* **2019**, *82*, 2443–2450. [CrossRef] [PubMed]
52. Li-Zhulanov, N.S.; Zakharenko, A.L.; Chepanova, A.A.; Patel, J.; Zafar, A.; Volcho, K.P.; Salakhutdinov, N.F.; Reynisson, J.; Leung, I.K.H.; Lavrik, O.I. A Novel Class of Tyrosyl-DNA Phosphodiesterase 1 Inhibitors That Contains the Octahydro-2H-chromen-4-ol Scaffold. *Molecules* **2018**, *23*, 2468. [CrossRef] [PubMed]
53. Salomatina, O.V.; Popadyuk, I.I.; Zakharenko, A.L.; Zakharova, O.D.; Fadeev, D.S.; Komarova, N.I.; Reynisson, J.; Arabshahi, H.J.; Chand, R.; Volcho, K.P.; et al. Novel semisynthetic derivatives of bile acids as effective tyrosyl-DNA phosphodiesterase 1 inhibitors. *Molecules* **2018**, *23*, 679. [CrossRef] [PubMed]
54. Zhang, Y.; Wang, Z.; Li, D.; Zang, W.; Zhu, H.; Wu, P.; Mei, Y.; Liang, Y. A polysaccharide from Antrodia cinnamomea mycelia exerts antitumor activity through blocking of TOP1/TDP1-mediated DNA repair pathway. *Int. J. Biol. Macromol.* **2018**, *120*, 1551–1560. [CrossRef] [PubMed]
55. Tian, L.W.; Feng, Y.; Tran, T.D.; Shimizu, Y.; Pfeifer, T.; Vu, H.T.; Quinn, R.J. Achyrodimer F, a tyrosyl-DNA phosphodiesterase I inhibitor from an Australian fungus of the family Cortinariaceae. *Bioorg. Med. Chem. Lett.* **2017**, *27*, 4007–4010. [CrossRef] [PubMed]

56. Takagi, M.; Ueda, J.Y.; Hwang, J.H.; Hashimoto, J.; Izumikawa, M.; Murakami, H.; Sekido, Y.; Shin-ya, K. Tyrosyl-DNA phosphodiesterase 1 inhibitor from an anamorphic fungus. *J. Nat. Prod.* **2012**, *75*, 764–767. [CrossRef]
57. Antony, S.; Marchand, C.; Stephen, A.G.; Thibaut, L.; Agama, K.K.; Fisher, R.J.; Pommier, Y. Novel high-throughput electrochemiluminescent assay for identification of human tyrosyl-DNA phosphodiesterase (Tdp1) inhibitors and characterization of furamidine (NSC 305831) as an inhibitor of Tdp1. *Nucleic Acids Res.* **2007**, *35*, 4474–4484. [CrossRef]
58. Huang, Y.; Qureshi, I.A.; Chen, H. Effects of phosphatidylinositol 4,5-bisphosphate and neomycin on phospholipase D: Kinetic studies. *Mol. Cell. Biochem.* **1999**, *197*, 195–201. [CrossRef]
59. Liao, Z.; Thibaut, L.; Jobson, A.; Pommier, Y. Inhibition of human tyrosyl-DNA phosphodiesterase by aminoglycoside antibiotics and ribosome inhibitors. *Mol. Pharmacol.* **2006**, *70*, 366–372. [CrossRef]
60. Dexheimer, T.S.; Gediya, L.K.; Stephen, A.G.; Weidlich, I.; Antony, S.; Marchand, C.; Interthal, H.; Nicklaus, M.; Fisher, R.J.; Njar, V.C.; et al. 4-Pregnen-21-ol-3,20-dione-21-(4-bromobenzenesulfonate) (NSC 88915) and related novel steroid derivatives as tyrosyl-DNA phosphodiesterase (Tdp1) inhibitors. *J. Med. Chem.* **2009**, *52*, 7122–7131. [CrossRef]
61. Huang, S.N.; Pommier, Y.; Marchand, C. Tyrosyl-DNA Phosphodiesterase 1 (Tdp1) inhibitors. *Expert Opin. Ther. Pat.* **2011**, *21*, 1285–1292. [CrossRef] [PubMed]
62. The Natural Compound InterBioScreen Collection. Available online: https://www.ibscreen.com/natural-compounds (accessed on 26 November 2019).
63. Jung, J.-W.; Kim, N.-J.; Yun, H.; Han, Y.T. Recent Advances in Synthesis of 4-Arylcoumarins. *Molecules* **2018**, *23*, 2417. [CrossRef] [PubMed]
64. Zakharenko, A.; Khomenko, T.; Zhukova, S.; Koval, O.; Zakharova, O.; Anarbaev, R.; Lebedeva, N.; Korchagina, D.; Komarova, N.; Vasil'ev, V.; et al. Synthesis and biological evaluation of novel tyrosyl-DNA phosphodiesterase 1 inhibitors with a benzopentathiepine moiety. *Bioorg. Med. Chem.* **2015**, *23*, 2044–2052. [CrossRef] [PubMed]
65. Gillis, E.P.; Eastman, K.J.; Hill, M.D.; Donnelly, D.J.; Meanwell, N.A. Applications of Fluorine in Medicinal Chemistry. *J. Med. Chem.* **2015**, *58*, 8315–8359. [CrossRef] [PubMed]
66. Lountos, G.T.; Zhao, X.Z.; Kiselev, E.; Tropea, J.E.; Needle, D.; Pommier, Y.; Burke, T.R.; Waugh, D.S. Identification of a ligand binding hot spot and structural motifs replicating aspects of tyrosyl-DNA phosphodiesterase I (TDP1) phosphoryl recognition by crystallographic fragment cocktail screening. *Nucleic Acids Res.* **2019**, *47*, 10134–10150. [CrossRef] [PubMed]
67. Mooij, W.; Verdonk, M.L. General and targeted statistical potentials for protein–ligand interactions. *Proteins* **2005**, *61*, 272–287. [CrossRef]
68. Korb, O.; Stutzle, T.; Exner, T.E. Empirical scoring functions for advanced protein−ligand docking with PLANTS. *J. Chem. Inf. Model* **2009**, *49*, 84–96. [CrossRef]
69. Eldridge, M.D.; Murray, C.W.; Auton, T.R.; Paolini, G.V.; Mee, R.P. Empirical scoring functions: I. The development of a fast empirical scoring function to estimate the binding affinity of ligands in receptor complexes. *J. Comput. Aided Mol. Des.* **1997**, *11*, 425–445. [CrossRef]
70. Verdonk, M.L.; Cole, J.C.; Hartshorn, M.J.; Murray, C.W.; Taylor, R.D. Improved protein–ligand docking using GOLD. *Proteins* **2003**, *52*, 609–623. [CrossRef]
71. Jones, G.; Willett, P.; Glen, R.C.; Leach, A.R.; Taylor, R. Development and validation of a genetic algorithm for flexible docking. *J. Mol. Biol.* **1997**, *267*, 727–748. [CrossRef]
72. Zhu, F.; Logan, G.; Reynisson, J. Wine Compounds as a Source for HTS Screening Collections. A Feasibility Study. *Mol. Inf.* **2012**, *31*, 847–855. [CrossRef] [PubMed]
73. Eurtivong, C.; Reynisson, J. The Development of a Weighted Index to Optimise Compound Libraries for High Throughput Screening. *Mol. Inf.* **2018**, *38*, 100068. [CrossRef] [PubMed]
74. Mosmann, T.J. Rapid colorimetric assay for cellular growth and survival: Application to proliferation and cytotoxicity assays. *Immunol. Methods* **1983**, *65*, 55–63. [CrossRef]
75. Davies, D.R.; Interthal, H.; Champoux, J.J.; Hol, W.G.J. Insights into Substrate Binding and Catalytic Mechanism of Human Tyrosyl-DNA Phosphodiesterase (Tdp1) from Vanadate and Tungstate-inhibited Structures. *J. Mol. Biol.* **2002**, *324*, 917–932. [CrossRef]
76. Berman, H.M.; Westbrook, J.; Feng, Z.; Gilliland, G.; Bhat, T.; Weissig, H.; Shindyalov, I.N.; Bourne, P.E. The protein data bank. *Nucleic Acids Res.* **2000**, *28*, 235–242. [CrossRef]

77. Berman, H.; Henrick, K.; Nakamura, H. Announcing the worldwide protein data bank. *Nat. Struct. Mol. Biol.* **2003**, *10*, 980. [CrossRef]
78. *Scigress*, version FJ 2.6 (EU 3.1.7); Fujitsu Limited: Tokyo, Japan, 2008–2016.
79. *QikProp*, version 3.2; Schrödinger: New York, NY, USA, 2009.
80. Ioakimidis, L.; Thoukydidis, L.; Naeem, S.; Mirza, A.; Reynisson, J. Benchmarking the Reliability of QikProp. Correlation between Experimental and Predicted Values. *QSAR Comb. Sci.* **2008**, *27*, 445–456. [CrossRef]

International Journal of
Molecular Sciences

MDPI

Article

A Small Compound KJ-28d Enhances the Sensitivity of Non-Small Cell Lung Cancer to Radio- and Chemotherapy

Hwani Ryu [1,2], Hyo Jeong Kim [1], Jie-Young Song [1], Sang-Gu Hwang [1], Jae-Sung Kim [1], Joon Kim [2], Thi Hong Nhung Bui [3], Hyun-Kyung Choi [3,*] and Jiyeon Ahn [1,*]

1 Division of Radiation Biomedical Research, Korea Institute of Radiological & Medical Sciences (KIRAMS), Seoul 01812, Korea; hwanya85@kirams.re.kr (H.R.); hjkim@kirams.re.kr (H.J.K.); immu@kirams.re.kr (J.-Y.S.); sgh63@kirams.re.kr (S.-G.H.); jaesung@kirams.re.kr (J.-S.K.)

2 Department of Biology, Korea University, Seoul 02841, Korea; joonkim@korea.ac.kr

3 Department of Medicinal Chemistry, Jungwon University, Goesan 28024, Korea; buihongnhung103@gmail.com

* Correspondence: hkchoi45@jwu.ac.kr (H.-K.C.); ahnjy@kirams.re.kr (J.A.); Tel.: +82-2-970-1311 (J.A.)

Received: 10 October 2019; Accepted: 29 November 2019; Published: 29 November 2019

Abstract: We previously reported on a poly (ADP-ribose) polymerase (PARP) 1/2 inhibitor *N*-(3-(hydroxycarbamoyl)phenyl)carboxamide (designated KJ-28d), which increased the death of human ovarian cancer *BRCA1*-deficient SNU-251 cells. In the present study, we further investigated the antitumor activities of KJ-28d in *BRCA*-proficient non-small cell lung cancer (NSCLC) cells to expand the use of PARP inhibitors. KJ-28d significantly inhibited the growth of NSCLC cells in vitro and in vivo, and induced DNA damage and reactive oxygen species in A549 and H1299 cells. Combined treatment with KJ-28d and ionizing radiation led to increased DNA damage responses in A549 and H1299 cells compared to KJ-28d or ionizing radiation alone, resulting in apoptotic cell death. Moreover, the combination of KJ-28d plus a DNA-damaging therapeutic agent (carboplatin, cisplatin, paclitaxel, or doxorubicin) synergistically inhibited cell proliferation, compared to either drug alone. Taken together, the findings demonstrate the potential of KJ-28d as an effective anti-cancer therapeutic agent for *BRCA*-deficient and -proficient cancer cells. KJ-28d might have potential as an adjuvant when used in combination with radiotherapy or DNA-damaging agents, pending further investigations.

Keywords: poly (ADP-ribose) polymerase inhibitor; non-small cell lung cancer; DNA damage; radiotherapy; chemotherapy; combination therapy

1. Introduction

Lung cancer is one of the most commonly occurring cancers worldwide. Non-small cell lung cancer (NSCLC) accounts for more than 84% of all lung cancers [1]. Despite recent advances in our understanding of the molecular and genetic basis of lung cancer and improvements in therapy, the 5-year survival rate of patients with NSCLC in both Korea (23.7%) and the United States (23%) is lower than that for many other leading cancers [1,2]. The standard chemotherapy treatment used for NSCLC is either the platinum-based (cisplatin, carboplatin, etc.) or taxane-based (paclitaxel, docetaxel, etc.) regimens, or epidermal growth factor receptor (EGFR) inhibitors. However, because of the lack of response or resistance to those therapies, many patients with NSCLC consider new alternative therapies or receive combination therapies [3]. Therefore, the development of novel drugs or strategies of combination therapy with existing drugs is urgently required.

Ionizing radiation (IR) is directly cytotoxic through the induction of DNA single- or double-strand breaks (DSB) and indirectly cytotoxic through the generation of reactive oxygen species (ROS), leading to tumor cell death and thus cancer treatment. Similarly, platinum- or taxane-based chemotherapeutic agents inhibit cell mitosis through the induction of DNA damage. DNA lesions induced by IR or cytotoxic agents can be repaired by six major DNA repair pathways: base excision repair, nucleotide excision repair, direct repair (i.e., O^6-alkylguanine DNA alkyltransferase directly repairs DNA damaged lesions by removing O^6-methylguanine), mismatch repair, homologous recombination (HR), and non-homologous end-joining pathways. NSCLC is characterized by genomic instability with mutations and translocations in oncogenes, such as the *Ki-ras2 Kirsten rat sarcoma viral oncogene homolog proto-oncogene* (*KRAS*), *EGFR*, *ALK receptor tyrosine kinase* (*ALK*), *ataxia-telangiectasia mutated* (*ATM*) or *phosphatidylinositol-4,5-bisphosphate 3-kinase catalytic subunit alpha* (*PIK3CA*), and tumor suppressor genes, such as the *tumor protein p53* (*TP53*), *liver kinase B1* (*LKB1*), or *Kelch-like ECH-associated protein 1* (*KEAP1*) [4–11].

Poly (ADP-ribose) polymerases (PARPs) are a family of 18 protein members that function as catalytic enzymes by adding ADP-ribose polymers to lysine residues of themselves and target proteins, a process termed poly (ADP-ribosyl)ation (PARylation). Among the proteins of the PARP family, PARP-1 and -2 are abundant nuclear enzymes involved in DNA damage repair, which involves binding to DNA breaks and facilitating the localization of repair enzymes, such as breast cancer 1/2 (BRCA1/2), to the sites of DNA damage [12,13]. Olaparib, rucaparib, niraparib, and talazoparib targeting PARP1/2 have been approved by the United States Food and Drug Administration (FDA) for the treatment of breast or ovarian cancer in patients harboring HR mutations that induce synthetic lethality in the *BRCA1* or *BRCA2* gene.

Evidence is accumulating that PARP inhibitors (PARPi) have therapeutic efficacy in cancer cells with high genomic instability by inducing synthetic lethality of cells with deficient or insufficient DNA repair [14–16]. Briefly, in this scenario, IR, or radiotherapy or cytotoxic chemotherapeutic agents that induce severe DNA damage could lead to insufficient DNA repair in targeted tumor cells. Consecutively, cells that adopt the "BRCAness" state could be treated with PARP inhibitor (PARPi) [17–20].

We have previously identified a potential novel PARPi, termed *N*-(3-(hydroxycarbamoyl)phenyl) carboxamide (designated KJ-28d). KJ-28d inhibited PARP-1/2 activities and displayed significant antitumor activity in human ovarian cancer *BRCA1*-deficient (*BRCA1* mutation at 5564G>A) SNU-251 cells [21]. In this study, we further investigated the antitumor activity of KJ-28d in *BRCA*-proficient cell lines, as well as the combination of KJ-28d and DNA damage-inducing radiotherapy or cytotoxic chemotherapeutics in human NSCLC cells.

2. Results

2.1. *KJ-28d Inhibits Growth of Human NSCLC Cells In Vitro and In Vivo*

Among the reported novel PARP-1 inhibitors, we examined KJ-28d, as shown in Figure 1A, based on the prior demonstrations of its significant inhibitory activity against PARP-1, as well as antitumor activity in *BRCA*-deficient ovarian cancer cells [21]. Specifically, we assessed whether KJ-28d can inhibit the growth of *BRCA*-proficient cancer cells. Human NSCLC cells were treated with KJ-28d and growth inhibition was determined using the 3-(4,5-dimethylthiazol-2-yl)-2,5- diphenyltetrazolium bromide (MTT) assay. KJ-28d significantly inhibited the growth of A549, H1299, H460, and H1650 human NSCLC cells with a determined IC_{50} value, as shown in Figure 1B, and induced the sub-G1 phase (apoptotic cell) in A549 and H1299 cells, as shown in Supplementary Figure S1.

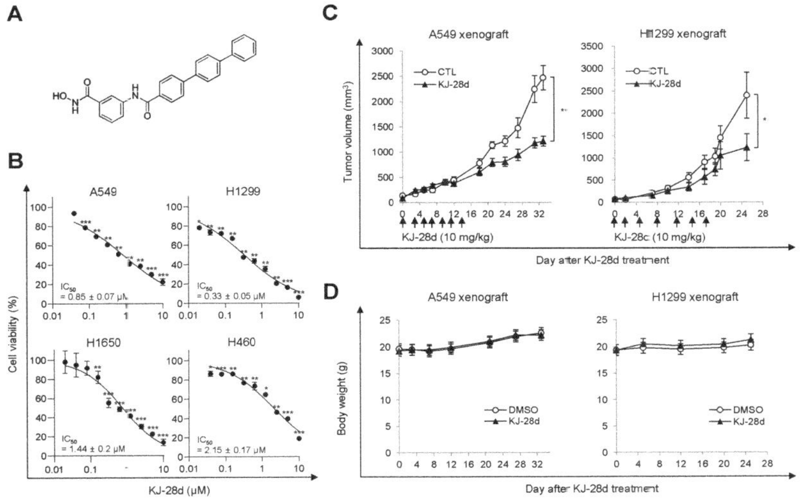

Figure 1. KJ-28d inhibits tumor growth of A549 and H1299 xenografts in nude mice. (**A**) The chemical structure of the KJ-28d compound. (**B**) A549, H1299, H1650, and H460 human non-small cell lung cancer (NSCLC) cells were treated with KJ-28d at the indicated concentrations for 5 days, and cell viabilities were determined by the MTT assay. Data are presented as means ± standard deviation (SD) from at least three independent experiments. * $p < 0.05$, ** $p < 0.01$, *** $p < 0.001$ versus DMSO-treated control. (**C**,**D**) A549 cells and H1299 cells were subcutaneously injected into the thigh of the right hind leg of BALB/c nu/nu mice ($n = 3$ per group, A549; $n = 4$ per group, H1299). Two weeks after tumor cell injection, KJ-28d (10 mg/kg) or DMSO (control) was intraperitoneally administered once every 2 or 3 days for seven times in total. (**C**) Longest (L) and shortest (W) tumor axes were measured, and tumor volume (mm^3) was calculated as $L \times W^2/2$. Data shown represent average tumor volume (* $p < 0.05$, ** $p < 0.01$). Results are shown as means ± SD. (**D**) The body weights of A549 and H1299 xenograft mice were determined once a week during the experiments. Data are shown as means ± SD.

We next determined whether the antitumor effect associated with the in vitro KJ-28d treatment could be translated into a similar effect in an in vivo xenograft mouse model. BALB/c-nu/nu mice were subcutaneously (*s.c.*) implanted with A549 or H1299 cells in the right hind leg, and when tumors were palpable (average diameter approximately 150 mm^3; 10 days post-implantation), mice were intraperitoneally (*i.p.*) administered a dose of 10 mg/kg KJ-28d or DMSO (control vehicle) once every 2 or 3 days for a total of seven times. KJ-28d treatment inhibited A549 or H1299 cell-derived tumor growth by 51% and 49%, respectively, as compared with the respective vehicle, as shown in Figure 1C. Additionally, to determine the toxicity of KJ-28d, we measured the body weight of mice. Mice treated with KJ-28d did not show any difference in body weight as compared with control mice, as shown in Figure 1D. The results suggested that KJ-28d has antitumor activity for NSCLC cells in vitro and in vivo.

2.2. KJ-28d Induces DNA Damage and Generation of ROS in NSCLC Cells

Since PARPi induce accumulation of DNA damage [22,23], we sought to determine whether KJ-28d could induce DNA damage in NSCLC cells. The DNA damage was measured in A549 and H1299 cells at different time points after KJ-28d treatment by detecting the phosphorylation on Ser139 of the H2AX (γ-H2AX) histone protein, which is an indicator of the presence of DNA double-strand damage. KJ-28d induced γ-H2AX in both cell types at the latest time point (24 h), as shown in Figure 2A. As treatment with KJ-28d resulted in inducing a DNA damage response, we also investigated whether KJ-28d could augment ROS generation in NSCLC cells. Flow cytometry analysis showed that treatment of A549 and H1299 cells with 5 μM KJ-28d led to distinctly increased ROS levels, which were then reduced following treatment with *N*-acetyl-L-cysteine (NAC), a general free radical scavenger, as shown in Figure 2B,C. These results suggested that KJ-28d is able to exhibit antitumor activities in NSCLC cells through the accumulation of DNA damage and the generation of ROS.

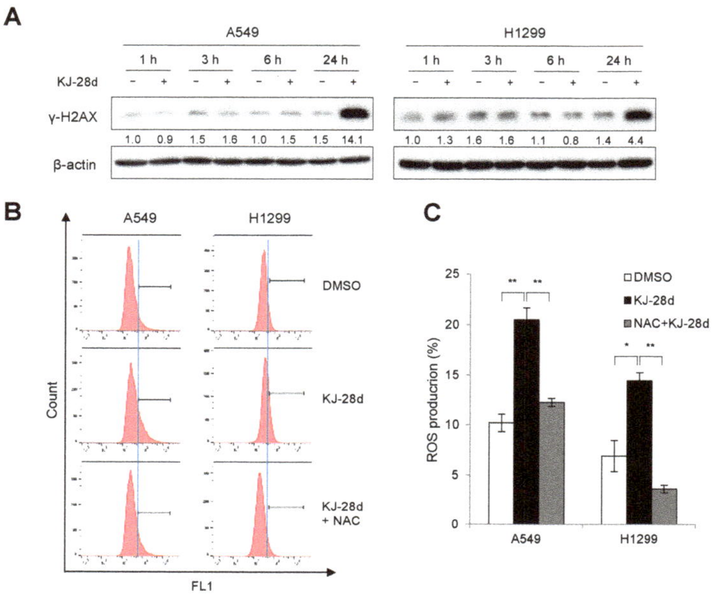

Figure 2. KJ-28d induces DNA damage and generation of reactive oxygen species (ROS). (**A**) A549 and H1299 cells were treated with 5 μM KJ-28d at indicated time points and immunoblotted for the detection of expression of γ-H2AX. (**B**,**C**) A549 and H1299 cells were treated with 5 mM NAC for 1 h, followed by KJ-28d for 24 h after incubation with 2′,7′-dichlorodihydrofluorescein diacetate (CM-H_2DCFHDA) for 30 min. Total cellular ROS production was measured using flow cytometry. Data are representative of three independent experiments (**B**). The bar graph shows the quantitative analysis of flow cytometer data (**C**). Data are presented as the mean ± SD of three independent experiments. * $p < 0.05$, ** $p < 0.01$ versus corresponding values.

2.3. *KJ-28d Potentiated Ionizing Radiation-Induced DNA Damage and Radiosensitized NSCLC Cells*

As IR induces severe DNA damage, which can lead to overloading DNA repair capacity, it has been reported that PARP inhibitors enhance IR-induced DNA damage [14,17,20,22]. To examine whether KJ-28d could induce increased DNA damage in combination with IR, DNA damage was measured in A549 and H1299 cells treated with KJ-28d and IR by detecting the presence of γ-H2AX. Immunoblot analysis revealed a significant increase in the phosphorylation levels of H2AX protein as compared with KJ-28d or IR alone. Similarly, we observed high levels of staining of γ-H2AX foci in A549 and H1299 cells treated with both KJ-28d and IR, as shown in Figure 3A–C. PARylation by PARP-1 catalytic activity is a post-translational modification involved in DNA damage repair. To determine whether KJ-28d suppresses cellular PARylation, H1299 cells were treated with the indicated concentrations of either KJ-28d or olaparib, and A549 cells were treated with 5 μM KJ-28d with or without IR. We observed that 10 μM KJ-28d and 5 and 10 μM olaparib inhibited protein PARylation in H1299, as shown in Figure 3D, and 5 μM of KJ-28d inhibited IR-induced PARylation in A549 cells, as shown in Figure 3E.

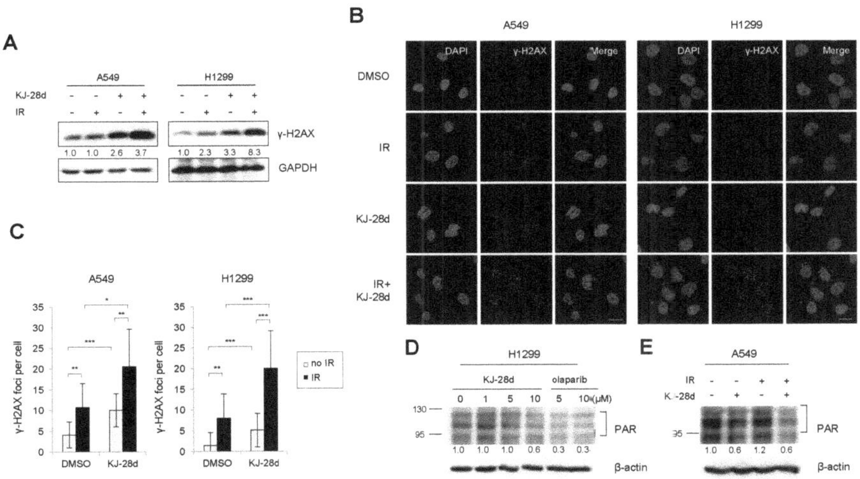

Figure 3. KJ-28d potentiates ionizing radiation (IR)-induced DNA damage responses. A549 and H1299 cells were treated with 5 μM KJ-28d 2 h before IR (4 Gy) and incubated for 24 h. The cell lysates were subjected to immunoblotting for detection of γ-H2AX (**A**), whereas cells were immunostained for γ-H2AX foci (red) and nuclei (DAPI: blue). Images were captured at 400× magnification. Scale bar: 20 μm (**B**). Quantification of the number of γ-H2AX foci per cell (**C**). Data represent the mean ± SD of three independent experiments. * $p < 0.05$, ** $p < 0.01$, *** $p < 0.001$ versus corresponding cells. (**D**) H1299 cells were treated with KJ-28d at indicated concentrations for 1 h. (**E**) A549 cells were treated with 5 μM KJ-28d and IR (4 Gy) and incubated for 1 h. The cell lysates were immunoblotted for the detection of expression of PAR. β-actin was used as a loading control.

Since KJ-28d potentiated IR-induced DNA damage in NSCLC cells, we further examined whether KJ-28d inhibited IR-induced cell growth. A549 and H1299 cells were treated with KJ-28d 2 h before IR. The clonogenic survival assay revealed that KJ-28d radiosensitized both cell lines, as shown in Figure 4A. Dose enhancement ratios (DER) of 0.75 μM KJ-28d-treated (at a surviving fraction of 0.37) to DMSO-treated cells were 1.5 and 1.23 in A549 and H1299 cells, respectively. We next determined whether treatment with KJ-28d could induce apoptotic cell death in A549 and H1299 cells. Apoptotic cell populations of these cell lines were detected using flow cytometer analysis with annexin V/propidium iodide (PI) staining. Following treatment with 5 μM of KJ-28d and IR, the number of A549 and H1299 cells undergoing both early-stage (annexin V-positive/PI-negative) and late-stage (annexin V-positive/PI-positive) apoptosis increased significantly by 1.5-fold compared to KJ-28d alone, respectively, as shown in Figure 4B. In addition, KJ-28d plus IR treatment increased the cleavage of caspase-3 in both cell lines, as shown in Figure 4C. Taken together, these results indicated that KJ-28d enhanced both IR-induced DNA damage and apoptotic cell death in A549 and H1299 human NSCLC cells.

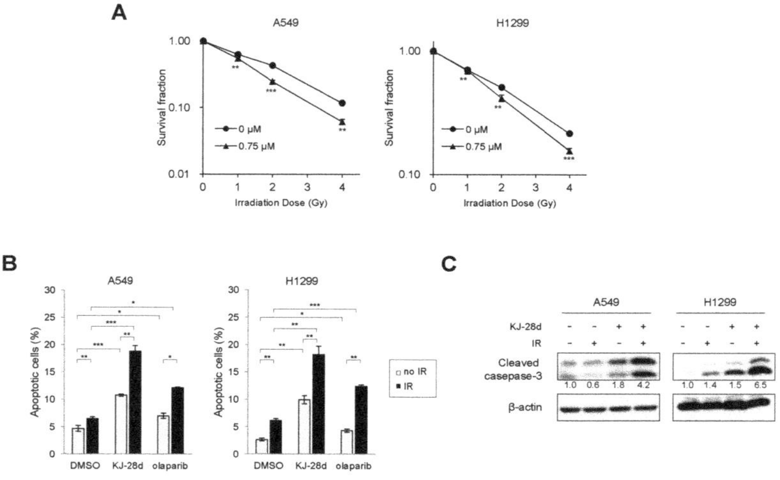

Figure 4. KJ-28d enhances the radiosensitivity of A549 and H1299 cells. (**A**) A549 and H1299 cells were treated with 0.75 μM KJ-28d for 2 h before IR (0, 1, 2, and 4 Gy). Clonogenic survival was measured 10 days after IR. Data are expressed as mean ± SD ($n = 3$) of the surviving fraction compared to non-irradiated cells. Colonies consisting of more than 50 cells were scored as survival colonies. (**B**) A549 and H1299 cells were treated with either 5 μM KJ-28d or 5 μM olaparib plus IR (4 Gy) for 48 h. Apoptotic cells were determined using the APC-conjugated annexin V/PI staining. Cell populations were gated into four groups, as described in Section 4. Bar graphs represent the mean percentage of early (annexin V-positive/PI-negative) and late apoptotic cells (annexin V-positive/PI-positive). Data represent the mean ± SD of three independent experiments. * $p < 0.05$, ** $p < 0.01$, *** $p < 0.001$ versus corresponding cells. (**C**) A549 and H1299 cells were treated with 5 μM KJ-28d plus IR (4 Gy), and the cell lysates were subjected to immunoblotting for detection of cleaved caspase-3. β-actin was used as a loading control.

2.4. The Combination of KJ-28d and DNA Damage-Inducing Chemotherapeutic Agents Synergistically Inhibits NSCLC Cell Growth

In preclinical and clinical studies of advanced NSCLC treatment, administration of PARPi in combination with DNA-damaging therapeutic agents, such as platinum-based compounds, taxane-based compounds, and topoisomerase inhibitors, has demonstrated enhanced cytotoxicity [24–26]. As KJ-28d enhanced radiosensitivity of A549 and H1299 cells, we expected that KJ-28d could increase cytotoxicity when combined with treatment with DNA DSB-inducing agents. A549 and H1299 cells were treated with KJ-28d and carboplatin, cisplatin, paclitaxel, or doxorubicin and evaluation of the synergistic effect of each pair-compound on cell viability was assessed using the MTT assay. Compared to treatment with KJ-28d or each compound alone, combination treatments showed a strong synergistic effect in both A549 and H1299 cells, decreasing cell viability in a dose-dependent manner, as shown in Figure 5A,C. To interpret the effects of all drug combinations, cell viability observed with each concentration of a KJ-28d-compound pair was converted to a combination index (CI) score using the CompuSyn software by Chou–Talalay [27]. CI scores were categorized as synergistic (CI < 0.9, green), additive (1.1 > CI ≥ 0.9, blue), or antagonistic (CI > 1.1, gray). We observed synergistic growth inhibition with a wide range of concentrations of KJ-28d and DSB-inducing agents in both cell lines, as shown in Figure 5B,D. Taken together, these results supported the conclusion that KJ-28d enhances the sensitivity of NSCLC cells to IR or chemotherapeutic agents of DNA-induced damage.

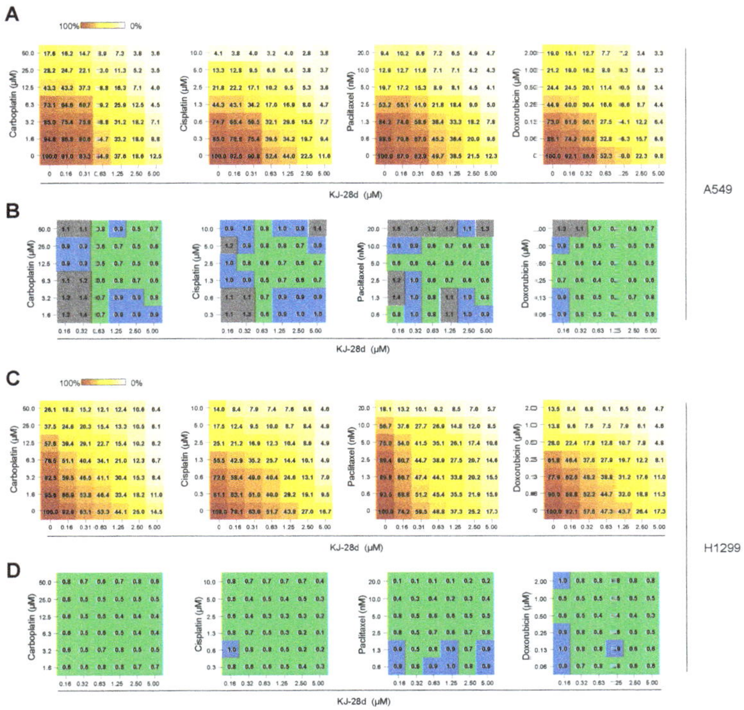

Figure 5. The combination of KJ-28d and DNA damage-inducing chemotherapeutic agents synergistically inhibits growth of A549 and H1299 cells. A549 (**A**) and H1299 (**C**) cells were treated with the single or combined administration of KJ-28d and carboplatin, cisplatin, paclitaxel, or doxorubicin at indicated concentrations. Cell viability was determined 5 days after the treatment by MTT assay. Relative viability (normalized to DMSO-treated cells) is shown for each combination at indicated concentrations. Data are from one representative experiment of three independently repeated experiments. (**B**,**D**) Summary of tables showing combination index (CI) scores of KJ-28d and each chemotherapeutic drug combined at indicated concentrations in A549 (**B**) and H1299 cells (**D**). CI scores were calculated using the CompuSyn software and categorized as synergistic (CI < 0.9, green), additive (1.1 > CI ≥ 0.9, blue), or antagonistic (CI ≥ 1.1, gray). Each CI score was one representative datum from treatment with the indicated concentrations of single- and paired compounds from more than three independent experiments.

3. Discussion

We previously identified KJ-28d as a novel PARP inhibitor that leads to increased cytotoxicity in human ovarian cancer *BRCA-1*-deficient SNU-251 (*BRCA1* mutation at 5564G>A) cells, as well as in triple-negative human ovarian cancer *BRCA1* heterozygous (*BRCA1*+/−) MDA-MB-231 cells [21]. In this study, we further investigated the antitumor activity of KJ-28d on *BRCA*-proficient NSCLC cancer cells, as well as a combination approach with DNA damage-inducing agents to evaluate the synergistic therapeutic efficacy in human NSCLC cells.

NSCLC is a type of cancer with a high mortality rate. Most patients with NSCLC receive treatment with platinum-based drugs, such as cisplatin or carboplatin, as first-line standard therapy. However, many patients who initially benefit from such chemotherapies gradually acquire chemoresistance. Thus, the development of novel therapies and strategies, including combination therapies, is needed. KJ-28d can inhibit the growth of various cancer cells, including NSCLC, breast, and colorectal cancer cells, as shown in Supplementary Figure S2. We specifically focused on NSCLC and revealed its antitumor effects on A549 and H1299 cells both in vitro and in vivo. A549 cells have a *KRAS* oncogenic mutation with c-Myc amplification, while H1299 cells are *TP53* mutant cells, as shown in Supplementary Table S1. Many studies have shown that KRAS-driven c-Myc amplification and TP53 correlate with genomic instability, thereby compromising DNA damage repair (DDR), causing the cells to be vulnerable to DDR inhibiting agents [28]. We observed that KJ-28d induced DNA damage and ROS generation in A549 and H1299 cells, which may facilitate cell death of NSCLC cells. The suppressed cellular PARylation might contribute to inducing DNA damage responses. Recent studies have suggested that PARPi can induce ROS generation due to DNA double stranded breaks. This is considered a characteristic of PARP inhibitors in DNA repair-deficient cells, as well as tumor suppressors or oncogene-mutated cells. Thus, PARPi may crosstalk with other signals [20,29]. In that regard, FDA-approved PARPi compounds have been studied to expand clinical use of PARPi in NSCLC [26,30,31].

Although the rationale for the use of PARPi was based on the reported induced synthetic lethality shown in *BRCA*-deficient cancer cells, preclinical and clinical studies suggested that PARPi could potentially be used as a combination partner with DNA-damaging agents in *BRCA*-proficient cancer cells [26,32,33]. Approximately 50% of patients with NSCLC receive radiotherapy during their treatment course [34]. Radiotherapy is an effective treatment modality, causing severe DNA damage. However, it can only treat tumors at defined doses because of the side effects to surrounding normal tissues. Although PARPi is not yet considered as a radiosensitizer in patients with ovarian cancer harboring *BRCA1/2* mutations in the clinic, the combination of radiotherapy with PARPi could provide promising synergistic therapeutic effects. For that reason, it has been studied in human NSCLC xenografts [35–37]. Olaparib is the first PARPi approved for the treatment of refractory ovarian cancer harboring *BRCA1* or *BRCA2* mutations. We have previously shown that treatment with KJ-28d induced more frequent apoptotic cell death in *BRCA* mutated ovarian cancer cells than olaparib did. It was noteworthy in this study that the combination of KJ-28d and IR also induced significantly more apoptotic cell deaths in NSCLC cells compared to combination treatment with olaparib and IR. Since we initially identified KJ-28d as a PARPi, further studies need to be conducted to elucidate other modes of action, as they may be cytotoxic. Platinum-based agents are widely used for a broad range of solid tumors, including NSCLC, and are the most commonly studied combination partners of PARPi [26,31,38]. Likewise, our results demonstrated that KJ-28d significantly enhanced the sensitivity of A549 and H1299 cells to carboplatin or cisplatin. In addition, we observed synergistic cell growth inhibitions of both cell populations when treated with a combination of KJ-28d with either paclitaxel or doxorubicin that constitute strong DNA-damaging therapeutic agents.

Besides the demonstrated inhibitory PARP activity of KJ-28d, the mechanisms responsible for KJ-28d-induced cytotoxicity in wild-type *BRCA* cancer cells have not been fully uncovered in this study. We suggest that KJ-28d might be involved in the antitumor activity exhibited in NSCLC cells with wild-type *BRCA* based on the experimentally exhibited markedly induced DNA DSBs and ROS generation at the latest examined time point (24 h). Structurally, KJ-28d contains the hydroxamic acid moiety found in histone deacetylase inhibitors (HDACi), including suberoylanilide hydroxamic acid. The hydroxamic acid moiety of HDACi acts as a chelator for zinc ions in the active site of histone deacetylases. To elucidate the possibility of HDAC inhibition, we examined the inhibitory activities of 1–11 HDAC isoforms using in vitro enzyme assays but observed little inhibitory activities of HDACs at KJ-28d concentrations under 5 μM, as shown in Supplementary Table S2. Further structure–activity relationship studies are required to understand the inhibitory activities of PARP-1/2 as a PARPi and to investigate the mechanisms associated with KJ-28d-induced growth inhibition and apoptotic cell death in NSCLC cells.

In summary, KJ-28d was cytotoxic to *BRCA*-proficient cancer cells, including NSCLC cells. Treatment with KJ-28d before IR led to increased DNA damage responses compared to treatment with KJ-28d or IR alone, resulting in the induction of apoptotic cell death. The combination of KJ-28d with carboplatin, cisplatin, paclitaxel, or doxorubicin considerably inhibited cell proliferation. In this context, KJ-28d might act as an effective anti-cancer therapeutic agent against both *BRCA*-deficient and -proficient cancer cells and might have further potential as an adjuvant when used in combination with radiotherapy or DNA-damaging agents. However, further investigations into the mechanisms of action of KJ-28d in cancer cells are warranted.

4. Materials and Methods

4.1. Cell Culture

Human lung cancer cells (A549, H1299, H460, and H1650) were obtained from American Type Culture Collection (ATCC, Manassas, VA, USA). Cells were maintained in Roswell Park Memorial Institute (RPMI) 1640 medium (Welgene, Gyeongsangbuk-do, Korea) supplemented with 10% fetal bovine serum (Welgene) and 100 units/mL penicillin–streptomycin solution (Gibco, Grand Island, NY, USA) at 37 °C in a humidified 5% CO_2 atmosphere.

4.2. Reagents

The PARPi olaparib (AZD2281) was purchased from Selleckchem (Houston, TX, USA). Doxorubicin was purchased from Sigma-Aldrich (St. Louis, MO, USA). All reagents, including carboplatin (Dong-A ST, Seoul, Korea), cisplatin (JW Pharmaceutical, Seoul, Korea), and paclitaxel (Samyang Biopharm, Gyeonggi-do, Korea) were dissolved in dimethyl sulfoxide (DMSO; Sigma-Aldrich). MTT was purchased from Amresco (Solon, OH, USA). Primary antibodies used in this study included the following; anti-cleaved caspase 3 (D175, #9661; Cell Signaling Technology; Danvers, MA, USA), phospho-histone H2AX (Ser 139, sc-517348), and glyceraldehyde 3-phosphate dehydrogenase (GAPDH, sc-47724; Santa Cruz Biotechnology; Dallas, TX, USA), and anti-β-actin (A5441; Sigma-Aldrich) antibodies.

4.3. Cell Viability Assay

Cell viability was assessed using the MTT colorimetric assay. Cells ($6 \sim 8 \times 10^2$ cells/well) were seeded into 96-well plates and treated with various concentrations of KJ-28d or a combination of anti-cancer drugs or IR. After 5 days of treatment, 10 μL MTT (0.5 mg/mL) was added, and cells were further incubated for 3 h. After removal of the supernatant, the resultant pellet was dissolved in DMSO. The absorbance of the resultant formazan was measured at 540 nm using a Multiskan EX plate reader (Thermo LabSystems, Waltham, MA, USA).

4.4. Tumor Xenograft Mouse Models

A549 and H1299 human lung cancer cell xenografts were established by *s.c.* implantation of 2×10^6 cultured cells into the thigh of the right hind leg of six-week-old mice. When tumor volumes had reached approximately 100 mm^3, KJ-28d (10 mg/kg) was administered *i.p.* once per 2 or 3 days for seven times in total. All animal experiments were reviewed and approved by the Institutional Animal Care and Use Committee of Korea Institute of Radiological and Medical Sciences (kirams2018-0063, 2018).

4.5. Detection of Intracellular ROS

Either DMSO- or KJ-28d-treated cells (5×10^5) were further treated with 10 μM CM-H_2DCFH-DA (Thermo Fisher Scientific, Rockford, IL, USA) for 30 min and then washed with phosphate-buffered saline (PBS). After trypsinization, cells were collected, washed, and resuspended in PBS. Inhibition of ROS was evaluated by treating cells with 5 mM NAC 2 h prior to KJ-28d treatment. Intracellular ROS levels were detected using a CyFlow cube 6 flow cytometer (Sysmex Partec, Gorlitz, Germany) at excitation/emission wavelengths of 488/525 nm.

4.6. Immunoblot Analysis

A549 and H1299 cells (1 × 10^6) were seeded onto a 60 mm dish. Cell lysates were prepared by extracting proteins with radioimmunoprecipitation assay (RIPA) lysis buffer (Millipore, Billerica, MA, USA) supplemented with a protease inhibitor cocktail (Thermo Fisher Scientific). Equal amounts of proteins were separated using SDS-PAGE on 8%–13% gels and transferred to nitrocellulose membranes (Bio-Rad, Hercules, CA, USA). Membranes were blocked with 5% skim milk in Tris-buffered saline-Tween 20 (TBST; 150 mM NaCl, 10 mM Tris, 0.2% Tween 20; Sigma-Aldrich), followed by overnight incubation with primary antibodies at 4 °C. Blots were developed using peroxidase-conjugated secondary antibody, and immunoreactive proteins were visualized using enhanced chemiluminescence reagents, according to the manufacturer's recommendations (Amersham, GE Healthcare, Buckinghamshire, UK). Protein bands were visualized using an ImageQuant LAS 4000 mini digital imaging system (GE Healthcare). Protein levels were analyzed using Image J software (National Institutes of Health, Bethesda, MD, USA). Experiments were repeated at least three times.

4.7. Clonogenic Assay

Cells were seeded on 60 mm culture dishes at various densities and then treated with DMSO or KJ-28d (0.75 μM). After 2 h, cells were treated with the indicated doses of 137Cs γ-radiation. After 10 d, colonies were fixed and stained with 1.5% methylene blue (Sigma-Aldrich) in methanol solution. Colonies containing >50 cells were counted. DER was calculated as the dose (Gy) of radiation that yielded a surviving fraction of 0.1 for DMSO-treated cells divided by that dose for KJ-28d-treated cells. The experiment was performed in triplicate.

4.8. Annexin V/PI-Based Flow Cytometric Analysis

Annexin V assays were performed according to the manufacturer's protocol (BD Pharmingen, San Diego, CA, USA). Briefly, 10,000 cells were plated into 60 mm plates and treated with KJ-28d (10 μM) for 48 h. Cells were harvested and incubated with 4 μL allophycocyanin (APC)-conjugated annexin V (20 μg/mL) and 4 μL PI (50 μg/mL) for 15 min. Fluorescence analyses were performed using flow cytometry (CyFlow cube 6). Cells were classified as early apoptotic (annexin V-positive/PI-negative), late apoptotic/necrotic (annexin V-positive/PI-positive), necrotic/dead (annexin V-negative/PI-positive), and live (annexin V-negative/PI-negative). Flow cytometry data were analyzed using FlowJo software (TreeStar Inc., Ashland, OR, USA).

4.9. Combination Index

CI scores were calculated using the CompuSyn software by Chou (CompuSyn Inc., Paramus, NJ, USA) [27] based on cell viability after treatment with single and paired drug concentrations. The CI equation for two drugs used is given below:

$$\mathrm{CI} = \frac{(\mathrm{D})\mathrm{A}}{(\mathrm{Dx})\mathrm{A}} + \frac{(\mathrm{D})\mathrm{B}}{(\mathrm{Dx})\mathrm{B}}$$

where (Dx)A is the concentration of drug A alone that inhibits x%, (Dx)B is the concentration of drug B alone that inhibits x%, (D)A or (D)B is the portion of drug A or drug B in the combination (D)A + (D)B that inhibits x%. Thus, (D)A + (D)B also inhibits x%.

4.10. Statistical Analyses

Results are shown as means ± SD. Data were analyzed using the two-tailed Student's *t*-test. Analysis of variance (ANOVA) and Tukey's post hoc test were used for 2 or 3 group comparisons. Differences between groups with p values < 0.05 were considered statistically significant.

Supplementary Materials: Supplementary materials can be found at http://www.mdpi.com/1422-0067/20/23/6026/s1.

Author Contributions: Conceptualization: J.A., H.-K.C., and J.-S.K.; funding acquisition: J.A.; investigation: H.R., H.J.K., and T.H.N.B.; resources: S.-G.H. and J.K.; writing—original draft: J.A. and H.R.; writing—review and editing: J.-Y.S. and H.-K.C.

Funding: This research was supported by the Basic Science Research Program through the National Research Foundation of Korea (NRF) funded by the Ministry of Education (NRF-2016R1D1A1B03936420; NRF-2016R1D1A3B03935674), and the Korea Institute of Radiological and Medical Sciences funded by the Ministry of Science and ICT, Republic of Korea (50531-2019; 50538-2019).

Conflicts of Interest: The authors declare that they have no conflicts of interest.

Abbreviations

PARP-1	Poly (ADP-ribose) polymerase-1
PARPi	PARP inhibitor(s)
PARylation	Poly ADP-ribosylation
NSCLC	non-small cell lung cancer
DSB	double-strand breaks
IR	ionizing radiation
ROS	reactive oxygen species
CI	combination index

References

1. American-Cancer-Society. *Cancer Facts & Figures 2019*; American Cancer Society: Atlanta, GA, USA, 2019.
2. Korea, N.C.C. *Annual Report of Cancer Statistics in Korea in 2016*; Ntational Cancer Center Korea: Gyeonggi-do, Korea, 2016. Available online: https://www.cancer.go.kr/lay1/bbs/S1T674C680/B/26/list.do (accessed on 29 November 2019).
3. Liu, C.Y.; Wang, C.L.; Li, S.H.; Hsu, P.C.; Chen, C.H.; Lin, T.Y.; Kuo, C.H.; Fang, Y.F.; Ko, H.W.; Yu, C.T.; et al. The efficacy of 40 mg versus dose de-escalation to less than 40 mg of afatinib (Giotrif) as the first-line therapy for patients with primary lung adenocarcinoma harboring favorable epidermal growth factor mutations. *Oncotarget* **2017**, *8*, 97602–97612. [CrossRef]
4. Jinesh, G.G.; Sambandam, V.; Vijayaraghavan, S.; Balaji, K.; Mukherjee, S. Molecular genetics and cellular events of K-Ras-driven tumorigenesis. *Oncogene* **2018**, *37*, 839–846. [CrossRef]
5. Chen, Y.; Chen, G.; Li, J.; Huang, Y.Y.; Li, Y.; Lin, J.; Chen, L.Z.; Lu, J.P.; Wang, Y.Q.; Wang, C.X.; et al. Association of Tumor Protein p53 and Ataxia-Telangiectasia Mutated Comutation With Response to Immune Checkpoint Inhibitors and Mortality in Patients With Non-Small Cell Lung Cancer. *JAMA Netw. Open* **2019**, *2*, e1911895. [CrossRef] [PubMed]
6. Fan, Y.; Zhu, X.; Xu, Y.; Lu, X.; Xu, Y.; Wang, M.; Xu, H.; Ding, J.; Ye, X.; Fang, L.; et al. Cell-Cycle and DNA-Damage Response Pathway Is Involved in Leptomeningeal Metastasis of Non-Small Cell Lung Cancer. *Clin. Cancer Res.* **2018**, *24*, 209–216. [CrossRef] [PubMed]
7. Alidousty, C.; Baar, T.; Martelotto, L.G.; Heydt, C.; Wagener, S.; Fassunke, J.; Duerbaum, N.; Scheel, A.H.; Frank, S.; Holz, B.; et al. Genetic instability and recurrent MYC amplification in ALK-translocated NSCLC: A central role of TP53 mutations. *J. Pathol.* **2018**, *246*, 67–76. [CrossRef] [PubMed]
8. Postel-Vinay, S.; Vanhecke, E.; Olaussen, K.A.; Lord, C.J.; Ashworth, A.; Soria, J.C. The potential of exploiting DNA-repair defects for optimizing lung cancer treatment. *Nat. Rev. Clin. Oncol.* **2012**, *9*, 144–155. [CrossRef] [PubMed]
9. Wang, Y.S.; Chen, J.; Cui, F.; Wang, H.; Wang, S.; Hang, W.; Zeng, Q.; Quan, C.S.; Zhai, Y.X.; Wang, J.W.; et al. LKB1 is a DNA damage response protein that regulates cellular sensitivity to PARP inhibitors. *Oncotarget* **2016**, *7*, 73389–73401. [CrossRef] [PubMed]
10. Singh, A.; Misra, V.; Thimmulappa, R.K.; Lee, H.; Ames, S.; Hoque, M.O.; Herman, J.G.; Baylin, S.B.; Sidransky, D.; Gabrielson, E.; et al. Dysfunctional KEAP1-NRF2 interaction in non-small-cell lung cancer. *PLoS Med.* **2006**, *3*, e420. [CrossRef] [PubMed]
11. Rojo de la Vega, M.; Chapman, E.; Zhang, D.D. NRF2 and the Hallmarks of Cancer. *Cancer Cell* **2018**, *34*, 21–43. [CrossRef]

12. Virag, L.; Szabo, C. The therapeutic potential of poly(ADP-ribose) polymerase inhibitors. *Pharmacol. Rev.* **2002**, *54*, 375–429. [CrossRef]
13. Dantzer, F.; Ame, J.C.; Schreiber, V.; Nakamura, J.; Menissier-de Murcia, J.; de Murcia, G. Poly(ADP-ribose) polymerase-1 activation during DNA damage and repair. *Methods Enzymol.* **2006**, *409*, 493–510. [PubMed]
14. Kalimutho, M.; Bain, A.L.; Mukherjee, B.; Nag, P.; Nanayakkara, D.M.; Harten, S.K.; Harris, J.L.; Subramanian, G.N.; Sinha, D.; Shirasawa, S.; et al. Enhanced dependency of KRAS-mutant colorectal cancer cells on RAD51-dependent homologous recombination repair identified from genetic interactions in Saccharomyces cerevisiae. *Mol. Oncol.* **2017**, *11*, 470–490. [CrossRef] [PubMed]
15. Kawahara, N.; Ogawa, K.; Nagayasu, M.; Kimura, M.; Sasaki, Y.; Kobayashi, H. Candidate synthetic lethality partners to PARP inhibitors in the treatment of ovarian clear cell cancer. *Biomed. Rep.* **2017**, *7*, 391–399. [CrossRef] [PubMed]
16. Sizemore, S.T.; Mohammad, R.; Sizemore, G.M.; Nowsheen, S.; Yu, H.; Ostrowski, M.C.; Chakravarti, A.; Xia, F. Synthetic Lethality of PARP Inhibition and Ionizing Radiation is p53-dependent. *Biomed. Rep.* **2017**, *7*, 391–399. [CrossRef] [PubMed]
17. Alli, E.; Sharma, V.B.; Sunderesakumar, P.; Ford, J.M. Defective repair of oxidative dna damage in triple-negative breast cancer confers sensitivity to inhibition of poly(ADP-ribose) polymerase. *Cancer Res.* **2009**, *69*, 3589–3596. [CrossRef] [PubMed]
18. Hastak, K.; Alli, E.; Ford, J.M. Synergistic chemosensitivity of triple-negative breast cancer cell lines to poly(ADP-Ribose) polymerase inhibition, gemcitabine, and cisplatin. *Cancer Res.* **2010**, *70*, 7970–7980. [CrossRef] [PubMed]
19. Turner, N.; Tutt, A.; Ashworth, A. Hallmarks of 'BRCAness' in sporadic cancers. *Nat. Rev. Cancer* **2004**, *4*, 814–819. [CrossRef]
20. Liu, Q.; Gheorghiu, L.; Drumm, M.; Clayman, R.; Eidelman, A.; Wszolek, M.F.; Olumi, A.; Feldman, A.; Wang, M.; Marcar, L.; et al. PARP-1 inhibition with or without ionizing radiation confers reactive oxygen species-mediated cytotoxicity preferentially to cancer cells with mutant TP53. *Oncogene* **2018**, *37*, 2793–2805. [CrossRef]
21. Ryu, H.; Ahn, J.; Choi, H.K. Novel Benzamide Derivatives: Synthesis and Bioactivity as Potent PARP-1 Inhibitors. *Bull. Korean Chem. Soc.* **2017**, *38*, 935–943. [CrossRef]
22. Nile, D.L.; Rae, C.; Hyndman, I.J.; Gaze, M.N.; Mairs, R.J. An evaluation in vitro of PARP-1 inhibitors, rucaparib and olaparib, as radiosensitisers for the treatment of neuroblastoma. *BMC Cancer* **2016**, *16*, 621. [CrossRef]
23. Gill, S.J.; Travers, J.; Pshenichnaya, I.; Kogera, F.A.; Barthorpe, S.; Mironenko, T.; Richardson, L.; Benes, C.H.; Stratton, M.R.; McDermott, U.; et al. Combinations of PARP Inhibitors with Temozolomide Drive PARP1 Trapping and Apoptosis in Ewing's Sarcoma. *PLoS ONE* **2015**, *10*, e0140988. [CrossRef] [PubMed]
24. Donawho, C.K.; Luo, Y.; Luo, Y.; Penning, T.D.; Bauch, J.L.; Bouska, J.J.; Bontcheva-Diaz, V.D.; Cox, B.F.; DeWeese, T.L.; Dillehay, L.E.; et al. ABT-888, an orally active poly(ADP-ribose) polymerase inhibitor that potentiates DNA-damaging agents in preclinical tumor models. *Clin. Cancer Res.* **2007**, *13*, 2728–2737. [CrossRef]
25. Palma, J.P.; Wang, Y.C.; Rodriguez, L.E.; Montgomery, D.; Ellis, P.A.; Bukofzer, G.; Niquette, A.; Liu, X.; Shi, Y.; Lasko, L.; et al. ABT-888 confers broad in vivo activity in combination with temozolomide in diverse tumors. *Clin. Cancer Res.* **2009**, *15*, 7277–7290. [CrossRef] [PubMed]
26. Ramalingam, S.S.; Blais, N.; Mazieres, J.; Reck, M.; Jones, C.M.; Juhasz, E.; Urban, L.; Orlov, S.; Barlesi, F.; Kio, E.; et al. Randomized, Placebo-Controlled, Phase II Study of Veliparib in Combination with Carboplatin and Paclitaxel for Advanced/Metastatic Non-Small Cell Lung Cancer. *Clin. Cancer Res.* **2017**, *23*, 1937–1944. [CrossRef] [PubMed]
27. Chou, T.C. Drug combination studies and their synergy quantification using the Chou-Talalay method. *Cancer Res.* **2010**, *70*, 440–446. [CrossRef] [PubMed]
28. Kastenhuber, E.R.; Lowe, S.W. Putting p53 in Context. *Cell* **2017**, *170*, 1062–1078. [CrossRef] [PubMed]
29. Yin, Z.X.; Hang, W.; Liu, G.; Wang, Y.S.; Shen, X.F.; Sun, Q.H.; Li, D.D.; Jian, Y.P.; Zhang, Y.H.; Quan, C.S.; et al. PARP-1 inhibitors sensitize HNSCC cells to APR-246 by inactivation of thioredoxin reductase 1 (TrxR1) and promotion of ROS accumulation. *Oncotarget* **2018**, *9*, 1885–1897. [CrossRef] [PubMed]

30. Reck, M.; Blais, N.; Juhasz, E.; Gorbunova, V.; Jones, C.M.; Urban, L.; Orlov, S.; Barlesi, F.; Kio, E.; Keilholz, U.; et al. Smoking History Predicts Sensitivity to PARP Inhibitor Veliparib in Patients with Advanced Non-Small Cell Lung Cancer. *J. Thorac. Oncol.* **2017**, *12*, 1098–1108. [CrossRef] [PubMed]
31. Mizugaki, H.; Yamamoto, N.; Nokihara, H.; Fujiwara, Y.; Horinouchi, H.; Kanda, S.; Kitazono, S.; Yagishita, S.; Xiong, H.; Qian, J.; et al. A phase 1 study evaluating the pharmacokinetics and preliminary efficacy of veliparib (ABT-888) in combination with carboplatin/paclitaxel in Japanese subjects with non-small cell lung cancer (NSCLC). *Cancer Chemother. Pharmacol.* **2015**, *76*, 1063–1072. [CrossRef] [PubMed]
32. Tuli, R.; Surmak, A.J.; Reyes, J.; Armour, M.; Hacker-Prietz, A.; Wong, J.; DeWeese, T.L.; Herman, J.M. Radiosensitization of Pancreatic Cancer Cells In Vitro and In Vivo through Poly (ADP-ribose) Polymerase Inhibition with ABT-888. *Transl. Oncol.* **2014**, *7*, 439–445. [CrossRef]
33. Tuli, R.; Shiao, S.L.; Nissen, N.; Tighiouart, M.; Kim, S.; Osipov, A.; Bryant, M.; Ristow, L.; Placencio-Hickok, V.; Hoffman, D.; et al. A phase 1 study of veliparib, a PARP-1/2 inhibitor, with gemcitabine and radiotherapy in locally advanced pancreatic cancer. *EBioMedicine* **2019**, *40*, 375–381. [CrossRef]
34. Tyldesley, S.; Boyd, C.; Schulze, K.; Walker, H.; Mackillop, W.J. Estimating the need for radiotherapy for lung cancer: An evidence-based, epidemiologic approach. *Int. J. Radiat. Oncol. Biol. Phys.* **2001**, *49*, 973–985. [CrossRef]
35. Wang, L.; Mason, K.A.; Ang, K.K.; Buchholz, T.; Valdecanas, D.; Mathur, A.; Buser-Doepner, C.; Toniatti, C.; Milas, L. MK-4827, a PARP-1/-2 inhibitor, strongly enhances response of human lung and breast cancer xenografts to radiation. *Invest. New Drugs* **2012**, *30*, 2113–2120. [CrossRef]
36. Senra, J.M.; Telfer, B.A.; Cherry, K.E.; McCrudden, C.M.; Hirst, D.G.; O'Connor, M.J.; Wedge, S.R.; Stratford, I.J. Inhibition of PARP-1 by olaparib (AZD2281) increases the radiosensitivity of a lung tumor xenograft. *Mol. Cancer Ther.* **2011**, *10*, 1949–1958. [CrossRef]
37. Albert, J.M.; Cao, C.; Kim, K.W.; Willey, C.D.; Geng, L.; Xiao, D.; Wang, H.; Sandler, A.; Johnson, D.H.; Colevas, A.D.; et al. Inhibition of poly(ADP-ribose) polymerase enhances cell death and improves tumor growth delay in irradiated lung cancer models. *Clin. Cancer Res.* **2007**, *13*, 3033–3042. [CrossRef]
38. Hu, X.L.; Feng, J.H.; Pham, T.A.; Ma, H.Y.; Ma, M.X.; Song, R.; Shen, W.; Xiong, F.; Zhang, X.Q.; Ye, W.C.; et al. Identification of amentoflavone as a potent highly selective PARP-1 inhibitor and its potentiation on carboplatin in human non-small cell lung cancer. *Phytomedicine* **2018**, *50*, 88–98. [CrossRef]

International Journal of *Molecular Sciences*

MDPI

Review

PARP-1-Associated Pathological Processes: Inhibition by Natural Polyphenols

Natalya V. Maluchenko [1,*], Alexey V. Feofanov [1,2] and Vasily M. Studitsky [1,3]

[1] Biology Faculty, Lomonosov Moscow State University, Lenin Hills 1/12, 119234 Moscow, Russia; avfeofanov@yandex.ru (A.V.F.); vasily.studitsky@fccc.edu (V.M.S.)
[2] Shemyakin-Ovchinnikov Institute of Bioorganic Chemistry, Russian Academy of Sciences, Mikluko-Maklaya Str., 16/10, 117997 Moscow, Russia
[3] Fox Chase Cancer Center, Cottman Avenue 333, Philadelphia, PA 19111, USA
[*] Correspondence: mal_nat@mail.ru; Tel.: +7-916-443-9114

Abstract: Poly (ADP-ribose) polymerase-1 (PARP-1) is a nuclear enzyme involved in processes of cell cycle regulation, DNA repair, transcription, and replication. Hyperactivity of PARP-1 induced by changes in cell homeostasis promotes development of chronic pathological processes leading to cell death during various metabolic disorders, cardiovascular and neurodegenerative diseases. In contrast, tumor growth is accompanied by a moderate activation of PARP-1 that supports survival of tumor cells due to enhancement of DNA lesion repair and resistance to therapy by DNA damaging agents. That is why PARP inhibitors (PARPi) are promising agents for the therapy of tumor and metabolic diseases. A PARPi family is rapidly growing partly due to natural polyphenols discovered among plant secondary metabolites. This review describes mechanisms of PARP-1 participation in the development of various pathologies, analyzes multiple PARP-dependent pathways of cell degeneration and death, and discusses representative plant polyphenols, which can inhibit PARP-1 directly or suppress unwanted PARP-dependent cellular processes.

Keywords: PARP-1; polyphenols; PARP-1 inhibitors

Citation: Maluchenko, N.V.; Feofanov, A.V.; Studitsky, V.M. PARP-1-Associated Pathological Processes: Inhibition by Natural Polyphenols. *Int. J. Mol. Sci.* **2021**, *22*, 11441. https://doi.org/10.3390/ijms222111441

Academic Editors: Ciro Isidoro, Konstantin Volcho and Olga Lavrik

Received: 28 September 2021
Accepted: 21 October 2021
Published: 23 October 2021

Publisher's Note: MDPI stays neutral with regard to jurisdictional claims in published maps and institutional affiliations.

1. Introduction

Poly (ADP-ribose) polymerase-1 (PARP-1) is a widespread nuclear protein with a spectrum of different activities due to its DNA-binding and enzymatic properties [1–5]. PARP-1 uses β-NAD+ as a substrate to synthesize branched polymers of ADP-ribose (PAR) and to covalently modifies more than 40 nuclear proteins and transcription factors, including PARP-1 itself. Under conditions of moderate genotoxic stress, the functioning of PARP-1 maintains integrity and activity of cell genome, while during severe genotoxic stress PARP-1 coordinates multiple pathways of cell death. General enhancement of PARP-1 activity is associated with the development of tumor, cardiovascular and neurodegenerative diseases, and pharmacological inhibition of PARP-1 is a promising strategy for their therapy. Several inhibitors of the enzymatic activity of PARP-1 (PARPi) are already used in clinical practice for the treatment of cancer [6,7]. A search for more active and less toxic PARPi, as well as for compounds that block the development of unwanted PARP-dependent cellular processes is in progress. In particular, a search for PARPi is carried out among natural compounds, since they might have a higher bioavailability, more effective cell penetration, higher pharmacological activity and fewer side effects than synthetic agents. Polyphenols are the largest and most studied group of plant metabolites, among which a considerable number of compounds were found to have therapeutic potential due to antiviral, antibacterial, antioxidant or antitumor activities. Some polyphenols were demonstrated to be effective PARPi or/and can affect signaling pathways that regulate cell survival under adverse conditions of oxidative/nitrosative or genotoxic stress. Many of these signaling pathways are closely related to molecular processes that are under the control of PARP-1.

Accordingly, in the first part of the review, data are systematized on PARP-1-dependent molecular mechanisms that contribute to the development of diseases and therefore are targets for therapeutic intervention; in the second part, the polyphenols are discussed, which affect PARP-1 or (and) the signaling pathways under its control.

2. Relationship of PARP-1 with Inflammatory and Metabolic Diseases

Hyperactivation of PARP-1 plays an important role in the development of diseases, which are associated with or caused by chronic inflammation [8]. These include diabetes, neurodegenerative disorders (Alzheimer's disease (AD), Parkinson's disease (PD)), and cardiovascular diseases (Figure 1) [9–12].

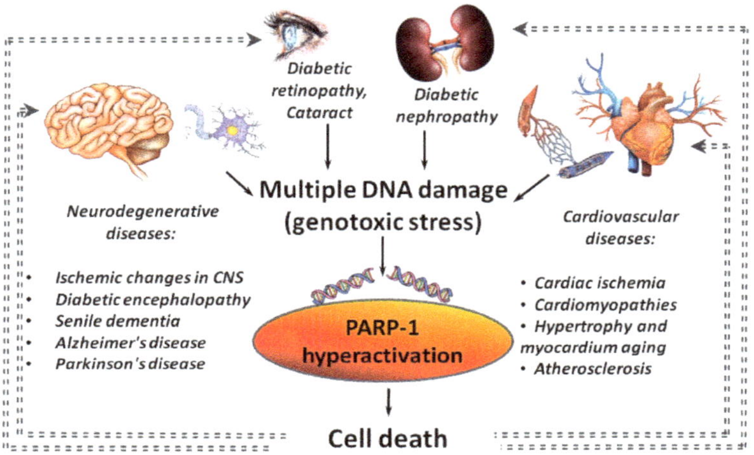

Figure 1. PARP-1 hyperactivation as an aggravating factor in the development of various diseases. Gray dotted lines indicate "vicious circles" when PARP-1 hyperactivation initiated by inflammation, cardiovascular, neurodegenerative or diabetic pathology leads to an increase in the severity of the disease.

Hyperactivation of PARP-1 was found in various cardiovascular diseases (ischemic heart disease, atherosclerosis, cardiomyopathies of various origins, hypertrophy and aging of the myocardium) [13,14] and in many models of a central nervous system (CNS) damage (stroke [15], traumatic brain injury [16], neurodegeneration [17] and senile dementia [18]). A negative role of the PARP-1 hyperactivation in cardiac ischemia/reperfusion is known for a long time [13,19]. It was shown that the strongest activation of PARP-1 is observed in a peri-infarction zone and areas of necrotic damage during a heart attack [20]. PARP-1 hyperactivation is involved in a cascade of events initiated by β-amyloid peptides (Aβ), the accumulation of which leads to the death of brain cells in AD. A significant increase in PARP-1 expression and accumulation of PAR polymers was found in the cerebral cortex at the early (3.5 months) and intermediate (6 months) stages of Aβ-aggregation in mouse models of AD [21]. Hyperglycemia is also associated with PARP-1 hyperactivation [22–24] that is usually an aggravating factor in the development of systemic diabetic dysfunctions. In particular, PARP-1 hyperactivation is involved in the death of insulin-producing pancreatic β-cells [25].

Inflammatory processes, hypoxia, hypo- and hyperglycemia are often accompanied by an uncontrolled increase in the levels of reactive oxygen (ROS) and nitrogen species (RNS), which cause DNA damage. As a consequence, an increase in PARP-1 activity is required for DNA repair. In turn, PARP-1 hyperactivation can initiate parthanatos—a programmed caspase-independent cell death (Figure 2) [26,27]. PAR and PARylated proteins that are formed in large quantities migrate from a nucleus to cytoplasm and cause the release of apoptosis-inducing factor (AIF) from mitochondria [26,28,29]. Released AIF is transported to a nucleus due to a nuclear localization signal (NLS) and activates endonucleases, which cause DNA fragmentation followed by cell death. PARP-1-mediated parthanatos is observed in neurons during PD, excitotoxicity of glutamate and cerebral ischemia [30,31].

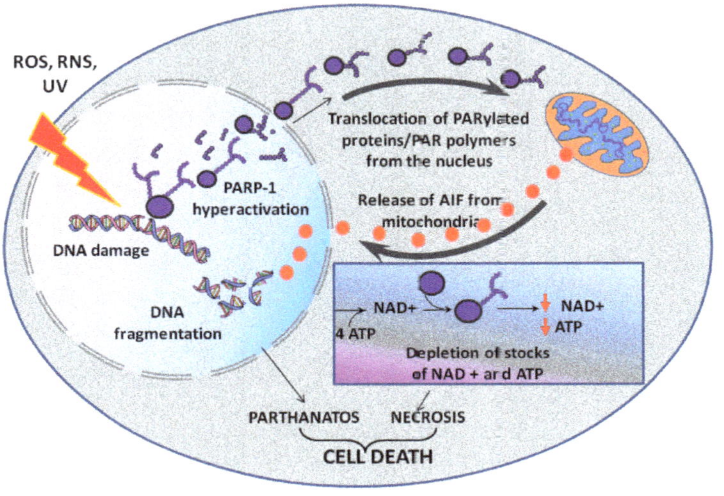

Figure 2. PARP-1 dependent cell death. UV—ultraviolet light, ROS—reactive oxygen species overproduced in oxidative stress, RNS—reactive nitrogen species (e.g., nitric oxide NO) overproduced in nitrosative stress. See text for detail.

The PARP-1-dependent cell death can also occur after PARP-1 hyperactivation due to an energy crisis caused by the depletion of cellular reserves of macroergic compounds [32,33]. A synthesis of a NAD+ molecule requires four ATP molecules, and intense consumption of NAD+ by PARP-1 can result in a rapid depletion of ATP and NAD+ stocks, lead to suppression of energy-dependent cellular processes followed by cell necrosis [34]. Suppression of energy-dependent processes is additionally enhanced by PAR metabolism. Free and protein-bound PAR is intensely cleaved by poly(ADP-ribose)glycohydrolase to ADP-ribose, which is then metabolized by NUDIX-hydrolases (NUDIX - NUcleoside DIphosphates linked to any other moiety X) to AMP [35]. A high AMP/ATP ratio is interpreted by a cell as an energy stress, and AMP-activated protein kinase corrects this apparent energy misbalance by blocking the mammalian target of rapamycin (mTOR) signaling pathway with a subsequent down-regulation of ATP consuming processes [36].

PARP-1 itself promotes the development of inflammatory processes by up-regulating expression of various inflammatory mediators such as tumor necrosis factor α (TNFα), inducible isoform of nitrite oxide synthase (iNOS), cyclooxygenase 2 (COX2), monocyte chemoattractant protein 1 (MCP1), interleukins 1β and 6 (IL-1β, IL-6). Here PARP-1 acts as a co-activator of transcription factors such as nuclear factor kappa-light-chain-enhancer of activated B cells (NF-κB), activator proteins 1 and 2 (AP1, AP2) that regulate immune and inflammatory responses (Figure 3) [37–40]. PARP-1 was shown to be acetylated at lysine residues (K498, K505, K508, K521, K524) by the p300/CREB-binding protein complex (CREB - cAMP-response element binding protein) and phosphorylated at Y829 by mitogen-activated protein kinases (MAPKs) in response to pro-inflammatory stimuli [37,38]. Modified in this way, PARP-1 stimulates transcription of NF-κB-dependent genes of inflammatory mediators (Figure 3) [37–40]. Interestingly, neither the enzymatic activity of PARP-1 nor its DNA-binding activity were required for full activation of NF-kB in response to various stimuli [37]. PAR polymers can act as alarmins releasing from a cell during stress and activating production of inflammatory cytokines by the cells of an innate immunity system [41].

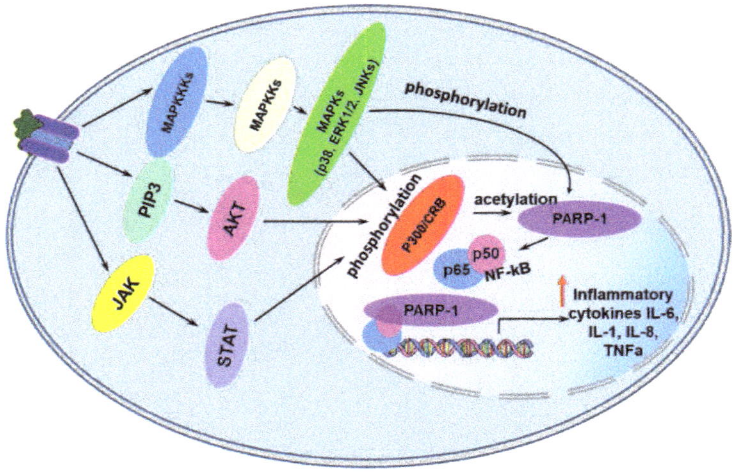

Figure 3. PARP-1-dependent transcriptional activation of genes encoding pro-inflammatory cytokines in eukaryotic cells. See text for details. Abbreviations: JAK—Janus kinase; PIP3—phosphatidylinositol (3,4,5)-trisphosphate; MAPKKKs—Mitogen-Activated Protein (MAP) kinase kinase kinases; MAPKKs—Mitogen-activated protein kinase kinases; MAPKs—mitogen-activated protein kinases; STAT—members of the signal transducer and activator of transcription protein family; AKT—subfamily of serine/threonine kinases; p38—p38 mitogen-activated protein kinases; JNK—c-Jun N-terminal kinases; ERKs—extracellular signal-regulated kinases; IL-l, IL6, IL-8—interleukins 1, 6, 8, p300/CRB—p300/CREB-binding protein complex, TNFα—tumor necrosis factor α.

During apoptosis, which may accompany the development of some pathologies, PARP-1 is cleaved by caspases 3 and 7 into DNA-binding and catalytically active fragments [42,43], but retains its ability to activate NF-κB and enhance transcription of inflammatory mediator genes [44]. The C-terminal fragment preserves catalytic activity, but is not stimulated by DNA damage. The N-terminal fragment remains associated with DNA injuries blocking access of repair factors to them [45,46].

As a co-activator of NF-κB and activator protein 1, PARP-1 was suggested to be responsible for accelerated aging during chronic inflammatory diseases [47].

An important role of PARP-1 in the development of inflammatory diseases was confirmed by experiments with PARP-1 knockout mice. These mice are better protected from diabetic and septic complications associated with inflammation such as multiple organ dysfunction syndromes [37,48].

In summary, PARP-1 hyperactivation, which occurs during oxidative/nitrosative stress, chronic inflammation and irreversible genotoxic damage, leads to massive cell death that at the level of the organism promotes development of metabolic syndrome, multiple organ dysfunction syndrome, cardiovascular and neurodegenerative diseases.

3. PARP-1 and Oncological Diseases

PARP-1 is involved in pathogenesis of oncological diseases in a complex way as described in several excellent comprehensive reviews [49–52]. Here we will only briefly describe the effect of PARP-1 on tumor cell metabolism, referring readers to the published reviews for more details.

In contrast to the negative role of severe hyperactivation of PARP-1 in inflammatory processes, a moderate activation of PARP-1 occurring during transformation of cells does not lead to a cell death. On the contrary, it contributes to cell survival. The pro-tumor activity of PARP-1 is mediated by PARP-1-dependent deregulation of factors involved in the cell cycle, mitosis, apoptosis and autophagy [53]. PARP-1 impedes with cell differentiation thus enhancing tumor malignancy [54], and moderate activation of PARP-1 caused by the accumulation of DNA damage during intensive cell division increases DNA repair efficiency and cell viability.

Malignancy of cancer cells and its ability to metastasize strongly depend on the tumor microenvironment [55]. PARP-1 plays an important role in the functioning of the tumor microenvironment, participating in angiogenesis as well as in the formation of a tumor-associated stroma [56]. PARP-1 is involved in the process of epithelial-mesenchymal transition (EMT) during the acquisition of the ability of tumor cells to metastasize [52]. PARP-1 knockdown leads to EMT reversal through inhibitory transcription factors such as Smad4, p65 and ZEB1 [57].

PARP-1 participates in several processes responsible for the resistance of tumor cells to therapy (Figure 4) [25,30,58–60]. As a key repair enzyme, PARP-1 ensures the stability of a tumor cell genome after treatment with DNA damaging chemotherapy agents [61–63]. PARP-1 is able to promote (directly or indirectly) epigenetic modifications, creating conditions for development of heterogeneity of tumor cells and formation of super-resistant clones in a heterogeneous population [64]. Another PARP-1-mediated mechanism of drug resistance is a non-lethal autophagy [65–67]. PARP-1 is also known to control the expression of heat shock protein 70 [68,69], which makes a significant contribution to the survival of tumor cells and their resistance to antitumor agents [70].

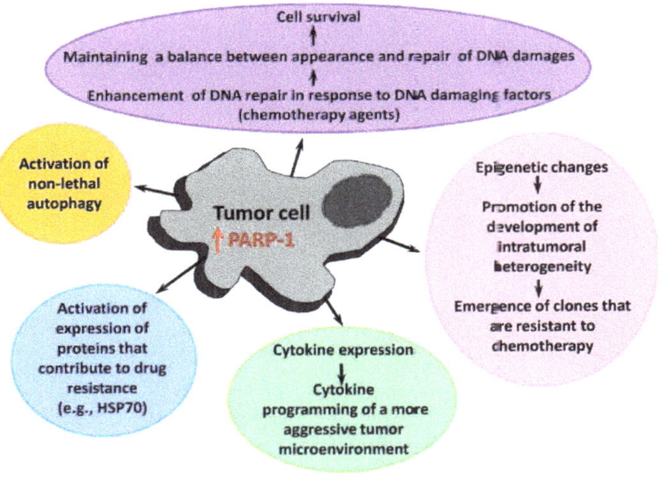

Figure 4. A role of PARP-1 in the tumor progression and development of its drug resistance. HSP70—heat shock protein 70.

In general, an increased level of PARP-1 expression is considered to be a prognostic marker associated with an aggressive phenotype of malignant tumors and a worse prognosis of patient survival [71–73].

4. Synthetic PARP-1 Inhibitors in Treatment of Diseases

PARPi are considered to be promising antitumor agents since the increased activity of PARP-1 is a key factor contributing to growth of tumors, to an increase in their malignancy and to the development of drug resistance [74]. Most PARPi that are currently in antitumor preclinical and clinical trials are nicotinamide mimetics. They act by competing with NAD+ for the binding to a catalytic domain of PARP-1 and suppressing PAR synthesis [74]. Several

PARPi are already used in clinical practice (Figure 5), and their combined administration with chemotherapy agents is promising for overcoming the drug resistance of tumor cells [75]. Inhibition of PARP1 is especially toxic to cells lacking functional forms of the tumor suppressors, breast cancer type 1 susceptibility protein (BRCA1) or breast cancer type 2 susceptibility protein (BRCA2) [76].

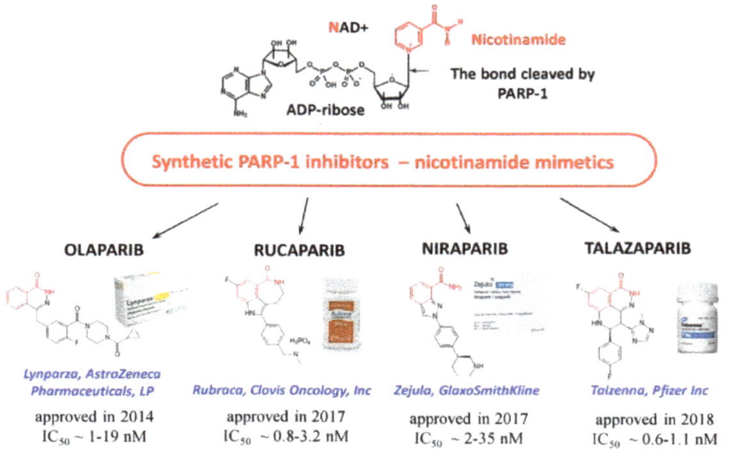

Figure 5. Synthetic PARPi approved for use in oncology. IC50 values (PARPi concentrations inducing 50% inhibition of PARP-1 activity) are cited from [19].

PARPi apparently can find wide application in the treatment of diseases related to inflammation. There are numerous examples of pharmacological or genetic inactivation of PARP-1 leading to a powerful anti-inflammatory effect that were demonstrated using different models of respiratory, gastrointestinal, osteochondral, cardiovascular and neurological pathologies (Table 1).

Table 1. Inhibition of PARP-1 in the treatment of non-cancer diseases.

Targeted Organ/System	Model	PARP-1 Inhibition Method	Effects	Ref.
The cardiovascular system	Various models of myocardial ischemia/reperfusion, models of acute coronary syndrome in mice and rats, as well as atherosclerotic vascular lesions	Genetic suppression (hereinafter—PARP-1$^{-/-}$ or activity inhibition with PARPi: 3-AB, TIQ-A, PJ-34, ABT-888, DPQ, INO-1001 or Doxycycline	↑TIMP-2, ↓NF-κB, ↓MCP-1, ↓ICAM-1, ↓TNF-α, ↓nitrotyrosine (marker of NO-dependent oxidative stress), ↓attraction of macrophages (MF), ↑ALDH2, ↓TC, ↓VLDL, ↓LDL, ↓ACAT-1, ↓caspase-3, ↑SMCs and collagen content, ↓atherosclerotic plaques, ↓MMPs, ↓ infarction zone, ↓CRP, ↓IL-6, ↓MPO activity, ↓neutrophil infiltration, ↓iNOS, ↓AIF nuclear translocation	[13]
Lungs, liver, gut, CNS	LPS-induced sepsis and endotoxic shock in animal model (mice, mini-pigs, rats)	PARP-1$^{-/-}$ or PARPi: PJ34, Olaparib, 3-AB or INO-1001	↓degree of organ inflammation ↓TNF-α, ↓IFN-γ, ↓iNOS, ↓IL-1β, ↓IL-6, ↑IL-10, ↓neutrophil infiltration, ↓increased vascular permeability in organs, ↓NO production, ↓lipid peroxidation, ↓MIP-1α↓MIP-2 (CXCL2), ↓MCP-1, ↓CXCL1 (mKC), ↑protective effect on membrane lipids	[40,77–79]

Table 1. *Cont.*

Targeted Organ/System	Model	PARP-1 Inhibition Method	Effects	Ref.
Gastrointestinal tract	Salmonella-induced sepsis in animal model (mice)	PARP-1$^{-/-}$	↓CXCL9, ↓Gbp2, CXCL10, ↓Iigp1, Cd274, ↓IFN-γ,	[80]
Gastrointestinal tract	TNBS-induced colitis in animal model (mice)	PARP-1$^{-/-}$	↓ICAM-1, ↓neutrophil infiltration, ↓lipid peroxidation,↓degree of nitrosative lesion.	[81]
CNS	Induced stroke in animal model (primates, mice, rats)	PARP-1$^{-/-}$or PARPi: benzamide, 3-AB, ISQ, DPQ, PHT, INH2BP, GPI-6150, PJ34, INO-10001, ONO-1924H, DR2313, GPH, MP-124 or JPI-289	↓PARP-1 activity, ↓PAR in the affected area, ↓inflammation and swelling of the brain,↓secondary neuronal damage	[82–85]
CNS	Alzheimer's disease: in vitro cellular models (human and rat cells treated with Aβ peptide); in vivo animal models (mice, rats with Aβ peptide)	PARPi: benzamide, Rukaparib, Veliparib, MC2050, PJ34, INO-1001, JPI-289, nicotinamide	↓neuroinflammation, ↓accumulation of Aβ plaques, ↑genes of antioxidant defense enzymes (*Sod1, Gpx1, Gpx4*), ↑genes regulating the mitochondriongenesis (*Mfn1, Mfn2, Dnm1l, Opa1, Fis1*), ↑mt-Nd1, ↑Sdha, ↑mt-Cytb, ↓membrane potential of mitochondria,↑Foxo1, ↓Nrf1, ↓Stat6, ↓NF-κB, ↓free radical concentration	[86–89]
CNS	Parkinson's disease: 6-OHDA-induced mice model of PD, MPTP-induced dopamine neurotoxicity, AIMP2 transgenic mice.	PARP-1$^{-/-}$ or PARPi: benzamide or Rukaparib	↓atrophy of dopaminergic (DA) neurons; ↓degeneration of DA neurons	[90–92]
Spinal cord	Spinal cord injuries in mice	PARPi: 3-AB or 5-AIQ	↓infiltration of the injured spinal cord with neutrophils, ↓cell apoptosis, ↓spinal cord injury	[93]
Diabetic multiple organ lesions	Mice, rats (by high-fat feeding and a single peritoneal dose of streptozotocin or obese animals with leptin resistant	PARPi: INO1001, MRL-45696 or JPI-289	↑SIRT1, ↑PGC-1α, ↓oxidative stress, ↓organ inflammation and fibrosis,↓TLR4, ↓NFκB signaling pathway.	[13,94–96]
Immune system	Arthritis in mice or rats	PARP-1$^{-/-}$ or PARPi: 3-AB	↓IL-17, ↓TNF-α, ↓IL-2, ↓MCP-1, ↓MIP-2, ↓VCAM-1, ↓ICAM-1 at the site of defeat, ↓iNOS, ↓COX-2, ↓MMP-2, ↓MMP-9	[8,97,98]

Abbreviations: AMPK—AMP-activated protein kinase; TNBS—trinitrobenzenesulfonic acid; PGC-1α—peroxisome proliferator-activated receptor gamma coactivator 1-α; TIMP-2—tissue inhibitor of metalloproteinases 2, ICAM-1—inter-cellular adhesion molecule 1, ALDH2 -aldehyde dehydrogenase,2, TC—total cholesterol, VLDL—very low density lipoproteins, LDL—low density lipoproteins, ACAT1—acetyl-CoA acetyltransferase 1, SMC—smooth muscle cell content, CRP—C-reactive protein, MPO—myeloperoxidase, MMPs—matrix metalloproteinases, MIP-1a and 2—macrophage inflammatory proteins 1a and 2, CXCLs—C-X-C motif chemokine ligands, GBP2- guanylate binding protein 2, IigP1—interferon-inducible GTPase 1, CD274—programmed death-ligand 1 (or PD-L1), Gpx1,4—glutathione peroxidase 1, 4, SOD1—superoxide dismutase 1, mt-Nd1—mitochondrially encoded NADH, Sdha—succinate dehydrogenase complex flavoprotein subunit A, mt-Cytb—mitochondrially encoded cytochrome B, FOXO1—forkhead box protein O1, Nrf1—nuclear respiratory factor 1, STAT6—signal transducer and activator of transcription 6, PGC-1α—peroxisome proliferator-activated receptor gamma coactivator, TLR4—toll-like receptor 4, VCAM-1—vascular cell adhesion molecule 1, ↑ - up-regulated, ↓- down-regulated.

Importantly, PARPi block NF-κB-mediated transcription of genes encoding pro-inflammatory cytokines, but do not reduce the transcription of anti-inflammatory cytokine-encoding genes of IL-10 and IL-13 [99]. At the same time, even clinically approved PARPi are characterized by side effects that demand the search for safer drugs that target PARP-1 [100].

5. Polyphenols as PARP-1 Inhibitors

An alternative to synthetic PARPi can be found among natural compounds—plant metabolites. Polyphenolic compounds (flavonoids and non-flavonoids, Figure 6), along with terpenoids and alkanoids, are the most common secondary plant metabolites. Flavonoids, which are subdivided into flavones, flavonols, flavanones, catechins (flavan-3-ols), isoflavonoids, and anthocyanidins (Figure 6) are the most widespread and studied natural polyphenols.

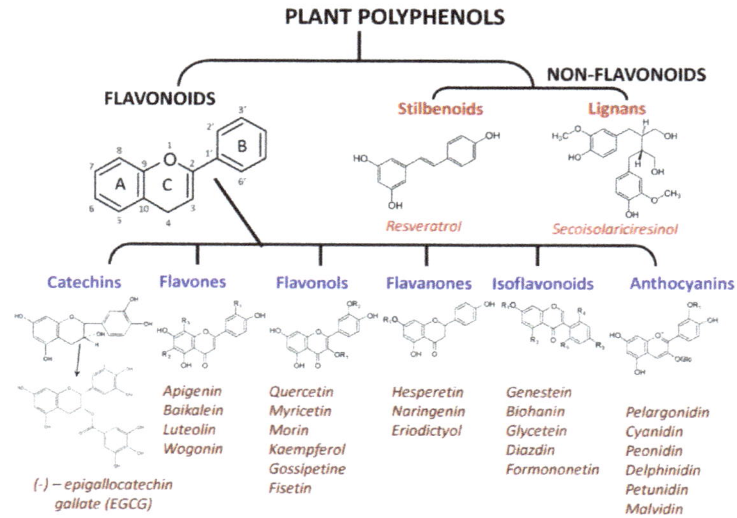

Figure 6. Plant polyphenols with known pharmacological properties.

Various types of plant polyphenols have different anti-inflammatory, antioxidant, anti-allergic, antiviral and/or antitumor activities [101,102]. Some natural polyphenols are epigenetically active compounds and may play an important role in the regulation of gene expression, including PARP-dependent genes [103]. Some of polyphenols were shown to have PARP-1 inhibiting activity [104–106]. Many polyphenols have high bioavailability, efficiently penetrate cells and induce biological effects at micromolar concentrations [107–110] that makes them good candidates for the search of new PARPi. Thus, a search in the flavonoid library led to the discovery of PARPi such as myricetin, quercetin, fisetin, tricetin, gossipetin and delphinidin [104]. Functional screening of the library of polyphenols used in traditional medicine resulted in identification of 11 compounds interacting with PARP-1 with the dissociation constants of the complexes ranging from 0.32 to 79 µM [111]. The most active PARPi among the polyphenols was 2″-hydroxygenkwanol A isolated from the plant *Daphne linearifolia* that has long been used to treat inflammation and rheumatism in Arab traditional medicine. This polyphenol is structurally similar to talazaparib, the strongest synthetic PARPi. Computer screening technologies also predict an existence of PARPi among polyphenols that may have affinities higher than clinically approved synthetic PARPi [112].

Below the features of PARP-1 inhibition by some representatives of polyphenols are considered in more detail.

5.1. Flavonols

Flavonols are the most abundant species of flavonoids existing in nature. Their distinct feature is the presence of 3-hydroxyflavone in the structure. Flavonols are often found as O-glycoside, glucuronide, methyl, and sulfate conjugates.

The flavonol quercetin (QC) is found in large quantities in plants, predominantly red and orange (sea buckthorn, cranberries, raspberries, blueberries, onions), as well as in food products such as buckwheat, tea, red wine, and olive oil. QC is usually found in plants in conjunction with glycosylated forms—isoquercetin and rutin [113]. These flavonols perform a wide range of physiological functions in plants, the most important of which is antioxidant. QC is able to inhibit PARP-1 in the micromolar concentration range, and its activity is approximately seven times higher than that of 3-aminobenzamide (3-AB) [114]. It was found that glycosylation improves the solubility of QC derivatives, but decreases their inhibitory activity. It was shown that QC at a concentration of more than 30 μM exhibits genotoxicity. Glycosylated analogs have less cyto- and genotoxicity, but this might be due to their lower cell permeability.

Regular consumption of citrus fruit reduces the risk of cancer, and this effect is likely associated with inhibition of PARP-1 by flavonols naringenin (NG), hesperetin (GP) and their O-glycoside forms naringin and hesperidin, which are contained in citrus fruit [115]. GP was found to be more active than QC and cytotoxic for both wild-type V79 cells and mutant cells deficient in the BRCA2 protein involved in DNA repair (100% cytotoxicity at 30 μM GP) [116]. In turn, QC selectively induces the death of BRCA2 mutant cells (40% cytotoxicity at 30 μM QC).

Glycosylated isoquercetin, rutin, naringin and hesperidin have less cytotoxicity than the corresponding non-glycosylated flavonoids, but at the same time exhibit selectivity towards BRCA2 mutant cells [115]. The death of more than 80% of the mutant cells was observed at 100 μM rutin and isoquercetin and at ~1 mM naringin and hesperidin.

In hepatocytes stimulated by the pro-inflammatory cytokine IL-1, QC reduces NO production through suppression of iNOS expression, which in turn, can block the enhancement of the inflammatory cascade [116]. Similarly, QC inhibits the LPS-induced iNOS gene expression in various models [117–119]. It is believed that the anti-inflammatory effect of QC is based on a combination of antioxidant and anti-PARP activities. Other flavonoids such as naringenin, apigenin, and resveratrol also block iNOS expression in macrophages [120,121].

Flavonoids are able to attenuate NAD+ depletion by inhibiting PARP-1 hyperactivation both in vitro and in vivo [104], therefore reducing the likelihood of cell death and exerting a pleiotropic protective effect at high glucose levels [122]. This may play an important role in preventing the development of diabetic complications caused by increased PARP-1 activity, including those associated with massive neuronal death [123]. The molecular mechanisms of the protective action of flavonoids in the suppression of diabetic complications remain the subject of active study [124]. QC can up-regulate expression of neural and synapse-associated proteins (nerve growth factor, brain-derived neurotrophic factor, post synaptic density 93 and 95 proteins) and inhibit neurodegeneration [125]. QC increases the level of SIRT1 (NAD+-dependent histone deacetylase) and inhibits the stress proteins of the endoplasmic reticulum (RNA-like endoplasmic reticulum kinase, inositol-requiring enzyme-1α, activating transcription factor 6α, eukaryotic initiation factor 2, binding immunoglobulin protein and protein disulfide isomerase). An increase in SIRT1 activity was shown to have a positive effect on the metabolism of mammals, leading to inhibition of aging and longevity [126,127]. PARP-1 knockout increases the NAD + content and, accordingly, SIRT1 activity in brown adipose tissue and muscles [128]. A similar effect is caused by PARPi. In aging rats, QC activates SIRT1, promotes monoamine synthesis and improves animal cognitive functions. QC improves learning and memory in diabetic rats [124,129,130].

5.2. Flavones

Flavones are flavonoids that have a 2-phenylchromen-4-one (2-phenyl-1-benzopyran-4-one) group. 4′-Methoxyflavone (4MF) and 3′, 4′-dimethoxyflavone (DMF) were reported to prevent parthanatos in cells treated with a DNA-alkylating agent and possess neuroprotective activity [131]. It was concluded that the anti-parthanatos effect of 4MF and DMF is related to the suppression of PARP-1 activity and is not associated with antioxidant or free radical scavenging properties. Both compounds almost equally prevented parthanatos in HeLa cells, but 4MF demonstrated higher neuroprotection than DMF.

It should be noted that some flavones (e.g., apigenin and luteolin) inhibit tankyrases (TNK), the proteins of the PARP family, which are attractive targets in cancer treatment [132]. The antitumor therapeutic potential of TNK is determined by their participation in telomere homeostasis, mitosis, and Wnt signaling pathways [133]. Interestingly, flavones do not contain the nicotinamide-like moiety that is characteristic for most PARP-1 inhibitors, and the flavone-based pharmacophore model was designed for TNK inhibitors [134,135].

5.3. Catechins

Flavan derivatives catechins include a wide variety of biologically active compounds. A feature of their structure is the absence of a double bond between the second and third carbon atoms leading to emergence of two chiral centers and four diastereoisomers. The trans and cis isomers are called catechins and epicatechins, respectively. Catechins are present in large quantities in tea leaves and cocoa beans. Green tea contains epigallocatechin gallate (EGCG), which is considered one of the most powerful dietary antioxidants [136]. During the production of black tea (enzymatic oxidation), catechin is oxidized to quinone, which is further condensed into several other chemical compounds, one of which is the theaflavin polyphenol (TF). Tea polyphenols affect regulatory systems of cells and may produce an inhibitory effect on various stages of carcinogenesis: inflammatory processes, cell transformation, proliferation, apoptosis, metastasis, and invasion [107,137–140]. It was found that EGCG and TF cause synthetic lethality in BRCA2-deficient cells through a PARP-dependent mechanism [141]. EGCG inhibits PARP-1 more effectively than TF, which is probably due to the presence of a haloyl group. Moreover PARP-1, the targets of tea polyphenols are histone deacetylases [142], transcription factors [143], DNA topoisomerase II [109] and ABC transporters responsible for the development of multidrug resistance [144,145].

Other catechins that affect PARP-1 include epicatechin, myricetin, epigallocatechin, catechin gallate, epicatechin gallate, gallocatechin, and gallocatechin gallate [146].

5.4. Resveratrol

The representative of stilbenoids, resveratrol (RSV), exists in the form of cis- and trans-stereoisomers and is often glycosylated. RSV is found in grapes, nuts and cocoa beans, as well as in berries, leaves and flowers of orchids, eucalyptus, gnetum and some other plants. Numerous studies have shown that the RSV containing extracts reduce thrombus formation, improve the rheological properties of blood, relax the vascular endothelium, lower cholesterol and triglyceride levels in the blood preventing atherosclerosis development, exhibit antioxidant and anti-inflammatory activity [147–149]. Such properties are strongly associated with the RSV-mediated blocking of the mTOR (mammalian target of rapamycin) signaling pathway [150]. mTOR is known to integrate various signaling pathways, including the pathways of insulin, growth factors, and mitogens. It functions as a sensor for redox status and cellular nutrient and energy levels. Dysregulation of mTOR pathway leads to the development of various metabolic and oncological diseases. The mTOR pathway can intersect with PARP-1 during partanotosis. In this case, SIRT1, involved in the regulation of the intracellular level of NAD +, can play an important role as a factor that binds PARP-1 and mTOR pathway [151]. It was demonstrated that RSV directly binds to PARP-1 and induces its dose-dependent inhibition (IC50 = 0.65 μM) [152]. Treatment of cells damaged by hyperglycemia with RSV reduces the production of ROS, improves the ratio of reduced/oxidized glutathione (GSH/GSSG), restoration mitochondrial membrane potential [153]. Studies on the suppression of metabolic stress leading to the onset of diabetic cataracts revealed that the administration of RSV led to a significant decrease in cataractogenesis. This effect may be associated with both the activation of antioxidant protection and the inhibition of PARP-1.

6. Conclusions

A moderate level of PARP-1 activation provides an effective reparation of DNA lesions supporting survival of cells under the action of genotoxic factors. A hyperactivation of PARP-1, which often occurs at the inflammation, hypoxia, hypo- and hyperglycemia, modulates or activates multiple cellular pathways leading to cell death or degeneration. Synthetic PARPi are already implemented in anticancer therapy, and might also be useful in treatment of metabolic syndrome, multiple organ dysfunction syndrome, diabetic complications, cardiovascular and neurodegenerative diseases.

Natural polyphenols capable of inhibiting PARP-1 directly or indirectly (Figure 7) can become a supplement or even an alternative to synthetic drugs, because besides a pronounced pharmacological activity they could have low systemic toxicity and minor side effects. As an adjunct to standard drug therapy, polyphenols can allow one to reduce a concentration of toxic drugs, providing a synergistic effect.

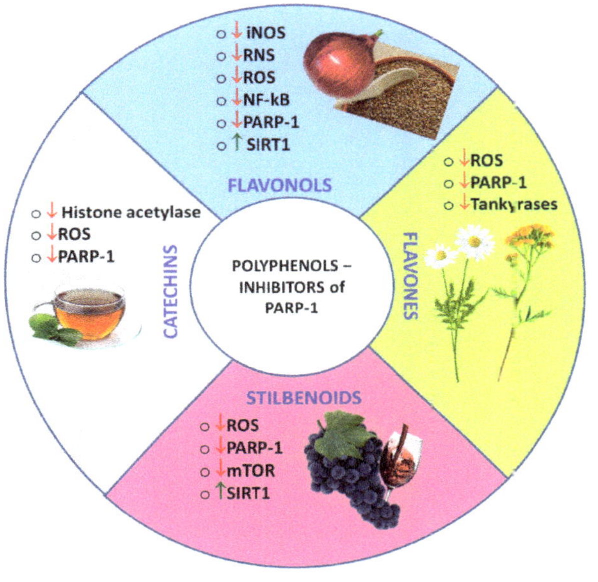

Figure 7. Classes of polyphenols, whose representatives were found to act as PARPi, and the observed polyphenol-induced regulatory effects. ↑ - up-regulated, ↓- down-regulated.

Extending a search for natural PARPi among the secondary plant metabolites, terpenoids should be also considered. Terpenoids, like polyphenols, have a wide spectrum of biological activities [154,155] and some of them were reported to be PARPi [21,140].

Author Contributions: Conceptualization, N.V.M., A.V.F. and V.M.S.; formal analysis N.V.M. and A.V.F.; writing—original draft N.V.M. and A.V.F.; writing—review and editing A.V.F. and V.M.S.; project administration V.M.S.; funding acquisition V.M.S.; supervision V.M.S. All authors have read and agreed to the published version of the manuscript.

Funding: Studies were financially supported by Russian Science Foundation (grant 21-74-20018).

Acknowledgments: The authors acknowledge the use of facilities of the Interdisciplinary Scientific and Educational School of Moscow University «Molecular Technologies of the Living Systems and Synthetic Biology».

Conflicts of Interest: The authors declare that they have no conflict of interest.

Abbreviations

PARP-1	poly (ADP-ribose) polymerase-1
PARPi	PARP inhibitors
AD	Alzheimer's disease
PD	Parkinson's disease
UV	ultraviolet light
ROS	reactive oxygen species overproduced in oxidative stress
RNS	reactive nitrogen species
TNFα	tumor necrosis factor α
iNOS	inducible isoform of nitrite oxide synthase
COX2	cyclooxygenase 2
MCP1	monocyte chemoattractantprotein 1
IL-1β, 6, 8	interleukins 1β, 6, 8
NF-κB	nuclear factor kappa-light-chain-enhancer of activated B cells
AP1, AP2	activator proteins 1 and 2
JAK	Janus kinase
PIP3	phosphatidylinositol (3,4,5)-trisphosphate
MAPKKKs	mitogen-activated protein (MAP) kinase kinase kinases
MAPKKs	mitogen-activated protein kinase kinases
MAPKs	mitogen-activated protein kinases
STAT	members of the signal transducer and activator of transcription protein family
AKT	subfamily of serine/threonine kinases
p38	p38 mitogen-activated protein kinases
JNK	c-Jun N-terminal kinases
ERKs	extracellular signal-regulated kinases
p300/CRBP	p300/CREB-binding protein complex
EMT	epithelial-mesenchymal transition
HSP70	heat shock protein 70
AIF	apoptosis inducing factor
HDAC	histone deacetylases
SIRT-1	sirtuin-1
AMPK	AMP-activated protein kinase
TNBS	trinitrobenzenesulfonic acid
PGC-1α	peroxisome proliferator-activated receptor gamma coactivator 1-α
TIMP-2	tissue inhibitor of metalloproteinases 2, I
ICAM-1	inter-cellular adhesion molecule 1
ALDH2	aldehyde dehydrogenase 2
TC	total cholesterol
VLDL	very low density lipoproteins
LDL	low density lipoproteins
ACAT1	acetyl-CoA acetyltransferase 1
SMC	smooth muscle cell content
CRP	C-reactive protein
MPO	myeloperoxidase
BRCA	BRCA1 (breast cancer type 1 susceptibility protein) and BRCA2 (breast cancer type 2 s
LPS	lipopolysaccharide
MMPs	matrix metalloproteinases
MIP-1a and 2	macrophage inflammatory proteins 1a and 2
CXCLs	C-X-C motif chemokine ligands
GBP2	guanylate binding protein 2
IigP1	interferon-inducible GTPase 1
CD274	programmed death-ligand 1 or PD-L1
Gpx1,4	glutathione peroxidase 1, 4

SOD1	superoxidedismutase 1
mt-Nd1	mitochondrially encoded NADH
Sdha	succinate dehydrogenasecomplexflavoproteinsubunit A
mt-Cytb	mitochondrially encoded cytochrome B
FOXO1	forkhead box protein O1
Nrf1	nuclear respiratory factor 1
STAT6	signal transducer and activator of transcription 6
TLR4	toll-like receptor 4
VCAM-1	vascular cell adhesion molecule 1
QC	quercetin
RSV	resveratrol
NG	naringenin
GP	hesperetin
4MF	4′-methoxyflavone
DMF	3′, 4′-dimethoxyflavone
3-AB	3-aminobenzamide
TNK	tankyrases
TF	theaflavin polyphenol
EGCG	epigallocatechin gallate

References

1. Gupte, R.; Liu, Z.; Kraus, W.L. PARPs and ADP-ribosylation: Recent advances linking molecular functions to biological out-comes. *Genes Dev.* **2017**, *31*, 101–126. [CrossRef]
2. Kraus, W.L. PARPs and ADP-ribosylation: 60 years on. *Genes Dev.* **2020**, *34*, 251–253. [CrossRef]
3. Eisemann, T.; Pascal, J.M. Poly(ADP-ribose) polymerase enzymes and the maintenance of genome integrity. *Cell. Mol. Life Sci.* **2019**, *77*, 19–33. [CrossRef]
4. Maluchenko, N.V.; Kulaeva, O.I.; Kotova, E.J.; Chupyrkina, A.A.; Nikitin, D.V.; Kirpichnikov, M.P.; Studitsky, V.M. Mo-lecular Mechanisms of Transcriptional Regulation by Poly(ADP-Ribose) Polymerase. *Mol. Biol.* **2015**, *49*, 1–15. [CrossRef]
5. Hottiger, M.O. Nuclear ADP-Ribosylation and Its Role in Chromatin Plasticity, Cell Differentiation, and Epigenetics. *Annu. Rev. Biochem.* **2015**, *84*, 227–263. [CrossRef] [PubMed]
6. Mateo, J.; Lord, C.J.; Serra, V.; Tutt, A.; Balmana, J.; Castroviejo-Bermejo, M.; Cruz, C.; Oaknin, A.; Kaye, S.B.; de Bono, J.S. A decade of clinical development of PARP inhibitors in perspective. *Annals of oncology: Off. J. Eur. Soc. Med. Oncol.* **2019**, *30*, 1437–1447. [CrossRef]
7. Curtin, N.J.; Szabo, C. Poly(ADP-ribose) polymerase inhibition: Past, present and future. *Nat. Rev. Drug Discov.* **2020**, *19*, 711–736. [CrossRef] [PubMed]
8. Brady, P.N.; Goel, A.; Johnson, M.A. Poly(ADP-Ribose) Polymerases in Host-Pathogen Interactions, Inflammation, and Immunity. *Microbiol. Mol. Biol. Rev.* **2019**, *83*, e00038-18. [CrossRef] [PubMed]
9. Vinuesa, A.; Pomilio, C.; Gregosa, A.; Bentivegna, M.; Presa, J.; Bellotto, M.; Saravia, F.; Beauquis, J. Inflammation and Insulin Resistance as Risk Factors and Potential Therapeutic Targets for Alzheimer's Disease. *Front. Neurosci.* **2021**, *15*. [CrossRef] [PubMed]
10. Hernández-Alonso, P.; Salas-Salvadó, J.; Baldrich-Mora, M.; Juanola-Falgarona, M.; Bulló, M. Beneficial Effect of Pistachio Consumption on Glucose Metabolism, Insulin Resistance, Inflammation, and Related Metabolic Risk Markers: A Randomized Clinical Trial. *Diabetes Care* **2014**, *37*, 3098–3105. [CrossRef]
11. Ministrini, S.; Carbone, F.; Montecucco, F. Updating concepts on atherosclerotic inflammation: From pathophysiology to treatment. *Eur. J. Clin. Investig.* **2020**, *51*, e13467. [CrossRef]
12. Ajala, O.N.; Everett, B.M. Targeting Inflammation to Reduce Residual Cardiovascular Risk. *Curr. Atheroscler. Rep.* **2020**, *22*, 1–9. [CrossRef] [PubMed]
13. Henning, R.J.; Bourgeois, M.; Harbison, R.D. Poly(ADP-ribose) Polymerase (PARP) and PARP Inhibitors: Mechanisms of Action and Role in Cardiovascular Disorders. *Cardiovasc. Toxicol.* **2018**, *18*, 493–506. [CrossRef] [PubMed]
14. Ordog, K.; Horvath, O.; Eros, K.; Bruszt, K.; Toth, S.; Kovacs, D.; Kalman, N.; Radnai, B.; Deres, L.; Gallyas, F.; et al. Mitochondrial protective effects of PARP-inhibition in hypertension-induced myocardial remodeling and in stressed cardiomyocytes. *Life Sci.* **2021**, *268*, 118936. [CrossRef] [PubMed]
15. Li, X.; Nemoto, M.; Xu, Z.; Yu, S.W.; Shimoji, M.; Andrabi, S.A.; Haince, J.F.; Poirier, G.G.; Dawson, T.M.; Dawson, V.L.; et al. Influence of duration of focal cerebral ischemia and neuronal nitric oxide synthase on translocation of apopto-sis-inducing factor to the nucleus. *Neuroscience* **2007**, *144*, 56–65. [CrossRef]
16. Zhang, X.; Chen, J.; Graham, S.H.; Du, L.; Kochanek, P.; Draviam, R.; Guo, F.; Nathaniel, P.D.; Szabo, C.; Watkins, S.; et al. Intranuclear localization of apoptosis-inducing factor (AIF) and large scale dna fragmentation after traumatic brain injury in rats and in neuronal cultures exposed to peroxynitrite. *J. Neurochem.* **2002**, *82*, 181–191. [CrossRef]

17. Yu, S.W.; Wang, H.; Dawson, T.M.; Dawson, V.L. Poly(ADP-ribose) polymerase-1 and apoptosis inducing factor in neuro-toxicity. *Neurobiol. Dis.* **2003**, *14*, 303–317. [CrossRef]
18. Koh, D.W.; Dawson, T.M.; Dawson, V.L. Poly(ADP-ribosyl)ation regulation of life and death in the nervous system. *Cell. Mol. Life Sci. CMLS* **2005**, *62*, 760–768. [CrossRef]
19. Antolin, A.A.; Ameratunga, M.; Banerji, U.; Clarke, P.A.; Workman, P.; Al-Lazikani, B. The kinase polypharmacology landscape of clinical PARP inhibitors. *Sci. Rep.* **2020**, *10*, 2585. [CrossRef]
20. Szabó, G.; Liaudet, L.; Hagl, S.; Szabó, C. Poly(ADP-ribose) polymerase activation in the reperfused myocardium. *Cardiovasc. Res.* **2004**, *61*, 471–480. [CrossRef]
21. Luccarini, I.; Pantano, D.; Nardiello, P.; Cavone, L.; Lapucci, A.; Miceli, C.; Nediani, C.; Berti, A.; Stefani, M.; Casamenti, F. The Polyphenol Oleuropein Aglycone Modulates the PARP1-SIRT1 Interplay: An In Vitro and In Vivo Study. *J. Alzheimer's Dis.* **2016**, *54*, 737–750. [CrossRef]
22. Andreone, T.; Meares, G.P.; Hughes, K.J.; Hansen, P.A.; Corbett, J.A. Cytokine-mediated beta-cell damage in PARP-1-deficient islets. *Am. J. Physiology. Endocrinol. Metab.* **2012**, *303*, E172–E179. [CrossRef]
23. Szkudelski, T. Streptozotocin-nicotinamide-induced diabetes in the rat. Characteristics of the experimental model. *Experiment. Biol. Med.* **2012**, *237*, 481–490. [CrossRef]
24. Pacher, P.; Szabo, C. Role of poly(ADP-ribose) polymerase-1 activation in the pathogenesis of diabetic complications: Endo-thelial dysfunction, as a common underlying theme. *Antioxid. Redox Signal.* **2005**, *7*, 1568–1580. [CrossRef]
25. Haider, T.; Pandey, V.; Banjare, N.; Gupta, P.N.; Soni, V. Drug resistance in cancer: Mechanisms and tackling strategies. *Pharmacol. Rep.* **2020**, *72*, 1125–1151. [CrossRef]
26. Fatokun, A.A.; Dawson, V.; Dawson, T.M. Parthanatos: Mitochondrial-linked mechanisms and therapeutic opportunities. *Br. J. Pharmacol.* **2014**, *171*, 2000–2016. [CrossRef] [PubMed]
27. Harrision, D.; Gravells, P.; Thompson, R.; Bryant, H.E. Poly(ADP-Ribose) Glycohydrolase (PARG) vs. Poly(ADP-Ribose) Polymerase (PARP)—Function in Genome Maintenance and Relevance of Inhibitors for Anti-cancer Therapy. *Front. Mol. Biosci.* **2020**, *7*, 191. [CrossRef] [PubMed]
28. Wang, Y.; Dawson, V.L.; Dawson, T.M. Poly(ADP-ribose) signals to mitochondrial AIF: A key event in parthanatos. *Exp. Neurol.* **2009**, *218*, 193–202. [CrossRef]
29. Mashimo, M.; Onishi, M.; Uno, A.; Tanimichi, A.; Nobeyama, A.; Mori, M.; Yamada, S.; Negi, S.; Bu, X.; Kato, J.; et al. The 89-kDa PARP1 cleavage fragment serves as a cytoplasmic PAR carrier to induce AIF-mediated apoptosis. *J. Biol. Chem.* **2021**, *296*, 100046. [CrossRef] [PubMed]
30. Santagostino, S.F.; Assenmacher, C.-A.; Tarrant, J.C.; Adedeji, A.O.; Radaelli, E. Mechanisms of Regulated Cell Death: Current Perspectives. *Vet. Pathol.* **2021**. [CrossRef]
31. Park, H.; Kam, T.-I.; Dawson, T.M.; Dawson, V.L. Poly (ADP-ribose) (PAR)-dependent cell death in neurodegenerative diseases. *Int. Rev. Cell Mol. Biol.* **2020**, *353*, 1–29. [CrossRef]
32. Andrabi, S.A.; Umanah, G.K.E.; Chang, C.; Stevens, D.A.; Karuppagounder, S.; Gagné, J.-P.; Poirier, G.G.; Dawson, V.L.; Dawson, T.M. Poly(ADP-ribose) polymerase-dependent energy depletion occurs through inhibition of glycolysis. *Proc. Natl. Acad. Sci. USA* **2014**, *111*, 10209–10214. [CrossRef]
33. Berger, N.A.; Sims, J.L.; Catino, D.M.; Berger, S.J. Poly(ADP-ribose) Polymerase Mediates the Suicide Response to Massive DNA Damage: Studies in Normal and DNA-repair Defective Cells. *Princess Takamatsu Symp.* **2020**, 219–226. [CrossRef]
34. Chiarugi, A. Poly(ADP-ribose) polymerase: Killer or conspirator? The 'suicide hypothesis' revisited. *Trends Pharmacol. Sci.* **2002**, *23*, 122–129. [CrossRef]
35. Formentini, L.; Macchiarulo, A.; Cipriani, G.; Camaioni, E.; Rapizzi, E.; Pellicciari, R.; Moroni, F.; Chiarugi, A. Poly(ADP-ribose) Catabolism Triggers AMP-dependent Mitochondrial Energy Failure. *J. Biol. Chem.* **2009**, *284*, 17668–17676. [CrossRef] [PubMed]
36. Carling, D.; Sanders, M.J.; Woods, A. The regulation of AMP-activated protein kinase by upstream kinases. *Int. J. Obes.* **2008**, *32*, S55–S59. [CrossRef] [PubMed]
37. Hassa, P.O.; Hottiger, M.O. The functional role of poly(ADP-ribose)polymerase 1 as novel coactivator of NF-kappaB in in-flammatory disorders. *Cell. Mol. Life Sci.* **2002**, *59*, 1534–1553. [CrossRef] [PubMed]
38. Zerfaoui, M.; Errami, Y.; Naura, A.S.; Suzuki, Y.; Kim, H.; Ju, J.; Liu, T.; Hans, C.P.; Kim, J.G.; Abd Elmageed, Z.Y. Poly(ADP-ribose) polymerase-1 is a determining factor in crm1-mediated nuclear export and retention of p65 NF-kappa B upon TLR4 stimulation. *J. Immunol.* **2010**, *185*, 1894–1902. [CrossRef]
39. Hassa, P.O.; Haenni, S.S.; Buerki, C.; Meier, N.I.; Lane, W.S.; Owen, H.; Gersbach, M.; Imhof, R.; Hottiger, M.O. Acetylation of PARP-1 by p300/CBP regulates coactivation of NF-kappa B-dependent transcription. *J. Biol. Chem.* **2005**, *280*, 40450–40464. [CrossRef]
40. Wasyluk, W.; Zwolak, A. PARP Inhibitors: An Innovative Approach to the Treatment of Inflammation and Metabolic Dis-orders in Sepsis. *J. Inflamm. Res.* **2021**, *14*, 1827–1844. [CrossRef]
41. Krukenberg, K.A.; Kim, S.; Tan, E.S.; Maliga, Z.; Mitchison, T.J. Extracellular Poly(ADP-Ribose) Is a Pro-inflammatory Signal for Macrophages. *Chem. Biol.* **2015**, *22*, 446–452. [CrossRef] [PubMed]
42. Schreiber, V.; Molinete, M.; Boeuf, H.; De Murcia, G.; Murcia, J.M.-D. The human poly(ADP-ribose) polymerase nuclear localization signal is a bipartite element functionally separate from DNA binding and catalytic activity. *EMBO J.* **1992**, *11*, 3263–3269. [CrossRef] [PubMed]

43. Scovassi, A.I.; Diederic, M. Modulation of poly(ADP-ribosylation) in apoptotic cells. *Biochem. Pharmacol.* **2004**, *68*, 1041–1047. [CrossRef] [PubMed]
44. Pétrilli, V.; Herceg, Z.; Hassa, P.; Patel, N.; Di Paola, R.; Cortes, U.; Dugo, L.; Filipe, H.-M.; Thiemermann, C.; Hottiger, M.; et al. Noncleavablepoly(ADP-ribose) polymerase-1 regulates the inflammation response in mice. *J. Clin. Invest.* **2004**, *114*, 1072–1081. [CrossRef]
45. Yung, T.M.C.; Satoh, M.S. Functional Competition between Poly(ADP-ribose) Polymerase and Its 24-kDa Apoptotic Fragment in DNA Repair and Transcription. *J. Biol. Chem.* **2001**, *276*, 11279–11286. [CrossRef]
46. Soldani, C.; Scovassi, A.I. Poly(ADP-ribose) polymerase-1 cleavage during apoptosis: An update. *Apoptosis* **2002**, *7*, 321–328. [CrossRef]
47. Boesten, D.M.; de Vos-Houben, J.M.; Timmermans, L.; den Hartog, G.J.; Bast, A.; Hageman, G.J. Accelerated aging during chronic oxidative stress: A role for PARP-1. *Oxidative Med. Cell. Longev.* **2013**, *2013*, 680414. [CrossRef]
48. Oliver, F.J.; Menissier-de Murcia, J.; Nacci, C.; Decker, P.; Andriantsitohaina, R.; Muller, S.; de la Rubia, G.; Stoclet, J.C.; de Murcia, G. Resistance to endotoxic shock as a consequence of defective NF-kappaB activation in poly (ADP-ribose) polymer-ase-1 deficient mice. *EMBO J.* **1999**, *18*, 4446–4454. [CrossRef]
49. Chan, C.Y.; Tan, K.V.; Cornelissen, B. PARP Inhibitors in Cancer Diagnosis and Therapy. *Clin. Cancer Res.* **2020**, *27*, 1585–1594. [CrossRef]
50. Singh, N.; Pay, S.L.; Bhandare, S.B.; Arimpur, U.; Motea, E.A. Therapeutic Strategies and Biomarkers to Modulate PARP Activity for Targeted Cancer Therapy. *Cancers* **2020**, *12*, 972. [CrossRef]
51. Kim, D.-S.; Camacho, C.V.; Kraus, W.L. Alternate therapeutic pathways for PARP inhibitors and potential mechanisms of resistance. *Exp. Mol. Med.* **2021**, *53*, 42–51. [CrossRef]
52. Kumar, M.; Jaiswal, R.K.; Yadava, P.K.; Singh, R.P. An assessment of poly (ADP-ribose) polymerase-1 role in normal and cancer cells. *BioFactors* **2020**, *46*, 894–905. [CrossRef]
53. Simbulan-Rosenthal, C.M.; Ly, D.H.; Rosenthal, D.S.; Konopka, G.; Luo, R.; Wang, Z.Q.; Schultz, P.G.; Smulson, M.E. Misregulation of gene expression in primary fibroblasts lacking poly(ADP-ribose) polymerase. *Proc. Natl. Acad. Sci. USA* **2000**, *97*, 11274–11279. [CrossRef]
54. Liu, Z.; Kraus, W.L. Catalytic-Independent Functions of PARP-1 Determine Sox2 Pioneer Activity at Intractable Genomic Loci. *Mol. Cell* **2017**, *65*, 589–603.e9. [CrossRef]
55. Neophytou, C.; Panagi, M.; Stylianopoulos, T.; Papageorgis, P. The Role of Tumor Microenvironment in Cancer Metastasis: Molecular Mechanisms and Therapeutic Opportunities. *Cancers* **2021**, *13*, 2053. [CrossRef]
56. Marti, J.M.; Fernandez-Cortes, M.; Serrano-Saenz, S.; Zamudio-Martinez, E.; Delgado-Bellido, D.; Garcia-Diaz, A.; Oliver, F.J. The Multifactorial Role of PARP-1 in Tumor Microenvironment. *Cancers* **2020**, *12*, 739. [CrossRef] [PubMed]
57. Kumar, M.; Jaiswal, R.K.; Prasad, R.; Yadav, S.S.; Kumar, A.; Yadava, P.K.; Singh, R.P. PARP-1 induces EMT in non-small cell lung carcinoma cells via modulating the transcription factors Smad4, p65 and ZEB1. *Life Sci.* **2021**, *269*, 118994. [CrossRef]
58. Balaji, S.; Kim, U.; Muthukkaruppan, V.; Vanniarajan, A. Emerging role of tumor microenvironment derived exosomes in therapeutic resistance and metastasis through epithelial-to-mesenchymal transition. *Life Sci.* **2021**, *280*, 119750. [CrossRef] [PubMed]
59. Vasan, N.; Baselga, J.; Hyman, D.M. A view on drug resistance in cancer. *Nature* **2019**, *575*, 299–309. [CrossRef]
60. Chatterjee, N.; Bivona, T.G. Polytherapy and Targeted Cancer Drug Resistance. *Trends Cancer* **2019**, *5*, 170–182. [CrossRef] [PubMed]
61. Yap, T.A.; Plummer, R.; Azad, N.S.; Helleday, T. The DNA Damaging Revolution: PARP Inhibitors and Beyond. *Am. Soc. Clin. Oncol. Educ. Book* **2019**, *39*, 185–195. [CrossRef] [PubMed]
62. de Haan, R.; Pluim, D.; van Triest, B.; van den Heuvel, M.; Peulen, H.; van Berlo, D.; George, J.; Verheij, M.; Schellens, J.H.M.; Vens, C. Improved pharmacodynamic (PD) assessment of low dose PARP inhibitor PD activity for radiotherapy and chem-otherapy combination trials. *Radiother. Oncol.* **2018**, *126*, 443–449. [CrossRef] [PubMed]
63. Matulonis, U.A.; Monk, B.J. PARP inhibitor and chemotherapy combination trials for the treatment of advanced malignancies: Does a development pathway forward exist? *Ann. Oncol.* **2017**, *28*, 443–447. [CrossRef]
64. Yang, X.-D.; Kong, F.-E.; Qi, L.; Lin, J.-X.; Yan, Q.; Loong, J.H.C.; Xi, S.-Y.; Zhao, Y.; Zhang, Y.; Yuan, Y.-F.; et al. PARP inhibitor Olaparib overcomes Sorafenib resistance through reshaping the pluripotent transcriptome in hepatocellular carcinoma. *Mol. Cancer* **2021**, *20*, 1–7. [CrossRef] [PubMed]
65. Huang, J.; Li, W.; Liao, W.; Hao, Q.; Tang, D.; Wang, D.; Wang, Y.; Ge, G. Green tea polyphenol epigallocatechin-3-gallate alleviates nonalcoholic fatty liver disease and ameliorates intestinal immunity in mice fed a high-fat diet. *Food Funct.* **2020**, *11*, 9924–9935. [CrossRef]
66. Karunarathna, K.H.T.; Mewan, K.M.; Weerasena, O.V.D.S.J.; Perera, S.A.C.N.; Edirisinghe, E.N.U. A functional molecular marker for detecting blister blight disease resistance in tea (*Camellia sinensis* L.). *Plant Cell Rep.* **2020**, *40*, 351–359. [CrossRef]
67. Yang, M.; Liu, L.; Xie, M.; Sun, X.; Yu, Y.; Kang, R.; Yang, L.; Zhu, S.; Cao, L.; Tang, D. Poly-ADP-ribosylation of HMGB1 regulates TNFSF10/TRAIL resistance through autophagy. *Autophagy* **2015**, *11*, 214–224. [CrossRef]
68. Tulin, A.; Spradling, A. Chromatin Loosening by Poly(ADP)-Ribose Polymerase (PARP) at Drosophila Puff Loci. *Science* **2003**, *299*, 560–562. [CrossRef]

69. Petesch, S.J.; Lis, J.T. Rapid, Transcription-Independent Loss of Nucleosomes over a Large Chromatin Domain at Hsp70 Loci. *Cell* **2008**, *134*, 74–84. [CrossRef]
70. Leu, J.I.; Pimkina, J.; Frank, A.; Murphy, M.E.; George, D.L. A small molecule inhibitor of inducible heat shock protein 70. *Mol. Cell* **2009**, *36*, 15–27. [CrossRef]
71. Rojo, F.; García-Parra, J.; Zazo, S.; Tusquets, I.; Ferrer-Lozano, J.; Menendez, S.; Eroles, P.; Chamizo, C.; Servitja, S.; Ramírez-Merino, N.; et al. Nuclear PARP-1 protein overexpression is associated with poor overall survival in early breast cancer. *Ann. Oncol.* **2011**, *23*, 1156–1164. [CrossRef]
72. Domagala, P.; Huzarski, T.; Lubinski, J.; Gugala, K.; Domagala, W. PARP-1 expression in breast cancer including BRCA1-associated, triple negative and basal-like tumors: Possible implications for PARP-1 inhibitor therapy. *Breast Cancer Res. Treat.* **2011**, *127*, 861–869. [CrossRef]
73. Ossovskaya, V.; Koo, I.C.; Kaldjian, E.P.; Alvares, C.; Sherman, B.M. Upregulation of Poly (ADP-Ribose) Polymerase-1 (PARP1) in Triple-Negative Breast Cancer and Other Primary Human Tumor Types. *Genes Cancer* **2010**, *1*, 812–821. [CrossRef]
74. Malyuchenko, N.V.; Kotova, E.Y.; Kulaeva, O.I.; Kirpichnikov, M.P.; Studitskiy, V.M. PARP1 Inhibitors: Antitumor drug design. *Acta Nat.* **2015**, *7*, 27–37. [CrossRef]
75. Singh, M.P.; Cho, H.J.; Kim, J.-T.; Baek, K.E.; Lee, H.G.; Kang, S.C.; Cho, K.; Baek, L. Morin Hydrate Reverses Cisplatin Resistance by Impairing PARP1/HMGB1-Dependent Autophagy in Hepatocellular Carcinoma. *Cancers* **2019**, *11*, 986. [CrossRef] [PubMed]
76. Lee, J.M.; Ledermann, J.A.; Kohn, E.C. PARP Inhibitors for BRCA1/2 mutation-associated and BRCA-like malignancies. *Ann. Oncol.* **2014**, *25*, 32–40. [CrossRef]
77. Liaudet, L.; Pacher, P.; Mabley, J.G.; Virag, L.; Soriano, F.G.; Hasko, G.; Szabo, C. Activation of poly(ADP-Ribose) poly-merase-1 is a central mechanism of lipopolysaccharide-induced acute lung inflammation. *Am. J. Respir. Crit. Care Med.* **2002**, *165*, 372–377. [CrossRef] [PubMed]
78. Zerfaoui, M.; Naura, A.S.; Errami, Y.; Hans, C.P.; Rezk, B.M.; Park, J.W.; Elsegeiny, W.; Kim, H.; Lord, K.; Kim, J.G.; et al. Effects of PARP-1 deficiency on airway inflammatory cell recruitment in response to LPS or TNF: Differential effects on CXCR2 ligands and Duffy antigen receptor for chemokines. *J. Leukoc. Biol.* **2009**, *86*, 1385–1392. [CrossRef] [PubMed]
79. Czapski, G.A.; Cakala, M.; Gajkowska, B.; Strosznajder, J.B. Poly(ADP-ribose) polymerase-1 inhibition protects the brain against systemic inflammation. *Neurochem. Int.* **2006**, *49*, 751–755. [CrossRef]
80. Altmeyer, M.; Barthel, M.; Eberhard, M.; Rehrauer, H.; Hardt, W.-D.; Hottiger, M.O. Absence of Poly(ADP-Ribose) Polymerase 1 Delays the Onset of Salmonella enterica Serovar Typhimurium-Induced Gut Inflammation. *Infect. Immun.* **2010**, *78*, 3420–3431. [CrossRef]
81. Zingarelli, B.; Szabó, C.; Salzman, A.L. Blockade of poly(ADP-ribose) synthetase inhibits neutrophil recruitment, oxidant generation, and mucosal injury in murine colitis. *Gastroenterology* **1999**, *116*, 335–345. [CrossRef]
82. Ullrich, O.; Diestel, A.; Eyupoglu, I.Y.; Nitsch, R. Regulation of microglial expression of integrins by poly(ADP-ribose) polymerase-1. *Nat. Cell Biol.* **2001**, *3*, 1035–1042. [CrossRef]
83. de la Lastra, C.A.; Villegas, I.; Sanchez-Fidalgo, S. Poly(ADP-ribose) polymerase inhibitors: New pharmacological functions and potential clinical implications. *Curr. Pharm. Des.* **2007**, *13*, 933–962. [CrossRef]
84. Matsuura, S.; Egi, Y.; Yuki, S.; Horikawa, T.; Satoh, H.; Akira, T. MP-124, a novel poly(ADP-ribose) polymerase-1 (PARP-1) inhibitor, ameliorates ischemic brain damage in a non-human primate model. *Brain Res.* **2011**, *1410*, 122–131. [CrossRef]
85. Kim, Y.; Kim, Y.S.; Noh, M.-Y.; Lee, H.; Joe, B.; Kim, H.Y.; Kim, J.; Kim, S.H.; Park, J. Neuroprotective effects of a novel poly (ADP-ribose) polymerase-1 inhibitor, JPI-289, in hypoxic rat cortical neurons. *Clin. Exp. Pharmacol. Physiol.* **2017**, *44*, 671–679. [CrossRef] [PubMed]
86. Thapa, K.; Khan, H.; Sharma, U.; Grewal, A.K.; Singh, T.G. Poly (ADP-ribose) polymerase-1 as a promising drug target for neurodegenerative diseases. *Life Sci.* **2020**, *267*, 118975. [CrossRef]
87. Martire, S.; Fuso, A.; Rotili, D.; Tempera, I.; Giordano, C.; De Zottis, I.; Muzi, A.; Vernole, P.; Graziani, G.; Lococo, E.; et al. PARP-1 Modulates Amyloid Beta Peptide-Induced Neuronal Damage. *PLoS ONE* **2013**, *8*, e72169. [CrossRef]
88. Bayrakdar, E.T.; Uyanikgil, Y.; Kanit, L.; Koylu, E.; Yalcin, A. Nicotinamide treatment reduces the levels of oxidative stress, apoptosis, and PARP-1 activity in Abeta(1-42)-induced rat model of Alzheimer's disease. *Free Radic. Res.* **2014**, *48*, 146–158. [CrossRef] [PubMed]
89. Czapski, G.A.; Cieslik, M.; Wencel, P.L.; Wojtowicz, S.; Strosznajder, R.P.; Strosznajder, J.B. Inhibition of poly(ADP-ribose) polymerase-1 alters expression of mitochondria-related genes in PC12 cells: Relevance to mitochondrial homeostasis in neurodegenerative disorders. *Biochim. Biophys. Acta Mol. Cell Res.* **2018**, *1865*, 281–288. [CrossRef]
90. Kim, T.W.; Cho, H.M.; Choi, S.Y.; Suguira, Y.; Hayasaka, T.; Setou, M.; Koh, H.C.; Hwang, E.M.; Park, J.Y.; Kang, S.J.; et al. (ADP-ribose) polymerase 1 and AMP-activated protein kinase mediate progressive dopaminergic neuronal degeneration in a mouse model of Parkinson's disease. *Cell Death Dis.* **2013**, *4*, e919. [CrossRef] [PubMed]
91. Cosi, C.; Marien, M. Decreases in mouse brain NAD+ and ATP induced by 1-methyl-4-phenyl-1,2,3,6-tetrahydropyridine (MPTP): Prevention by the poly(ADP-ribose) polymerase inhibitor, benzamide. *Brain Res.* **1998**, *809*, 58–67. [CrossRef]
92. Lee, Y.; Karuppagounder, S.S.; Shin, J.H.; Lee, Y.I.; Ko, H.S.; Swing, D.; Jiang, H.; Kang, S.U.; Lee, B.D.; Kang, H.C. Parthanatos mediates AIMP2-activated age-dependent dopaminergic neu-ronal loss. *Nat. Neurosci.* **2013**, *16*, 1392–1400. [CrossRef]

93. Genovese, T.; Mazzon, E.; Muià, C.; Patel, N.S.A.; Threadgill, M.D.; Bramanti, P.; De Sarro, A.; Thiemermann, C.; Cuzzocrea, S. Inhibitors of Poly(ADP-Ribose) Polymerase Modulate Signal Transduction Pathways and Secondary Damage in Experimental Spinal Cord Trauma. *J. Pharmacol. Exp. Ther.* **2004**, *312*, 449–457. [CrossRef] [PubMed]
94. Waldman, M.; Nudelman, V.; Shainberg, A.; Abraham, N.G.; Kornwoski, R.; Aravot, D.; Arad, M.; Hochhauser, E. PARP-1 inhibition protects the diabetic heart through activation of SIRT1-PGC-1α axis. *Exp. Cell Res.* **2018**, *373*, 112–118. [CrossRef] [PubMed]
95. Jang, H.R.; Lee, K.; Jeon, J.; Kim, J.-R.; Lee, J.E.; Kwon, G.Y.; Kim, Y.-G.; Kim, D.J.; Ko, J.-W.; Huh, W. Poly (ADP-Ribose) Polymerase Inhibitor Treatment as a Novel Therapy Attenuating Renal Ischemia-Reperfusion Injury. *Front. Immunol.* **2020**, *11*, 564288. [CrossRef]
96. Ji, Z.; Zeng, L.; Cheng, Y.; Liang, G. Poly(ADP-ribose) polymerases-1 inhibitor MRL-45696 alleviates DNA damage after myocardial ischemia-reperfusion in diabetic rats. *Nan Fang Yike Da Xuexue Bao* **2018**, *38*, 830–835.
97. Liu, Z.; Wang, H.; Wang, S.; Gao, J.; Niu, L. PARP-1 inhibition attenuates the inflammatory response in the cartilage of a rat model of osteoarthritis. *Bone Jt. Res.* **2021**, *10*, 401–410. [CrossRef] [PubMed]
98. Ahmad, S.F.; Attia, S.M.; Zoheir, K.M.; Ashour, A.E.; Bakheet, S.A. Attenuation of the progression of adjuvant-induced arthritis by 3-aminobenzamide treatment. *Int. Immunopharmacol.* **2014**, *19*, 52–59. [CrossRef]
99. Idowu, S.O.; Fatokun, A.A. Artificial Intelligence (AI) to the Rescue: Deploying Machine Learning to Bridge the Biorelevance Gap in Antioxidant Assays. *SLAS Technol. Transl. Life Sci. Innov.* **2020**, *26*, 16–25. [CrossRef] [PubMed]
100. Nilov, D.; Maluchenko, N.; Kurgina, T.; Pushkarev, S.; Lys, A.; Kutuzov, M.; Gerasimova, N.; Feofanov, A.; Švedas, V.; Lavrik, O.; et al. Molecular Mechanisms of PARP-1 Inhibitor 7-Methylguanine. *Int. J. Mol. Sci.* **2020**, *21*, 2159. [CrossRef] [PubMed]
101. Arora, I.; Sharma, M.; Sun, L.Y.; Tollefsbol, T.O. The Epigenetic Link between Polyphenols, Aging and Age-Related Diseases. *Genes* **2020**, *11*, 1094. [CrossRef]
102. Arora, I.; Sharma, M.; Tollefsbol, T.O. Combinatorial Epigenetics Impact of Polyphenols and Phytochemicals in Cancer Pre-vention and Therapy. *Int. J. Mol. Sci.* **2019**, *20*, 4567. [CrossRef] [PubMed]
103. Sharma, V.; Pandey, S.N.; Khawaja, H.; Brown, K.J.; Hathout, Y.; Chen, Y.-W. PARP1 Differentially Interacts with Promoter region of DUX4 Gene in FSHD Myoblasts. *J. Genet. Syndr. Gene Ther.* **2016**, *7*, 1–9. [CrossRef] [PubMed]
104. Geraets, L.; Moonen, H.J.J.; Brauers, K.; Wouters, E.F.M.; Bast, A.; Hageman, G.J Dietary Flavones and Flavonoles Are Inhibitors of Poly(ADP-ribose)polymerase-1 in Pulmonary Epithelial Cells. *J. Nutr.* **2007**, *137*, 2190–2195. [CrossRef]
105. Maeda, J.; Roybal, E.J.; Brents, C.A.; Uesaka, M.; Aizawa, Y.; Kato, T.A. Natural and glucosyl flavonoids inhibit poly(ADP-ribose) polymerase activity and induce synthetic lethality in BRCA mutant cells. *Oncol. Rep.* **2013**, *31*, 551–556. [CrossRef] [PubMed]
106. Narwal, M.; Haikarainen, T.; Fallarero, A.; Vuorela, P.M.; Lehtiö, L. Screening and Structural Analysis of Flavones Inhibiting Tankyrases. *J. Med. Chem.* **2013**, *56*, 3507–3517. [CrossRef]
107. Qadir, M.I. Role of Green Tea Flavonoids and Other Related Contents in Cancer Prevention. *Crit. Rev. Eukaryot. Gene Expr.* **2017**, *27*, 163–171. [CrossRef]
108. Arts, I.C.W. A Review of the Epidemiological Evidence on Tea, Flavonoids, and Lung Cancer. *J. Nutr.* **2008**, *138*, 1561S–1566S. [CrossRef]
109. Amin, A.; Buratovich, M. The anti-cancer charm of flavonoids: A cup-of-tea will do! *Recent Pat. Anti-Cancer Drug Discov.* **2007**, *2*, 109–117. [CrossRef]
110. Bandele, O.J.; Clawson, S.J.; Osheroff, N. Dietary Polyphenols as Topoisomerase II Poisons: B Ring and C Ring Substituents Determine the Mechanism of Enzyme-Mediated DNA Cleavage Enhancement. *Chem. Res. Toxicol.* **2008**, *21*, 1253–1260. [CrossRef]
111. Piaz, F.D.; Ferro, P.; Vassallo, A.; Vasaturo, M.; Forte, G.; Chini, M.G.; Bifulco, G.; Tosco, A.; De Tommasi, N. Identification and mechanism of action analysis of the new PARP-1 inhibitor 2″-hydroxygenkwanol A. *Biochim. Biophys. Acta Gen. Subj.* **2015**, *1850*, 1806–1814. [CrossRef]
112. Inyang, O.K.; Omotuyi, O.I.; Ogunleye, A.J.; Eniafe, G.O.; Adewumi, B.; Metibemu, D.S. Molecular Interaction and Inhibitory Potential of Polyphenol on DNA Repair Pathway in Small Cell Lung Cancer: A Computational Study. *J. Anal. Pharm. Res.* **2017**, *6*, 00178–00186. [CrossRef]
113. Formica, J.V.; Regelson, W. Review of the biology of Quercetin and related bioflavonoids. *Food Chem. Toxicol. Int. J. Publ. Br. Ind. Biol. Res. Assoc.* **1995**, *33*, 1061–1080. [CrossRef]
114. Engen, A.; Maeda, J.; Wozniak, D.E.; Brents, C.A.; Bell, J.J.; Uesaka, M.; Aizawa, Y.; Kato, T.A. Induction of cytotoxic and genotoxic responses by natural and novel quercetin glycosides. *Mutat. Res. Toxicol. Environ. Mutagen.* **2015**, *784–785*, 15–22. [CrossRef] [PubMed]
115. Su, C.; Haskins, A.H.; Omata, C.; Aizawa, Y.; Kato, T.A. PARP Inhibition by Flavonoids Induced Selective Cell Killing to BRCA2-Deficient Cells. *Pharmaceuticals* **2017**, *10*, 80. [CrossRef]
116. Li, Y.; Yao, J.; Han, C.; Yang, J.; Chaudhry, M.T.; Wang, S.; Liu, H.; Yin, Y. Quercetin, Inflammation and Immunity. *Nutrients* **2016**, *8*, 167. [CrossRef]
117. Cui, S.; Wu, Q.; Wang, J.; Li, M.; Qian, J.; Li, S. Quercetin inhibits LPS-induced macrophage migration by suppressing the iNOS/FAK/paxillin pathway and modulating the cytoskeleton. *Cell Adhes. Migr.* **2018**, *13*, 1–12. [CrossRef]
118. Xue, F.; Nie, X.; Shi, J.; Liu, Q.; Wang, Z.; Li, X.; Zhou, J.; Su, J.; Xue, M.; Chen, W.-D.; et al. Quercetin Inhibits LPS-Induced Inflammation and ox-LDL-Induced Lipid Deposition. *Front. Pharmacol.* **2017**, *8*, 40. [CrossRef]

119. Li, C.; Zhang, W.-J.; Frei, B. Quercetin inhibits LPS-induced adhesion molecule expression and oxidant production in human aortic endothelial cells by p38-mediated Nrf2 activation and antioxidant enzyme induction. *Redox Biol.* **2016**, *9*, 104–113. [CrossRef]
120. Raso, G.M.; Meli, R.; Di Carlo, G.; Pacilio, M.; Di Carlo, R. Inhibition of inducible nitric oxide synthase and cyclooxygenase-2 expression by flavonoids in macrophage J774A.1. *Life Sci.* **2001**, *68*, 921–931. [CrossRef]
121. Lim, H.; Heo, M.Y.; Kim, H.P. Flavonoids: Broad Spectrum Agents on Chronic Inflammation. *Biomol. Ther.* **2019**, *27*, 241–253. [CrossRef]
122. Ikeda, A.; Iso, H.; Yamagishi, K.; Iwasaki, M.; Yamaji, T.; Miura, T.; Sawada, N.; Inoue, M.; Tsugane, S. Plasma tea catechins and risk of cardiovascular disease in middle-aged Japanese subjects: The JPHC study. *Atherosclerosis* **2018**, *277*, 90–97. [CrossRef]
123. Yu, S.-W.; Wang, H.; Poitras, M.F.; Coombs, C.; Bowers, W.J.; Federoff, H.J.; Poirier, G.G.; Dawson, T.M.; Dawson, V.L. Mediation of Poly(ADP-Ribose) Polymerase-1-Dependent Cell Death by Apoptosis-Inducing Factor. *Science* **2002**, *297*, 259–263. [CrossRef]
124. Oh, C.M.; Oh, I.H.; Choe, B.K.; Yoon, T.Y.; Choi, J.M.; Hwang, J. Consuming Green Tea at Least Twice Each Day Is As-sociated with Reduced Odds of Chronic Obstructive Lung Disease in Middle-Aged and Older Korean Adults. *J. Nutr.* **2018**, *148*, 70–76. [CrossRef]
125. Kuriyama, S. The Relation between Green Tea Consumption and Cardiovascular Disease as Evidenced by Epidemiological Studies. *J. Nutr.* **2008**, *138*, 1548S–1553S. [CrossRef]
126. Fatokun, A.A.; Tome, M.; Smith, R.A.; Darlington, L.G.; Stone, T.W. Protection by the flavonoids quercetin and luteolin against peroxide- or menadione-induced oxidative stress in MC3T3-E1 osteoblast cells. *Nat. Prod. Res.* **2015**, *29*, 1127–1132. [CrossRef] [PubMed]
127. Uberti, D.; Ferrari-Toninelli, G.; Bonini, S.A.; Sarnico, I.; Benarese, M.; Pizzi, M.; Benussi, L.; Ghidoni, R.; Binetti, G.; Spano, P.; et al. Blockade of the tumor necrosis factor-related apoptosis inducing ligand death receptor DR5 prevents beta-amyloid neurotoxicity. *Neuropsychopharmacol. Off. Publ. Am. Coll. Neuropsychopharmacol.* **2007**, *32*, 872–880. [CrossRef]
128. Huang, Y.; Xing, K.; Qiu, L.; Wu, Q.; Wei, H. Therapeutic implications of functional tea ingredients for ameliorating inflam-matory bowel disease: A focused review. *Crit. Rev. Food Sci. Nutr.* **2021**, 1–15. [CrossRef]
129. Li, X.; Yu, C.; Guo, Y.; Bian, Z.; Si, J.; Yang, L.; Chen, Y.; Ren, X.; Jiang, G.; Chen, J.; et al. Tea consumption and risk of ischaemic heart disease. *Heart* **2017**, *103*, 783–789. [CrossRef] [PubMed]
130. Mineharu, Y.; Koizumi, A.; Wada, Y.; Iso, H.; Watanabe, Y.; Date, C.; Yamamoto, A.; Kikuchi, S.; Inaba, Y.; Toyoshima, H.; et al. Coffee, green tea, black tea and oolong tea consumption and risk of mortality from cardiovascular disease in Japanese men and women. *J. Epidemiol. Community Health* **2009**, *65*, 230–240. [CrossRef] [PubMed]
131. Fatokun, A.A.; O Liu, J.; Dawson, V.L.; Dawson, T.M. Identification through high-throughput screening of 4′-methoxyflavone and 3′,4′-dimethoxyflavone as novel neuroprotective inhibitors of parthanatos. *Br. J. Pharmacol.* **2013**, *169*, 1263–1278. [CrossRef] [PubMed]
132. Lin, C.M.; Chen, H.H.; Lin, C.A.; Wu, H.C.; Sheu, J.J.; Chen, H.J. Apigenin-induced lysosomal degradation of beta-catenin in Wnt/beta-catenin signaling. *Sci. Rep.* **2017**, *7*, 372. [CrossRef] [PubMed]
133. Lehtiö, L.; Chi, N.-W.; Krauss, S. Tankyrases as drug targets. *FEBS J.* **2013**, *280*, 3576–3593. [CrossRef]
134. Zhang, X.H.; Gao, Y.T.; Rashid, A.; Deng, J.; Liu, E.J.; Wu, K.; Sun, L.; Cheng, J.R.; Gridley, G.; Hsing, A.W. Tea consumption and risk of biliary tract cancers and gallstone disease: A population-based case-control study in Shanghai, China. *ZhonghuaZhongliu Za Zhi* **2005**, *27*, 667–671.
135. Alam, S.; Khan, F. 3D-QSAR, Docking, ADME/Tox studies on Flavone analogs reveal anticancer activity through Tankyrase inhibition. *Sci. Rep.* **2019**, *9*, 5414. [CrossRef]
136. Stewart, A.J.; Mullen, W.; Crozier, A. On-line high-performance liquid chromatography analysis of the antioxidant activity of phenolic compounds in green and black tea. *Mol. Nutr. Food Res.* **2004**, *49*, 52–60. [CrossRef]
137. Larsen, C.A.; Dashwood, R.H.; Bisson, W.H. Tea catechins as inhibitors of receptor tyrosine kinases: Mechanistic insights and human relevance. *Pharmacol. Res.* **2010**, *62*, 457–464. [CrossRef]
138. Fang, J.; Sureda, A.; Silva, A.S.; Khan, F.; Xu, S.; Nabavi, S.M. Trends of tea in cardiovascular health and disease: A critical review. *Trends Food Sci. Technol.* **2019**, *88*, 385–396. [CrossRef]
139. Wang, X.; Liu, F.; Li, J.; Yang, X.; Chen, J.; Cao, J.; Wu, X.; Lu, X.; Huang, J.; Li, Y.; et al. Tea consumption and the risk of atherosclerotic cardiovascular disease and all-cause mortality: The China-PAR project. *Eur. J. Prev. Cardiol.* **2020**, *27*, 1956–1963. [CrossRef]
140. Roman, G.C.; Jackson, R.E.; Gadhia, R.; Roman, A.N.; Reis, J. Mediterranean diet: The role of long-chain omega-3 fatty acids in fish; polyphenols in fruits, vegetables, cereals, coffee, tea, cacao and wine; probiotics and vitamins in prevention of stroke, age-related cognitive decline, and Alzheimer disease. *Rev. Neurol.* **2019**, *175*, 724–741. [CrossRef]
141. Alqahtani, S.; Welton, K.; Gius, J.P.; Elmegerhi, S.; Kato, T.A. The Effect of Green and Black Tea Polyphenols on BRCA2 Deficient Chinese Hamster Cells by Synthetic Lethality through PARP Inhibition. *Int. J. Mol. Sci.* **2019**, *20*, 1274. [CrossRef]
142. Bassett, S.A.; Barnett, M.P.G. The Role of Dietary Histone Deacetylases (HDACs) Inhibitors in Health and Disease. *Nutrients* **2014**, *6*, 4273–4301. [CrossRef] [PubMed]
143. Park, S. Polyphenol Compound as a Transcription Factor Inhibitor. *Nutrients* **2015**, *7*, 8987–9004. [CrossRef] [PubMed]
144. Qian, F.; Wei, D.; Zhang, Q.; Yang, S. Modulation of P-glycoprotein function and reversal of multidrug resistance by (−)-epigallocatechin gallate in human cancer cells. *Biomed. Pharmacother.* **2005**, *59*, 64–69. [CrossRef]

145. Albassam, A.A.; Markowitz, J.S. An Appraisal of Drug-Drug Interactions with Green Tea (*Camellia sinensis*). *Planta Med.* **2017**, *83*, 496–508. [CrossRef]
146. Formentini, L.; Arapistas, P.; Pittelli, M.; Jacomelli, M.; Pitozzi, V.; Menichetti, S.; Romani, A.; Giovannelli, L.; Moroni, F.; Chiarugi, A. Mono-galloyl glucose derivatives are potent poly(ADP-ribose) glycohydrolase (PARG) inhibitors and partially reduce PARP-1-dependent cell death. *Br. J. Pharmacol.* **2008**, *155*, 1235–1249. [CrossRef]
147. Singh, A.P.; Singh, R.; Verma, S.S.; Rai, V.; Kaschula, C.H.; Maiti, P.; Gupta, S.C. Health benefits of resveratrol: Evidence from clinical studies. *Med. Res. Rev.* **2019**, *39*, 1851–1891. [CrossRef] [PubMed]
148. de Vries, K.; Strydom, M.; Steenkamp, V. A Brief Updated Review of Advances to Enhance Resveratrol's Bioavailability. *Molecules* **2021**, *26*, 4367. [CrossRef]
149. Fraiz, G.M.; da Conceição, A.R.; Vilela, D.L.D.S.; Rocha, D.M.U.P.; Bressan, J.; Hermsdorff, H.H.M. Can resveratrol modulate sirtuins in obesity and related diseases? A systematic review of randomized controlled trials. *Eur. J. Nutr.* **2021**, *60*, 2961–2977. [CrossRef] [PubMed]
150. Cole, L.; Kurscheid, S.; Nekrasov, M.; Domaschenz, R.; Vera, D.L.; Dennis, J.H.; Tremethick, D.J. Multiple roles of H2A.Z in regulating promoter chromatin architecture in human cells. *Nat. Commun.* **2021**, *12*, 1–15. [CrossRef] [PubMed]
151. Pan, Y.-R.; Song, J.-Y.; Fan, B.; Wang, Y.; Che, L.; Zhang, S.-M.; Chang, Y.-X.; He, C.; Li, G.-Y. mTOR may interact with PARP-1 to regulate visible light-induced parthanatos in photoreceptors. *Cell Commun. Signal.* **2020**, *18*, 1–17. [CrossRef] [PubMed]
152. Alharbi, H.; Alshehri, A.S.; Ahmad, M.; Guo, W.W. Promising anti- cervical carcinoma and inflammatory agent, Resveratrol targets poly (ADP-ribose) polymerase 1 (PARP-1) induced premature ovarian failure with a potent enzymatic modulatory activity. *J. Reprod. Immunol.* **2021**, *144*, 103272. [CrossRef]
153. Fatokun, C.; Girma, G.; Abberton, M.; Gedil, M.; Unachukwu, N.; Oyatomi, O.; Yusuf, M.; Rabbi, I.; Boukar, O. Genetic di-versity and population structure of a mini-core subset from the world cowpea (*Vigna unguiculata* (L.) Walp.) germplasm col-lection. *Sci. Rep.* **2018**, *8*, 16035. [CrossRef]
154. Jaeger, R.; Cuny, E. Terpenoids with Special Pharmacological Significance: A Review. *Nat. Prod. Commun.* **2016**, *11*, 1373–1390. [CrossRef] [PubMed]
155. Manayi, A.; Nabavi, S.M.; Daglia, M.; Jafari, S. Natural terpenoids as a promising source for modulation of GABAergic system and treatment of neurological diseases. *Pharmacol. Rep.* **2016**, *68*, 671–679. [CrossRef] [PubMed]

International Journal of
Molecular Sciences

MDPI

Review

Role of Nucleotide Excision Repair in Cisplatin Resistance

Mingrui Duan [1], Jenna Ulibarri [1], Ke Jian Liu [2] and Peng Mao [1,*]

[1] Comprehensive Cancer Center, Department of Internal Medicine, University of New Mexico, Albuquerque, NM 87131, USA; mduan@salud.unm.edu (M.D.); eriellako@unm.edu (J.U.)

[2] Department of Pharmaceutical Sciences, College of Pharmacy, University of New Mexico, Albuquerque, NM 87131, USA; KLiu@salud.unm.edu

* Correspondence: PMao@salud.unm.edu

Received: 31 October 2020; Accepted: 30 November 2020; Published: 4 December 2020

Abstract: Cisplatin is a chemotherapeutic drug used for the treatment of a number of cancers. The efficacy of cisplatin relies on its binding to DNA and the induction of cytotoxic DNA damage to kill cancer cells. Cisplatin-based therapy is best known for curing testicular cancer; however, treatment of other solid tumors with cisplatin has not been as successful. Pre-clinical and clinical studies have revealed nucleotide excision repair (NER) as a major resistance mechanism against cisplatin in tumor cells. NER is a versatile DNA repair system targeting a wide range of helix-distorting DNA damage. The NER pathway consists of multiple steps, including damage recognition, pre-incision complex assembly, dual incision, and repair synthesis. NER proteins can recognize cisplatin-induced DNA damage and remove the damage from the genome, thereby neutralizing the cytotoxicity of cisplatin and causing drug resistance. Here, we review the molecular mechanism by which NER repairs cisplatin damage, focusing on the recent development of genome-wide cisplatin damage mapping methods. We also discuss how the expression and somatic mutations of key NER genes affect the response of cancer cells to cisplatin. Finally, small molecules targeting NER factors provide important tools to manipulate NER capacity in cancer cells. The status of research on these inhibitors and their implications in cancer treatment will be discussed.

Keywords: DNA damage; chemotherapy; XPA; ERCC1-XPF; CSB

---

1. Introduction

Cis-diamminedichloridoplatinum(II) (best known as cisplatin) and its derivatives, such as carboplatin and oxaliplatin, are platinum (Pt)-based chemotherapeutics [1]. Since its approval by the FDA (Food and Drug Administration) in 1978, cisplatin has been widely used to treat a number of cancers, including testicular, ovarian, bladder, lung cancer, and others [2]. Cisplatin exerts its anti-cancer activity by inducing cytotoxic DNA lesions after activation in cells by a series of aquation reactions [1]. The aquated cisplatin is highly reactive and prone to interact with a wide range of cellular substrates, including DNA. The activated cisplatin preferentially binds to two purines on the same DNA strand, causing 1,2-intrastrand crosslinks of purine bases such as Pt-d(GpG) and to a lesser extent, Pt-d(ApG) adducts [3]. These two adducts represent about 90% and 10% of cisplatin lesions, respectively [1]. Additionally, cisplatin also binds guanines on the two opposing DNA strands, inducing the formation of interstrand crosslink G-Pt-G. However, the frequency of interstrand lesion is much lower than intrastrand crosslinks. Cisplatin-induced DNA crosslinks can strongly inhibit replicative DNA polymerases and induce apoptosis [4], which explains why cisplatin selectively kills fast-proliferating cancer cells. Additionally, cisplatin adducts block elongating RNA polymerases and inhibit gene transcription [5], which also contributes to cisplatin-induced cell death (Figure 1).

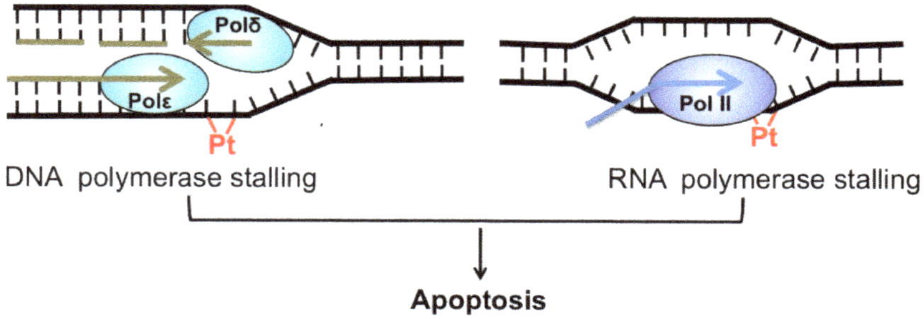

Figure 1. Cytotoxicity of cisplatin. Cisplatin-induced DNA adduct (red) blocks replicative DNA polymerase (**left**) and RNA polymerase (**right**). Both replication and transcription stalling can trigger apoptosis.

The most impressive clinical success of cisplatin is the cure of over 80% patients with metastatic testicular germ cell tumors (TGCT) [6], a phenomenon that has been attributed to the intrinsic defects in DNA repair in TGCT cells [7,8]. However, the same clinical benefits are not observed for other solid tumors. Some tumor cells, including colorectal, lung, and prostate cancer cells, are intrinsically resistant to cisplatin [9]. Other tumors are originally sensitive to cisplatin, but develop resistance during the process of treatment, when tumor cells acquire further mutations to adapt to cisplatin damage. The acquired cisplatin resistance is particularly common for advanced ovarian cancer patients [10], and there is an urgent need to develop new strategies to reverse drug resistance. Resistance to platinum drugs such as cisplatin is associated with a number of cellular mechanisms, including low drug uptake through reduced expression of the copper transporter protein Ctr1 [11], inhibition of cisplatin potency by proteins such as glutathione (GSH) [12], replicative bypass of cisplatin-induced DNA damage by translesion synthesis DNA polymerases [13], and removal of cisplatin adducts by DNA repair to abolish cytotoxicity [14].

Nucleotide excision repair (NER) plays a critical role in removing helix-distorting DNA damage, including cisplatin-induced intrastrand crosslinks [15–17]. Previous studies in testicular tumors (i.e., TGCT) have shown low NER activity in TGCT cell extracts and low expression of several key NER proteins [7,18]. The intrinsic NER defects in TGCT cells are correlated with high cisplatin sensitivity and high cure rate among TGCT patients [19]. The impressive findings in TGCT triggered investigations in other solid tumors such as lung and ovarian cancer, in order to understand how NER contributes to cisplatin resistance in these tumors. Through extensive studies, it is now evident that elevated NER capacity in cancer cells is generally correlated with drug resistance [20,21]. The expression levels of several NER proteins (e.g., XPC, XPA, ERCC1-XPF, and XPG) and somatic mutations in XPD significantly affect cellular sensitivity to cisplatin [22–24]. Recent CRISPR-Cas9 screening also identified transcription-coupled NER (TC-NER) factors in resolving cisplatin-induced transcription stalling to avoid apoptosis [25]. Hence, the status of these critical NER factors represents a prospective biomarker to predict the outcome of cisplatin treatment. Furthermore, manipulation of cellular NER capacity with specific inhibitors is a potentially useful strategy to cope with cisplatin resistance.

2. Mechanism of NER in the Repair of Cisplatin Adducts

NER targets a wide range of mutagenic and cytotoxic DNA lesions. Common substrates for NER include ultraviolet (UV)-induced pyrimidine dimers, DNA adducts formed by Benzo[a]pyrene (BaP) in cigarette smoke, intrastrand crosslinks, and other helix-distorting DNA lesions [26]. The structural alteration caused by these bulky lesions interferes with DNA replication and gene transcription, and thus, needs to be repaired correctly before leading to permanent mutations or cell death. DNA helix-distorting lesions can be recognized by NER surveillance proteins such as Xeroderma pigmentosum complementation group C (XPC) and the UV-damaged DNA binding protein complex

(UV-DDB) [26]. Once bound to damaged DNA, XPC recruits transcription factor II H (TFIIH), a 10-subunit protein complex consisting of two important DNA helicases, XPB and XPD. XPB and XPD, together with XPA and replication protein A (RPA), separate the two DNA strands around the damage site, creating a pre-incision DNA bubble that is recognized by repair endonucleases ERCC1-XPF and XPG. XPF and XPG cleave the damaged DNA strand on the 5′ and 3′ side relative to the damage, releasing an oligonucleotide of ~30 nucleotide (nt) containing the lesion (Figure 2). The resulting gap is filled with the activity of DNA polymerase using the undamaged strand as the synthesis template. Finally, DNA ligase I or ligase IIIα is recruited to seal the DNA backbone [26]. As damage recognition by XPC and UV-DDB can occur across the whole genome, this NER subpathway is known as global genomic NER (GG-NER). GG-NER is responsible for the removal for the majority of bulky lesions residing in the genome. NER can also be activated through the stalling of RNA polymerases. When the elongating RNA Pol II is blocked by DNA damage, the prolonged arrest of Pol II serves as a strong signal to activate transcription-coupled NER (TC-NER) [27]. TC-NER is a highly efficient repair mechanism that specifically removes transcription-impeding damage from the transcribed strand (TS) to allow transcription resumption. XPC and UV-DDB are not required for TC-NER. Instead, the stalled RNA Pol II recruits the Cockayne syndrome group B (CSB) protein, a master regulator of TC-NER [28]. CSB coordinates assembly of the downstream NER core factors for dual incision and repair synthesis. Mechanistically, GG-NER and TC-NER mainly differ in damage recognition but share the same set of core NER enzymes for the "cut-and-patch"-type repair process (Figure 2).

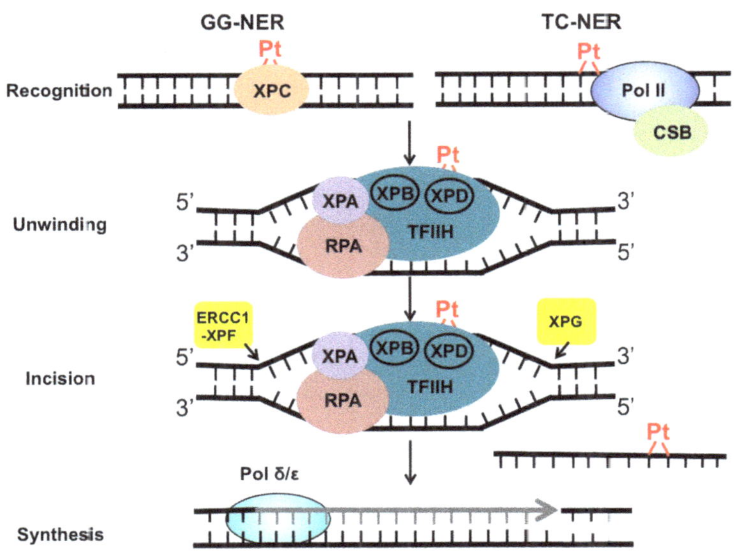

Figure 2. Repair of cisplatin-induced DNA damage by NER. Cisplatin damage can be recognized by surveillance protein XPC or elongating RNA Pol II, to initiate GG-NER or TC-NER. After damage recognition, the two subpathways utilize the same set of repair enzymes to conduct DNA unwinding, dual incision, and repair synthesis and ligation.

Previous studies using mammalian cell extracts as well as reconstituted NER system with purified proteins have shown high activity of NER toward cisplatin-induced intrastrand crosslinks [15,29]. These in vitro cisplatin repair experiments used chemically synthesized oligonucleotides containing site-specific cisplatin adducts. Incubation of the DNA substrates with whole cell extracts or purified NER enzymes resulted in 25–29 nt excision products, due to the dual incision by XPF and XPG [15,29]. Using these well-defined NER systems, the abundance of the excision products can be detected to investigate NER activity for different types of cisplatin adducts. For example, experiments conducted in HeLa cell extracts have shown that NER enzymes exhibit slightly higher repair activity for the intrastrand crosslink Pt-d(ApG) relative to Pt-d(GpG), suggesting the local DNA sequence for the crosslink can affect repair efficiency [15]. Additionally, intrastrand adducts induced by cisplatin and its derivatives, such as oxaliplatin and JM216, can all be repaired by NER proteins with similar repair kinetics [16]. In contrast, the interstrand crosslink G-Pt-G cannot be repaired by NER [15], consistent with findings showing that interstrand crosslinks are repaired by the Fanconi anemia (FA) pathway [30]. Furthermore, published data also uncovered the regulation of cisplatin damage repair by non-NER proteins. In this regard, in vitro studies showed that the high mobility group (HMG)-domain proteins suppress excision repair of the main cisplatin adduct, Pt-d(GpG) [15,29]. HMG proteins bind to cisplatin-damaged DNA by recognizing the altered DNA structure [31], and thus, shield cisplatin adducts from NER [32]. As a result, cisplatin may cause more cytotoxicity in cancer cells with high expression of HMG proteins [33].

Repair of cisplatin damage by NER is also confirmed by experiments conducted in cells. Compared to biochemical analysis, cellular repair studies provided the advantage of analyzing both GG-NER and TC-NER in resolving cisplatin-induced DNA damage. Particularly, the development of high-throughput damage sequencing methods such as Damage-seq and eXcision Repair-seq (XR-seq) has significantly improved the understanding of cisplatin damage distribution and repair kinetics at the genome scale [17,34]. Damage-seq was built on the observation that cisplatin damage blocks replicative DNA polymerases. This method first uses a cisplatin damage-specific antibody to enrich DNA fragments containing cisplatin adducts. The damage-containing single stranded DNA is used as the template to synthesize a new strand with a replicative DNA polymerase, so that the newly synthesized strand terminates at the site of cisplatin damage. By mapping the location of the 3′ end of the nascent strand, this method allows for identification of the damage site [17]. On the other hand, XR-seq is used for high-resolution mapping of NER activity in cisplatin-treated cells. In XR-seq, the ~30-nt single-stranded excision repair fragments generated during NER are purified from cells and ligated with sequencing adaptors for next-generation sequencing [34]. By mapping the sequencing reads to the human genome and counting the number of XR-seq reads in different genomic regions, this method offers a genome-wide repair profile for cisplatin adducts as well as other bulky lesions [17]. The XR-seq data have revealed highly variable repair efficiency for cisplatin damage across the genome. Several factors, including transcription and chromatin states, can affect the repair efficiency. Particularly, the transcribed strand of active genes display robust repair by TC-NER at the early repair time points, whereas cisplatin damage located in heterochromatin is poorly repaired [17], likely due to the low access of cisplatin damage to GG-NER proteins. In contrast, Damage-seq data revealed that formation of cisplatin damage is largely uniform across the genome and mainly dictated by the underlying DNA sequences [17]. Thus, these new methods provide high-resolution maps of cisplatin damage formation and repair kinetics at the genome level, and highlight the important roles for both GG-NER and TC-NER in resolving cytotoxic damage induced by cisplatin. Indeed, as discussed below, both NER subpathways have a significant impact on the cisplatin response in cancer cells.

3. Roles of GG-NER Factors in Cisplatin Resistance

XPC: XPC is an important damage recognition factor in GG-NER [26]. XPC physically interacts with RAD23B for DNA binding and the prevention of proteasomal degradation [35]. In vitro binding data indicate that XPC-RAD23B binds with high affinity (K_D ~1−3 nM) to DNA fragments containing a single cisplatin intrastrand crosslink or a single UV-induced 6,4-photoproduct [36]. The wide range of structurally diverse lesions recognized by XPC-RAD23B is accomplished by its strong binding to the thermodynamically destabilized DNA double helix caused by bulky lesions [37]. Binding of XPC to damaged DNA is critical for the recruitment of the next NER factor, TFIIH, via physical interaction with the TFIIH subunit p62 [38,39]. Hence, damage recognition by XPC represents an important rate-limiting step that largely dictates cellular GG-NER capacity. Consistent with this notion, studies performed in cancer cell lines have shown a correlation between XPC protein levels and cisplatin resistance. Overexpression of XPC in several types of cancer cells (e.g., lung, colorectal, and gastric cancers) leads to elevated cisplatin resistance [23,24,40]. On the other hand, inhibition of XPC protein production by siRNA knockdown increases cisplatin sensitivity. While the observations in cancer cell lines show a clear correlation between cisplatin sensitivity and XPC protein levels, whether or not resistance in cisplatin-treated tumors is associated with upregulated XPC has not been analyzed. Damage recognition by XPC is assisted by UV-DDB in some cases. For example, recognition of UV-induced cyclobutane pyrimidine dimer (CPD), which is less helix-distorting than 6,4-photoproduct, requires cooperation between XPC and UV-DDB [26]. UV-DDB does not appear to be required for the repair of cisplatin adducts, and overexpression of UV-DDB surprisingly sensitizes ovarian cancer cells to cisplatin by promoting apoptosis [41]. This suggests that UV-DDB may play a different role from XPC in cellular response to cisplatin damage.

TFIIH: Following damage recognition, the TFIIH complex is recruited to DNA to promote the subsequent repair steps. The DNA helicases XPB and XPD in TFIIH unwind damaged DNA to promote the assembly of the pre-incision complex [26] (see Figure 2). XPB is important for both transcription initiation and repair. The ATPase, but not DNA helicase activity of XPB, is required for NER [42]. In contrast, XPD appears to mainly function in NER [43]. XPD mutations that disrupt its helicase activity can significantly affect cellular NER capacity [43]. As a result, germline mutations in the *XPD* gene are associated with three severe human diseases: Xeroderma pigmentosum (XP), XP combined with Cockayne syndrome (XP/CS), and Trichothiodystrophy (TTD) [44]. Both XP and CS are known genetic disorders caused by NER defects, while TTD is linked with defective transcription [44]. Interestingly, somatic XPD mutations are frequently found in human cancers, particularly in bladder tumors [45]. A number of bladder cancer-derived XPD somatic mutations can significantly inhibit NER and strongly increase cisplatin cytotoxicity in human cells [22]. Data generated in a preclinical bladder cancer mouse model showed that XPD-deficient tumors are significantly more sensitive to cisplatin relative to XPD-proficient tumors [22]. Hence, these analyses suggest a potential role for XPD somatic mutations as a predictive marker of cisplatin treatment in bladder cancer. Furthermore, clinical studies performed in two cohorts of muscle-invasive bladder cancer patients revealed the association between XPD somatic mutations and cisplatin response [46,47]. In both studies, patients with XPD-mutated tumors had better overall survival after receiving platinum-based chemotherapy compared to XPD wild-type tumors.

XPA: XPA is a key scaffold protein required for both GG-NER and TC-NER. XPA itself does not possess enzymatic activities, but it plays a critical role in assembling the pre-incision complex on the damaged DNA. The N- and C-termini of XPA protein are disordered and physically interact with a variety of NER factors such as XPC, UV-DDB, TFIIH, ERCC1-XPF, and RPA [48]. The interaction between XPA and TFIIH allows the recruitment of XPA to the single-strand and double-strand DNA junction (ss/dsDNA) [49] around the damaged site created by TFIIH. Binding to the repair bubble allows XPA to properly position NER factors such as the endonuclease XPF in the right place for DNA incision [26]. As a critical regulator of NER, the XPA protein level in tumors is an important predictor for cisplatin response. XPA protein levels are significantly low in metastatic testicular tumor cells, which is consistent with positive response to cisplatin and excellent prognosis for TGCT patients [7]. In fact, supplementing cell extracts isolated from testicular tumor cells with purified XPA protein can restore the repair activity for cisplatin damage [7], implying that the low repair capacity is due to insufficient XPA protein in these cancer cells. The predictive role for XPA in cisplatin therapy is further confirmed by a survey including 207 patients diagnosed with germ cell tumors (GCTs) [50]. This study showed that GCTs patients with low *XPA* gene expression had significantly better overall survival than patients with high *XPA* expression after platinum treatment. These published data and the fact that XPA plays a central role in NER suggest that targeting XPA may be an effective strategy to improve cisplatin treatment.

ERCC1-XPF: ERCC1-XPF is a structure-specific endonuclease that nicks the damaged strand 5′ to the lesion during NER (Figure 2). XPF is the catalytically active subunit in the heterodimer [26]. ERCC1 does not have enzymatic activity, but it is important for recognizing ss/dsDNA junctions and interacting with XPA [51,52]. The physical interaction between ERCC1 and XPA is important for recruiting XPF to the unwound DNA to incise the damaged strand [51]. High ERCC1 expression has been linked with poor response to cisplatin therapy in a number of tumors, including non-small-cell lung carcinoma [53], head and neck [54], and ovarian cancer [55]. In contrast, low ERCC1 expression is associated with high cisplatin sensitivity, as shown in testicular tumors [18]. One molecular mechanism that suppresses *ERCC1* gene expression is DNA methylation in its promoter. Hypermethylation has been found in the *ERCC1* gene promoter in some glioma cells and is associated with suppressed *ERCC1* gene expression [56]. However, the prevalence of *ERCC1* promoter methylation in other tumor types is unknown. Furthermore, knockdown of either ERCC1 or XPF with siRNA in lung cancer cells resulted in defective repair of both intrastrand and interstrand cisplatin crosslinks, which is accompanied with increased cellular cytotoxicity [57]. Although these studies suggest that ERCC1 is an important biomarker, the reliability of ERCC1 expression alone in predicting treatment outcome has been challenged by other studies. For example, a survey in lung cancer patients showed no clear correlation between *ERCC1* mRNA levels and patient survival after cisplatin therapy [58]. While the *ERCC1* mRNA may not directly reflect its protein levels, another large-scale study using specimens obtained from 494 non-small-cell lung cancer patients failed to validate the predictive effect of immunostaining for ERCC1 protein [59], which suggests that ERCC1 protein levels alone cannot accurately predict clinical outcomes. Therefore, more detailed studies are needed to determine if the predictive role for ERCC1 only applies to a subset of tumors, or if ERCC1 should be considered together with other NER factors. One technical challenge for the analysis of ERCC1 protein in tumors is the presence of four ERCC1 isoforms. Commercial ERCC1 antibodies recognize all different isoforms [59]; however, only one isoform, ERCC1-202, is active in DNA repair [60]. Hence, a more specific antibody targeting this isoform is required to revalidate the correlation between tumor ERCC1 protein levels and cisplatin effectiveness.

XPG: XPG is the endonuclease that incises the damaged strand 3′ to the lesion during NER. XPG is recruited to the damaged DNA through interaction with TFIIH and it nicks DNA after 5′ incision by XPF [26]. Previous studies have shown that loss of heterozygosity of the *XPG* gene locus and reduction in *XPG* gene expression are associated with better survival for ovarian cancer patients treated with cisplatin [61]. Interestingly, the *XPG* gene promoter contains a putative CpG island and analysis of specimens obtained from ovarian cancer patients identified DNA methylation in ~20% of the tumor samples, but not in the matched blood DNA [62], suggesting that XPG expression may be influenced by epigenetic mechanisms in tumors. However, whether the methylation in *XPG* promoter is associated with better cisplatin response has not been investigated.

4. Roles of TC-NER Factors in Cisplatin Resistance

CSB: Cockayne syndrome group B (CSB) protein is the central regulator of human TC-NER [27]. CSB deficiency leads to impaired TC-NER and is associated with Cockayne syndrome, a severe neurodegenerative disorder. CSB is one of the earliest responders that bind damage-stalled RNA Pol II to initiate repair. While the detailed mechanism for CSB in initiating TC-NER remains elusive, new studies in budding yeast suggest that Rad26, a CSB homolog, utilizes the ATPase activity to evict the transcription elongation factor Spt4–Spt5 from the stalled Pol II, and thus, switches Pol II from elongation to repair [63,64]. The cellular TC-NER capacity has been shown to influence cisplatin sensitivity in cancer cells. Specifically, the reduction in TC-NER by knockdown of CSB was shown to significantly sensitize prostate and colorectal carcinoma cell lines and increase apoptosis upon cisplatin treatment, even in the absence of p53 and DNA mismatch repair [65]. The prominent cisplatin sensitivity in CSB-deficient cells is likely due to transcription blockage by cisplatin-induced DNA crosslinks [66], and the failed rescue of arrested RNA Pol II, because the prolonged blockage of Pol II has been shown as a strong signal for apoptosis [67]. CSB protein is overexpressed in a number of cancer cell lines collected from different tissues, which appears to reduce apoptosis and promote cell proliferation [68], consistent with the role for CSB in stimulating transcription elongation [69]. Considering the crucial role of CSB in TC-NER, the increased CSB expression may be linked with elevated TC-NER capacity in cancer cells. However, this hypothesis has not been tested in tumor cells. Further studies characterizing how CSB is upregulated and how its upregulation is related to cisplatin resistance will provide new insights into the connection between TC-NER and drug resistance. On the other hand, cisplatin-induced transcription stalling is harmful to non-dividing cells such as terminally differentiated healthy cells. The essential role for TC-NER in rapidly resolving transcription stress plays an important role in protecting healthy cells from the side effects of cisplatin. Indeed, it has been shown that CSB deficiency inhibits the repair of cisplatin damage in the sensory hair cells in the organ of Corti, and thus, predisposes mice to hearing loss after cisplatin treatment [70].

CSA: Like CSB, mutations in Cockayne syndrome group A (CSA) cause deficient TC-NER and neurological degeneration [71]. The CSA protein contains seven WD40 motifs that are required for protein–protein interaction. CSA functions as a component of the DDB1–CUL4-based E3 ubiquitin ligase complex to ubiquitylate TC-NER proteins, including CSB and the stalled RNA Pol II [72]. Additionally, CSA cooperates with CSB to recruit XPA Binding Protein 2 (XAB2), the nucleosomal binding protein HMGN1, and elongation factor TFIIS to UV-stalled RNA Pol II [73]. A published study using CRISPR-Cas9 screening identified CSA as one prominent target that protects colon adenocarcinoma cells against the clinically applied platinum drug oxaliplatin [25]. The same study also showed that CSB, ERCC1, and XPF are important for cell survival upon oxaliplatin treatment [25]. In addition to CSB and CSA, previous studies in human cells have revealed another two important TC-NER factors—UV Stimulated Scaffold Protein A (UVSSA) and Ubiquitin Specific Peptidase 7 (USP7) [71]. UVSSA interacts with RNA Pol II and recruits the deubiquitylase USP7 to the stalled Pol II to stabilize the CSB protein [74,75]. However, the roles for UVSSA and USP7 in the repair of cisplatin damage have not been characterized.

5. NER Inhibitors and Their Roles in Suppressing DNA Repair

Given the important role of NER in the removal of cisplatin adducts, many efforts have been made to develop small molecules targeting different NER proteins [76]. These inhibitors are useful tools for basic research as well as clinical applications to sensitize tumors to cisplatin and its derivatives.

A previous small molecule screening identified spironolactone (SP), an anti-aldosterone drug used for the treatment of heart failure and hypertension, as a specific inhibitor of XPB [77]. Studies conducted in several cell lines indicate that SP reduces cellular XPB protein by inducing its degradation [78,79]. XPB is a critical DNA helicase in TFIIH and functions in both transcription initiation and NER [26]. The rapid degradation of XPB protein in SP-treated cells has been shown to inhibit NER and increase cell sensitivity to DNA damaging agents such as UV and cisplatin [77], suggesting that SP might enhance cancer chemotherapy when combined with platinum drugs. Mechanistically, SP was shown to induce XPB phosphorylation on Ser90 by the kinase CDK7. The phosphorylated XPB is prone to poly-ubiquitylation by the SCF^{FBXL18} E3 ubiquitin ligase [78,79], followed by proteasomal degradation. SP-induced XPB degradation is reversible and XPB protein levels can be fully restored after removing the drug from the growth media.

New inhibitors for XPA have also been reported previously. A computer-aided screening using the XPA protein structure and a virtual library of small molecules identified 63 putative inhibitors targeting the DNA binding domain of XPA [80]. Biochemical analysis has revealed that one of the candidate inhibitors, X80, inhibits binding of XPA protein to single-stranded DNA and double-stranded DNA containing a cisplatin lesion [80]. Additionally, inhibitors targeting XPA–ERCC1 interaction have been identified. XPA interacts with ERCC1 to recruit ERCC1-XPF to damaged DNA. A previous study showed that the cell cycle checkpoint abrogator UCN-01 significantly weakens the interaction between

XPA and ERCC1 and increases the cytotoxicity of cisplatin [81]. A virtual screening also identified putative inhibitors that disrupt XPA–ERCC1 interaction [82].

The repair endonuclease XPF can also be targeted to inhibit NER. Indeed, a high-throughput screen against ERCC1-XPF identified two compounds that exhibit robust inhibitory effect on the endonuclease activity at nanomolar concentrations [83]. The inhibition appears to be highly specific for XPF's enzymatic activity, because they did not inhibit other endonucleases or DNA binding by ERCC1-XPF. Furthermore, the two compounds significantly potentiated sensitivity to cisplatin in cancer cells, and one of them has been shown to enhance cisplatin antitumor activity in a lung cancer xenograft model [83]. Another study using virtual screening based on the known XPF protein structure led to the identification of a compound labeled NSC 130813 that can disrupt ERCC1–XPF interaction and increase cytotoxicity to cisplatin [84]. A recent study followed this identified inhibitor and developed derivatives with enhanced inhibitory activity. One of the derivatives has been shown to be a potent inhibitor for ERCC1-XPF and suppress cellular NER activity upon UV irradiation [85].

While these small molecules can inhibit NER at different repair steps and represent important progress in developing new therapeutics to improve cisplatin efficacy, none of them have been clinically tested.

6. Discussion and Future Directions

The central role of NER in the repair of cisplatin-induced DNA damage makes a strong case for NER factors functioning as both predictive biomarkers and potential therapeutic targets to reverse drug resistance. As shown in Table 1, previous studies have demonstrated that many NER factors are associated with cisplatin resistance. Current evidence built on the clinical success in the treatment of testicular cancer patients with platinum-based therapy strongly indicates a correlation between defective NER and positive treatment response. Among NER factors analyzed in testis tumor cells, XPA, ERCC1, and XPF exhibited the most significant reduction compared to other types of cells [7]. These three core NER factors function downstream of damage recognition and are required for both GG-NER and TC-NER. Defective expression of the three factors likely inhibits two key steps in the NER process: the assembly of the pre-incision NER complex and repair incision on the 5′ side of the cisplatin damage (Figure 2). Although damage recognition by either XPC-RAD23B or stalled RNA Pol II may still occur, the low availability of downstream factors likely blocks processing of the damage, causing replication fork stalling and transcription stalling, both of which can lead to strong apoptosis. Hence, analyzing NER protein levels in tumors cells represents a feasible strategy to predict drug response and treatment efficacy. Whether one specific NER factor is sufficient to predict treatment outcomes or several NER factors should be considered simultaneously is currently an open question. Adding to this strategy is the identification of somatic mutations in important NER genes. Sequencing of tumor genomes has shown prevalent somatic mutations in *XPD* gene in bladder cancer patients [22]. These mutations likely disrupt XPD helicase activity and contribute to increased genome instability and tumorigenesis [86]. On the other hand, XPD somatic mutations have been shown to sensitize tumor cells to toxic cisplatin damage and represent a potential biomarker for positive cisplatin response [22]. Genome sequencing of more tumor samples offers the opportunity to identify new NER gene mutations that may help treatment decision.

Table 1. NER factors with known functions in cisplatin resistance.

Name	Function	Role in Cisplatin Resistance	References
XPC	Damage recognition	Overexpression of XPC in cancer cells increases cisplatin resistance.	[23,24,40]
UV-DDB	Damage recognition	Overexpression of DDB2 sensitizes ovarian cancer cells to cisplatin.	[41]
XPB	3′-5′ helicase	Degradation of XPB by spironolactone increases cisplatin sensitivity.	[77]

Table 1. *Cont.*

Name	Function	Role in Cisplatin Resistance	References
XPD	5′-3′ helicase	Somatic mutations of XPD increase cisplatin sensitivity in cells and in bladder cancer patients.	[22,46,47]
XPA	Scaffolding protein	Low XPA expression correlates with increased cisplatin sensitivity in testicular tumors.	[7,50]
ERCC1	Partner of XPF	ERCC1 expression in tumor cells may predict cisplatin efficacy.	[18,53,54]
XPF	Endonuclease (5′ to lesion)	Low XPF expression correlates with high cisplatin sensitivity.	[18]
XPG	Endonuclease (3′ to lesion)	Low XPG expression is associated with good cisplatin response in ovarian patients.	[61]
CSB	TC-NER initiation	Knockdown of CSB sensitizes cancer cells to cisplatin.	[25,65]
CSA	TC-NER; E3 ligase protein	Knockout of CSA gene sensitizes human cells to oxaliplatin.	[25]

While acquired cisplatin resistance is widely observed for different solid tumors, whether and how the resistance is correlated with changes in NER factors are still poorly understood. It has been shown that the overexpression of NER genes in cancer cells using expression vectors can increase cisplatin resistance (e.g., [23,24]); however, it is unknown if upregulation of these same NER genes occurs in tumors during cancer treatment. If cisplatin treatment indeed activates NER gene expression in tumors, it is important to determine what NER genes are most frequently upregulated and how their expression is elevated. Systematic analysis of protein and mRNA levels of NER factors at different stages of drug treatment may hold the key to uncover treatment-induced changes of specific NER proteins. Published studies have shown that DNA methylation plays an important role in regulating the transcription of several NER genes such as *ERCC1* and *XPG* [56,62]. Monitoring the DNA methylation status of these genes during drug treatment should also provide insights into how cisplatin changes the expression of NER genes.

Funding: Research related to this review is supported by NIH grants (R21ES029302 to P.M. and R01ES030993 to K.J.L.), a pilot grant from the UNM Center for Metals in Biology and Medicine (P20GM130422), the UNM Comprehensive Cancer Center Support Grant NCI P30CA118100, and the UNM Analytical and Translational Genomics Shared Resource.

Acknowledgments: We thank Amelia Hodges and Eric Burns for reading the manuscript.

Conflicts of Interest: The authors declare no conflict of interest.

References

1. Dasari, S.; Tchounwou, P.B. Cisplatin in cancer therapy: Molecular mechanisms of action. *Eur. J. Pharmacol.* **2014**, *740*, 364–378. [CrossRef] [PubMed]
2. Galluzzi, L.; Senovilla, L.; Vitale, I.; Michels, J.; Martins, I.; Kepp, O.; Castedo, M.; Kroemer, G. Molecular mechanisms of cisplatin resistance. *Oncogene* **2012**, *31*, 1869–1883. [CrossRef] [PubMed]
3. Fichtinger-Schepman, A.M.J.; Van der Veer, J.L.; Den Hartog, J.H.J.; Lohman, P.H.M.; Reedijk, J. Adducts of the antitumor drug cis-diamminedichloroplatinum(II) with DNA: Formation, identification, and quantitation. *Biochemistry* **1985**, *24*, 707–713. [CrossRef] [PubMed]
4. Siddik, Z.H. Cisplatin: Mode of cytotoxic action and molecular basis of resistance. *Oncogene* **2003**, *22*, 7265–7279. [CrossRef] [PubMed]
5. Todd, R.C.; Lippard, S.J. Inhibition of transcription by platinum antitumor compounds. *Metallomics* **2009**, *1*, 280–291. [CrossRef]
6. Horwich, A.; Shipley, J.; Huddart, R. Testicular germ-cell cancer. *Lancet* **2006**, *367*, 754–765. [CrossRef]
7. Köberle, B.; Masters, J.R.; Hartley, J.A.; Wood, R.D. Defective repair of cisplatin-induced DNA damage caused by reduced XPA protein in testicular germ cell tumours. *Curr. Biol.* **1999**, *9*, 273–276. [CrossRef]

8. Einhorn, L.H. Curing metastatic testicular cancer. *PNAS* **2002**, *99*, 4592–4595. [CrossRef]
9. Galluzzi, L.; Vitale, I.; Michels, J.; Brenner, C.; Szabadkai, G.; Harel-Bellan, A.; Castedo, M.; Kroemer, G. Systems biology of cisplatin resistance: Past, present and future. *Cell Death Dis.* **2014**, *5*, e1257. [CrossRef]
10. Pokhriyal, R.; Hariprasad, R.; Kumar, L.; Hariprasad, G. Chemotherapy Resistance in Advanced Ovarian Cancer Patients. *Biomark. Cancer* **2019**, *11*. [CrossRef]
11. Ishida, S.; Lee, J.; Thiele, D.J.; Herskowitz, I. Uptake of the anticancer drug cisplatin mediated by the copper transporter Ctr1 in yeast and mammals. *Proc. Natl. Acad. Sci. USA* **2002**, *99*, 14298–14302. [CrossRef] [PubMed]
12. Chen, H.H.W.; Kuo, M.T. Role of Glutathione in the Regulation of Cisplatin Resistance in Cancer Chemotherapy. *Met. Based Drugs* **2010**, *2010*. [CrossRef] [PubMed]
13. Vaisman, A.; Varchenko, M.; Umar, A.; Kunkel, T.A.; Risinger, J.I.; Barrett, J.C.; Hamilton, T.C.; Chaney, S.G. The Role of hMLH1, hMSH3, and hMSH6 Defects in Cisplatin and Oxaliplatin Resistance: Correlation with Replicative Bypass of Platinum-DNA Adducts. *Cancer Res.* **1998**, *58*, 3579–3585. [PubMed]
14. Rocha, C.R.R.; Silva, M.M.; Quinet, A.; Cabral-Neto, J.B.; Menck, C.F.M. DNA repair pathways and cisplatin resistance: An intimate relationship. *Clinics (Sao Paulo)* **2018**, *73*. [CrossRef] [PubMed]
15. Zamble, D.B.; Mu, D.; Reardon, J.T.; Sancar, A.; Lippard, S.J. Repair of Cisplatin–DNA Adducts by the Mammalian Excision Nuclease. *Biochemistry* **1996**, *35*, 10004–10013. [CrossRef] [PubMed]
16. Reardon, J.T.; Vaisman, A.; Chaney, S.G.; Sancar, A. Efficient nucleotide excision repair of cisplatin, oxaliplatin, and Bis-aceto-ammine-dichloro-cyclohexylamine-platinum(IV) (JM216) platinum intrastrand DNA diadducts. *Cancer Res.* **1999**, *59*, 3968–3971. [PubMed]
17. Hu, J.; Lieb, J.D.; Sancar, A.; Adar, S. Cisplatin DNA damage and repair maps of the human genome at single-nucleotide resolution. *Proc. Natl. Acad. Sci. USA* **2016**, *113*, 11507–11512. [CrossRef]
18. Welsh, C.; Day, R.; McGurk, C.; Masters, J.R.W.; Wood, R.D.; Köberle, B. Reduced levels of XPA, ERCC1 and XPF DNA repair proteins in testis tumor cell lines. *Int. J. Cancer* **2004**, *110*, 352–361. [CrossRef]
19. Masters, J.R.W.; Köberle, B. Curing metastatic cancer: Lessons from testicular germ-cell tumours. *Nat. Rev. Cancer* **2003**, *3*, 517–525. [CrossRef]
20. Ferry, K.V.; Hamilton, T.C.; Johnson, S.W. Increased nucleotide excision repair in cisplatin-resistant ovarian cancer cells: Role of ercc1–xpf. *Biochem. Pharmacol.* **2000**, *60*, 1305–1313. [CrossRef]
21. Rosell, R.; Taron, M.; Barnadas, A.; Scagliotti, G.; Sarries, C.; Roig, B. Nucleotide Excision Repair Pathways Involved in Cisplatin Resistance in Non-Small-Cell Lung Cancer. *Cancer Control.* **2003**, *10*, 297–305. [CrossRef] [PubMed]
22. Li, Q.; Damish, A.W.; Frazier, Z.; Liu, D.; Reznichenko, E.; Kamburov, A.; Bell, A.; Zhao, H.; Jordan, E.J.; Gao, S.P.; et al. ERCC2 Helicase Domain Mutations Confer Nucleotide Excision Repair Deficiency and Drive Cisplatin Sensitivity in Muscle-Invasive Bladder Cancer. *Clin. Cancer Res.* **2019**, *25*, 977–988. [CrossRef] [PubMed]
23. Zhang, Y.; Cao, J.; Meng, Y.; Qu, C.; Shen, F.; Xu, L. Overexpression of xeroderma pigmentosum group C decreases the chemotherapeutic sensitivity of colorectal carcinoma cells to cisplatin. *Oncol. Lett.* **2018**, *15*, 6336–6344. [CrossRef] [PubMed]
24. Pajuelo-Lozano, N.; Bargiela-Iparraguirre, J.; Dominguez, G.; Quiroga, A.G.; Perona, R.; Sanchez-Perez, I. XPA, XPC, and XPD Modulate Sensitivity in Gastric Cisplatin Resistance Cancer Cells. *Front. Pharmacol.* **2018**, *9*. [CrossRef] [PubMed]
25. Slyskova, J.; Sabatella, M.; Ribeiro-Silva, C.; Stok, C.; Theil, A.F.; Vermeulen, W.; Lans, H. Base and nucleotide excision repair facilitate resolution of platinum drugs-induced transcription blockage. *Nucleic Acids Res.* **2018**, *46*, 9537–9549. [CrossRef]
26. Schärer, O.D. Nucleotide Excision Repair in Eukaryotes. *Cold Spring Harb. Perspect. Biol.* **2013**, *5*. [CrossRef]
27. Hanawalt, P.C.; Spivak, G. Transcription-coupled DNA repair: Two decades of progress and surprises. *Nat. Rev. Mol. Cell Biol.* **2008**, *9*, 958–970. [CrossRef]
28. Gregersen, L.H.; Svejstrup, J.Q. The Cellular Response to Transcription-Blocking DNA Damage. *Trends Biochem. Sci.* **2018**, *43*, 327–341. [CrossRef]
29. Huang, J.C.; Zamble, D.B.; Reardon, J.T.; Lippard, S.J.; Sancar, A. HMG-domain proteins specifically inhibit the repair of the major DNA adduct of the anticancer drug cisplatin by human excision nuclease. *Proc. Natl. Acad. Sci. USA* **1994**, *91*, 10394–10398. [CrossRef]

30. Deans, A.J.; West, S.C. DNA interstrand crosslink repair and cancer. *Nat. Rev. Cancer* **2011**, *11*, 467–480. [CrossRef]
31. Ohndorf, U.-M.; Rould, M.A.; He, Q.; Pabo, C.O.; Lippard, S.J. Basis for recognition of cisplatin-modified DNA by high-mobility-group proteins. *Nature* **1999**, *399*, 708–712. [CrossRef]
32. Awuah, S.G.; Riddell, I.A.; Lippard, S.J. Repair shielding of platinum-DNA lesions in testicular germ cell tumors by high-mobility group box protein 4 imparts cisplatin hypersensitivity. *PNAS* **2017**. [CrossRef] [PubMed]
33. Zamble, D.B.; Mikata, Y.; Eng, C.H.; Sandman, K.E.; Lippard, S.J. Testis-specific HMG-domain protein alters the responses of cells to cisplatin. *J. Inorg. Biochem.* **2002**, *91*, 451–462 [CrossRef]
34. Hu, J.; Adar, S.; Selby, C.P.; Lieb, J.D.; Sancar, A. Genome-wide analysis of human global and transcription-coupled excision repair of UV damage at single-nucleotide resolution. *Genes Dev.* **2015**, *29*, 948–960. [CrossRef] [PubMed]
35. Ng, J.M.Y.; Vermeulen, W.; van der Horst, G.T.J.; Bergink, S.; Sugasawa, K.; Vrieling, H.; Hoeijmakers, J.H.J. A novel regulation mechanism of DNA repair by damage-induced and RAD23-dependent stabilization of xeroderma pigmentosum group C protein. *Genes Dev.* **2003**, *17*, 1630–1645. [CrossRef] [PubMed]
36. Hey, T.; Lipps, G., Sugasawa, K.; Iwai, S.; Hanaoka, F.; Krauss, G. The XPC−HR23B Complex Displays High Affinity and Specificity for Damaged DNA in a True-Equilibrium Fluorescence Assay. *Biochemistry* **2002**, *41*, 6583–6587. [CrossRef] [PubMed]
37. Min, J.-H.; Pavletich, N.P. Recognition of DNA damage by the Rad4 nucleotide excision repair protein. *Nature* **2007**, *449*, 570–575. [CrossRef]
38. Okuda, M.; Nakazawa, Y.; Guo, C.; Ogi, T.; Nishimura, Y. Common TFIIH recruitment mechanism in global genome and transcription-coupled repair subpathways. *Nucleic Acids Res.* **2017**, *45*, 13043–13055. [CrossRef]
39. Yokoi, M.; Masutani, C.; Maekawa, T.; Sugasawa, K.; Ohkuma, Y.; Hanaoka, F. The Xeroderma Pigmentosum Group C Protein Complex XPC-HR23B Plays an Important Role in the Recruitment of Transcription Factor IIH to Damaged DNA. *J. Biol. Chem.* **2000**, *275*, 9870–9875. [CrossRef]
40. Teng, X.; Fan, X.-F.; Li, Q.; Liu, S.; Wu, D.-Y.; Wang, S.-Y.; Shi, Y.; Dong, M. XPC inhibition rescues cisplatin resistance via the Akt/mTOR signaling pathway in A549/DDP lung adenocarcinoma cells. *Oncol. Rep.* **2019**, *41*, 1875–1882. [CrossRef]
41. Barakat, B.M.; Wang, Q.-E.; Han, C.; Milum, K.; Yin, D.-T.; Zhao, Q.; Wani, G.; Arafa, E.-S.A.; El-Mahdy, M.A.; Wani, A.A. Overexpression of DDB2 enhances the sensitivity of human ovarian cancer cells to cisplatin by augmenting cellular apoptosis. *Int J. Cancer* **2010**, *127*, 977–988. [CrossRef] [PubMed]
42. Coin, F.; Oksenych, V.; Egly, J.-M. Distinct roles for the XPB/p52 and XPD/p44 subcomplexes of TFIIH in damaged DNA opening during nucleotide excision repair. *Mol. Cell* **2007**, *26*, 245–256. [CrossRef] [PubMed]
43. Kuper, J.; Braun, C.; Elias, A.; Michels, G.; Sauer, F.; Schmitt, D.R.; Poterszman, A.; Egly, J.-M.; Kisker, C. In TFIIH, XPD Helicase Is Exclusively Devoted to DNA Repair. *PLoS Biol.* **2014**, *12*, e1001954. [CrossRef]
44. Lehmann, A.R. The xeroderma pigmentosum group D (XPD) gene: One gene, two functions, three diseases. *Genes Dev.* **2001**, *15*, 15–23. [CrossRef]
45. Guo, G.; Sun, X.; Chen, C.; Wu, S.; Huang, P.; Li, Z.; Dean, M.; Huang, Y.; Jia, W.; Zhou, Q.; et al. Whole-genome and whole-exome sequencing of bladder cancer identifies frequent alterations in genes involved in sister chromatid cohesion and segregation. *Nat. Genet.* **2013**, *45*, 1459–1463. [CrossRef]
46. Van Allen, E.M.; Mouw, K.W.; Kim, P.; Iyer, G.; Wagle, N.; Al-Ahmadie, H.; Zhu, C.; Ostrovnaya, I.; Kryukov, G.V.; O'Connor, K.W.; et al. Somatic ERCC2 mutations correlate with cisplatin sensitivity in muscle-invasive urothelial carcinoma. *Cancer Discov.* **2014**, *4*, 1140–1153. [CrossRef]
47. Liu, D.; Plimack, E.R.; Hoffman-Censits, J.; Garraway, L.A.; Bellmunt, J.; Van Allen, E.; Rosenberg, J.E. Clinical Validation of Chemotherapy Response Biomarker ERCC2 in Muscle-Invasive Urothelial Bladder Carcinoma. *JAMA Oncol.* **2016**, *2*, 1094–1096. [CrossRef]
48. Sugitani, N.; Sivley, R.M.; Perry, K.E.; Capra, J.A.; Chazin, W.J. XPA: A key scaffold for human nucleotide excision repair. *DNA Repair* **2016**, *44*, 123–135. [CrossRef]
49. Yang, Z.; Roginskaya, M.; Colis, L.C.; Basu, A.K.; Shell, S.M.; Liu, Y.; Musich, P.R.; Harris, C.M.; Harris, T.M.; Zou, Y. Specific and Efficient Binding of Xeroderma Pigmentosum Complementation Group A to Double-Strand/Single-Strand DNA Junctions with 3′- and/or 5′-ssDNA Branches. *Biochemistry* **2006**, *45*, 15921–15930. [CrossRef]

50. Cierna, Z.; Miskovska, V.; Roska, J.; Jurkovicova, D.; Pulzova, L.B.; Sestakova, Z.; Hurbanova, L.; Machalekova, K.; Chovanec, M.; Rejlekova, K.; et al. Increased levels of XPA might be the basis of cisplatin resistance in germ cell tumours. *BMC Cancer* **2020**, *20*, 17. [CrossRef]
51. Tsodikov, O.V.; Ivanov, D.; Orelli, B.; Staresincic, L.; Shoshani, I.; Oberman, R.; Schärer, O.D.; Wagner, G.; Ellenberger, T. Structural basis for the recruitment of ERCC1-XPF to nucleotide excision repair complexes by XPA. *EMBO J.* **2007**, *26*, 4768–4776. [CrossRef] [PubMed]
52. Tsodikov, O.V.; Enzlin, J.H.; Schärer, O.D.; Ellenberger, T. Crystal structure and DNA binding functions of ERCC1, a subunit of the DNA structure-specific endonuclease XPF–ERCC1. *PNAS* **2005**, *102*, 11236–11241. [CrossRef] [PubMed]
53. Olaussen, K.A.; Dunant, A.; Fouret, P.; Brambilla, E.; André, F.; Haddad, V.; Taranchon, E.; Filipits, M.; Pirker, R.; Popper, H.H.; et al. DNA repair by ERCC1 in non-small-cell lung cancer and cisplatin-based adjuvant chemotherapy. *N. Engl. J. Med.* **2006**, *355*, 983–991. [CrossRef]
54. Jun, H.J.; Ahn, M.J.; Kim, H.S.; Yi, S.Y.; Han, J.; Lee, S.K.; Ahn, Y.C.; Jeong, H.-S.; Son, Y.-I.; Baek, J.-H.; et al. ERCC1 expression as a predictive marker of squamous cell carcinoma of the head and neck treated with cisplatin-based concurrent chemoradiation. *Br. J. Cancer* **2008**, *99*, 167–172. [CrossRef]
55. Du, P.; Wang, Y.; Chen, L.; Gan, Y.; Wu, Q. High ERCC1 expression is associated with platinum-resistance, but not survival in patients with epithelial ovarian cancer. *Oncol. Lett.* **2016**, *12*, 857–862. [CrossRef]
56. Chen, H.-Y.; Shao, C.-J.; Chen, F.-R.; Kwan, A.-L.; Chen, Z.-P. Role of ERCC1 promoter hypermethylation in drug resistance to cisplatin in human gliomas. *Int. J. Cancer* **2010**, *126*, 1944–1954. [CrossRef]
57. Arora, S.; Kothandapani, A.; Tillison, K.; Kalman-Maltese, V.; Patrick, S.M. Downregulation of XPF–ERCC1 enhances cisplatin efficacy in cancer cells. *DNA Repair (Amst)* **2010**, *9*, 745–753. [CrossRef]
58. Booton, R.; Ward, T.; Ashcroft, L.; Morris, J.; Heighway, J.; Thatcher, N. ERCC1 mRNA Expression Is Not Associated with Response and Survival after Platinum-Based Chemotherapy Regimens in Advanced Non-Small Cell Lung Cancer. *J. Thorac. Oncol.* **2007**, *2*, 902–906. [CrossRef]
59. Friboulet, L.; Olaussen, K.A.; Pignon, J.-P.; Shepherd, F.A.; Tsao, M.-S.; Graziano, S.; Kratzke, R.; Douillard, J.-Y.; Seymour, L.; Pirker, R.; et al. ERCC1 Isoform Expression and DNA Repair in Non–Small-Cell Lung Cancer. *N. Eng. J. Med.* **2013**, *368*, 1101–1110. [CrossRef]
60. Friboulet, L.; Postel-Vinay, S.; Sourisseau, T.; Adam, J.; Stoclin, A.; Ponsonnailles, F.; Dorvault, N.; Commo, F.; Saulnier, P.; Salome-Desmoulez, S.; et al. ERCC1 function in nuclear excision and interstrand crosslink repair pathways is mediated exclusively by the ERCC1-202 isoform. *Cell Cycle* **2013**, *12*, 3298–3306. [CrossRef]
61. Walsh, C.S.; Ogawa, S.; Karahashi, H.; Scoles, D.R.; Pavelka, J.C.; Tran, H.; Miller, C.W.; Kawamata, N.; Ginther, C.; Dering, J.; et al. ERCC5 is a novel biomarker of ovarian cancer prognosis. *J. Clin. Oncol. Off. J. Am. Soc. Clin. Oncol.* **2008**, *26*, 2952–2958. [CrossRef] [PubMed]
62. Sabatino, M.A.; Marabese, M.; Ganzinelli, M.; Caiola, E.; Geroni, C.; Broggini, M. Down-regulation of the Nucleotide Excision Repair gene XPG as a new mechanism of drug resistance in human and murine cancer cells. *Mol. Cancer* **2010**, *9*, 259. [CrossRef] [PubMed]
63. Wang, D. A panorama of transcription-coupled repair in yeast chromatin. *Proc. Natl. Acad. Sci. USA* **2020**, *117*, 20991–20993. [CrossRef] [PubMed]
64. Duan, M.; Selvam, K.; Wyrick, J.J.; Mao, P. Genome-wide role of Rad26 in promoting transcription-coupled nucleotide excision repair in yeast chromatin. *Proc. Natl. Acad. Sci. USA* **2020**, 202003868. [CrossRef]
65. Stubbert, L.J.; Smith, J.M.; McKay, B.C. Decreased transcription-coupled nucleotide excision repair capacity is associated with increased p53- and MLH1-independent apoptosis in response to cisplatin. *BMC Cancer* **2010**, *10*, 207. [CrossRef]
66. Damsma, G.E.; Alt, A.; Brueckner, F.; Carell, T.; Cramer, P. Mechanism of transcriptional stalling at cisplatin-damaged DNA. *Nat. Struct. Mol. Biol.* **2007**, *14*, 1127–1133. [CrossRef]
67. Ljungman, M.; Zhang, F. Blockage of RNA polymerase as a possible trigger for u.v. light-induced apoptosis. *Oncogene* **1996**, *13*, 823–831.
68. Caputo, M.; Frontini, M.; Velez-Cruz, R.; Nicolai, S.; Prantera, G.; Proietti-De-Santis, L. The CSB repair factor is overexpressed in cancer cells, increases apoptotic resistance, and promotes tumor growth. *DNA Repair (Amst)* **2013**, *12*, 293–299. [CrossRef]
69. Selby, C.P.; Sancar, A. Cockayne syndrome group B protein enhances elongation by RNA polymerase II. *Proc. Natl. Acad. Sci. USA* **1997**, *94*, 11205–11209. [CrossRef]

70. Rainey, R.N.; Ng, S.; Llamas, J.; van der Horst, G.T.J.; Segil, N. Mutations in Cockayne Syndrome-Associated Genes (Csa and Csb) Predispose to Cisplatin-Induced Hearing Loss in Mice. *J. Neurosci.* **2016**, *36*, 4758–4770. [CrossRef]
71. Vermeulen, W.; Fousteri, M. Mammalian Transcription-Coupled Excision Repair. *Cold Spring Harb. Perspect. Biol.* **2013**, *5*, a012625. [CrossRef] [PubMed]
72. Saijo, M. The role of Cockayne syndrome group A (CSA) protein in transcription-coupled nucleotide excision repair. *Mech. Ageing Dev.* **2013**, *134*, 196–201. [CrossRef] [PubMed]
73. Fousteri, M.; Vermeulen, W.; Van Zeeland, A.A.; Mullenders, L.H.F. Cockayne Syndrome A and B Proteins Differentially Regulate Recruitment of Chromatin Remodeling and Repair Factors to Stalled RNA Polymerase II In Vivo. *Mol. Cell* **2006**, *23*, 471–482. [CrossRef] [PubMed]
74. Nakazawa, Y.; Sasaki, K.; Mitsutake, N.; Matsuse, M.; Shimada, M.; Nardo, T.; Takahashi, Y.; Ohyama, K.; Ito, K.; Mishima, H.; et al. Mutations in UVSSA cause UV-sensitive syndrome and impair RNA polymerase IIo processing in transcription-coupled nucleotide-excision repair. *Nat. Genet.* **2012**, *44*, 586–592. [CrossRef]
75. Schwertman, P.; Lagarou, A.; Dekkers, D.H.W.; Raams, A.; van der Hoek, A.C.; Laffeber, C.; Hoeijmakers, J.H.J.; Demmers, J.A.A.; Fousteri, M.; Vermeulen, W.; et al. UV-sensitive syndrome protein UVSSA recruits USP7 to regulate transcription-coupled repair. *Nat. Genet.* **2012**, *44*, 598–602. [CrossRef]
76. Barakat, K.; Tuszynski, J. Nucleotide Excision Repair Inhibitors: Still a Long Way to Go. *New Res. Dir. DNA Repair* **2013**. [CrossRef]
77. Alekseev, S.; Ayadi, M.; Brino, L.; Egly, J.-M.; Larsen, A.K.; Coin, F. A Small Molecule Screen Identifies an Inhibitor of DNA Repair Inducing the Degradation of TFIIH and the Chemosensitization of Tumor Cells to Platinum. *Chem. Biol.* **2014**, *21*, 398–407. [CrossRef]
78. Ueda, M.; Matsuura, K.; Kawai, H.; Wakasugi, M.; Matsunaga, T. Spironolactone-induced XPB degradation depends on CDK7 kinase and SCFFBXL18 E3 ligase. *Genes Cells* **2019**, *24*, 284–296. [CrossRef]
79. Kemp, M.G.; Krishnamurthy, S.; Kent, M.N.; Schumacher, D.L.; Sharma, P.; Excoffon, K.J.D.A.; Travers, J.B. Spironolactone Depletes the XPB Protein and Inhibits DNA Damage Responses in UVB-Irradiated Human Skin. *J. Invest. Dermatol.* **2019**, *139*, 448–454. [CrossRef]
80. Neher, T.M.; Shuck, S.C.; Liu, J.; Zhang, J.-T.; Turchi, J.J. Identification of novel small molecule inhibitors of the XPA protein using in silico based screening. *ACS Chem. Biol.* **2010**, *5*, 953–965. [CrossRef]
81. Jiang, H.; Yang, L.-Y. Cell Cycle Checkpoint Abrogator UCN-01 Inhibits DNA Repair: Association with Attenuation of the Interaction of XPA and ERCC1 Nucleotide Excision Repair Proteins. *Cancer Res.* **1999**, *59*, 4529–4534. [PubMed]
82. Barakat, K.H.; Jordheim, L.P.; Perez-Pineiro, R.; Wishart, D.; Dumontet, C.; Tuszynski, J.A. Virtual Screening and Biological Evaluation of Inhibitors Targeting the XPA-ERCC1 Interaction. *PLoS ONE* **2012**, *7*, e51329. [CrossRef] [PubMed]
83. Arora, S.; Heyza, J.; Zhang, H.; Kalman-Maltese, V.; Tillison, K.; Floyd, A.M.; Chalfin, E.M.; Bepler, G.; Patrick, S.M. Identification of small molecule inhibitors of ERCC1-XPF that inhibit DNA repair and potentiate cisplatin efficacy in cancer cells. *Oncotarget* **2016**, *7*, 75104–75117. [CrossRef] [PubMed]
84. Jordheim, L.P.; Barakat, K.H.; Heinrich-Balard, L.; Matera, E.-L.; Cros-Perrial, E. Bouledrak, K.; El Sabeh, R.; Perez-Pineiro, R.; Wishart, D.S.; Cohen, R.; et al. Small molecule inhibitors of ERCC1-XPF protein-protein interaction synergize alkylating agents in cancer cells. *Mol. Pharmacol.* **2013**, *84*, 12–24. [CrossRef]
85. Elmenoufy, A.H.; Gentile, F.; Jay, D.; Karimi-Busheri, F.; Yang, X.; Soueidan, O.M.; Weilbeer, C.; Mani, R.S.; Barakat, K.H.; Tuszynski, J.A.; et al. Targeting DNA Repair in Tumor Cells via Inhibition of ERCC1-XPF. *J. Med. Chem.* **2019**, *62*, 7684–7696. [CrossRef]
86. Kim, J.; Mouw, K.W.; Polak, P.; Braunstein, L.Z.; Kamburov, A.; Tiao, G.; Kwiatkowski, D.J.; Rosenberg, J.E.; Van Allen, E.M.; D'Andrea, A.D.; et al. Somatic *ERCC2* mutations are associated with a distinct genomic signature in urothelial tumors. *Nat. Genet.* **2016**, *48*, 600–606. [CrossRef]

International Journal of
Molecular Sciences

MDPI

Review

Fused in Sarcoma (FUS) in DNA Repair: Tango with Poly(ADP-ribose) Polymerase 1 and Compartmentalisation of Damaged DNA

Maria V. Sukhanova [1,*], Anastasia S. Singatulina [1], David Pastré [2] and Olga I. Lavrik [1]

1 Institute of Chemical Biology and Fundamental Medicine, SB RAS, 630090 Novosibirsk, Russia; lasty@ngs.ru (A.S.S.); lavrik@niboch.nsc.ru (O.I.L.)

2 Laboratoire Structure-Activité des Biomolécules Normales et Pathologiques, INSERM U1204, Université Paris-Saclay, 91025 Evry, France; david.pastre@univ-evry.fr

* Correspondence: mary@niboch.nsc.ru; Tel.: +7-383-363-5196

Received: 12 August 2020; Accepted: 21 September 2020; Published: 24 September 2020

Abstract: The fused in sarcoma (FUS) protein combines prion-like properties with a multifunctional DNA/RNA-binding domain and has functions spanning the regulation of RNA metabolism, including transcription, pre-mRNA splicing, mRNA transport and translation. In addition to its roles in RNA metabolism, FUS is implicated in the maintenance of DNA integrity. In this review, we examine the participation of FUS in major DNA repair pathways, focusing on DNA repair associated with poly(ADP-ribosyl)ation events and on how the interaction of FUS with poly(ADP-ribose) may orchestrate transient compartmentalisation of DNA strand breaks. Unravelling how prion-like RNA-binding proteins control DNA repair pathways will deepen our understanding of the pathogenesis of some neurological diseases and cancer as well as provide the basis for the development of relevant innovative therapeutic technologies. This knowledge may also extend the range of applications of poly(ADP-ribose) polymerase inhibitors to the treatment of neurodegenerative diseases related to RNA-binding proteins in the cell, e.g., amyotrophic lateral sclerosis and frontotemporal lobar degeneration.

Keywords: fused in sarcoma; DNA repair; poly(ADP-ribose) polymerase; poly(ADP-ribose); protein phase separation

---

1. Introduction

Throughout their lifespan organisms are constantly exposed to genotoxic agents, both exogenous and endogenous. A rough assessment has revealed that up to 70,000 DNA damage events occur per human cell per day [1]. Under these conditions, preserving cell genome integrity is one of the most important challenges faced by multicellular organisms. Notably, unrepaired DNA damage contributes to the appearance of a pathological mutation [2–4]. The maintenance of genome stability is achieved by the machineries associated with the DNA damage response (DDR) and the cell death pathways that carry out detection of DNA damage or a signal of its presence to orchestrate DNA repair and induce cell death upon massive DNA damage, respectively [5,6]. For a long time, it has been believed that RNA-binding proteins (RBPs) interfere indirectly with the DDR through post-transcriptional regulation of gene expression [7–10]. Nonetheless, more and more studies are revealing the direct functions of many RBPs in the presence of DNA damage, e.g., sensing, signalling, and repair [11–13]. In this regard, poly(ADP-ribose) polymerases (PARPs), which are DNA damage sensors, may provide the basis for the control of RBPs over DNA repair and accordingly aroused special interest recently [14–19]. Indeed, DNA damage followed by PARPs' activation is accompanied with protein poly(ADP-ribosyl)ation (PARylation) through covalent attachment of ADP-ribose moieties and the formation of an ADP-ribose chain: poly(ADP-ribose) (PAR) [20–22]. PAR is a polymeric molecule that shares several features with RNA, including a high negative charge and structural diversity (size, chain length, and branching complexity) [23–25]. A large body of evidence suggests that many RBPs interact with PAR and/or undergo post-translational modification through PARylation during genotoxic stress [26,27]. Moreover, RBPs contain low-complexity regions with self-adhesive properties abundantly as compared to other proteins. Low-complexity domains (LCDs), also called intrinsically disordered regions, are considered key components of membrane-less assemblies such as DNA damage foci, P-bodies or stress granules formed under stressful conditions in the cell [28–31]. PAR, in turn, may serve as a scaffolding factor for these RBP-containing assemblies [16]. Just as many RBPs, fused in sarcoma (FUS) combines self-adhesive LCDs with multifunctional DNA/RNA-binding domains. FUS is involved in the regulation of RNA metabolism, including transcription, pre-mRNA splicing, mRNA transport and translation [32]. In addition to its role in RNA metabolism, FUS was recently implicated in the maintenance of DNA integrity in response to DNA damage [33,34]. In particular, FUS is mainly a nuclear protein [35] that interacts with DNA repair factors and is associated with DNA damage–induced formation of nuclear foci [17,36–40]. Nonetheless, the exact functions of FUS in DNA repair remain unclear. In this review, we highlight the role of FUS in the maintenance of genome integrity, focusing on PARylation events and on how the FUS–PAR interaction may be connected with DNA repair.

2. Intrinsically Disordered Regions and Prion-Like Properties of FUS

The protein called fused in sarcoma (FUS; also known as translocated in liposarcoma; TLS) was first identified in human myxoid liposarcomas ~30 years ago, suggesting that this protein plays a critical part in cancers [41,42]. In liposarcoma, a chromosomal translocation leads to the fusion of two genes, *FUS* and UPR-regulated CCAAT/enhancer-binding protein homologous protein, resulting in the synthesis of a chimeric protein that acts as a transcription factor enhancing cell proliferation and promoting tumour progression [41]. FUS, along with Ewing's sarcoma (EWS) and TATA-binding protein–associated factor 15 (TAF15), belongs to the FET family of RBPs that are highly conserved and perform functions primarily related to RNA metabolism [43,44]. FUS is a 526-amino-acid-long protein that possesses a serine/tyrosine/glycine/glutamine (SYGQ)-rich N terminus of low complexity, three arginine/glycine/glycine (RGG)-rich regions (named RGG1–3), a conserved RNA recognition motif, a zinc finger motif and a proline-tyrosine nuclear localisation signal at the C-terminus [45,46] (Figure 1). The C-terminal domains of FUS comprising the RNA recognition motif, RGG and zinc finger motifs mainly participate in the binding of FUS to RNA, DNA and PAR [47–51]. On the other hand, FUS's unstructured N-terminal domain of low complexity is mainly associated with FUS self-interactions caused by homotypic multivalent interactions [52,53].

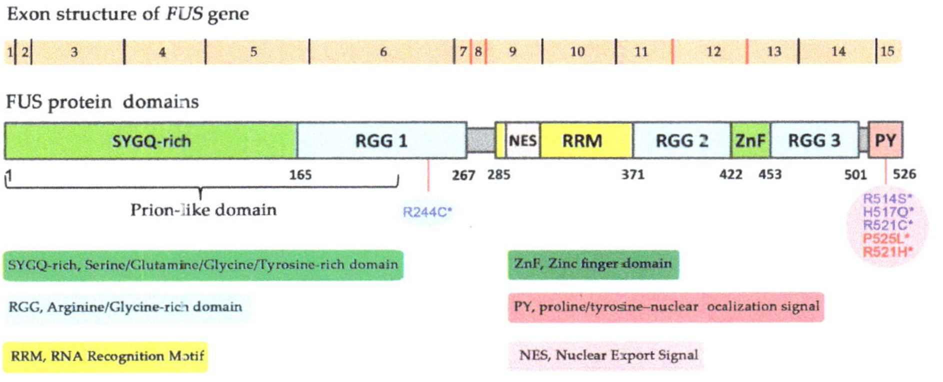

Figure 1. A schematic diagram of exon structure of the *FUS* gene and domain structure of the FUS protein [54]. * Mutations identified in patients with familial amyotrophic lateral sclerosis (fALS) and implicated in DNA repair and DDR [37,40].

3. Higher-Order Assembly and Phase Separation of the FUS Protein for the Formation of Membrane-Less Assemblies in the Cell

Notably, through its long N-terminal LCD with prion-like properties, higher-order multimolecular assembly of FUS either alone or in the presence of RNA gives rise to diverse structures including aggregates, hydrogels, amyloid fibrils and liquid droplets in vivo that have been the subject of intense research since pathological mutations in *FUS* were directly associated with two major neurodegenerative diseases, amyotrophic lateral sclerosis (ALS) and frontotemporal lobar degeneration (FTLD), and the identification of cytoplasmic inclusions of FUS in neurons of the affected patients [41,52,53,55–57]. In agreement with its ability to aggregate, among more than 200 yeast-prion-like proteins that have been identified in the human proteome [58], FUS has been ranked 15th for its prion-like properties and 1st among RBPs [59]. In particular, it has been shown that SYGQ- and glycine-rich regions at the N terminus of FUS have prion-like properties and accordingly play an important role in FUS aggregation. In line with this notion, a truncated FUS protein lacking the N terminus is not able to form droplets or aggregates both in vitro and in vivo [17,56,60–64]. Weak homotypic multivalent intermolecular interactions occur between N-terminal LCDs thereby resulting in FUS self-assembly into liquid-like dynamical compartments as a single component. Heterotypic interaction with other proteins and/or nucleic acids can lead to heterogeneous higher-order structures giving rise to higher complexity in terms of composition, shape and dynamics and most probably biological functions [52,55,64,65]. In this way, it has been suggested that phase transition, in particular liquid–liquid phase separation (LLPS) of protein or protein–nucleic acid mixtures underlies the emergence of membrane-less compartments such as nucleoli, Cajal bodies, gemini of Cajal bodies, Nuage bodies, speckles, paraspeckles, DNA damage foci, stress granules and P-bodies in the cell [28–30,66]. In the emerging field of phase separation biology, FUS has received even more attention since pathological mutations in the LCD that impair FUS were shown to trigger a phase transition from a reversible liquid-like droplet or gel-like state to irreversible solid-like states possibly promoting the formation of cytoplasmic inclusions of FUS found in ALS and FTLD [66]. Taking into account that FUS undergoes phase separation and interacts with other macromolecules such as RNA, DNA or PAR, FUS is regarded as an important player in the creation of membrane-less compartments in vivo under physiological and stressful conditions [67,68]. Indeed, nuclear FUS has been detected in association with DNA damage foci [17,37], paraspeckles [69] and SMN1 bodies (Gems) [70], whereas cytoplasmic FUS is recruited into stress granules or P-bodies [71–80] and neuronal RNA granules [81,82] (Figure 2).

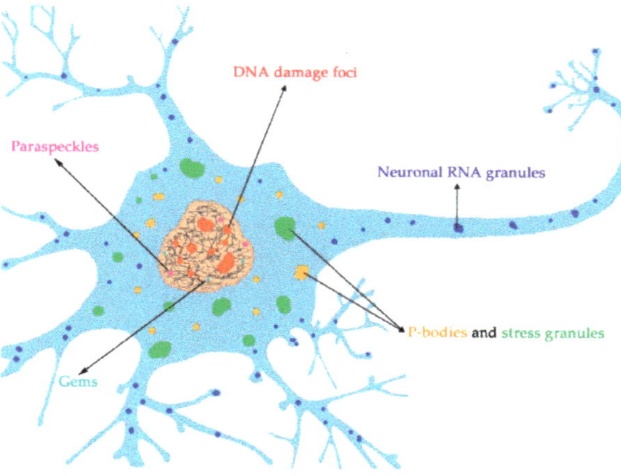

Figure 2. Schematic illustration of a neuronal cell and membrane-less compartments generated with the participation of FUS in the nucleus and cytoplasm. We corrected the Figure 2 (Please, see the attached file).

Accordingly, the functions of FUS in DNA repair may be related to its capacity to induce the formation of dynamic compartments that regulate DNA repair through protein phase separation. Although numerous studies have shown that FUS can be recruited to a region containing DNA damage in the nucleus, whether FUS directly affects the efficiency of DNA repair, by promoting the emergence of DNA repair foci, is still an open question.

4. Direct Functions of FUS in DNA Repair

As mentioned above, FUS mainly features a nuclear localisation [83], while a smaller proportion of FUS is found in the cytoplasm under physiological conditions [48,82]. For a long time, FUS has been known to be involved in the regulation of RNA metabolism [43,84]. FUS binds preferentially to a nascent GU-rich mRNA transcript and has been identified as a component of membrane-less organelles associated with RNA processing such as SMN1 bodies (Gems) in the nucleus and RNA granules or stress granules in the cytoplasm [70].

More recently, the recruitment of FUS to DNA damage–induced foci in the nucleus has drawn attention to its involvement in DNA repair processes [17,38,40,51]. DNA damage repair proceeds through one of five major pathways: direct repair of certain types of UV light–induced photo-lesions or methylated bases; homologous recombination (HR) or nonhomologous end-joining (NHEJ) to repair double-strand breaks (DSBs); nucleotide excision repair of bulky lesions; base excision repair (BER) of damaged bases, apurinic/apyrimidinic sites and single-strand breaks (SSBs); and mismatch repair of unpaired bases [85]. Numerous studies indicate that FUS's functions in the DDR are associated with DNA strand break signalling and the repair of oxidative DNA damage types such as oxidised DNA bases and DNA SSBs and DSBs [33,34]; therefore, here we focus on the possible participation of FUS in HR, NHEJ or BER pathways (Figures 3 and 4).

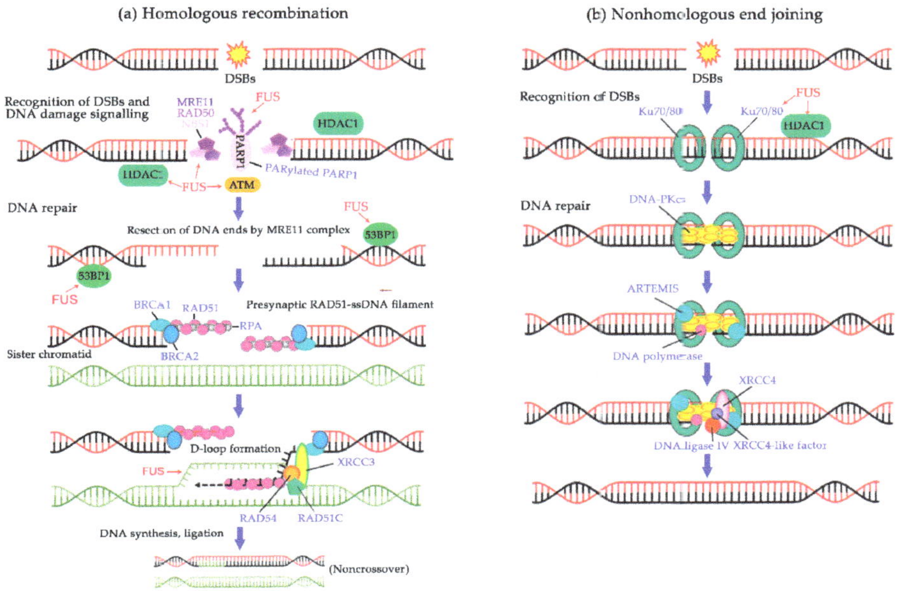

Figure 3. The involvement of FUS in double-strand DNA break (DSB) repair pathways. Schematic overview of HR, NHEJ illustrating basic steps of these pathways along with the proteins implicated in each step [86]. (**a**) The influence of FUS on HR. Simplified scheme for HR. The Mre11/Rad50/NBS1 complex starts resection on the DSBs to generate single stranded DNA. The ssDNA is first coated by Replication protein A (RPA), which is subsequently replaced by Rad51 with the help of BRCA1 and BRCA2. These Rad51 nucleoprotein filaments mediate strand invasion on the homologous template. The invading 3′-end of ssDNA serves as a primer for DNA synthesis. D-loop strands extended by DNA repair synthesis dissociate from their sister chromatid complements and reanneal with their original complementary strands. Additional DNA synthesis in the reannealed DNA duplex and ligation of the remaining single strand nicks complete the repair in the case of synthesis-dependent strand-annealing model of D-loop resolution, forming non-crossover products. FUS interacts with histone deacetylase 1 (HDAC1) and PARylated PARP1, binds to a D-loop intermediate and affects DSB-dependent accumulation of ATM (Ataxia-telangiectasia mutated serine-proteins kinase), NBS1 (Nijmegen breakage syndrome 1) and 53BP1 (p53-binding protein 1) [37,49]. (**b**) The effect of FUS on NHEJ. Simplified scheme for NHEJ. The two broken DNA ends are processed by the action of the end-binding Ku70/80 heterodimer complex, DNA-dependent protein kinase, catalytic subunits, (DNA-PKcs), artemis, DNA polymerase and ligated by DNA Ligase IV-XRCC4 complex. FUS interacts with HDAC1 and affects accumulation of Ku70/80 at DSBs [37].

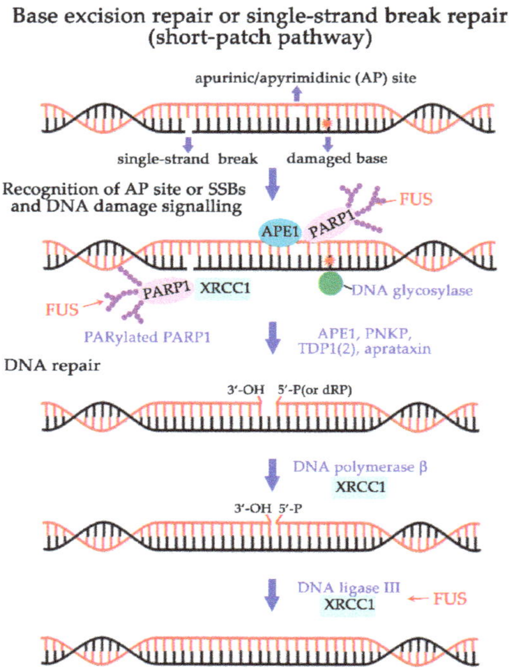

Figure 4. The involvement of FUS in base excision repair (BER) or single-strand break (SSB) repair pathways. Simplified scheme for BER/SSBR short-patch pathway [87]. Monofunctional DNA glycosylases catalyse the removal of the damaged base through cleavage of the *C1′-N*-glycosylic bond, leaving an AP site. AP endonuclease1 (APE1) cleaves AP site, DNA polymerase β inserts a single nucleotide and removes the 5′-deoxyribose phosphate (dRP), and the resulting nick is sealed by DNA ligase III-XRCC1 complex. In the case of SSB, the 5′-and 3′-termini containing blocking modifications can be converted to 5′-phosphate (P) and 3′-hydroxyl (OH) moieties by APE1, polynucleotide kinase 3′-phosphatase (PNKP), tyrosyl-DNA phosphodiesterase 1/2 (TDP1, TDP2) and/or aprataxin. AP sites, SSBs, arising endogenously or exogenously are bound by PARP1, which is then activated and autoPARylated. AutoPARylated PARP1 recruits repair proteins, in particular XRCC1 to the SSB [87]. The impact of FUS on BER is as follows: FUS interacts with PARylated PARP1 and the DNA ligase III–XRCC1 complex, thereby stimulating the ligation of DNA ends [40].

4.1. A FUS Gene Knockdown in Mice and Cells Impairs HR

One of the first pieces of evidence that FUS may play an important part in the maintenance of genome stability has been provided by experiments with FUS-deficient mice and cells originating from the knockout animals. The first *FUS* knockout mice were generated by disruption of a region (homologous to the 12th exon of the human gene) encoding a domain including the zinc finger motif (Figure 1), resulting in a lack of normal transcripts or protein expression in mice [88]. The homozygous mouse pups fail to suckle, and most of them die shortly, within 16 h, after birth. *FUS*$^{-/-}$ primary fibroblasts or B lymphocytes derived from the knockout mice are characterised by genomic instability, and the lymphocyte proliferative response to mitogens is significantly affected. Those authors have suggested that FUS acts as a modulator or effector of gene expression by binding to RNA and thereby participates indirectly in the cellular response to DNA damage or mitogenic stimuli [88]. Another strain of FUS-deficient mice has been created via disruption of a region homologous to the 8th exon of the human gene [89] (Figure 1). These *FUS*$^{-/-}$ animals manifest complete male sterility and reduced fertility of females; moreover, these mice and their fibroblasts ($FUS^{-/-}$)are sensitive to ionising radiation. Detailed analyses have revealed that the FUS deficiency causes a defect in homologous pairing and synapsis during HR, thus leading to degeneration of spermatocytes (Figure 3a). According to measurements of homologous DNA-pairing activities in cell extracts, the contribution of FUS to ATP-independent annealing of complementary single-stranded DNAs and D-loop formation in superhelical double-stranded DNA has uncovered its role in homologous pairing [49]. Therefore, FUS may contribute to meiotic HR through interaction with a D-loop intermediate (Figure 3a). Meiotic HR is a programmed event, but HR is also one of two major DSB repair pathways, and homologous DNA pairing is an essential step in the repair pathway [90]. Consequently, defects in the repair of DNA damage produced by ionising radiation in FUS knockout animals and cells may be explained by impaired DSB repair via HR.

4.2. Deficiency of FUS or Mutations of FUS Affect the Repair of DNA Strand Breaks

Because FUS deficiency increases genome instability in animal and cell models, FUS has been the subject of further research on its role in the repair processes. As mentioned above, eukaryotic cells have evolved two pathways to repair DSBs, namely, HR and NHEJ (Figure 3a,b) [86,90]. Other experimental pieces of direct evidence for the participation of FUS in DSB repair come from examination of the effects of a knockdown of FUS by a small interfering RNA or expression of fALS-associated mutant FUS versions (R244C, R514S, H517Q or R521C) in murine primary neuronal culture and/or human osteosarcoma U2OS cells [37]. Upon DNA damage, FUS depletion leads to a decrease in the level of H2AX phosphorylation, an impairment of DSB repair foci creation and deficient accumulation of DDR proteins such as p53-binding protein 1 (53BP1), Nijmegen breakage syndrome 1 (NBS1), phospho-ATM (ataxia telangiectasia mutated) and Ku70 (Figure 3a,b). The role of FUS in DSB repair is further underscored by results of FUS depletion: decreased HR and NHEJ efficiency and elevating the number of DNA damages in primary neuronal culture [37]. Besides, FUS directly associates with chromatin [91] and with a remodelling factor such as histone deacetylase 1 (HDAC1) [37]. HDAC1 plays an important part in the promotion of DNA DSB repair in post-mitotic neurons [92]. Furthermore, the expression of FUS mutants defective in their interaction with HDAC1 impairs both NHEJ and HR pathways in the $U2OS^{-GFP}$ cell line [37]. These observations indicate an impaired capacity to repair a DNA DSB when expression levels of FUS are low or FUS is mutated in the cell. The functions of FUS in NHEJ and HR repair pathways are therefore clearly non-transcriptional.

In addition to its functions in DSB repair, FUS is involved in DNA SSB repair in HEK293 cells [40]. FUS co-immunoprecipitates with BER players such as XRCC1 and DNA ligase III (Lig III) and stimulates Lig III activity via direct interaction. Moreover, a CRISPR/Cas9-mediated knockout of FUS in HEK293 cells reduces the efficiency of SSB repair [40]. Defective DNA repair of oxidative DNA damage is observed in induced pluripotent stem cell lines derived from ALS patients carrying either the R521H or P525L mutation of FUS (Figure 1), which promotes the formation of cytosolic FUS aggregates at the expense of the pool of nuclear FUS [66,93].

These findings therefore define FUS as a novel participant of DNA break repair processes that plays an upstream role in DSB signalling; these data also point to a functional link to HR, NHEJ and BER/SSB repair via interaction with DNA intermediates and/or repair factors. In the case of DSB repair (Figure 3a,b), FUS facilitates the initial recruitment of DNA damage signalling proteins to DNA lesions and regulates HDAC's chromatin-remodelling activity [37] In the case of SSB repair (Figure 4), FUS interacts with the XRCC1–DNA ligase III complex and stimulates the SSB ligation step [40]. On the other hand, we cannot rule out the possibility that FUS has functions downstream of ATM or DNA-PK in response to DNA breaks. This is because FUS phosphorylation drives its translocation to the cytoplasm under genotoxic stress thereby possibly affecting FUS functions in RNA metabolism [36,39].

4.3. FUS Is Connected to DNA Strand Breaks by PARPs' Signalling Activities

Although in vitro studies indicate that FUS directly binds to single- or double-stranded DNA or G-quadruplexes in telomeres [49,94], how FUS is recruited to the sites of DNA damage in the cell has been an open question. FUS recruitment to DNA repair sites has been analysed owing to the development of the laser micro-irradiation technique able to generate a spatially controlled DNA damage region in the cell nucleus [37,38,40]. The first evidence that FUS can be translocated directly to DNA damage foci was obtained in cells subjected to 405 nm laser micro-irradiation. Moreover, FUS recruitment to DSBs is specifically dependent on the activity of PARP1, thus pointing to an interaction between FUS and the PAR produced by PARP1 at DNA damage sites [51]. Later, PAR-dependent FUS accumulation at damaged-DNA sites has been detected in response to oxidative DNA damage induced by a UVA (320–400 nm) laser, suggesting the recruitment of FUS not only to DSBs but also to SSBs generated directly or indirectly during the repair of oxidised bases in DNA [38]. Therefore, DNA damage–dependent activation of PARPs and the synthesis of PAR are some of the ways in which FUS can be recruited to sites of DNA damage [37,38,40]. PARPs, primarily PARP1 and PARP2, can directly recognise damaged DNA, and their binding to it results in PARP activation [95,96]. PARP enzymes use NAD^+ as a substrate and catalyse the transfer of an ADP-ribose residue from NAD^+ to target protein-acceptors, leading finally to the synthesis of ADP-ribose chains (PAR) attached to proteins, mainly PARPs themselves [21] (Figure 5).

PAR chains is commonly linked to Lys/Arg or Glu/Asp residues attached to the C1″ or C1″,C2″, C3″ atoms of ADP-ribose with the formation of ketoamine and carboxyl ester, respectively (Figure 5). Moreover, amino acid specificity of PARP1 (PARP2) from Glu/Asp to Ser residue can be changed under influence of other protein factors such as histone PARylation factor 1 [95,97].

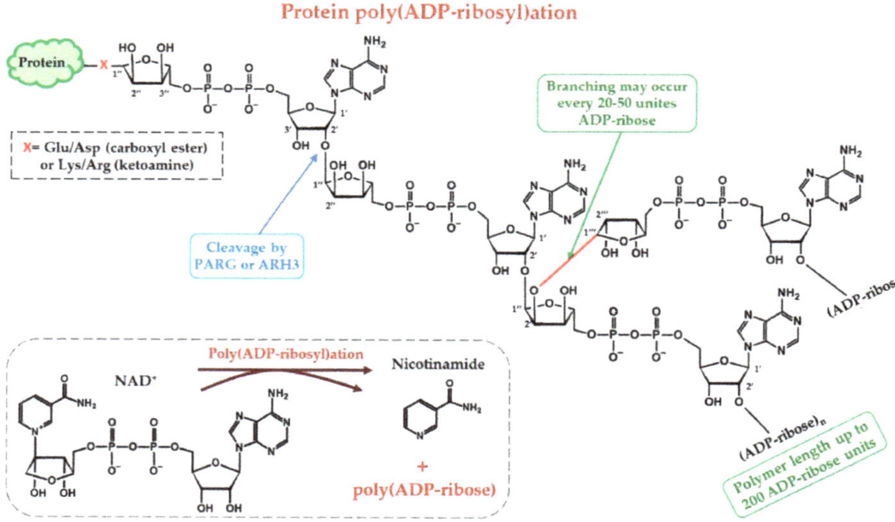

Figure 5. A diagram of the structure of PAR and protein PARylation. The scheme illustrates chemical structure of NAD^+ and a chain of ADP-ribose units linked by the $\alpha(2'\text{-}1'')$ *O*-glycosidic bond between ribose residues (linear chain) and by the $\alpha(2''\text{-}1''')$ glycosidic bond between two nicotinamide-proximal ribose residues (branching). PARG and ARH3 is the main enzymes that degrade PAR and possesses exo- and endoglycosidase activities hydrolysing the glycosidic bonds between ribose units of PAR [98,99].

Protein PARylation is a reversible process mainly due to poly(ADP-ribose) glycohydrolase (PARG) activity and ADP-ribosyl-acceptor hydrolases 3 (ARH3), which catalyse the cleavage between ADP-ribose structural units at the terminal position and inside the polymer, thereby releasing ADP-ribose or oligo(ADP-ribose), respectively [98,99]. Thus, the action of PAR-degrading enzymes makes protein PARylation a dynamic and reversible post-translational modification and plays an important role in the regulation of DNA repair [98–100]. Protein PARylation, PAR length and formation of protein-free PAR molecules all contribute to the regulation of DNA repair, in particular, through the binding of proteins to PAR or their PARylation [16,21,25,100]. Therefore, PAR (either attached to proteins or free), because of its biochemical properties and tight regulation of its synthesis and degradation by PARP and PARG/ARH3 activities, is considered a critical factor that orchestrates reversible assembly of DNA repair compartments [16,25,99,101]. It is commonly accepted now that PARPs and PARylation regulate DSB and BER/SSB repair [102,103]. Recent research showed that FUS binds to PAR non-covalently and/or can be PARylated in vivo and in vitro [17,26,27,40,101]. Consistently with this notion, the FUS interaction with PAR after DNA damage constitutes the missing link between FUS and DNA repair events [17,38,40,51,56,101]. In line with this view, PAR synthesis at DNA damage sites induces the relocation of FUS to DNA damage foci [17,38,40,51]. Furthermore, inhibition of PARP activity impairs FUS accumulation at sites of laser-induced damage in the cell, and it is likely that the absence of PAR prevents FUS from being directed to DNA damage sites [17,38,51].

Consequently, FUS may take part in DNA strand break repair in a PAR-dependent manner, although little is known about the functional significance of FUS–PAR interactions in DNA repair [40,104]. One supposition is the unusual capacity of FUS to form a liquid-like compartment making it an ideal organiser of DNA repair compartments [17,101,104] able to concentrate damaged DNA at the early stages of the DNA strand break response in order to undergo spatially controlled PAR-dependent phase separation [101,104].

4.4. DNA Damage Sensing by the Phase Separation of FUS?

Even though the nucleus, unlike the cytoplasm, does not contain membrane-bound organelles, the nucleus is a highly organised structure with separate proteinaceous and nucleic-acid–rich subnuclear compartments that have special morphology and specific composition of protein–nucleic acid complexes and are functionally specialised [105]. It was recently proposed that many DNA- or RNA-containing structures in the nucleus are assembled through a physicochemical process called LLPS [30,105–107]. In fact, LLPS of protein or protein–nucleic acid mixtures is now regarded as the principal mechanism behind the formation of protein-rich membrane-less compartments in the cell [29,105–108]. In eukaryotes, an abundant group of RBPs—that are intrinsically disordered proteins (IDPs) with prion-like properties—represents key factors that contribute to the creation of these compartments in the cell through the phase separation mechanism [30,64,65,67]. Among these proteins is FUS, which is presently intensively studied due to its ability to undergo reversible phase separation in vitro and in vivo (thereby generating protein-rich droplets, hydrogels and amyloid aggregates) and due to its link with major neurodegenerative diseases such as ALS [56,64,79,109,110]. Moreover, other biomolecules, such as RNA and PAR, influence the phase behaviour of FUS and can drive phase separation [17,56,101,111]. For example, functional interactions between RNA and FUS are assumed to play a key role in the dynamics of compartments among which we can find mRNA-rich stress granules [71,74,78,83,112]. The similarity to its association with RNA and the evidence of FUS condensation at a damaged-DNA site mediated by its interaction with PAR both in vitro and in cell systems have been reported [17,38,51,56,101]. *In vitro,* FUS has been found to interact with purified PAR directly, and its binding to PAR promotes LLPS of FUS and possibly its aggregation [16,17,38,56,101]. In addition, the appearance of FUS-rich assemblies in nuclear damage regions observed in vivo depends on PAR synthesised at DNA damage sites [17,56]. Accordingly, the interaction of FUS with PAR and the ensuing phase separation may be an important process underlying the formation of DNA repair compartments [17,56,101]. Nevertheless, the exact molecular mechanism(s) responsible for the creation of these compartments and their possible functions are difficult to address in a cellular context. At least three mechanisms have been considered to explain the formation of protein-rich DNA compartments in the nucleus, namely, (i) a cooperative binding of proteins to specific sites along DNA without phase separation, (ii) polymer–polymer phase separation in which proteins form molecular 'bridges' between different binding sites and (iii) LLPS of IDPs and nucleic acids [106,113]. So far, it is not clear what types of mechanisms take place in the context of a compartment generated to orchestrate DNA repair after FUS is directed to DNA damage sites upon PARP1 activation. Single-molecule experiments have partly shed light on this important issue. The atomic force microscopy single-molecule technique has been used to analyse the assemblies orchestrated by FUS at DNA damage sites after PARP1 activation in a reconstituted molecular system that includes mRNA to mimic mRNA targets of FUS in the nucleus, damaged DNA, PARP1 to recognise damage sites in DNA, NAD^+ to trigger the synthesis of PAR by PARP1 and PARG to hydrolyse PAR [101]. Indeed, in the presence of free PAR, FUS can assemble into large PAR-containing aggregates (Figure 6a). Of note, when auto-PARylated PARP1 is still complexed with a DNA damage site, the FUS interaction with PAR gives rise to large aggregates in which DNA damage sites are condensed (Figure 6b).

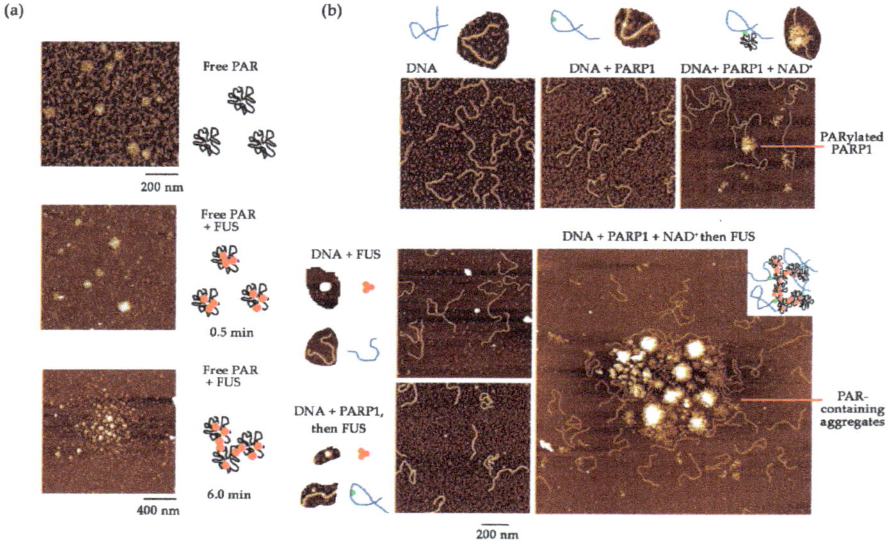

Figure 6. The formation of large DNA-rich assemblies in the presence of FUS after PAR synthesis by PARP1 [101]. (**a**) Representative AFM images of PAR:FUS complexes at different incubation times. PAR: 1 μM; FUS: 40 nM. (**b**) Representative AFM images of 1200-bp nicked DNA (1.25 nM) after incubation with PARP1 (3 nM) for 5 min in the presence or absence of NAD^+ (0.3 mM) followed by the addition of FUS (40 nM) and incubation for 1 min.

Thus, damaged DNA is rapidly confined to multi-protein complexes assembled after the FUS interaction with PARylated PARP1, thereby initiating the spatial segregation of damaged DNA (from intact DNA) into dynamical compartments [101]. The following sequence of events may then take place: (i) the site-specific interaction of PARP1 with damaged DNA triggers PAR synthesis to direct FUS to DNA damage sites, (ii) FUS induces the compartmentalisation of damaged DNA if PARylated PARP1 still bound to damaged DNA [101,114]. In agreement with the latter point, FUS fails to support the assembly of the DNA-rich compartments in mixtures of damaged DNA and free PAR [101]. Moreover, PAR hydrolysis by PARG induces the disassembly of DNA-rich compartments and a release of FUS. Due to the PARG/ARH3 activities, the formation of damaged-DNA–rich compartments by FUS is dynamic and reversible, which is important for the turnover of DNA repair. Taken together, these observations suggest that FUS-mediated DNA repair compartments form not only through the LLPS capacity of FUS but may also be due to mechanisms underlying the binding of PARP1 to damaged DNA and then bridging by FUS after PAR synthesis (Figure 7). Thus, PARP1 binding to damaged DNA proceeds without phase separation, then the interaction of FUS with PARylated PARPs bridges damage sites in close spatial proximity to each other.

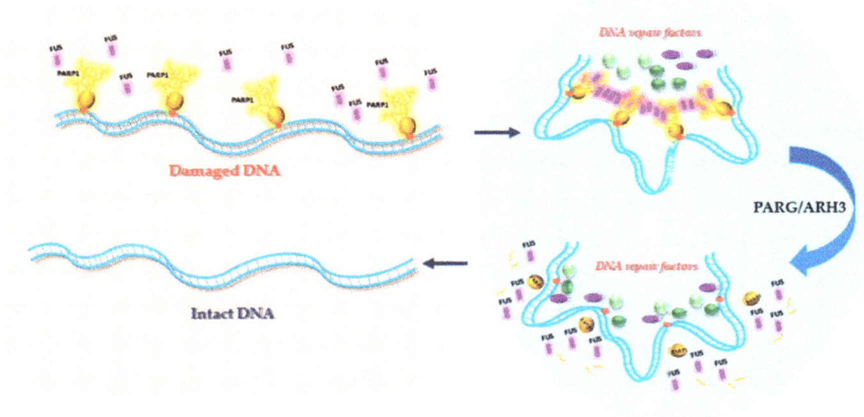

Figure 7. Creation of a repair compartment driven by the interaction of FUS with PAR. The compartment is a cluster of PARylated PARP1 bound by FUS that is concentrated to create a genomic region that is active in terms of DNA repair and may concentrate damaged DNA with subsequent recruitment of DNA repair proteins. PARG dissociates damaged DNA compartments by hydrolyzing PAR.

Because some of the DNA repair factors interact with PAR [25,99,115–117], they can be in turn directed to the damaged-DNA–rich FUS/PAR compartments. Such a function in the organisation of DNA damage repair has previously been attributed to PAR itself [116,118] but the presence of an IDP like FUS may be necessary to increase the capacity of PAR to recruit DNA repair factors and/or to concentrate the DNA damage sites within the compartment. Consequently, the FUS interaction with PAR not only results in the emergence of damaged-DNA–rich compartments but also can stimulate the assembly of the relevant DNA machineries. Altogether, these findings provide new insights into DNA repair regulation at least for SSB repair and BER, because these processes are highly dependent on PARP1 activation [102,103]. The interaction of key proteins of BER/SSB repair pathway (Figure 4) such as AP endonuclease1, XRCC1 and DNA polymerase β with PAR was recently demonstrated [117]. FUS directly affects the efficiency of a repair process, for example, nick sealing by DNA ligase III at the last step of the repair pathways [40]. The spatiotemporal control of FUS/PAR-rich damaged-DNA compartments is most likely complicated and highly regulated, with a probable critical role of additional protein factors and phosphatases and/or kinases that remain to be identified. The hydrolysis of these compartments by PARG causes their dissociation and ensures reversibility of the whole repair process.

Thus, FUS may generate dynamic compartments in which damaged DNA accumulates, and this event should facilitate the recognition of DNA lesions by DNA repair proteins because of the relative increase in concentrations of damaged DNA and of repair factors within the compartments. The increase in the local concentration of DNA lesions and repair factors should accelerate DNA repair and its turnover in the cell.

5. Conclusions

Lately, a number of fundamental discoveries were made concerning the functions of RBPs in the DDR; they may profoundly change the concept of the regulation and organisation of DNA repair processes in the cell. Notably, the recent finding that RBPs with prion-like domains undergo phase separation points to the participation of RBPs in the formation of DNA damage–induced compartments [17,19,56,101,104]. In this regard, the interaction of RBPs with the PAR produced in response to DNA damage is currently receiving increasing attention because PAR not only contributes to the recruitment of RBPs to a DNA repair site but also promotes their LLPS.

Since the discovery of the PAR-dependent recruitment of FUS to DNA repair foci in 2013 [51], there has been a substantial increase in the number of research articles confirming the role of FUS in DNA repair, thereby nurturing special interest in the question which molecular mechanisms may enable FUS to play a part in the DDR [33,34]. FUS seems to be involved in the DDR through an interaction with DNA repair intermediates, DNA repair factors and DNA damage signalling molecules like PAR [33,34]. Furthermore, FUS, having a self-adhesive LCD, can undergo phase separation [53,55,56,65,109,110]. Therefore, FUS–PAR interactions during DNA repair are possibly directly related to the assembly of damaged DNA with repair proteins and of transient repairosome compartments, which may carry out specific functions and implement spatiotemporal control over the DNA repair process [101]. To date, it is still unknown how FUS modulates DNA repair through compartmentalisation. Understanding the biological role of FUS in the generation of the repairosome and in its molecular composition and functions is necessary to clarify diverse characteristics of DNA repair–regulatory processes.

Further research into FUS functions and into the other protein members of the FET family, EWSR1 and TAF15, may advance the present knowledge on the mechanisms of prion-like-RBP–dependent regulation of DNA repair processes through the formation of compartments in human cells. Future investigation of how these RBPs orchestrate DNA repair pathways will deepen our understanding of the response of the cancer cell to genotoxic stress and will elucidate the mutations in these proteins associated with neurological diseases as well as will lay the foundation for the development of relevant innovative preventive or therapeutic modalities. Notably, in neurodegenerative diseases, FUS forms cytoplasmic inclusions that can be toxic by themselves or may impair the DNA-related function of nuclear FUS in ALS or FTLD patients. Given that PARPs' or PARG activities have been demonstrated to interfere with nucleocytoplasmic shuttling of FUS in the cytoplasm upon genotoxic stress [101,119], these data may extend the range of applications of PARPs or PARG inhibitors from cancer to neurodegenerative diseases [120].

Author Contributions: Writing-original draft preparation: M.V.S., D.P., O.I.L.; Visualisation: A.S.S., M.V.S.; Conceptualisation and Supervision: M.V.S., O.I.L. All authors have read and agreed to the published version of the manuscript.

Funding: This research was funded by the Russian Science Foundation, grant number 20-14-00086 (to M.S.), and by a Russian-state–funded budget project, grant number AAAA-A17-117020210022-4.

Conflicts of Interest: The authors declare no conflict of interest.

References

1. Lindahl, T.; Barnes, D.E. Repair of endogenous DNA damage. *Cold Spring Harb. Symp. Quant. Biol.* **2000**, *65*, 127–133. [CrossRef]
2. Torgovnick, A.; Schumacher, B. DNA repair mechanisms in cancer development and therapy. *Front. Genet.* **2015**, *6*, 157. [CrossRef]
3. Dietlein, F.; Thelen, L.; Reinhardt, H.C. Cancer-specific defects in DNA repair pathways as targets for personalized therapeutic approaches. *Trends Genet.* **2014**, *30*, 326–339. [CrossRef]
4. Tubbs, A.; Nussenzweig, A. Endogenous DNA damage as a source of genomic instability in cancer. *Cell* **2017**, *168*, 644–656. [CrossRef]
5. Ciccia, A.; Elledge, S.J. The DNA damage response: Making it safe to play with knives. *Mol. Cell* **2010**, *40*, 179–204. [CrossRef]
6. Sirbu, B.M.; Cortez, D. DNA damage response: Three levels of DNA repair regulation. *Cold Spring Harb. Perspect. Biol.* **2013**, *5*, a012724. [CrossRef]
7. Matsuoka, S.; Ballif, B.A.; Smogorzewska, A.; McDonald, E.R.; Hurov, K.E.; Luo, J.; Bakalarski, C.E.; Zhao, Z.; Solimini, N.; Lerenthal, Y.; et al. ATM and ATR substrate analysis reveals extensive protein networks responsive to DNA damage. *Science* **2007**, *316*, 1160–1166. [CrossRef]

8. Dutertre, M.; Sanchez, G.; Barbier, J.; Corcos, L.; Auboeuf, D. The emerging role of pre-messenger RNA splicing in stress responses: Sending alternative messages and silent messengers. *RNA Biol.* **2011**, *8*, 740–747. [CrossRef]
9. Dutertre, M.; Vagner, S. DNA-damage response RNA-binding proteins (DDRBPs): Perspectives from a new class of proteins and their RNA targets. *J. Mol. Biol.* **2017**, *429*, 3139–3145. [CrossRef]
10. Paulsen, R.D.; Soni, D.V.; Wollman, R.; Hahn, A.T.; Yee, M.C.; Guan, A.; Hesley, J.A ; Miller, S.C.; Cromwell, E.F.; Solow-Cordero, D.E.; et al. A genome-wide siRNA screen reveals diverse cellular processes and pathways that mediate genome stability. *Mol. Cell* **2009**, *35*, 228–239. [CrossRef]
11. Adamson, B.; Smogorzewska, A.; Sigoillot, F.D.; King, R.W.; Elledge, S.J. A genome-wide homologous recombination screen identifies the RNA-binding protein RBMX as a component of the DNA-damage response. *Nat. Cell Biol.* **2012**, *14*, 318–328. [CrossRef]
12. Dutertre, M.; Lambert, S.; Carreira, A.; Amor-Guéret, M.; Vagner, S. DNA damage: RNA-binding proteins protect from near and far. *Trends Biochem. Sci.* **2014**, *39*, 141–149. [CrossRef]
13. Kai, M. Roles of RNA-binding proteins in DNA damage response. *Int. J. Mol. Sci.* **2016**, *17*, 310. [CrossRef]
14. Leung, A.K. Poly (ADP-ribose): An organizer of cellular architecture. *J. Cell Biol.* **2014**, *205*, 613–619. [CrossRef]
15. Hong, Z.; Jiang, J.; Ma, J.; Dai, S.; Xu, T.; Li, H.; Yasui, A. The role of hnRPUL1 involved in DNA damage response is related to PARP1. *PLoS ONE* **2013**, *8*, e60208. [CrossRef]
16. Leung, A.K. Poly (ADP-ribose): A dynamic trigger for biomolecular condensate formation. *Trends Cell Biol.* **2020**, *30*, 370–383. [CrossRef]
17. Altmeyer, M.; Neelsen, K.J.; Teloni, F.; Pozdnyakova, I.; Pellegrino, S.; Grøfte, M.; Rask, M.-B.D.; Streicher, W.; Jungmichel, S.; Nielsen, M.L.; et al. Liquid demixing of intrinsically disordered proteins is seeded by poly (ADP-ribose). *Nat. Commun.* **2015**, *6*, 8088. [CrossRef]
18. Sun, X.; Fu, K.; Hodgson, A.; Wier, E.M.; Wen, M.G.; Kamenyeva, O.; Xia, X.; Koo, L.Y.; Wan, F. Sam68 is required for DNA damage responses via regulating poly(ADP-ribosyl)ation. *PLoS Biol.* **2016**, *14*, e1002543. [CrossRef]
19. Jang, Y.; Elsayed, Z.; Eki, R.; He, S.; Du, K.P.; Abbas, T.; Kai, M. Intrinsically disordered protein RBM14 plays a role in generation of RNA: DNA hybrids at double-strand break sites. *Proc. Natl. Acad. Sci. USA* **2020**, *117*, 5329–5338. [CrossRef]
20. Althaus, F.R.; Kleczowska, H.E.; Malanga, M.; Muntener, C.R.; Pleschke, J.M.; Ebner, M.; Auer, B. Poly ADP-ribosylation: A DNA break signal mechanism. *Mol. Cell Biochem.* **1999**, *193*, 5–11. [CrossRef]
21. D'Amours, D.; Desnoyers, S.; D'Silva, I.; Poirier, G.G. Poly(ADP-ribosyl)ation reactions in the regulation of nuclear functions. *Biochem. J.* **1999**, *342*, 249–268. [CrossRef] [PubMed]
22. Caldecott, K.W. Protein ADP-ribosylation and the cellular response to DNA strand breaks. *DNA Repair* **2014**, *19*, 108–113. [CrossRef]
23. Kiehlbauch, C.C.; Aboul-Ela, N.; Jacobson, E.L.; Ringer, D.P.; Jacobson, M.K. High resolution fractionation and characterization of ADP-ribose polymers. *Anal. Biochem.* **1993**, *208*, 26–34. [CrossRef] [PubMed]
24. Alvarez-Gonzalez, R.; Jacobson, M.K. Characterization of polymers of adenosine diphosphate ribose generated in vitro and in vivo. *Biochemistry* **1987**, *26*, 3218–3224. [CrossRef]
25. Teloni, F.; Altmeyer, M. Readers of poly (ADP-ribose): Designed to be fit for purpose. *Nucleic Acids Res.* **2015**, *44*, 993–1006. [CrossRef]
26. Gagne, J.P.; Isabelle, M.; Lo, K.S.; Bourassa, S.; Hendzel, M.J.; Dawson, V.L.; Dawson, T.M.; Poirier, G.G. Proteome-wide identification of poly (ADP-ribose) binding proteins and poly (ADP-ribose)-associated protein complexes. *Nucleic Acids Res.* **2008**, *36*, 6959–6976. [CrossRef]
27. Jungmichel, S.; Rosenthal, F.; Altmeyer, M.; Lukas, J.; Hottiger, M.O.; Nielsen, M.L. Proteome-wide identification of poly(ADP-Ribosyl) ation targets in different genotoxic stress responses. *Mol. Cell* **2013**, *52*, 272–285. [CrossRef]
28. Babu, M.M. The contribution of intrinsically disordered regions to protein function, cellular complexity, and human disease. *Biochem. Soc. Trans.* **2016**, *44*, 1185–1200. [CrossRef]
29. Boeynaems, S.; Alberti, S.; Fawzi, N.L.; Mittag, T.; Polymenidou, M.; Rousseau, F.; Schymkowitz, J.; Shorter, J.; Wolozin, B.; Van Den Bosch, L.; et al. Protein phase separation: A new phase in cell biology. *Trends Cell Biol.* **2018**, *28*, 420–435. [CrossRef]
30. Darling, A.L.; Liu, Y.; Oldfield, C.J.; Uversky, V.N. Intrinsically Disordered Proteome of Human Membrane-Less Organelles. *Proteomics* **2018**, *18*, 1700193. [CrossRef]

31. Jaunmuktane, Z.; Brandner, S. Invited Review: The role of prion-like mechanisms in neurodegenerative diseases. Neuropath. *Appl. Neuro.* **2019**. [CrossRef]
32. Yang, S.; Warraich, S.T.; Nicholson, G.A.; Blair, I.P. Fused in sarcoma/translocated in liposarcoma: A multifunctional DNA/RNA binding protein. *Int. J. Biochem. Cell Biol.* **2010**, *42*, 1408–1411. [CrossRef]
33. Sama, R.R.K.; Ward, C.L.; Bosco, D.A. Functions of FUS/TLS from DNA repair to stress response: Implications for ALS. *ASN Neuro* **2014**, *6*, 1759091414544472. [CrossRef]
34. Wang, H.; Hegde, M.L. New mechanisms of DNA repair defects in fused in sarcoma–associated neurodegeneration: Stage set for DNA repair-based therapeutics? *J. Exp. Neurosci.* **2019**, *13*, 1179069519856358. [CrossRef]
35. Aman, P.; Panagopoulos, I.; Lassen, C.; Fioretos, T.; Mencinger, M.; Toresson, H.; Höglund, M.; Forster, A.; Rabbitts, T.H.; Ron, D.; et al. Expression Patterns of the human sarcoma-associated genes FUS and EWS and the genomic structure of FUS. *Genomics* **1996**, *37*, 1–8. [CrossRef]
36. Gardiner, M.; Toth, R.; Vandermoere, F.; Morrice, N.A.; Rouse, J. Identification and characterization of FUS/TLS as a new target of ATM. *Biochem. J.* **2008**, *415*, 297–307. [CrossRef]
37. Wang, W.Y.; Pan, L.; Su, S.C.; Quinn, E.J.; Sasaki, M.; Jimenez, J.C.; Mackenzie, I.R.; Huang, E.J.; Tsai, L.H. Interaction of FUS and HDAC1 regulates DNA damage response and repair in neurons. *Nat. Neurosci.* **2013**, *16*, 1383. [CrossRef]
38. Rulten, S.L.; Rotheray, A.; Green, R.L.; Grundy, G.J.; Moore, D.A.; Gomez-Herreros, F.; Hafezparast, M.; Caldecott, K.W. PARP-1 dependent recruitment of the amyotrophic lateral sclerosis-associated protein FUS/TLS to sites of oxidative DNA damage. *Nucleic Acids Res.* **2014**, *42*, 307–314. [CrossRef]
39. Deng, Q.; Holler, C.J.; Taylor, G.; Hudson, K.F.; Watkins, W.; Gearing, M.; Ito, D.; Murray, M.E.; Dickson, D.W.; Seyfried, N.T.; et al. FUS is phosphorylated by DNA-PK and accumulates in the cytoplasm after DNA damage. *J. Neurosci.* **2014**, *34*, 7802–7813. [CrossRef] [PubMed]
40. Wang, H.; Guo, W.; Mitra, J.; Hegde, P.M.; Vandoorne, T.; Eckelmann, B.J.; Mitra, S.; Tomkinson, A.E.; Van Den Bosch, L.; Hegde, M.L. Mutant FUS causes DNA ligation defects to inhibit oxidative damage repair in Amyotrophic Lateral Sclerosis. *Nat. Commun.* **2018**, *9*, 3683. [CrossRef]
41. Crozat, A.; Åman, P.; Mandahl, N.; Ron, D. Fusion of CHOP to a novel RNA-binding protein in human myxoid liposarcoma. *Nature* **1993**, *363*, 640. [CrossRef] [PubMed]
42. Rabbitts, T.H.; Forster, A.; Larson, R.; Nathan, P. Fusion of the dominant negative transcription regulator CHOP with a novel gene FUS by translocation t (12; 16) in malignant liposarcoma. *Nat. Genet.* **1993**, *4*, 175–180. [CrossRef] [PubMed]
43. Law, W.J.; Cann, K.L.; Hicks, G.G. TLS, EWS and TAF15: A model for transcriptional integration of gene expression. *Brief. Funct. Genomics* **2006**, *5*, 8–14. [CrossRef]
44. Kovar, H. Dr. Jekyll and Mr. Hyde: The two faces of the FUS/EWS/TAF15 protein family. *Sarcoma* **2011**, *2011*, 837474. [CrossRef]
45. Iko, Y.; Kodama, T.S.; Kasai, N.; Oyama, T.; Morita, E.H.; Muto, T.; Okumura, M.; Fujii, R.; Takumi, T.; Tate, S.-I.; et al. Domain architectures and characterization of an RNA-binding protein, TLS. *J. Biol. Chem.* **2004**, *279*, 44834–44840. [CrossRef]
46. Lee, B.J.; Cansizoglu, A.E.; Süel, K.E.; Louis, T.H.; Zhang, Z.; Chook, Y.M. Rules for nuclear localization sequence recognition by karyopherin β2. *Cell* **2006**, *126*, 543–558. [CrossRef] [PubMed]
47. Prasad, D.D.; Ouchida, M.; Lee, L.; Rao, V.N.; Reddy, E.S. TLS/FUS fusion domain of TLS/FUS-erg chimeric protein resulting from the t (16; 21) chromosomal translocation in human myeloid leukemia functions as a transcriptional activation domain. *Oncogene* **1994**, *9*, 3717–3729. [PubMed]
48. Zinszner, H.; Sok, J.; Immanuel, D.; Yin, Y.; Ron, D. TLS (FUS) binds RNA in vivo and engages in nucleo-cytoplasmic shuttling. *J. Cell Sci.* **1997**, *110*, 1741–1750.
49. Baechtold, H.; Kuroda, M.; Sok, J.; Ron, D.; Lopez, B.S.; Akhmedov, A.T. Human 75-kDa DNA-pairing protein is identical to the pro-oncoprotein TLS/FUS and is able to promote D-loop formation. *J. Biol. Chem.* **1999**, *274*, 34337–34342. [CrossRef]
50. Perrotti, D.; Bonatti, S.; Trotta, R.; Martinez, R.; Skorski, T.; Salomoni, P.; Grassilli, E.; Lozzo, R.V.; Cooper, D.R.; Calabretta, B. TLS/FUS, a pro-oncogene involved in multiple chromosomal translocations, is a novel regulator of BCR/ABL-mediated leukemogenesis. *EMBO J.* **1998**, *17*, 4442–4455. [CrossRef]

51. Mastrocola, A.S.; Kim, S.H.; Trinh, A.T.; Rodenkirch, L.A.; Tibbetts, R.S. The RNA binding protein fused in sarcoma (FUS) functions downstream of PARP in response to DNA damage. *J. Biol. Chem.* **2013**, *288*, 24731–24741. [CrossRef] [PubMed]
52. Kwon, I.; Kato, M.; Xiang, S.; Wu, L.; Theodoropoulos, P.; Mirzaei, H.; Han, T.; Xie, S.; Corden, J.L.; McKnight, S.L. Phosphorylation-regulated binding of RNA polymerase II to fibrous polymers of low-complexity domains. *Cell* **2013**, *155*, 1049–1060. [CrossRef] [PubMed]
53. Kato, M.; Han, T.W.; Xie, S.; Shi, K.; Du, X.; Wu, L.C.; Mirzaei, H.; Goldsmith, E.J.; Longgood, J.; Pei, J.; et al. Cell-free formation of RNA granules: Low complexity sequence domains form dynamic fibers within hydrogels. *Cell* **2012**, *149*, 753–767. [CrossRef] [PubMed]
54. Shang, Y.; Huang, E.J. Mechanisms of FUS mutations in familial amyotrophic lateral sclerosis. *Brain Res.* **2016**, *1647*, 65–78. [CrossRef] [PubMed]
55. Han, T.W.; Kato, M.; Xie, S.; Wu, L.C.; Mirzaei, H.; Pei, J.; Chen, M.; Xie, Y.; Allen, J.; Xiao, G.; et al. Cell-free formation of RNA granules: Bound RNAs identify features and components of cellular assemblies. *Cell* **2012**, *149*, 768–779. [CrossRef] [PubMed]
56. Patel, A.; Lee, H.O.; Jawerth, L.; Maharana, S.; Jahnel, M.; Hein, M.Y.; Stoynov, S.; Mahamid, J.; Saha, S.; Franzmann, T.M.; et al. A liquid-to-solid phase transition of the ALS protein FUS accelerated by disease mutation. *Cell* **2015**, *162*, 1066–1077. [CrossRef]
57. Webber, C.J.; Lei, S.E.; Wolozin, B. The pathophysiology of neurodegenerative disease: Disturbing the balance between phase separation and irreversible aggregation. Dancing Protein Clouds: Intrinsically Disordered Proteins in Health and Disease. *Prog. Mol. Biol. Transl. Sci.* **2020**, *174*, 187–223.
58. Franzmann, T.M.; Alberti, S. Prion-like low-complexity sequences: Key regulators of protein solubility and phase behavior. *J. Biol. Chem.* **2019**, *294*, 7128–7136. [CrossRef]
59. King, O.D.; Gitler, A.D.; Shorter, J. The tip of the iceberg: RNA-binding proteins with prion-like domains in neurodegenerative disease. *Brain Res.* **2012**, *1462*, 61–80. [CrossRef]
60. Sun, Z.; Diaz, Z.; Fang, X.; Hart, M.P.; Chesi, A.; Shorter, J.; Gitler, A.D. Molecular determinants and genetic modifiers of aggregation and toxicity for the ALS disease protein FUS/TLS. *PLoS Biol.* **2011**, *9*, e1000614. [CrossRef]
61. Li, P.; Banjade, S.; Cheng, H.C.; Kim, S.; Chen, B.; Guo, L.; Llaguno, M.; Hollingsworth, J.V.; King, D.S.; Banani, S.F.; et al. Phase transitions in the assembly of multivalent signalling proteins. *Nature* **2012**, *483*, 336. [CrossRef]
62. Burke, K.A.; Janke, A.M.; Rhine, C.L.; Fawzi, N.L. Residue-by-residue view of in vitro FUS granules that bind the C-terminal domain of RNA polymerase II. *Mol. Cell* **2015**, *60*, 231–241. [CrossRef] [PubMed]
63. Lin, Y.; Protter, D.S.; Rosen, M.K.; Parker, R. Formation and maturation of phase-separated liquid droplets by RNA-binding proteins. *Mol. Cell* **2015**, *60*, 208–219. [CrossRef]
64. Murthy, A.C.; Dignon, G.L.; Kan, Y.; Zerze, G.H.; Parekh, S.H.; Mittal, J.; Fawzi, N.L. Molecular interactions underlying liquid− liquid phase separation of the FUS low-complexity domain. *Nat. Struct. Mol. Biol.* **2019**, *26*, 637–648. [CrossRef]
65. Wang, J.; Choi, J.M.; Holehouse, A.S.; Lee, H.O.; Zhang, X.; Jahnel, M.; Maharana, S.; Lemaitre, R.; Pozniakovsky, A.; Drechsel, D.; et al. A molecular grammar governing the driving forces for phase separation of prion-like RNA binding proteins. *Cell* **2018**, *174*, 688–699. [CrossRef] [PubMed]
66. Dormann, D.; Haass, C. Fused in sarcoma (FUS): An oncogene goes awry in neurodegeneration. *Mol. Cell Neurosci.* **2013**, *56*, 475–486. [CrossRef] [PubMed]
67. Gomes, E.; Shorter, J. The molecular language of membraneless organelles. *J. Biol. Chem.* **2019**, *294*, 7115–7127. [CrossRef] [PubMed]
68. Lin, Y.; Currie, S.L.; Rosen, M.K. Intrinsically disordered sequences enable modulation of protein phase separation through distributed tyrosine motifs. *J. Biol. Chem.* **2017**, *292*, 19110–19120. [CrossRef]
69. Nishimoto, Y.; Nakagawa, S.; Hirose, T.; Okano, H.J.; Takao, M.; Shibata, S.; Suyama, S.; Kuwako, K.-I.; Imai, T.; Murayama, S.; et al. The long non-coding RNA nuclear-enriched abundant transcript 1-2 induces paraspeckle formation in the motor neuron during the early phase of amyotrophic lateral sclerosis. *Mol. Brain* **2013**, *6*, 31. [CrossRef]
70. Yamazaki, T.; Chen, S.; Yu, Y.; Yan, B.; Haertlein, T.C.; Carrasco, M.A.; Tapia, J.C.; Zhai, B.; Das, R.; Lalancette-Hebert, M.; et al. FUS-SMN protein interactions link the motor neuron diseases ALS and SMA. *Cell Rep.* **2012**, *2*, 799–806. [CrossRef]

71. Sama, R.R.K.; Ward, C.L.; Kaushansky, L.J.; Lemay, N.; Ishigaki, S.; Urano, F.; Bosco, D.A. FUS/TLS assembles into stress granules and is a prosurvival factor during hyperosmolar stress. *J. Cell. Physiol.* **2013**, *228*, 2222–2231. [CrossRef] [PubMed]
72. Gal, J.; Zhang, J.; Kwinter, D.M.; Zhai, J.; Jia, H.; Jia, J.; Zhu, H. Nuclear localization sequence of FUS and induction of stress granules by ALS mutants. *Neurobiol. Aging* **2011**, *32*, 2323.e27–2323.e40. [CrossRef] [PubMed]
73. Dormann, D.; Rodde, R.; Edbauer, D.; Bentmann, E.; Fischer, I.; Hruscha, A.; Than, M.E.; Mackenzie, I.R.A.; Capell, A.; Schmid, B.; et al. ALS-associated fused in sarcoma (FUS) mutations disrupt Transportin-mediated nuclear import. *EMBO J.* **2010**, *29*, 2841–2857. [CrossRef] [PubMed]
74. Bosco, D.A.; Lemay, N.; Ko, H.K.; Zhou, H.; Burke, C.; Kwiatkowski, T.J.; Sapp, P.; McKenna-Yasek, D.; Brown, R.H.; Hayward, L.J. Mutant FUS proteins that cause amyotrophic lateral sclerosis incorporate into stress granules. *Hum. Mol. Genet.* **2010**, *19*, 4160–4175. [CrossRef]
75. Ito, D.; Seki, M.; Tsunoda, Y.; Uchiyama, H.; Suzuki, N. Nuclear transport impairment of amyotrophic lateral sclerosis-linked mutations in FUS/TLS. *Ann. Neurol.* **2010**, *69*, 152–162. [CrossRef]
76. Vance, C.; Scotter, E.L.; Nishimura, A.L.; Troakes, C.; Mitchell, J.C.; Kathe, C.; Urwin, H.; Manser, C.; Miller, C.C.; Hortobágyi, T.; et al. ALS mutant FUS disrupts nuclear localization and sequesters wild-type FUS within cytoplasmic stress granules. *Hum. Mol. Genet.* **2013**, *22*, 2676–2688. [CrossRef]
77. Baron, D.M.; Kaushansky, L.J.; Ward, C.L.; Sama, R.R.K.; Chian, R.-J.; Boggio, K.J.; Quaresma, A.J.C.; Nickerson, J.A.; Bosco, D.A. Amyotrophic lateral sclerosis-linked FUS/TLS alters stress granule assembly and dynamics. *Mol. Neurodegener.* **2013**, *8*, 30. [CrossRef]
78. Shelkovnikova, T.A.; Robinson, H.K.; Connor-Robson, N.; Buchman, V.L. Recruitment into stress granules prevents irreversible aggregation of FUS protein mislocalized to the cytoplasm. *Cell Cycle* **2013**, *12*, 3383–3391. [CrossRef]
79. Shelkovnikova, T.A.; Robinson, H.K.; Southcombe, J.A.; Ninkina, N.; Buchman, V.L. Multistep process of FUS aggregation in the cell cytoplasm involves RNA-dependent and RNA-independent mechanisms. *Hum. Mol. Genet.* **2014**, *23*, 5211–5226. [CrossRef]
80. Sun, S.; Ling, S.C.; Qiu, J.; Albuquerque, C.P.; Zhou, Y.; Tokunaga, S.; Li, H.; Qiu, H.; Bui, A.; Yeo, G.W.; et al. ALS-causative mutations in FUS/TLS confer gain and loss of function by altered association with SMN and U1-snRNP. *Nat. Commun.* **2015**, *6*, 6171. [CrossRef]
81. Yoshimura, A.; Fujii, R.; Watanabe, Y.; Okabe, S.; Fukui, K.; Takumi, T. Myosin-Va facilitates the accumulation of mRNA/protein complex in dendritic spines. *Curr. Biol.* **2006**, *16*, 2345–2351. [CrossRef] [PubMed]
82. Fujii, R.; Takumi, T. TLS facilitates transport of mRNA encoding an actin-stabilizing protein to dendritic spines. *J. Cell Sci.* **2005**, *118*, 5755–5765. [CrossRef] [PubMed]
83. Andersson, M.K.; Ståhlberg, A.; Arvidsson, Y.; Olofsson, A.; Semb, H.; Stenman, G.; Nilsson, O.; Åman, P. The multifunctional FUS, EWS and TAF15 proto-oncoproteins show cell type-specific expression patterns and involvement in cell spreading and stress response. *BMC Cell Biol.* **2008**, *9*, 37. [CrossRef] [PubMed]
84. Zhou, Y.; Liu, S.; Öztürk, A.; Hicks, G.G. FUS-regulated RNA metabolism and DNA damage repair: Implications for amyotrophic lateral sclerosis and frontotemporal dementia pathogenesis. *Rare Dis.* **2014**, *2*, e1003895. [CrossRef]
85. Schärer, O.D. Chemistry and biology of DNA repair. Angew. *Chem. Int. Ed.* **2003**, *42*, 2946–2974. [CrossRef]
86. Ceccaldi, R.; Rondinelli, B.; D'Andrea, A.D. Repair pathway choices and consequences at the double-strand break. *Trends Cell Biol.* **2016**, *26*, 52–64. [CrossRef]
87. Beard, W.A.; Horton, J.K.; Prasad, R.; Wilson, S.H. Eukaryotic base excision repair: New approaches shine light on mechanism. *Annu. Rev. Biochem.* **2019**, *88*, 137–162. [CrossRef]
88. Hicks, G.G.; Singh, N.; Nashabi, A.; Mai, S.; Bozek, G.; Klewes, L.; Arapovic, D.; White, E.K.; Koury, M.J.; Oltz, E.M.; et al. Fus deficiency in mice results in defective B-lymphocyte development and activation, high levels of chromosomal instability and perinatal death. *Nat. Genet.* **2000**, *24*, 175. [CrossRef]
89. Kuroda, M.; Sok, J.; Webb, L.; Baechtold, H.; Urano, F.; Yin, Y.; Chung, P.; de Rooij, D.G.; Akhmedov, A.; Ashley, T.; et al. Male sterility and enhanced radiation sensitivity in TLS−/−mice. *EMBO J.* **2000**, *19*, 453–462. [CrossRef]
90. Guirouilh-Barbat, J.; Lambert, S.; Bertrand, P.; Lopez, B.S. Is homologous recombination really an error-free process? *Front. Genet.* **2014**, *5*, 175. [CrossRef]

91. Yang, L.; Gal, J.; Chen, J.; Zhu, H. Self-assembled FUS binds active chromatin and regulates gene transcription. *Proc. Natl. Acad. Sci. USA* **2014**, *111*, 17809–17814. [CrossRef]
92. Dobbin, M.M.; Madabhushi, R.; Pan, L.; Chen, Y.; Kim, D.; Gao, J.; Ahanonu, B.; Pao, P.-C.; Qiu, Y.; Zhao, Y.; et al. SIRT1 collaborates with ATM and HDAC1 to maintain genomic stability in neurons. *Nat. Neurosci.* **2013**, *16*, 1008. [CrossRef]
93. Guerrero, E.N.; Wang, H.; Mitra, J.; Hegde, P.M.; Stowell, S.E.; Liachko, N.F.; Kraemer, B.C.; Garruto, R.M.; Rao, K.S.; Hegde, M.L. TDP-43/FUS in motor neuron disease: Complexity and challenges. *Prog. Neurobiol.* **2016**, *145–146*, 78–97. [CrossRef]
94. Takahama, K.; Takada, A.; Tada, S.; Shimizu, M.; Sayama, K.; Kurokawa, R.; Oyoshi, T. Regulation of telomere length by G-quadruplex telomere DNA-and TERRA-binding protein TLS/FUS. *Chem. Biol.* **2013**, *20*, 341–350. [CrossRef]
95. Alemasova, E.E.; Lavrik, O.I. Poly(ADP-ribosyl)ation by PARP1: Reaction mechanism and regulatory proteins. *Nucleic Acids Res.* **2019**, *47*, 3811–3827. [CrossRef]
96. Ali, S.O.; Khan, F.A.; Galindo-Campos, M.A.; Yélamos, J. Understanding specific functions of PARP-2: New lessons for cancer therapy. *Am. J. Cancer Res.* **2016**, *6*, 1842–1863.
97. Suskiewicz, M.J.; Palazzo, L.; Hughes, R.; Ahel, I. Progress and outlook in studying the substrate specificities of PARPs and related enzymes. *FEBS J.* **2020**. [CrossRef]
98. Davidovic, L.; Vodenicharov, M.; Affar, E.B.; Poirier, G.G. Importance of poly(ADP-ribose) glycohydrolase in the control of poly(ADP-ribose) metabolism. *Exp. Cell Res.* **2001**, *268*, 7–13. [CrossRef]
99. Mashimo, M.; Moss, J. Functional role of ADP-ribosyl-acceptor hydrolase 3 in poly(ADP-ribose) polymerase-1 response to oxidative stress. *Curr. Protein Pept. Sc.* **2016**, *17*, 633–640. [CrossRef]
100. Kamaletdinova, T.; Fanaei-Kahrani, Z.; Wang, Z.Q. The Enigmatic Function of PARP1: From PARylation Activity to PAR Readers. *Cells* **2019**, *8*, 1625. [CrossRef]
101. Singatulina, A.S.; Hamon, L.; Sukhanova, M.V.; Desforges, B.; Joshi, V.; Bouhss, A.; Lavrik, O.I.; Pastré, D. PARP-1 activation directs FUS to DNA damage sites to form PARG-reversible compartments enriched in damaged DNA. *Cell Rep.* **2019**, *27*, 1809–1821. [CrossRef] [PubMed]
102. Martin-Hernandez, K.; Rodriguez-Vargas, J.M.; Schreiber, V.; Dantzer, F. Expanding functions of ADP-ribosylation in the maintenance of genome integrity. *Semin. Cell Dev. Biol.* **2017**, *63*, 92–101. [CrossRef]
103. Khodyreva, S.N.; Lavrik, O.I. Poly(ADP-Ribose) polymerase 1 as a key regulator of DNA repair. *Mol. Biol.* **2016**, *50*, 580–595. [CrossRef]
104. Lenzken, S.C.; Levone, B.R.; Filosa, G.; Antonaci, M.; Conte, F.; Kizilirmak, C.; Mühlemann, O. FUS-dependent phase separation initiates double-strand break repair. *bioRxiv* **2019**, 798884. [CrossRef]
105. Sleeman, J.E.; Trinkle-Mulcahy, L. Nuclear bodies: New insights into assembly/dynamics and disease relevance. *Curr. Opin. Cell Biol.* **2014**, *28*, 76–83. [CrossRef]
106. Erdel, F.; Rippe, K. Formation of chromatin subcompartments by phase separation. *Biophys. J.* **2018**, *114*, 2262–2270. [CrossRef]
107. Yoshizawa, T.; Nozawa, R.S.; Jia, T.Z.; Saio, T.; Mori, E. Biological phase separation: Cell biology meets biophysics. *Biophys. Rev.* **2020**, *12*, 519–539. [CrossRef]
108. Yoo, H.; Triandafillou, C.; Drummond, D.A. Cellular sensing by phase separation: Using the process, not just the products. *J. Boil. Chem.* **2019**, *294*, 7151–7159. [CrossRef]
109. Murakami, T.; Qamar, S.; Lin, J.Q.; Schierle, G.S.; Rees, E.; Miyashita, A.; Costa, A.R.; Dodd, R.B.; Chan, F.T.; Michel, C.H.; et al. ALS/FTD Mutation-induced phase transition of FUS liquid droplets and reversible hydrogels into irreversible hydrogels impairs RNP granule function. *Neuron* **2015**, *88*, 678–690. [CrossRef]
110. Murray, D.T.; Kato, M.; Lin, Y.; Thurber, K.R.; Hung, I.; McKnight, S.L.; Tycko, R. Structure of FUS protein fibrils and its relevance to self-assembly and phase separation of low-complexity domains. *Cell* **2017**, *171*, 615–627. [CrossRef]
111. Niaki, A.G.; Sarkar, J.; Cai, X.; Rhine, K.; Vidaurre, V.; Guy, B.; Hurst, M.; Lee, J.C.; Koh, H.R.; Guo, L.; et al. Loss of dynamic RNA interaction and aberrant phase separation induced by two distinct types of ALS/FTD-Linked FUS mutations. *Mol. Cell* **2020**, *77*, 82–94. [CrossRef]
112. Bentmann, E.; Neumann, M.; Tahirovic, S.; Rodde, R.; Dormann, D.; Haass C. Requirements for stress granule recruitment of fused in sarcoma (FUS) and TAR DNA-binding protein of 43 kDa (TDP-43). *J. Biol. Chem.* **2012**, *287*, 23079–23094. [CrossRef]

113. Weber, S.C. Evidence for and against liquid-liquid phase separation in the Nucleus. *Non-Coding RNA* **2019**, *5*, 50.
114. Sukhanova, M.V.; Abrakhi, S.; Joshi, V.; Pastre, D.; Kutuzov, M.M.; Anarbaev, R.O.; Pastre, D.; Lavrik, O.I. Single molecule detection of PARP1 and PARP2 interaction with DNA strand breaks and their poly (ADP-ribosyl) ation using high-resolution AFM imaging. *Nucleic Acids Res.* **2016**, *44*, e60. [CrossRef]
115. Pleschke, J.M.; Kleczkowska, H.E.; Strohm, M.; Althaus, F.R. Poly (ADP-ribose) binds to specific domains in DNA damage checkpoint proteins. *J. Biol. Chem.* **2000**, *275*, 40974–40980. [CrossRef]
116. Fisher, A.E.; Hochegger, H.; Takeda, S.; Caldecott, K.W. Poly (ADP-ribose) polymerase 1 accelerates single-strand break repair in concert with poly(ADP-ribose) glycohydrolase. *Mol. Cell Biol.* **2007**, *27*, 5597–5605. [CrossRef]
117. Moor, N.A.; Vasil'eva, I.A.; Kuznetsov, N.A.; Lavrik, O.I. Human apurinic/apyrimidinic endonuclease 1 is modified in vitro by poly (ADP-ribose) polymerase 1 under control of the structure of damaged DNA. *Biochimie* **2020**, *168*, 144–155. [CrossRef]
118. Liu, C.; Vyas, A.; Kassab, M.A.; Singh, A.K.; Yu, X. The role of poly ADP-ribosylation in the first wave of DNA damage response. *Nucleic Acids Res.* **2017**, *45*, 8129–8141. [CrossRef]
119. Naumann, M.; Pal, A.; Goswami, A.; Lojewski, X.; Japtok, J.; Vehlow, A.; Naujock, M.; Günther, R.; Jin, M.; Stanslowsky, N. Impaired DNA damage response signaling by FUS-NLS mutations leads to neurodegeneration and FUS aggregate formation. *Nature Commun.* **2018**, *9*, 335. [CrossRef]
120. McGurk, L.; Rifai, O.M.; Bonini, N.M. Poly(ADP-ribosylation) in age-related neurological disease. *Trends Genet.* **2019**, *35*, 601–613. [CrossRef]

International Journal of
Molecular Sciences

MDPI

Review

Inhibition of DNA Repair in Cancer Therapy: Toward a Multi-Target Approach

Samuele Lodovichi [1], Tiziana Cervelli [2], Achille Pellicioli [1,*] and Alvaro Galli [2,*]

1 Bioscience Department, University of Milan, Via Celoria 26, 20131 Milan, Italy; 87samuele@gmail.com
2 Yeast Genetics and Genomics Group, Laboratory of Functional Genetics and Genomics, Institute of Clinical Physiology CNR, Via Moruzzi 1, 56125 Pisa, Italy; tizicerv@ifc.cnr.it
* Correspondence: achille.pellicioli@unimi.it (A.P.); alvaro.galli@ifc.cnr.it (A.G.)

Received: 4 August 2020; Accepted: 8 September 2020; Published: 12 September 2020

Abstract: Alterations in DNA repair pathways are one of the main drivers of cancer insurgence. Nevertheless, cancer cells are more susceptible to DNA damage than normal cells and they rely on specific functional repair pathways to survive. Thanks to advances in genome sequencing, we now have a better idea of which genes are mutated in specific cancers and this prompted the development of inhibitors targeting DNA repair players involved in pathways essential for cancer cells survival. Currently, the pivotal concept is that combining the inhibition of mechanisms on which cancer cells viability depends is the most promising way to treat tumorigenesis. Numerous inhibitors have been developed and for many of them, efficacy has been demonstrated either alone or in combination with chemo or radiotherapy. In this review, we will analyze the principal pathways involved in cell cycle checkpoint and DNA repair focusing on how their alterations could predispose to cancer, then we will explore the inhibitors developed or in development specifically targeting different proteins involved in each pathway, underscoring the rationale behind their usage and how their combination and/or exploitation as adjuvants to classic therapies could help in patients clinical outcome.

Keywords: cell cycle checkpoint; DNA repair; cancer therapy; DNA repair inhibitors; synthetic lethality

1. Introduction

Cells ability to faithfully repair DNA from insults from endogenous or exogenous sources is essential to maintain viability. Different kinds of damage arise from different sources and require specific DNA repair pathways, allowing recovery of the original sequence. Damage left unrepaired could be inherited after cell division, causing permanent genetic alteration. Accumulation of these mutations leads to cell senescence or apoptosis and may predispose to cancer development.

A complex network of sensors and effectors regulates DNA repair, which starts from the recognition of the damage on DNA to the choice of the most feasible repair pathway, depending on the type of damage and cell cycle phase. Among the various DNA repair pathways, we can distinguish: the base excision repair (BER) and the nucleotide excision repair (NER) primarily involved in removing small, non-helix-distorting base lesions from the genome and to repair bulky helix-distorting lesions and inter-strands crosslinks in G0/G1 phase, respectively [1,2]; the Fanconi anemia (FA) pathway acting during DNA replication for the repair of inter-strand crosslinks [3]; the mismatch repair (MMR) which is a system recognizing and repairing erroneous insertion, deletion, and mis-incorporation of bases [4]; the translesion DNA synthesis (TLS) which is a process that allows DNA replication machinery to replicate past DNA lesions by switching to specialized DNA polymerases [5]; the error-prone non-homologous end joining (NHEJ), which repairs DNA double strand breaks (DSBs) without the requirement of complementary strand [6]; the homologous recombination (HR) which faithfully repairs

DSBs taking advantage of the sequence present on the complementary strand of the homologous chromosome or chromatid, acting in S and G2 cell cycle phases [7]. Moreover, DSBs can also be repaired by the annealing of short (1-few bp) or long (>100 bp) complementary sequences through the highly mutagenic MMEJ (microhomology-mediated end-joining, also known as alternative non-homologous end-joining or Alt-NHEJ), and single strand annealing (SSA), respectively [8,9]. Despite the specificity of each pathway for particular DNA damage and cell cycle phase, none of them are mutually exclusive and they form a network that involves proteins determining the repair outcomes. Mutations affecting the numerous players of these mechanisms could lead to an accumulation of unfaithfully repaired DNA, thus predisposing to cancer development. Moreover, alterations in processes that support genome integrity in normal cells, allow cancer cells to acquire aggressiveness and facilitate the emergence of resistance to DNA damaging cancer treatments [10]. Nevertheless, understanding the emerging role of DNA repair pathway deficiency in cancer allows the development of specific drugs targeting the remaining functional pathways, on which cancer cells are strongly dependent for survival and proliferation. This kind of approach is based on the concept of synthetic lethality, which is a genetic interaction between two genes. If the mutation occurs in only one of the two genes, cell viability is not affected; whereas when two genes are simultaneously mutated, cell death occurs [11]. The discovery of many synthetic lethal interactions between proteins involved in DNA repair allowed the development of personalized therapeutic treatments that target specific DNA repair enzymes to kill cancer cells.

Thanks to the advancement of next-generation sequencing technologies, we now have a clearer indication of which DNA repair-related genes are mutated in each specific cancer: for example, some breast and ovarian cancers are characterized by mutations in BRCA1/BRCA2, while alterations in MMR genes such as MSH2 or MLH1 are related to colorectal cancer [12]. Putting this information together with new synthetic lethality interactions, discovered by novel CRISPR/Cas9 screening approaches in human cells [13,14], will allow the development of the most effective therapeutic approaches by production of drugs targeting DNA repair enzymes specific for each pathological situation.

In this review, we will give a brief introduction to DNA damage checkpoints and repair focusing on different players involved in each pathway and describe how their alterations could impact cancer development and aggressiveness. Then, we will describe how these players can be targeted in cancer therapy and the most encouraging drugs under study, focusing on the rationale behind their usage to improve efficacy of classic therapy or induce specific synthetic lethal interactions.

2. DNA Damage Checkpoint Promotes DNA Damage Recognition and Cell Cycle Arrest: The First Barrier to Cancer Progression

When DNA damage arises, a complex network of signaling cascades named DNA damage checkpoint (DDC) are activated to recognize DNA damage and arrest the cell cycle, thus ensuring enough time to cells to repair the lesion or eventually leading to senescence/apoptosis. These processes are tightly regulated to guarantee cell survival by promoting genomic stability and reducing the possibility that lesions are inherited, after cell division. Similar to other cellular signaling cascades, DDC is driven by protein phosphorylation and at its core there are three upstream kinases: ATM, ATR, and DNA-PKs [15] (Figure 1); moreover, recent evidence also suggests that non-coding (nc)RNA molecules and the RNA interference (RNAi) factors Drosha and Dicer have a role in this mechanism [16,17].

Each kinase is recruited to DNA lesions by specific co-factors: ATM is recruited by the C terminus of NBS1 a component of the MRE11-RAD50-NBS1 (MRN) complex, DNA-PKs is recruited by Ku80/Ku70 heterodimer and ATR is recruited by its stable binding partner ATRIP-RPA. In general, the first two kinases are involved in the initial recognition of DSBs, then, ATM promotes either HR or NHEJ, while DNA-PK activity is restricted to NHEJ. Instead, ATR is recruited to all kind of lesions that generate ssDNA coated by replication protein A (RPA) such as: DSBs after end resection, bulky lesions when the damaged strand is excised by NER and stalled replication forks when the double helix is opened and/or processed by nucleases. ATR is also involved in the recognition of particular DNA damage such as UV-damaged DNA [18]. Following DNA damage recognition, all the three kinases determine cell cycle arrest by interacting with specific substrates (Figure 1). ATR and ATM phosphorylate and activate the tumor suppressor checkpoint kinase 1 (CHK1). CHK1 regulates the G2/M checkpoint by activating kinase WEE1, which in turn phosphorylates the Cyclin-dependent kinase 1 (CDK1) reducing its activity and preventing entry into mitosis; CHK1 also regulates S phase checkpoint by promoting CDC25A phosphatase degradation, whose activity is essential to remove inhibitory phosphate groups from kinases CDK4 and CDK2 and guarantee cell cycle progression [19].

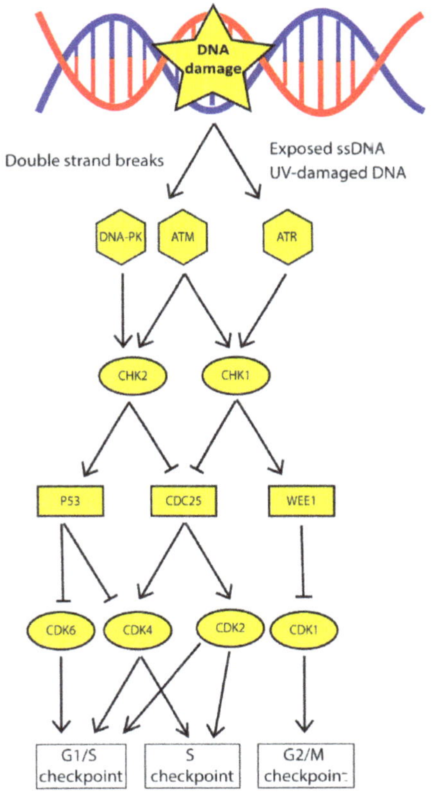

Figure 1. DDC signaling cascade determines arrest in specific phases of the cell cycle. ATM, DNA-PK, and ATR phosphorylate the downstream kinases CHK2 and CHK1; in turn they activate P53 and WEE1 that, by inhibiting CDK6/4 and CDK1 determine arrest in G1 and G2/S, respectively. CHK1 and CHK2 also inhibit CDC25 phosphatase that does not remove inhibitory phosphate groups from CDK2/4, therefore determining arrest in the S phase.

Due to its activity among replication fork collapse, ATR-CHK1 axis protects cells from replication stress and mitotic catastrophe that could arise after uncontrolled division [20]. ATM and DNA-PK also phosphorylate the tumor suppressor checkpoint kinase 2 (CHK2); CHK2 stabilizes P53, which inhibits CDK6 and CDK4 leading to cell cycle arrest in G1/S phase [21].

Thanks to its inhibition of cell cycle progression, the activation of DDC is considered the first barrier to tumor progression and the loss of one or more DDC-involved proteins determines accumulation of un-repaired DNA lesions which greatly contribute to cancer initiation and progression. Indeed, DDC helps cells recover from DNA injuries and overexpression of its players is frequent in cancer resistant to DNA damaging agents.

ATM is a well-recognized tumor suppressor and its alterations are common in several cancers including breast, gastric, colorectal, and prostate cancer [22]. Mutations in DNA-PK have been found associated to several cancers such as prostate cancer and melanoma [23,24]. Finally, ATR mutations promote development of melanoma and oropharyngeal cancer syndrome [25,26].

After initial recognition of DNA damage and cell cycle arrest, cells undergo specific DNA repair pathway depending on the cell cycle phase and type of damage.

3. Defects in DSB Repair Greatly Promotes Cancer Development

DSBs are fatal DNA lesions that must be quickly repaired to prevent dangerous chromosomal rearrangements events. Several DSBs sensors must be located on the lesion to recognize it and activate downstream effectors. The initial proteins recruited to the DSBs are the Poly (ADP-ribose) polymerase 1 (PARP1) and Ku70/Ku80 complex: the first one catalyzes the formation and attachment of mono and/or poly (ADP-ribose) polymers to itself and target proteins, promoting chromatin decondensation and allowing binding of successive players; the second protein recognizes and binds to the DSBs ends recruiting DNA-PKs and NHEJ factors [27,28] (Figure 2).

Following this initial recognition, the MRN complex, composed of RAD50, NBS1, and MRE11, binds to the lesion and recruits ATM that promotes the checkpoint arrest. ATM also phosphorylates the histone H2AX, producing gamma-H2AX (γH2AX), which is essential for the recruitment of multiple factors, including the mediator of DNA damage checkpoint 1 (MDC1). This is accompanied by simultaneous accumulation of E3 ubiquitin-protein ligase RNF8, which further promotes chromatin relaxation allowing recruitment of additional DNA repair players [29].

At this point, the choice between HR and NHEJ is strongly dependent on cell cycle phase and resection process. In G1, DSB end resection is limited by several negative regulators, such as the Ku70/Ku80 heterodimer that is steadily bound to the broken ends of the DSB, the tumor suppressor p53-binding protein 1 (TP53BP1) that acts as a scaffold to regulate a network of proteins that prevent resection and several other factors that act as barriers [30]; activity of these proteins promote NHEJ [31,32].

Conversely, in S and G2 phases, HR is favored by DSB end resection stimulators mainly by CtBP-interacting protein (CtIP) in association with BRCA1 and MRN complex; CtIP recruitment relies on ATR and on the activity of E3 ubiquitin-protein ligase RNF138, which also promotes ubiquitination of Ku80 leading to its removal from DSB ends [33,34].

In HR (Figure 2A), the endonuclease activity of MRE11 nicks the strand several nucleotides away from the break and then resects the DNA towards the DSB; this "short-range" resection activity requires interaction with CtIP [35]. This process is thought to displace Ku70/Ku80 complex allowing access of "long range" resection factors, such as endonucleases DNA2 and EXO1, the Bloom syndrome helicase (BLM) and Werner syndrome ATP-dependent helicase (WRN) [36]. The ssDNA formed is rapidly coated by the replication protein A (RPA) that protects it from degradation; afterwards, RPA displacement is mediated by BRCA2 and PALB2 that also promote RAD51 binding to the ssDNA. Then, RAD51 initiates the search for homologous sequences and invasion of the complementary strand; this activity is stimulated by RAD52 as well, which also mediates the DNA-DNA interaction necessary for annealing of the complementary DNA strands [37–39]. After strand invasion, the replicative DNA polymerases (POL δ, ε) or translesion DNA polymerases (POL η, κ) extend the DNA strand generating a displacement loop (D-loop). At this point, D-loop structures can be solved by three pathways: double-strand break repair (DSBR), synthesis-dependent strand annealing (SDSA) or break induced replication (BIR); the first allowing the formation of crossover and non-crossover products, the second allowing only the formation of non-crossover and the third generating half-crossover products with loss of heterozygosis (LOH), frequently promoting mutagenesis [40]. It has been recently shown that TP53BP1 has also a role in regulating the last steps of HR by limiting helicases activity that prevents D-loop stability, therefore favoring crossover and BIR events [41].

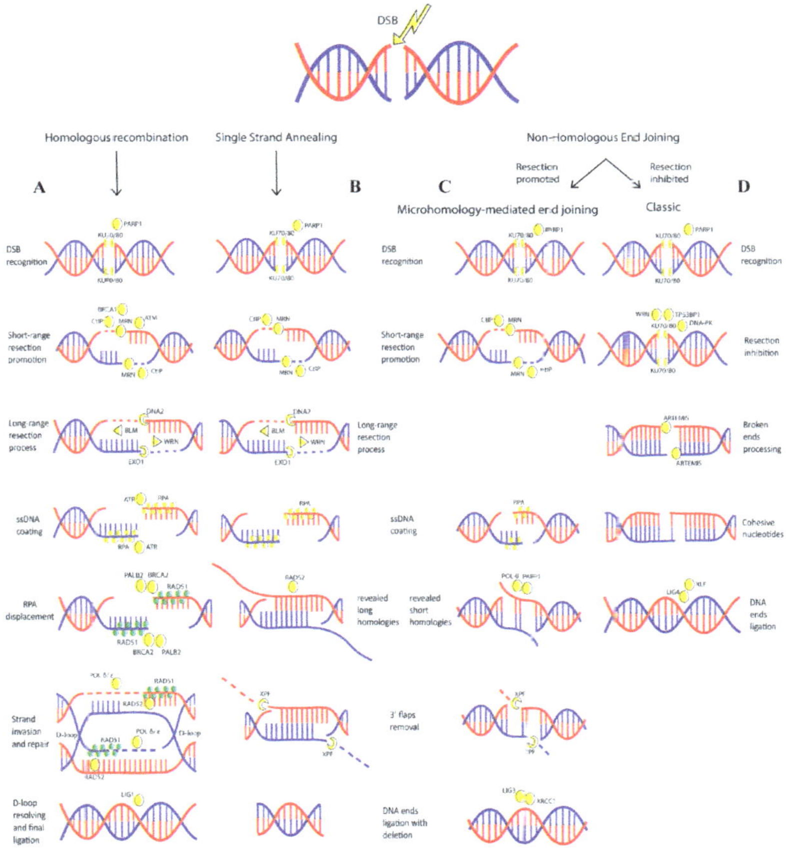

Figure 2. Several DSB repair pathways can be used depending on cell cycle phase and presence of homology on DNA strand. (**A**) In the S/G2 phase, when the homologous chromatid is present, HR error-free DSB repair pathway is favored: initial resection performed by the combined action of CtIP, MRN complex and BRCA1, promotes this pathway and allow access of long-range resection nucleases DNA2 and EXO1. WRN and BLM helicases unwind the two DNA strands; the ssDNA formed is rapidly stabilized by RPA and bound by ATR. Thereafter, RPA is displaced by RAD51 aided by BRCA2/PALB2 activity. RAD51 promotes search and invasion of complementary strand and RAD52 determines interaction between the complementary strands; finally, POL δ/ε re-synthetizes the damaged strand and LIG1 performs the final ligation. (**B**) Similar to HR, SSA requires long range resection of the DSB, but differently from HR this resection process exposes long regions of homology; DNA strands can bind together, also favored by the activity of RAD52, the overhanging 3′ flaps are removed by XLF nucleases and the DNA ends are ligated resulting in long deletion. (**C**) MMEJ pathway has several steps in common with SSA; the differences are that there is no long-range resection and that homology between strands are short (few base pairs). Short homology is revealed by the activity of Pol θ-associated helicase together with PARP1 and, as in SSA, the 3′flaps are removed by XLF nuclease; in the last step, XRCC1 and LIG3 catalyze DNA ends ligation resulting in short deletion. (**D**) c- NHEJ is favored by resection inhibition performed by Ku70/80 heterodimer and several other factors modulated by TP53BP1. Ku70/80 activates DNA-PK kinase which recruits ARTEMIS nucleases to the DSB; this promotes the processing of the broken ends until cohesive nucleotides are found, then XLF and LIG4 catalyze the final ligation.

In classic NHEJ (c-NHEJ) (Figure 2D), the binding of Ku70/Ku80 heterodimer to the DSB ends, together with TP53BP1, WRN helicase and several other barriers prevents end resection and recruits DNA-PK kinase. DNA-PK recruits the endonuclease ARTEMIS, which processes the broken ends until it finds cohesive nucleotides. In the last step, NHEJ factor 1 (XLF) interacts with DNA ligase 4 (LIG4) to catalyze the DSB ligation. Alternatively, DSBs can be repaired by SSA or MMEJ (Figure 2B,C) that, depending on the extension of end resection, results in DNA deletion. Both MMEJ and SSA also require ATM signaling [42] In SSA the resection process reveals flanking homologous sequences (>100 bp) that are annealed together by RAD52 and any gaps are filled by DNA polymerases; in MMEJ the Pol θ-associated helicase functions together with PARP1 to displace RPA from ssDNA, revealing short internal microhomologies (few bps) on the ssDNA ends and stabilizing their interaction. Then, the MRN complex recruits X-ray repair cross-complementing protein 1 (XRCC1) which form a complex with DNA ligase 3 (LIG3) to catalyze DNA ends ligation [43]. The choice between NHEJ and MMEJ is also dependent on WRN, which suppresses the recruitment of MRE11 and CtIP on the DSBs, thus promoting c-NHEJ [44]. Another critical step in MMEJ and SSA is the removal of the non-homologous 3′ ssDNA tail, which is mediated mainly by the XPF flap nucleases [45].

Considering the importance of DSB repair, alteration of the players is frequently related to cancer predisposition and development.

Accordingly, a comprehensive analysis across 33 cancer types identified HR pathway as the most frequently altered DNA repair pathway, particularly in ovarian cancer [46]. The most frequent mutated HR genes are *BRCA1* and *BRCA2*, followed by *RAD51*, *BLM*, and *RAD50* [46,47]. Germline mutations in *BRCA1* and *BRCA2* are related to the majority of hereditary breast and ovarian cancer (HBOC); however, beside these genes, several other low penetrance genes responsible of HBOC have been identified, such as *PALB2* and *RAD51* [48], and also mutations in genes coding for the components of the MRN complex [49,50].

Besides HBOC, mutations or alterations of HR-related genes are responsible for predisposition to other cancer types; for example, *RAD51* overexpression has been associated with poor prognosis in patients with solid malignancies [51]; mutations of *MRE11* are related to sporadic gastric cancer and neuroblastoma [52,53], while *RAD50* mutations are related to leukemia and endometrial carcinoma [54,55]; moreover, *NBS1* has emerged as a prostate and lung cancer-susceptibility gene [56,57].

Regarding cancer predisposition caused by alteration of c-NHEJ, TP53BP1 is the most interesting protein; when this protein is downregulated, it determines resistance to PARP inhibitors in breast and ovarian BRCA1 deficient cancer and to chemotherapeutic agents in colorectal cancer cells by reducing the protein level of the ATM-CHK2 pathway [58,59]. In general, by playing a pivotal role in the choice of DSB repair pathway, it has been extensively demonstrated that aberrant expression of the TP53BP1 protein contributes to tumor development [60].

Beside this, so far, few cancers are associated with the downregulation or alteration of genes involved in c-NHEJ. Only rare mutations of genes encoding for Ku70/Ku80, LIG4, ARTEMIS and XLF have been found in colon and endometrial cancer [24].

MMEJ and SSA are intrinsically mutagenic, generating deletions and causing genomic instability found in many human cancers. Initial DSB end resection is favored by CtIP, suggesting that this gene has oncogenic potential by promoting these pathways; accordingly, CtIP inactivation suppresses mammary tumorigenesis caused by p53 deficiency in mouse model [61]. Moreover, MMEJ relies on Pol-θ and elevated *POLQ* expression (encoding for Pol-θ) has been described in numerous cancer types, including breast and ovarian cancer [62].

Finally, having a role either in HR and NHEJ, alterations of DNA helicases such as: BLM, WRN, and REQLs predispose to tumorigenesis, in general their upregulation is involved in cell proliferation and resistance to DNA damaging agents, conversely their downregulation leads to genomic aberrations [63]. Particularly, *DNA2* alterations are related to genome instability by an enhanced end resection activity, caused by its overexpression in early stages of cancer [64]. Mutations of the *WRN* gene cause Werner syndrome, which is characterized by genetic instability and hematological disease [65], this helicase is often highly expressed in chronic myeloid leukemia determining increased cell survival through NHEJ [66].

4. Single Base Variations and Mismatched Base Pairs Are Repaired by Specific Pathways Linked to Cancer

BER is a pathway involved in repair of small base lesions such as oxidation, alkylation, and deamination (Figure 3A). The steps of the mechanism are: DNA damage recognition, excision of the base by a DNA glycosylase to generate an abasic (AP) site, and cleavage of the AP site by the AP endonuclease APE1 to form a DNA single strand break (SSB) [67]. When the SSB is formed, one of the first proteins activated is PARP1, which, acting as a SSB sensor, promotes recruitment of other enzymes involved in the repair process, such as XRCC1 [68,69]. Finally, the gap is filled by DNA polymerase β and the complex XRCC1 and LIG3 seals the nick, or as an alternative, two to 10 nucleotides are removed, a new nucleotide chain is synthetized by DNA polymerase and the final ligation step is performed by LIG1 (Figure 3A).

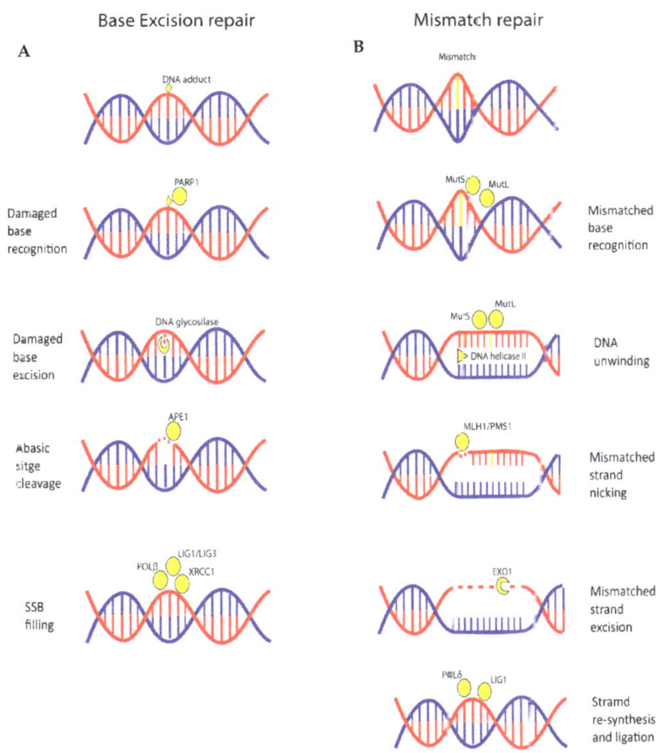

Figure 3. Small base lesions are repaired by base excision repair; mismatched base pairs and insertion/deletion loops are corrected by mismatch repair pathway. (**A**) DNA lesion is recognized by PARP1 which recruits a DNA glycosylase that removes the damaged base; then, APE1 digests the abasic site determining a SSB that is repaired by the combined action of LIG1/3, XRCC1 and Pol-β. (**B**) The mismatched base is recognized by the complexes MutS and MutL that, in turn, recruit the DNA helicases II to unwind the DNA strands; thereafter, the complex formed by MLH1/PMS1 digests DNA several bases from the mismatch, forming a nick. The EXO1 nuclease digests the DNA from the nick towards the mismatch; finally, the removed strand is re-synthetized by Pol-δ and ligated by LIG1.

MMR is a pathway responsible for correcting mismatched base pairs and insertion/deletion loops (IDLs) that may occur during DNA replication [70] (Figure 3B). The central players are MutS proteins that form two heterodimers, MutSα (MSH2-MSH6) and MutSβ (MSH2-MSH3), and MutL proteins that form three heterodimers, MutLα (MLH1-PMS2), MutLβ (MLH1-MLH3), and MutLγ (MLH1-PMS1). MutS dimer recognizes the mismatched base on the daughter strand and binds the damaged DNA; MutL recruits the DNA Helicase II to separate the two strands. Then, the entire complex slides along the DNA, unwinding the strand that must be excised. Thereafter, the endonuclease activity of MLH1-PMS1, activated by the PCNA clamp produces a nick of the DNA strand [71]. Subsequently, the nicked DNA strand with the mismatch is excised by EXO1. Finally, POL δ synthesizes a new fragment and DNA ligase I (LIG1) catalyzes strand ligation [72].

There are few relations between BER and cancer; alterations in this pathway have been associated with colon and breast cancer [73,74]. Particularly, Pol-β expression seems to have a protective role against breast and lung carcinomas by impairing cancer cell metastasis due to an increased DNA de-methylation [75]; accordingly, Pol-β deficiency is associated with aggressive breast cancer [76]. Moreover, variants of this protein affecting its fidelity are associated to prostate and colon cancer [77,78] and could drive cellular transformation leading to cancer onset [79]. Instead, germline mutations in MMR genes are associated with hereditary non-polyposis colorectal cancer (HNPCC), also known as Lynch Syndrome, an autosomal dominant disease. Moreover, mutations in all the MMR genes, *MLH1* and *MSH2, PMS2* and *MSH6* have been found related to breast cancer susceptibility [50,80–84].

5. The DNA Crosslinks Repair Deals with NER and FA Pathways with Implication for Cancer

Inter-strand crosslinks (ICLs) covalently link the two strands of the DNA double helix and they arise from exposure to chemicals such as DNA damaging agents used in cancer therapy.

In quiescent cells (G0/G1), an ICL is recognized by NER pathway, which is also involved in the repair of intra-strand crosslinks and UV-damage [85] (Figure 4). In NER, two mechanisms of DNA damage sensing are known: one pathway recognizes damage all over the genome, and the other one recognizes damage in the transcribed strand of active genes that cause blockage of the RNA polymerase II. In the first case, detection is performed by the binding of the XPC complex (XPC, HR23B and CENT2) to the non-damaged strand, assisted by PARP1 activity [86]. In the second case, arrest of RNA polymerase II is the initial signal that promotes recruitment of the Cockayne syndrome group A (CSA) and group B (CSB) proteins. At this point, both pathways require XPA that acts as a scaffold organizing the other components of the pathway [87] and the transcriptional factor IIH complex (TFIIH) which first unwinds the DNA around the damage by the helicase activity of XPB and XPD. Then, DNA damage is removed by the excision repair protein ERCC-1-XPF and XPG endonuclease activity, generating a gap that is filled by DNA polymerase and by the DNA ligase activity of XRCC1-LIG3 or LIG1 [88].

In S phase, ICL lesions cause stalling of DNA replication forks that are recognized by the FA pathway (Figure 4). The anchor complex is the one recognizing the stalled replication fork; when activated, the anchor complex recruits a core complex which attaches a single ubiquitin to both FANCD2 and FANCI that together form the ID2 complex. This ubiquitination is essential for the activation of the nucleases involved in the unhooking of the DNA strands that is carried out by SLX4 and XPF [89], and also for the recruitment of the TLS polymerases that replace the stalled DNA polymerases and bypass the lesion via TLS; this pathway is promoted by ubiquitination of TLS regulator PCNA, which is performed by FANCD2 [90]. Among TLS polymerases, Pol ξ, is the one involved in the lesion bypass step of ICL, while REV1 facilitates polymerase switching and coordinates extension steps [91]. FANCD2 deubiquitination is then performed by deubiquitinating complex USP1/UAF1. Finally, after DNA unhooking, the downstream players of the pathway coordinate HR to repair the DSB formed; these players are FANCS (BRCA1), FANCD1 (BRCA2), FANCN (PALB2), and FANCO (RAD51); they are referred to as downstream components of the FA pathway since their activation depends on the mono-ubiquitination step (Figure 4) [92].

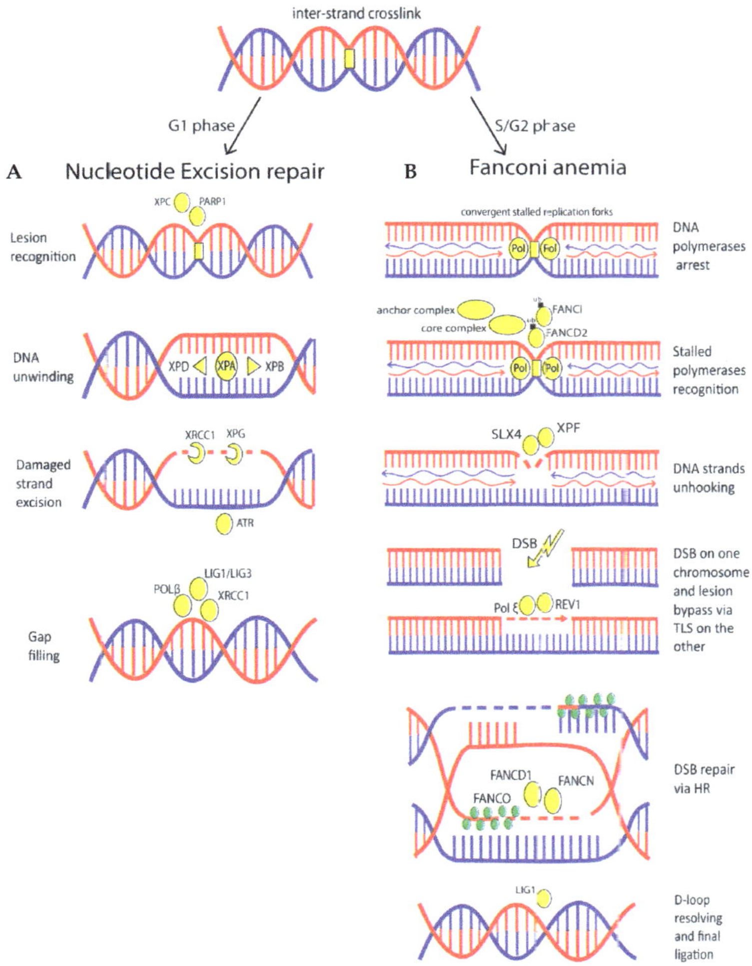

Figure 4. Bulky lesions are repaired by nucleotide excision repair, inter-strand crosslinks formed in S/G2 phase, are repaired by Fanconi Anemia pathway. (**A**) In NER, the bulky lesion is recognized by PARP1 and XPC: they recruit XPA that acts as a scaffold organizing the helicases XPB and XPD that unwind the DNA around the damage. Then, several bases are removed by ERCC-1-XPF and XPG endonuclease; the gap generated is filled by Pol-β and ligated by XRCC1, LIG3/LIG1. (**B**) Convergent arrested-replication forks are recognized by the anchor complex; this recruits the core complex which attaches a single ubiquitin residue to FANCI and FANCD2. After, SLX4 and XPF nucleases unhook the DNA strand determining a DSB on one strand. The strand without the DSB is repaired by TLS by Pol ξ assisted by REV1; the other strand is repaired via HR by FANCS (BRCA1), FANCD1 (BRCA2), FANCN (PALB2) and FANCO (RAD51).

Germline mutations of the FA-related genes determine development of FA which is a genetic disease causing an impaired response to DNA damage. The majority of the affected people develop cancer, most often leukemia [93]. FA disease can develop also by mutations of HR-related genes such as *BRCA1*, *BRCA2*, and *PALB2*; hence, the FA pathway is often called the FA/HR pathway. Alterations of FA-related proteins are also associated to breast and ovarian cancer development; mutations of

FANCM, a component of the anchor complex, and FANCA, which is part of the core complex, are both correlated to increased risk of breast cancer [94].

6. Classic Anti-Cancer Therapy Relies on DNA Damage Induction to Which Cancer Cells Are More Sensitive

Cancer cells deficient in DNA damage checkpoint and repair pathways are particularly sensitive to DNA damage upon which chemo and radiation cancer therapies rely.

In chemotherapy, the most common agents used are alkylating agents, which cause mainly intra-strand and few inter-strand crosslinks by adding alkyl groups to DNA [95]. Currently, platinum-based chemotherapeutic drugs, called "alkylating-like", are more frequently used because they form DNA crosslinks without having an alkyl group. Another group of DNA damaging agents used are inhibitors of topoisomerases, affecting the activity of two enzymes catalyzing the breaking and rejoining of the phosphodiester backbone of DNA strands during the normal cell cycle, topoisomerase I and topoisomerase II. Topoisomerase I inhibitors stabilize the cleavable complex of topoisomerases I, preventing DNA relegation and thus inducing DNA strand breaks; some examples are topotecan and camptothecin. Drugs against topoisomerase II consist of chemicals which target the complex DNA-topoisomerases II and promote cleavage activity or prevent re-ligation of the DNA, and inhibitors that reduce the turn-over of the enzyme; some examples are etoposide and doxorubicin [96].

A third group are cytotoxic antibiotics that act by different mechanisms of action and share the ability to interrupt cell division. Among this group, important examples are bleomycin and mitomycin C that damage DNA by producing free radicals and DNA inter-strand crosslinks, respectively [97].

Anti-metabolites are a group of molecules with a structure similar to nucleotides, but with altered chemical groups. These drugs exert their effect by either blocking the enzymes required for DNA synthesis or becoming incorporated into DNA and promoting DNA damage; some examples are capecitabine and gemcitabine [98].

Radiotherapy induces DNA damage using ionizing radiation (IR), either directly by ionizing the DNA strands or indirectly by producing free radicals as a result of ionized water molecules [99].

These treatments have proven to be effective against cancer, but resistance development is a common problem. Various factors drive drug resistance, such as increase of drug efflux, over-expression of DNA repair proteins or inhibition of proteins involved in apoptotic pathway [100]. To overcome this, it is now evident that combination with specific drugs targeting cancer deregulated pathways can greatly help, for this reason several compounds inhibiting different players in DDC and DNA repair are under study in pre-clinical and in clinical trials, alone or in combination with classic therapy.

7. Targeting DDC Is One of the Most Efficient Ways to Tackle Cancer Cells

Cancer cells accumulate DNA damage in an increased fashion as compared to normal cells and rely on DDC to avoid that excessive damage is inherited after cell division, for this reason several inhibitors targeting this pathway have been developed or are in development (Table 1).

As previously mentioned, DNA damage determines cell cycle arrest before entering mitosis, if this arrest is prolonged enough, cells go through the apoptotic pathway. Sometimes, cancer cells can avoid mitotic arrest and inappropriately enter mitosis before completion of the S and G2 phase; this frequently determines an improper distribution of chromosomes leading to mitotic catastrophe, which is an onco-suppressive signaling cascade preceding cell death via apoptosis, senescence or necrosis, in the first or subsequent cell divisions [101]. Inhibition of proteins that control the S/G2 checkpoint promotes this pathway and in solid cancer this is the main cell death-leading pathway induced by IR treatment [102,103].

ATM mutations could induce synthetic lethality in association with other mutations or specific drugs; this could be exploited to develop more accurate therapies [22]. Since ATM inhibition determines uncontrolled cell entry into mitosis by removing several checkpoints control, the use of ATM inhibitors in cancer therapy has been reported to cause sensitization to radiotherapy. In addition, ATM inhibition enhances cell sensitivity to topoisomerase I and II inhibitors such as camptothecin, etoposide, and doxorubicin [104]. Several ATM inhibitors have been used in cancer therapy. AZD0156 is a strong radio-sensitizer that recently has been demonstrated to inhibit tumor growth after radiation treatment of lung xenograft [105]; in addition, a combinatory effect with PARP inhibitors (olaparib) is under study since sensitivity to these drugs is related to ATM deficiency [106]. Currently, AZD0156 is in phase 1 trial (NCT02588105). A phase I clinical trial with the inhibitor AZD1390 (NCT03423628) (Table 1), which has been proved to radio-sensitize brain tumor models in pre-clinical studies, has also been initiated [107]. Several other compounds have been produced such as KU-55933, KU-60019 and KU-59403 that are in pre-clinical studies, therefore targeting ATM is a promising strategy for cancer treatment [108].

DNA-PK inhibition promotes radio-sensitivity in gastric cancer [109]; accordingly, its overexpression is associated to radio-resistance in thyroid, cervix, and prostate cancer [110] (Table 1). DNA-PKs deficiency can sensitize cells to DNA-damaging agents such as topoisomerase I and II inhibitors. This has led to the development of several DNA-PK inhibitors such as M3814, which has been demonstrated to increase efficacy of topoisomerase II inhibition in ovarian cancer models [111]. A clinical trial is ongoing to study efficacy of this drug in combination with radiotherapy and capecitabine, a drug that interferes with DNA and RNA metabolism (NCT03770689). Similarly, the DNA-PK inhibitor VX-984 enhances radio-sensitivity of glioblastoma and a clinical trial has been completed to assess safety of this drug in combination with chemotherapy (NCT02644278) [112]. Other drugs, currently under study, are CC-115 and MSC2490484A (NCT02516813) (NCT02833883), whose efficacy has been suggested in ATM-deficient cells and in combination with radiotherapy [113] (Table 1).

Differently from the other kinases, ATR does not participate in the initial recognition phases of DNA damage, but its activity is essential to prevent replication fork collapse, thus ensuring faithful duplication by binding to ssDNA and to arrest cell cycle in G2/M phase by interacting with CHK1 [114]. Since cancer cells are often defective in G1 phase checkpoint due to mutation of p53 and Rb tumor suppressor genes, they arrive in S phase with replication stress that could stall replication fork; therefore, cells rely on S-G2 phase checkpoint to avoid cell death due to excessive DSBs. Thus, a deficiency of ATR activity determines sensitivity to DNA damage agents such as IR, DNA cross-linking agents, and topoisomerase poisons [115]; its inhibition combined with PARP inhibitors also reduces tumor growth in BRCAness models [116] (Table 1). Currently, the ATR inhibitors M6620, AZD6738, BAY1895344, and VE-821 have been shown to increase lethality of cancer cell lines ATM, defective such as gastric cancer, non-small cell lung cancer and chronic lymphocytic leukemia, strongly suggesting a synthetic lethality interaction between the ATM and ATR pathways [117,118] (Table 1). A synthetic lethality interaction has been observed in triple-negative breast cancer cell lines also by combining AZD6738 with WEE1 inhibition, and with CHK1 inhibitor AZD7762, since combination of ATR and CHK1 inhibition leads to ssDNA accumulation and replication stress [119,120] (Table 1). In addition, since ATR inhibition enhances replication stress, as anticipated in pre-clinical trials, several studies analyze cancer cell sensitization by these drugs to DNA damaging agents such as cisplatin, gemcitabine, and radiation. In particular, M6620 was studied in combination with radiation in patients with brain metastases (NCT02589522) and with cisplatin for head and neck squamous cell carcinoma (NCT02567422) (Table 1); AZD6738 in combination with radiotherapy in patients with solid tumors, VE-821 in combination with radiation and chemotherapeutics in pancreatic cancer patients [121,122]. The effect of BAY1895344 in combination with chemotherapeutics or radiotherapy on tumor cell growth and survival was studied in cancer xenograft models that carry DNA damage repair deficiencies [123].

Targeting DDC pathway is an attractive strategy in cancer therapy. In fact, it has proven to be particularly effective in combination with radiotherapy determining cell death by induction of mitotic catastrophe pathway. Moreover, as DDC is upstream to several DNA repair pathways, different synthetic lethality combinations, currently under study, could be exploited taking advantage of this network.

Table 1. DDC-related proteins targeted in cancer therapy.

DDC Pathway					
Protein	**Drug Name**	**Combined Treatment**	**Possible Synthetic Lethal Interaction**	**Clinical Trial**	**References**
ATM	AZD0156	Ionizing radiation Topoisomerase I and II inhibitors	PARP1 inhibition	NCT02588105	[105–108]
	AZD1390			NCT03423628	
	KU-55933				
	KU-60019				
	KU-59403				
ATR	M6620	Ionizing radiation Cisplatin	ATM inhibition	NCT02589522 NCT02567422	[115–123]
	AZD6738	Ionizing radiation	ATM inhibition WEE1 inhibition		
	BAY1895344	Ionizing radiation Cisplatin	ATM inhibition PARP1 inhibition		
	VE-821	Ionizing radiation Anti-metabolites	ATM inhibition CHK1 inhibition		
DNA-PK	M3814	Topoisomerase I and II inhibitors Anti-metabolites Ionizing radiation	ATM inhibition	NCT03770689	[109–113]
	VX-984	Ionizing radiation Topoisomerase I and II inhibitors		NCT02644278	
	CC-115	Ionizing radiation		NCT02516813	
	MSC2490484A	Ionizing radiation		NCT02833883	
CHK1	AZD7762	Anti-metabolites	ATR inhibition	NCT00937664	[120]
	MK-8776		FANCD2 inhibition		[124]
WEE1	AZD1775	Cisplatin Ionizing radiation	PARP1 inhibition	NCT03028766	[119]

For each target it is reported the name of the drugs currently in analysis; combination of target inhibition and classic anti-cancer therapy under study; possible synthetic lethality interaction; ongoing clinical trials with the drugs; references.

8. Combination of DSB Inducers and DSB-Repair Proteins Inhibition Is Greatly Effective

A large number of DNA damaging agents exert their activity by inducing DSBs; therefore, combining these agents with compounds inhibiting DSB repair-involved players produced promising results. Particularly, inhibition of HR-involved proteins combined with DNA damage induction forces cells to use more error-prone DNA repair pathways, thus determining accumulation of DNA aberrations. Nevertheless, upregulation of mutagenic repair pathways could promote therapy resistance ultimately leading to disease progression [125]; for this reason, compounds inhibiting NHEJ or MMEJ players are in development to help in overcoming this problem and improve clinical outcomes of classic therapies.

Regarding HR, several inhibitors have been developed, targeting MRN complex and among them, mirin is the most studied (Table 2).

Mirin inhibits MRE11-associated exonuclease activity preventing MRN-dependent ATM activation. This inhibitor potentially sensitizes different cancers to genotoxic agents, such as malignant glioma, prostate cancer, and neuroblastoma [24,126,127]. In addition, several studies suggest that MRN defects could synergize with PARP inhibition, indicating a possible synthetic lethality interaction that could be exploited in specific cancer therapy [128].

As overexpression of *RAD51* stimulates cancer aggressiveness, it has been shown that lowering its expression or activity (through inhibition) can sensitize cancer cells to chemotherapeutics [113]. To date, different classes of RAD51 inhibitors which take advantage of different inhibition mechanisms have been developed (Table 2): some inhibitors interfere with RAD51 ssDNA binding ability and could lead to the formation of toxic RAD51 complexes; other inhibitors interfere with RAD51 ability to interact with other DNA repair factors such as BRCA2 and finally drugs that inhibit the RAD51 capacity to migrate through the DNA strand and form foci after DNA damage. B02 is a RAD51 inhibitor which interferes with ssDNA binding ability and has proved to sensitize cancer cells to several DNA damage agents in preclinical studies also in combination with PARP inhibitors [129–131]. The RAD51 inhibitor RI-1 destabilizes RAD51 oligomerization/filament formation on DNA strand leading to the formation of cytotoxic complexes. This compound specifically works on those cancer cells where *RAD51* is overexpressed, sensitizing them to chemo/radiotherapy and PARP inhibitors [132,133]. IBR120 is an inhibitor that disrupts RAD51 interaction with BRCA2 sensitizing cancer cells to IR [134,135]. Finally, CYT-0851 is a compound that reduced RAD51 migration to damaged DNA ends [125]. A clinical trial to test this inhibitor has already started, (NCT03997968) (Table 2). Anyhow, even if the other inhibitors are not in clinical trials, they proved to have great potential in pre-clinical studies.

Since BRCA1 and BRCA2 are large proteins that interact with several partners with their functional domains and act as hubs in DNA damage signaling/repair, it is difficult to design specific compounds to inhibit their activities [136]. Some BRCA1/BRCA2 inhibitors acting on the BRCT domain, which is essential for the interaction with DNA repair partners, such as RAD51, determine an increase of cell sensitivity to drugs such as PARP inhibitors [137,138]. Moreover, small molecules acting on the RING domain of BRCA1 have been developed and this could provide novel ways to inhibit HR [24].

Table 2. HR repair proteins targeted in cancer therapy.

			HR		
Protein	**Drug Name**	**Combined Treatment**	**Possible Synthetic Lethal Interaction**	**Clinical Trial**	**References**
MRE11	Mirin		PARP1 inhibition		[126–128]
RAD51	B02	Ionizing radiation Topoisomerases I and II inhibitors	PARP1 inhibition		[129–132]
	RI-1	Cisplatin Ionizing radiation			[132,133]
	IBR120	Ionizing radiation			[134,135]
	CYT-0851			NCT03997968	[125]
BRCA1			PARP1 inhibition		[24,137,138]
BRCA2					
RAD52	F79 6-OH-dopa D-I03	Cisplatin	PARP1 inhibition BRCA2 inhibition BRCA1/PALB2 inhibition		[139–142]
	A5MP AICAR/ZMP		BRCA1 inhibition		
	NP-004255 F779-0434		BRCA2 inhibition		

For each target it is reported: name of the drug currently in analysis; combination of target inhibition and classic anti-cancer therapy under study; possible synthetic lethality interaction; ongoing clinical trials with the drug; references.

RAD52 is another therapeutic target in breast and ovarian cancer because its depletion is synthetically lethal to cells BRCA2 or BRCA1/PALB2 deficient [139,140]. Different groups have developed several RAD52 inhibitors (F79, 6-OH-dopa, A5MP, AICAR/ZMP, D-I03, NP-004255, F779-0434) that act either by inhibiting oligomerization or blocking ssDNA binding activities of the protein, similar to some RAD51 inhibitors [141] (Table 1). Recently, to potentiate the effect of RAD52 drugs and reduce resistance, a new strategy named "dual synthetic lethality" has been proposed with the idea to combine them with PARP inhibitors, in BRCA1-deficient cancer cells [142].

Regarding NHEJ, several chemical compounds have been developed for cancer therapy, but they are all in preclinical studies. These studies are analyzing how their inhibition could influence cancer development and drugs resistance (Table 3). It has been reported that Ku70/80 inhibition sensitizes cells to radiation treatment, XLF inhibition contributes to overcome chemoresistance in colorectal cancer cells, and LIG IV silencing results in increased cellular sensitivity to chemotherapeutic alkylating agents [143–145].

Some studies are developing small ligands able to inhibit TP53BP1 by binding to its interacting domain and competing with regular substrates, making it unable to coordinate downstream effectors [146,147], moreover a synthetic lethal interaction between TP53BP1 loss and treatment with ATR inhibitors and cisplatin has been demonstrated [148] (Table 3).

MMEJ pathway is still far from being fully characterized and several groups are working to identify drugs able to inhibit Pol-θ activity and to exploit this pathway in cancer therapy [149]. Recently, it has been discovered that Pol-θ inhibition sensitizes cells to replication stress caused by agents such as topoisomerases poisons or ATR inhibitors, suggesting novel cancer treatments [150] (Table 3). In addition, thanks to a CRISPR genetic screen, 140 genes whose deletion is synthetically lethal with *POLQ* mutations have been discovered. Among them, several HR and NHEJ-involved genes and components of the TP53BP1 pathway have been identified. Moreover, in the same study, it has been discovered that 30% of breast cancer cases deposited in TCGA (the Cancer Genome Atlas) have alteration in at least one of these 140 genes; the proteins encoded by these genes could be potential targets to develop new strategies for cancer patients who have Pol-θ alteration [151]. Moreover, knockdown of Pol-θ in ovarian cancer cell lines deficient for FA factor FANCD2 enhances cell death [152]. Finally, the antibiotic novobicin is now under study as specific Pol-θ inhibitor, and preliminary results showed that this drug is able to kill HR-deficient tumor cells due to accumulation of toxic RAD51 foci [153] (Table 3).

Table 3. NHEJ repair proteins targeted in cancer therapy.

NHEJ				
Protein	**Drug Name**	**Combined Treatment**	**Possible Synthetic Lethal Interaction**	**References**
TP53BP1	i53 UNC2170	Cisplatin	ATR inhibition	[146–148]
KU70/80		Ionizing radiation		[143]
LIG4		Alkyating agents		[145]
XLF	G3	Cisplatin Anti-metabolities	PARP1 inhibition	[144]
Pol-θ	Novobicin	Topoisomerase I and II inhibitors	ATR inhibition	[150]
			FANCD2 inhibition	[152,153]

For each target it is reported: name of the drugs currently in analysis; combination of target inhibition and classic anti-cancer therapy under study; possible synthetic lethality interaction; ongoing clinical trials with the drug; references.

9. Targeting DNA Helicases Increases Sensitivity to Several Treatments

BLM is one of the helicases that have an essential role in HR pathway, promoting DNA end resection, RAD51 filament and D-loop formation [154]. The inhibitor ML216 targets this helicase (Table 4), by competing for ATP binding and for BLM binding to DNA; anyhow, this compound showed poor aqueous solubility and cell permeability, therefore it will require further optimization [155].

DNA2 helicase allows cancer cells to resist DNA replication stresses induced by chemotherapy or radiotherapy, by promoting flap removal during DNA replication, DNA resection during repair, and stabilization and restart of stalled replication forks [156]. C5, a potent DNA2 inhibitor, has been identified by a high throughput screening; this inhibitor sensitizes cells to camptothecin and PARP inhibitors, making it interesting for cancer therapy [157] (Table 4).

WRN downregulation leads to mitotic catastrophe and cancer cell death [158]. Interestingly, evidences show that treatment with a WRN inhibitor (NSC 19630) induces apoptosis in leukemia cell lines [159]. NSC617145 is an improved derivative of the previous drug, which likely traps WRN on the DNA substrate, leading to sensitization of cancer cells to DNA-damaging agents. This drug sensitizes cells deficient of the FA and NHEJ pathways to mitomycin C, suggesting the possibility of a synthetic lethality approach [160] (Table 4). Recently, new small molecule WRN inhibitors, identified by a high throughput screen, have been shown to reduce proliferation of cancer cells [161].

These data suggest that DNA helicases are potential targets for cancer therapy to be exploited as novel synthetic lethal interactions. Several studies are now ongoing to identify new drugs that could provide alternative and useful strategies to sensitize several cancers to DNA damage agents [155,162] (Table 4).

Table 4. Helicases and nucleases targeted in cancer therapy.

Helicases/Nucleases				
Protein	**Drug Name**	**Combined Treatment**	**Possible Synthetic Lethal Interaction**	**References**
BLM	ML216			[155]
DNA2	C5	Topoisomerase I and II inhibitors	PARP1 inhibition	[157]
WRN	NSC 19630			[159–161]
	NSC617145	Mitomycin C	FANCD2/DNA-PK inhibition	

For each target it is reported: name of the drugs currently in analysis; combination of target inhibition and classic anti-cancer therapy under study; possible synthetic lethality interaction; references.

10. Targeting Repair of Single Base Lesions and DNA Crosslinks Improves Efficiency of Cancer Treatments

A strong association between APE1 overexpression and cancer progression has been reported in BER, most likely due to an increased resistance to DNA damaging agents such as platinum compounds [163,164]. For this reason, several studies are investigating the possibility to identify bioactive compounds able to inhibit APE1 anti-proliferative activity and consequently sensitize cells to DNA damaging agents, such as bleomycin [165,166]. The APE1 inhibitor E3330 has been proved to reduce the migration of human breast cancer cells, and it thus might have therapeutic potential in metastatic breast cancer [167] (Table 5). Strikingly, it has been shown that treatment with PARP inhibitor olaparib reduces APE1 expression and this depletion increases sensitivity to olaparib treatment in breast cancer cells (Table 5) [168].

Table 5. BER/MMR-related proteins targeted in cancer therapy.

BER/MMR				
Protein	**Drug Name**	**Combined Treatment**	**Possible Synthetic Lethal Interaction**	**References**
APE1	E3330	Bleomycin	PARP1 inhibition	[163–168]
POL-β	NSC666715	Alkylating agents	MSH2/MLH1 inhibition	[169,170]

For each target it is reported: name of the drugs currently in analysis; combination of target inhibition and classic anti-cancer therapy under study; possible synthetic lethality interaction; references.

NER has a pivotal role in resolving helix-distortion DNA lesions; thus, it would be expected that depletion of proteins involved in this pathway would increase sensitivity to DNA damaging agents (Table 6). Most of the current data confirm this assumption, even if some contrasting evidences exist. Single nucleotide polymorphisms (SNPs) of NER genes could be associated to chemotherapy response in lung cancer patients and, in gastric cancer cells, low XPC expression together with the absence of translocation of XPA and XPD to the nucleus increase sensitivity to cisplatin [171,172]. However, in contrast with previous data, high XPC expression correlates with longer survival time in colon cancer patients sensitizing colon cancer cell lines to cisplatin and radiation [173]. Low expression of ERCC1 correlates with increased sensitivity to platinum in several tumor types and its depletion sensitizes cells to PARP inhibitors [174–176]. Compounds targeting ERCC1/XPF complex are being developed to sensitize cancer cells to DNA damage-based chemotherapy [177,178].

Table 6. NER/FA-related proteins targeted in cancer therapy.

NER/FA				
Protein	**Drug Name**	**Combined Treatment**	**Possible Synthetic Lethal Interaction**	**References**
ERCC1	NSC16168	Cisplatin	PARP1 inhibition	[174–178]
FANCS	Phenylbutyrate	Cisplatin		[179]
FANCD2	Curcumin	Cisplatin	ATM inhibition	[124]
	MLN4924	DNA cross-linking agents	CHK1 inhibition	[180–182]
USP1/UAF1	ML323	Topoisomerase I and II inhibitors	PARP1 inhibition	[183,184]

For each target it is reported: name of the drugs currently in analysis; combination of target inhibition and classic anti-cancer therapy under study; possible synthetic lethality interaction; references.

Considering the importance of FA pathway in resolving ICLs, synthetic lethal interactions with this pathway for the development of inhibitors have been explored (Table 6).

A siRNA-based screening identified ATM as synthetically lethal in Fanconi anemia deficient cell lines, indicating that this gene could be targeted concomitantly with a FA pathway inhibitor [185]. Moreover, combination of CHK1 inhibitor with FANCD2 depletion hyper-sensitizes lung cancer cells to gemcitabine [124]. Several molecules that inhibit specific components of the FA pathway have been identified acting at different levels; for instance, phenylbutyrate sensitizes head and neck cancer cells to cisplatin by reducing the expression of FANCS [179], curcumin acts on the mono-ubiquitination step of FANCD2 sensitizing ovarian and breast tumor cell lines to cisplatin and glioma cell lines to alkylating agents, also MLN4924 suppresses FANCD2 mono-ubiquitination sensitizing cells to DNA ICLs agents [180–182] (Table 6). The USP1/UAF1 complex has been identified as a druggable target: ML323 inhibits it, leading to sensitization of colorectal cancer, non-small cell lung cancer, and osteosarcoma cells to DNA-damaging chemotherapeutics and PARP inhibitors (Table 6); inhibiting this complex also compromises TLS due to reduced deubiquitination of PCNA, thus this drug could act on FA by simultaneously targeting two steps [183,184].

These data suggest that among intra and inter-strand DNA crosslinks-repair involved proteins, the most promising are the ones involved in FA pathway that could increase efficacy of the currently used anti-cancer treatments based on DNA crosslinker such as platinum compounds and mitomycin C.

11. Targeting MMR and Synthetic Lethality Interactions: Possible New Combination Therapies

Defects in MMR genes confer resistance to platinum-based chemotherapy and re-expression of these components restores sensitivity to these compounds [169]. Although not many specific drugs that inhibit MMR genes have been proposed, it has been shown that Pol-β inhibition is synthetically lethal with MSH2/MLH1 deficiencies, likely because this inhibition determines accumulation of unrepaired oxidized nucleotides, which increases the amount of mismatches. Thus, Pol-β inhibitors (NSC666715) are promising agents for treatment of MMR-deficient cancers in combination with DNA-alkylating agents such as temozolomide (Table 5) [170].

12. PARP1 Inhibition Gave the Most Important Results in the Synthetic Lethality Field Promoting Development of New Drugs and Treatments

PARP1 is one of the most studied DNA-repair involved proteins having a role in different DNA repair pathways such as BER, NER, NHEJ, and HR. In addition, PARP1 also regulates chromatin decondensation and cell cycle arrest [186]. Several inhibitors of this protein have been developed, because PARP1 depletion is synthetically lethal together with deficiency in BRCA1 or BRCA2 [187,188]. The synthetic lethality interaction is likely due to accumulation of DNA damage caused by the combined depletion of BER or NER and HR repair; NER/BER depletion determines SSB accumulation which eventually generates DSB after replication fork collapse. HR depletion due to BRCA1/2 mutation prevents DSB repair, leading to cancer cell death [189]. Thus, PARP inhibitors have been proved to be greatly effective in breast and ovarian cancers carrying mutations in HR-related genes. To date, four PARP inhibitors have been approved by Food and Drug Administration (FDA) for cancer therapy: olaparib, rucaparib, niraparib, talazoparib, and other inhibitors are under study [190,191]. Several clinical trials are ongoing to approve these drugs for several cancer types such as colorectal, melanoma, and prostate cancers (NCT00535353, NCT00804908, NCT03732820) and to assess efficacy of new inhibitors in combination with DNA damaging agents (NCT01127178, NCT01311713). Combinatorial treatments with these drugs and classic treatments have been explored with almost any chemotherapeutic agents and with radiotherapy [192], recently prompting the development of inhibitors labeled with therapeutic isotopes, that have proven to be particularly effective in brain tumor models [193,194]. Moreover, discovering that PARP inhibitors present antiangiogenic activity promoted development of targeted therapies combining PARP inhibition with anti-angiogenic agents to improve treatment of ovarian cancer [195,196]. As shown in this review, several synthetically lethal interactions that exploit the combination of DNA damage induced by PARP inhibitors and inhibition of several DNA-repair related proteins are under study, particularly those ones involved in HR (Figure 5).

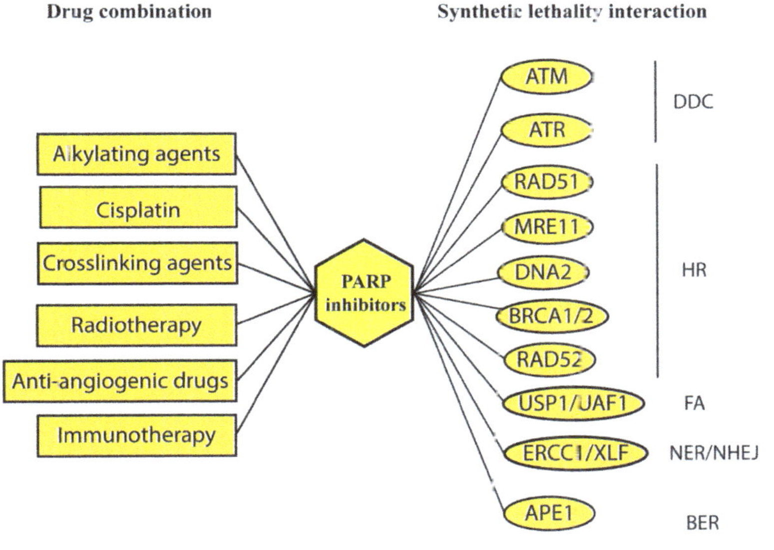

Figure 5. Field of PARP inhibitors in cancer therapy widely explored numerous combinations. Combination of PARP inhibitors with classic chemotherapeutic agents (alkylating agents, cisplatin, cross-linking agents), radiotherapy, targeted therapy (anti-angiogenic drugs) and immunotherapy. Numerous synthetic lethality interactions have been explored: from initial DNA damage recognition (DDC) to DNA repair, most promising results have been obtained combining PARP inhibition with DSB repair inhibition (HR, FA, NHEJ).

Despite their efficacy, resistance occurrence, often caused by a recovery of BRCA-related activity, is a common problem [197]. Several studies have investigated the network of PARP1 interactors in order to identify novel targets in combination with PARP inhibition, also exploring pathways not directly related to DNA repair. The final aim of these investigations is to reduce drug resistance, improve efficacy of currently used PARP inhibitors and eventually develop new compounds. Our group has identified novel PARP1 interactors by using a yeast wide-screening assay and selected by an integrative computational analysis the most relevant interactors that have been validated in cell-based assays [198,199]. Within the field of PARP inhibitors improvement, some studies are analyzing the landscape of secondary targets of these drugs to modulate efficacy and side-effects in the clinical therapy [200,201].

13. Conclusions

It is now evident that targeted inhibition of DNA repair and DDC proteins has numerous applications in cancer treatment. The kinases involved in the first phases of DNA damage recognition and cell cycle arrest such as: ATM, ATR, DNA-PK, and CHK1, have been demonstrated to be efficiently targeted and several inhibitors are currently being tested. Importantly, identifying new targets involved in specific pathways and new drugs could allow the development of more personalized therapeutic approaches that may result in reducing the possibility of side-effects and the occurrence of drug resistance.

Another promising application is the combinatorial treatment which takes advantage of simultaneous inhibition of related pathways, leading to an increase of efficacy as compared to the drugs alone. Recently, there is an emerging paradigm that DNA damage, which frequently occurs in cancer cells, leads to the expression of interferon and chemokines activating immune cells in the tumor microenvironment [202]; therefore, immunotherapy combined with drugs inhibiting DNA-repair proteins is now one of the most promising therapies. Numerous ongoing clinical trials are investigating efficacy of these treatments; for instance, PARP inhibitors combined with immune inhibitors (see clinical trials NCT03834519, NCT03851614, NCT03602859, and NCT03308942). Moreover, the efficacy of PARP inhibitors in HR-deficient cancers underlines the importance of the discovery of new synthetically lethal interactions and how these could be exploited in specific cancers, hence promoting development of personalized therapies. The progression in CRISPR/Cas9-based screening has made possible systematic analysis for synthetic lethal drug targets in human cancers leading to the identification of novel genetic cancer targets and eventually the development of effective drugs specific for different pathogenic conditions [203,204].

Finally, DNA damage is repaired by a protein network that is deeply connected; uncovering these connections might generate novel targets to increase efficacy of current in use therapies and overcome resistance development.

Author Contributions: All the authors were involved in curation of the manuscript. S.L. and A.G. collect data and wrote the article. A.P. contributed to designing and editing of the manuscript. S.L. was also involved in preparing the figures and tables. T.C. reviewed the manuscript and contributed to organize the conclusions. All authors have read and agreed to the published version of the manuscript.

Funding: This work was supported by grants assigned to AG by the "Fondazione Pisa" (127/16) and to AP by Associazione Italiana Ricerca sul Cancro (AIRC_IG19917). SL was supported by a fellowship from Fondazione Umberto Veronesi (Post-Doctoral Fellowship 2020).

Acknowledgments: The authors wish to thank all the members of the laboratories for suggestions and comments. Special thanks to Marcella Simili and Michael Minks for critical reading and language editing.

Conflicts of Interest: The authors declare no conflict of interest.

References

1. Beard, W.A.; Horton, J.K.; Prasad, R.; Wilson, S.H. Eukaryotic Base Excision Repair: New Approaches Shine Light on Mechanism. *Annu. Rev. Biochem.* **2019**, *88*, 137–162. [CrossRef] [PubMed]
2. Spivak, G. Nucleotide excision repair in humans. *DNA Repair. (Amst.)* **2015**, *36*, 13–18. [CrossRef] [PubMed]
3. Ceccaldi, R.; Sarangi, P.; D'Andrea, A.D. The Fanconi anaemia pathway: New players and new functions. *Nat. Rev. Mol. Cell Biol.* **2016**, *17*, 337–349. [CrossRef] [PubMed]
4. Larrea, A.A.; Lujan, S.A.; Kunkel, T.A. SnapShot: DNA mismatch repair. *Cell* **2010**, *141*, 730.e1. [CrossRef]
5. Waters, L.S.; Minesinger, B.K.; Wiltrout, M.E.; D'Souza, S.; Woodruff, R.V.; Walker, G.C. Eukaryotic translesion polymerases and their roles and regulation in DNA damage tolerance. *Microbiol. Mol. Biol. Rev.* **2009**, *73*, 134–154. [CrossRef] [PubMed]
6. Yang, K.; Guo, R.; Xu, D. Non-homologous end joining: Advances and frontiers. *Acta Biochim. Biophys. Sin. (Shanghai)* **2016**, *48*, 632–640. [CrossRef]
7. Wright, W.D.; Shah, S.S.; Heyer, W.D. Homologous recombination and the repair of DNA double-strand breaks. *J. Biol. Chem.* **2018**, *293*, 10524–10535. [CrossRef] [PubMed]
8. McVey, M.; Lee, S.E. MMEJ repair of double-strand breaks (director's cut): Deleted sequences and alternative endings. *Trends Genet.* **2008**, *24*, 529–538. [CrossRef]
9. Bhargava, R.; Onyango, D.O.; Stark, J.M. Regulation of Single-Strand Annealing and its Role in Genome Maintenance. *Trends Genet.* **2016**, *32*, 566–575. [CrossRef]
10. Jeggo, P.A.; Lobrich, M. How cancer cells hijack DNA double-strand break repair pathways to gain genomic instability. *Biochem. J.* **2015**, *471*, 1–11. [CrossRef]
11. O'Neil, N.J.; Bailey, M.L.; Hieter, P. Synthetic lethality and cancer. *Nat. Rev. Genet.* **2017**, *18*, 613–623. [CrossRef] [PubMed]
12. Hoadley, K.A.; Yau, C.; Hinoue, T.; Wolf, D.M.; Lazar, A.J.; Drill, E.; Shen, R.; Taylor, A.M.; Cherniack, A.D.; Thorsson, V.; et al. Cell-of-Origin Patterns Dominate the Molecular Classification of 10,000 Tumors from 33 Types of Cancer. *Cell* **2018**, *173*, 291–304. [CrossRef] [PubMed]
13. Horlbeck, M.A.; Gilbert, L.A.; Villalta, J.E.; Adamson, B.; Pak, R.A.; Chen, Y.; Fields, A.P.; Park, C.Y.; Corn, J.E.; Kampmann, M.; et al. Compact and highly active next-generation libraries for CRISPR-mediated gene repression and activation. *Elife* **2016**, *5*, e19760. [CrossRef] [PubMed]
14. Wang, T.; Yu, H.; Hughes, N.W.; Liu, B.; Kendirli, A.; Klein, K.; Chen, W.W.; Lander, E.S.; Sabatini, D.M. Gene Essentiality Profiling Reveals Gene Networks and Synthetic Lethal Interactions with Oncogenic Ras. *Cell* **2017**, *168*, 890–903. [CrossRef]
15. Blackford, A.N.; Jackson, S.P. ATM, ATR, and DNA-PK: The Trinity at the Heart of the DNA Damage Response. *Mol. Cell* **2017**, *66*, 801–817. [CrossRef]
16. Burger, K.; Schlackow, M.; Gullerova, M. Tyrosine kinase c-Abl couples RNA polymerase II transcription to DNA double-strand breaks. *Nucleic Acids Res.* **2019**, *47*, 3467–3484. [CrossRef]
17. Pong, S.K.; Gullerova, M. Noncanonical functions of microRNA pathway enzymes-Drosha, DGCR8, Dicer and Ago proteins. *FEBS Lett.* **2018**, *592*, 2973–2986. [CrossRef]
18. Falck, J.; Coates, J.; Jackson, S.P. Conserved modes of recruitment of ATM, ATR and DNA-PKcs to sites of DNA damage. *Nature* **2005**, *434*, 605–611. [CrossRef]
19. Zhang, Y.; Hunter, T. Roles of Chk1 in cell biology and cancer therapy. *Int. J. Cancer* **2014**, *134*, 1013–1023. [CrossRef]
20. Zannini, L.; Delia, D.; Buscemi, G. CHK2 kinase in the DNA damage response and beyond. *J. Mol. Cell Biol.* **2014**, *6*, 442–457. [CrossRef]
21. Wagner, S.A.; Oehler, H.; Voigt, A.; Dalic, D.; Freiwald, A.; Serve, H.; Beli, P. ATR inhibition rewires cellular signaling networks induced by replication stress. *Proteomics* **2016**, *16*, 402–416. [CrossRef] [PubMed]
22. Choi, M.; Kipps, T.; Kurzrock, R. ATM Mutations in Cancer: Therapeutic Implications. *Mol. Cancer Ther.* **2016**, *15*, 1781–1791. [CrossRef]
23. Goodwin, J.F.; Knudsen, K.E. Beyond DNA repair: DNA-PK function in cancer. *Cancer Discov.* **2014**, *4*, 1126–1139. [CrossRef]
24. Sishc, B.J.; Davis, A.J. The Role of the Core Non-Homologous End Joining Factors in Carcinogenesis and Cancer. *Cancers (Basel)* **2017**, *9*, 81. [CrossRef]

25. Tanaka, A.; Weinel, S.; Nagy, N.; O'Driscoll, M.; Lai-Cheong, J.E.; Kulp-Shorten, C.L.; Knable, A.; Carpenter, G.; Fisher, S.A.; Hiragun, M.; et al. Germline mutation in ATR in autosomal- dominant oropharyngeal cancer syndrome. *Am. J. Hum. Genet.* **2012**, *90*, 511–517. [CrossRef] [PubMed]
26. Chen, C.F.; Ruiz-Vega, R.; Vasudeva, P.; Espitia, F.; Krasieva, T.B.; de Feraudy, S.; Tromberg, B.J.; Huang, S.; Garner, C.P.; Wu, J.; et al. ATR Mutations Promote the Growth of Melanoma Tumors by Modulating the Immune Microenvironment. *Cell Rep.* **2017**, *18*, 2331–2342. [CrossRef] [PubMed]
27. Ray Chaudhuri, A.; Nussenzweig, A. The multifaceted roles of PARP1 in DNA repair and chromatin remodelling. *Nat. Rev. Mol. Cell Biol.* **2017**, *18*, 610–621. [CrossRef]
28. Yang, G.; Liu, C.; Chen, S.H.; Kassab, M.A.; Hoff, J.D.; Walter, N.G.; Yu, X. Super-resolution imaging identifies PARP1 and the Ku complex acting as DNA double-strand break sensors. *Nucleic Acids Res.* **2018**, *46*, 3446–3457. [CrossRef] [PubMed]
29. Mian, E.; Wiesmuller, L. Phenotypic Analysis of ATM Protein Kinase in DNA Double-Strand Break Formation and Repair. *Methods Mol. Biol.* **2017**, *1599*, 317–334. [CrossRef] [PubMed]
30. Marini, F.; Rawal, C.C.; Liberi, G.; Pellicioli, A. Regulation of DNA Double Strand Breaks Processing: Focus on Barriers. *Front. Mol. Biosci.* **2019**, *6*, 55. [CrossRef]
31. Zhao, X.; Wei, C.; Li, J.; Xing, P.; Li, J.; Zheng, S.; Chen, X. Cell cycle-dependent control of homologous recombination. *Acta Biochim. Biophys. Sin. (Shanghai)* **2017**, *49*, 655–668. [CrossRef]
32. Gupta, A.; Hunt, C.R.; Chakraborty, S.; Pandita, R.K.; Yordy, J.; Ramnarain, D.B.; Horikoshi, N.; Pandita, T.K. Role of 53BP1 in the regulation of DNA double-strand break repair pathway choice. *Radiat. Res.* **2014**, *181*, 1–8. [CrossRef] [PubMed]
33. Peterson, S.E.; Li, Y.; Wu-Baer, F.; Chait, B.T.; Baer, R.; Yan, H.; Gottesman, M.E.; Gautier, J. Activation of DSB processing requires phosphorylation of CtIP by ATR. *Mol. Cell* **2013**, *49*, 657–667. [CrossRef] [PubMed]
34. Ismail, I.H.; Gagne, J.P.; Genois, M.M.; Strickfaden, H.; McDonald, D.; Xu, Z.; Poirier, G.G.; Masson, J.Y.; Hendzel, M.J. The RNF138 E3 ligase displaces Ku to promote DNA end resection and regulate DNA repair pathway choice. *Nat. Cell Biol.* **2015**, *17*, 1446–1457. [CrossRef] [PubMed]
35. Reginato, G.; Cannavo, E.; Cejka, P. Physiological protein blocks direct the Mre11-Rad50-Xrs2 and Sae2 nuclease complex to initiate DNA end resection. *Genes Dev.* **2017**, *31*, 2325–2330. [CrossRef]
36. Daley, J.M.; Jimenez-Sainz, J.; Wang, W.; Miller, A.S.; Xue, X.; Nguyen, K.A.; Jensen, R.B.; Sung, P. Enhancement of BLM-DNA2-Mediated Long-Range DNA End Resection by CtIP. *Cell Rep.* **2017**, *21*, 324–332. [CrossRef]
37. Buisson, R.; Masson, J.Y. PALB2 self-interaction controls homologous recombination. *Nucleic Acids Res.* **2012**, *40*, 10312–10323. [CrossRef]
38. Liu, J.; Doty, T.; Gibson, B.; Heyer, W.D. Human BRCA2 protein promotes RAD51 filament formation on RPA-covered single-stranded DNA. *Nat. Struct. Mol. Biol.* **2010**, *17*, 1260–1262. [CrossRef]
39. Honda, M.; Okuno, Y.; Yoo, J.; Ha, T.; Spies, M. Tyrosine phosphorylation enhances RAD52-mediated annealing by modulating its DNA binding. *EMBO J.* **2011**, *30*, 3368–3382. [CrossRef]
40. Piazza, A.; Shah, S.S.; Wright, W.D.; Gore, S.K.; Koszul, R.; Heyer, W.D. Dynamic Processing of Displacement Loops during Recombinational DNA Repair. *Mol. Cell* **2019**, *73*, 1255–1266 e1254. [CrossRef]
41. Ferrari, M.; Rawal, C.C.; Lodovichi, S.; Vietri, M.Y.; Pellicioli, A. Rad9/53BP1 promotes DNA repair via crossover recombination by limiting the Sgs1 and Mph1 helicases. *Nat. Commun.* **2020**, *11*, 3181. [CrossRef] [PubMed]
42. Truong, L.N.; Li, Y.; Shi, L.Z.; Hwang, P.Y.; He, J.; Wang, H.; Razavian, N.; Berns, M.W.; Wu, X. Microhomology-mediated End Joining and Homologous Recombination share the initial end resection step to repair DNA double-strand breaks in mammalian cells. *Proc. Natl. Acad. Sci. USA* **2013**, *110*, 7720–7725. [CrossRef]
43. Rodgers, K.; McVey, M. Error-Prone Repair of DNA Double-Strand Breaks. *J. Cell Physiol.* **2016**, *231*, 15–24. [CrossRef] [PubMed]
44. Shamanna, R.A.; Lu, H.; de Freitas, J.K.; Tian, J.; Croteau, D.L.; Bohr, V.A. WRN regulates pathway choice between classical and alternative non-homologous end joining. *Nat. Commun.* **2016**, *7*, 13785. [CrossRef] [PubMed]
45. Faridounnia, M.; Folkers, G.E.; Boelens, R. Function and Interactions of ERCC1-XPF in DNA Damage Response. *Molecules* **2018**, *23*, 3205. [CrossRef]

46. Knijnenburg, T.A.; Wang, L.; Zimmermann, M.T.; Chambwe, N.; Gao, G.F.; Cherniack, A.D.; Fan, H.; Shen, H.; Way, G.P.; Greene, C.S.; et al. Genomic and Molecular Landscape of DNA Damage Repair Deficiency across The Cancer Genome Atlas. *Cell Rep.* **2018**, *23*, 239–254. [CrossRef]
47. Heeke, A.L.; Pishvaian, M.J.; Lynce, F.; Xiu, J.; Brody, J.R.; Chen, W.J.; Baker, T.M.; Marshall, J.L.; Isaacs, C. Prevalence of Homologous Recombination-Related Gene Mutations Across Multiple Cancer Types. *JCO Precis. Oncol.* **2018**, *2018*. [CrossRef]
48. Nielsen, F.C.; van Overeem Hansen, T.; Sorensen, C.S. Hereditary breast and ovarian cancer: New genes in confined pathways. *Nat. Rev. Cancer* **2016**, *16*, 599–612. [CrossRef]
49. Kleibl, Z.; Kristensen, V.N. Women at high risk of breast cancer: Molecular characteristics, clinical presentation and management. *Breast* **2016**, *28*, 136–144. [CrossRef]
50. Maresca, L.; Lodovichi, S.; Lorenzoni, A.; Cervelli, T.; Monaco, R.; Spugnesi, L.; Tancredi, M.; Falaschi, E.; Zavaglia, K.; Landucci, E.; et al. Functional Interaction Between BRCA1 and DNA Repair in Yeast May Uncover a Role of RAD50, RAD51, MRE11A, and MSH6 Somatic Variants in Cancer Development. *Front. Genet.* **2018**, *9*, 397. [CrossRef]
51. Gachechiladze, M.; Skarda, J.; Soltermann, A.; Joerger, M. RAD51 as a potential surrogate marker for DNA repair capacity in solid malignancies. *Int. J. Cancer* **2017**, *141*, 1286–1294. [CrossRef] [PubMed]
52. Kim, H.S.; Kim, J.W.; Hwang, I.G.; Lee, H.S.; Kim, W.H. Expression of DNA Damage Response Markers in Early-Onset or Familial Gastric Cancers. *Asian Pac. J. Cancer Prev.* **2019**, *20*, 1369–1376. [CrossRef] [PubMed]
53. Takagi, M.; Yoshida, M.; Nemoto, Y.; Tamaichi, H.; Tsuchida, R.; Seki, M.; Uryu, K.; Nishii, R.; Miyamoto, S.; Saito, M.; et al. Loss of DNA Damage Response in Neuroblastoma and Utility of a PARP Inhibitor. *J. Natl. Cancer Inst.* **2017**, *109*. [CrossRef] [PubMed]
54. Simonetti, G.; Padella, A.; do Valle, I.F.; Fontana, M.C.; Fonzi, E.; Bruno, S.; Baldazzi, C.; Guadagnuolo, V.; Manfrini, M.; Ferrari, A.; et al. Aneuploid acute myeloid leukemia exhibits a signature of genomic alterations in the cell cycle and protein degradation machinery. *Cancer* **2019**, *125*, 712–725. [CrossRef] [PubMed]
55. Garcia-Sanz, P.; Trivino, J.C.; Mota, A.; Perez Lopez, M.; Colas, E.; Rojo-Sebastian, A.; Garcia, A.; Gatius, S.; Ruiz, M.; Prat, J.; et al. Chromatin remodelling and DNA repair genes are frequently mutated in endometrioid endometrial carcinoma. *Int. J. Cancer* **2017**, *140*, 1551–1563. [CrossRef]
56. Zhen, J.T.; Syed, J.; Nguyen, K.A.; Leapman, M.S.; Agarwal, N.; Brierley, K.; Llor, X.; Hofstatter, E.; Shuch, B. Genetic testing for hereditary prostate cancer: Current status and limitations. *Cancer* **2018**, *124*, 3105–3117. [CrossRef]
57. Kaluzna, E.M.; Rembowska, J.; Ziolkowska-Suchanek, I.; Swiatek-Koscielna, B.; Gabryel, P.; Dyszkiewicz, W.; Nowak, J.S. Heterozygous p.I171V mutation of the NBN gene as a risk factor for lung cancer development. *Oncol. Lett.* **2015**, *10*, 3300–3304. [CrossRef]
58. Hurley, R.M.; Wahner Hendrickson, A.E.; Visscher, D.W.; Ansell, P.; Harrell, M.I.; Wagner, J.M.; Negron, V.; Goergen, K.M.; Maurer, M.J.; Oberg, A.L.; et al. 53BP1 as a potential predictor of response in PARP inhibitor-treated homologous recombination-deficient ovarian cancer. *Gynecol. Oncol.* **2019**, *153*, 127–134. [CrossRef]
59. Yao, J.; Huang, A.; Zheng, X.; Liu, T.; Lin, Z.; Zhang, S.; Yang, Q.; Zhang, T.; Ma, H. 53BP1 loss induces chemoresistance of colorectal cancer cells to 5-fluorouracil by inhibiting the ATM-CHK2-P53 pathway. *J. Cancer Res. Clin. Oncol.* **2017**, *143*, 419–431. [CrossRef]
60. Mirza-Aghazadeh-Attari, M.; Mohammadzadeh, A.; Yousefi, B.; Mihanfar, A.; Karimian, A.; Majidinia, M. 53BP1: A key player of DNA damage response with critical functions in cancer. *DNA Repair (Amst)* **2019**, *73*, 110–119. [CrossRef]
61. Reczek, C.R.; Shakya, R.; Miteva, Y.; Szabolcs, M.; Ludwig, T.; Baer, R. The DNA resection protein CtIP promotes mammary tumorigenesis. *Oncotarget* **2016**, *7*, 32172–32183. [CrossRef] [PubMed]
62. Wood, R.D.; Doublie, S. DNA polymerase theta (POLQ), double-strand break repair, and cancer. *DNA Repair (Amst)* **2016**, *44*, 22–32. [CrossRef] [PubMed]
63. Brosh, R.M., Jr. DNA helicases involved in DNA repair and their roles in cancer. *Nat. Rev. Cancer* **2013**, *13*, 542–558. [CrossRef] [PubMed]
64. Pawlowska, E.; Szczepanska, J.; Blasiak, J. DNA2-An Important Player in DNA Damage Response or Just Another DNA Maintenance Protein? *Int. J. Mol. Sci.* **2017**, *18*, 1562. [CrossRef]

65. Moser, M.J.; Bigbee, W.L.; Grant, S.G.; Emond, M.J.; Langlois, R.G.; Jensen, R.H.; Oshima, J.; Monnat, R.J., Jr. Genetic instability and hematologic disease risk in Werner syndrome patients and heterozygotes. *Cancer Res.* **2000**, *60*, 2492–2496.
66. Sallmyr, A.; Tomkinson, A.E.; Rassool, F.V. Up-regulation of WRN and DNA ligase IIIalpha in chronic myeloid leukemia: Consequences for the repair of DNA double-strand breaks. *Blood* **2008**, *112*, 1413–1423. [CrossRef]
67. Lee, T.H.; Kang, T.H. DNA Oxidation and Excision Repair Pathways. *Int. J. Mol. Sci.* **2019**, *20*, 6092. [CrossRef]
68. Breslin, C.; Hornyak, P.; Ridley, A.; Rulten, S.L.; Hanzlikova, H.; Oliver, A.W.; Caldecott, K.W. The XRCC1 phosphate-binding pocket binds poly (ADP-ribose) and is required for XRCC1 function. *Nucleic Acids Res.* **2015**, *43*, 6934–6944. [CrossRef]
69. Hanzlikova, H.; Gittens, W.; Krejcikova, K.; Zeng, Z.; Caldecott, K.W. Overlapping roles for PARP1 and PARP2 in the recruitment of endogenous XRCC1 and PNKP into oxidized chromatin. *Nucleic Acids Res.* **2017**, *45*, 2546–2557. [CrossRef]
70. Liu, D.; Keijzers, G.; Rasmussen, L.J. DNA mismatch repair and its many roles in eukaryotic cells. *Mutat. Res.* **2017**, *773*, 174–187. [CrossRef]
71. Kunkel, T.A.; Erie, D.A. Eukaryotic Mismatch Repair in Relation to DNA Replication. *Annu. Rev. Genet.* **2015**, *49*, 291–313. [CrossRef] [PubMed]
72. Chakraborty, U.; Alani, E. Understanding how mismatch repair proteins participate in the repair/anti-recombination decision. *FEMS Yeast Res.* **2016**, *16*, 1–12. [CrossRef] [PubMed]
73. Farrington, S.M.; Tenesa, A.; Barnetson, R.; Wiltshire, A.; Prendergast, J.; Porteous, M.; Campbell, H.; Dunlop, M.G. Germline susceptibility to colorectal cancer due to base-excision repair gene defects. *Am. J. Hum. Genet.* **2005**, *77*, 112–119. [CrossRef] [PubMed]
74. Sahadevan, M.; Lee, O.; Muzzio, M.; Phan, B.; Jacobs, L.; Khouri, N.; Wang, J.; Hu, H.; Stearns, V.; Chatterton, R.T. The relationship of single-strand breaks in DNA to breast cancer risk and to tissue concentrations of oestrogens. *Biomarkers* **2017**, *22*, 689–697. [CrossRef]
75. Wang, M.; Long, K.; Li, E.; Li, L.; Li, B.; Ci, S.; He, L.; Pan, F.; Hu, Z.; Guo, Z. DNA polymerase beta modulates cancer progression via enhancing CDH13 expression by promoter demethylation. *Oncogene* **2020**, *39*, 5507–5519. [CrossRef]
76. Abdel-Fatah, T.M.; Russell, R.; Agarwal, D.; Moseley, P.; Abayomi, M.A.; Perry, C.; Albarakati, N.; Ball, G.; Chan, S.; Caldas, C.; et al. DNA polymerase beta deficiency is linked to aggressive breast cancer: A comprehensive analysis of gene copy number, mRNA and protein expression in multiple cohorts. *Mol. Oncol* **2014**, *8*, 520–532. [CrossRef]
77. Alnajjar, K.S.; Negahbani, A.; Nakhjiri, M.; Krylov, I.S.; Kashemirov, B.A.; McKenna, C.E.; Goodman, M.F.; Sweasy, J.B. DNA Polymerase beta Cancer-Associated Variant I260M Exhibits Nonspecific Selectivity toward the beta-gamma Bridging Group of the Incoming dNTP. *Biochemistry* **2017**, *56*, 5449–5456. [CrossRef]
78. Mahmoud, M.M.; Schechter, A.; Alnajjar, K.S.; Huang, J.; Towle-Weicksel, J.; Eckenroth, B.E.; Doublie, S.; Sweasy, J.B. Defective Nucleotide Release by DNA Polymerase beta Mutator Variant E288K Is the Basis of Its Low Fidelity. *Biochemistry* **2017**, *56*, 5550–5559. [CrossRef]
79. Zhou, T.; Pan, F.; Cao, Y.; Han, Y.; Zhao, J.; Sun, H.; Zhou, X.; Wu, X.; He, L.; Hu, Z.; et al. R152C DNA Pol beta mutation impairs base excision repair and induces cellular transformation. *Oncotarget* **2016**, *7*, 6902–6915. [CrossRef]
80. Malik, S.S.; Masood, N.; Asif, M.; Ahmed, P.; Shah, Z.U.; Khan, J.S. Expressional analysis of MLH1 and MSH2 in breast cancer. *Curr. Probl. Cancer* **2019**, *43*, 97–105. [CrossRef]
81. Wang, S.M.; Jiang, B.; Deng, Y.; Huang, S.L.; Fang, M.Z.; Wang, Y. Clinical significance of MLH1/MSH2 for stage II/III sporadic colorectal cancer. *World J. Gastrointest. Oncol.* **2019**, *11*, 1065–1080. [CrossRef] [PubMed]
82. Liccardo, R.; Della Ragione, C.; Mitilini, N.; De Rosa, M.; Izzo, P.; Duraturo, F. Novel variants of unknown significance in the PMS2 gene identified in patients with hereditary colon cancer. *Cancer Manag. Res.* **2019**, *11*, 6719–6725. [CrossRef] [PubMed]
83. Roberts, M.E.; Jackson, S.A.; Susswein, L.R.; Zeinomar, N.; Ma, X.; Marshall, M.L.; Stettner, A.R.; Milewski, B.; Xu, Z.; Solomon, B.D.; et al. MSH6 and PMS2 germ-line pathogenic variants implicated in Lynch syndrome are associated with breast cancer. *Genet. Med.* **2018**, *20*, 1167–1174. [CrossRef] [PubMed]

84. Maresca, L.; Spugnesi, L.; Lodovichi, S.; Cozzani, C.; Naccarato, A.G.; Tancredi, M.; Collavoli, A.; Falaschi, E.; Rossetti, E.; Aretini, P.; et al. MSH2 role in BRCA1-driven tumorigenesis: A preliminary study in yeast and in human tumors from BRCA1-VUS carriers. *Eur. J. Med. Genet.* **2015**, *58*, 531–539. [CrossRef] [PubMed]
85. Hashimoto, S.; Anai, H.; Hanada, K. Mechanisms of interstrand DNA crosslink repair and human disorders. *Genes Environ.* **2016**, *38*, 9. [CrossRef]
86. Robu, M.; Shah, R.G.; Purohit, N.K.; Zhou, P.; Naegeli, H.; Shah, G.M. Poly(ADP-ribose) polymerase 1 escorts XPC to UV-induced DNA lesions during nucleotide excision repair. *Proc. Natl. Acad. Sci. USA* **2017**, *114*, E6847–E6856. [CrossRef]
87. Sugitani, N.; Sivley, R.M.; Perry, K.E.; Capra, J.A.; Chazin, W.J. XPA: A key scaffold for human nucleotide excision repair. *DNA Repair (Amst)* **2016**, *44*, 123–135. [CrossRef]
88. Park, J.M.; Kang, T.H. Transcriptional and Posttranslational Regulation of Nucleotide Excision Repair: The Guardian of the Genome against Ultraviolet Radiation. *Int. J. Mol. Sci.* **2016**, *17*, 1840. [CrossRef]
89. Klein Douwel, D.; Boonen, R.A.; Long, D.T.; Szypowska, A.A.; Raschle, M.; Walter, J.C.; Knipscheer, P. XPF-ERCC1 acts in Unhooking DNA interstrand crosslinks in cooperation with FANCD2 and FANCP/SLX4. *Mol. Cell* **2014**, *54*, 460–471. [CrossRef]
90. Howlett, N.G.; Harney, J.A.; Rego, M.A.; Kolling, F.W., IV; Glover, T.W. Functional interaction between the Fanconi Anemia D2 protein and proliferating cell nuclear antigen (PCNA) via a conserved putative PCNA interaction motif. *J. Biol. Chem.* **2009**, *284*, 28935–28942. [CrossRef]
91. Jo, U.; Kim, H. Exploiting the Fanconi Anemia Pathway for Targeted Anti-Cancer Therapy. *Mol. Cells* **2015**, *38*, 669–676. [CrossRef]
92. Michl, J.; Zimmer, J.; Tarsounas, M. Interplay between Fanconi anemia and homologous recombination pathways in genome integrity. *EMBO J.* **2016**, *35*, 909–923. [CrossRef] [PubMed]
93. Alter, B.P. Fanconi anemia and the development of leukemia. *Best Pract. Res. Clin. Haematol.* **2014**, *27*, 214–221. [CrossRef] [PubMed]
94. Niraj, J.; Farkkila, A.; D'Andrea, A.D. The Fanconi Anemia Pathway in Cancer. *Annu. Rev. Cancer Biol.* **2019**, *3*, 457–478. [CrossRef]
95. Kondo, N.; Takahashi, A.; Ono, K.; Ohnishi, T. DNA damage induced by alkylating agents and repair pathways. *J. Nucleic Acids* **2010**, *2010*, 543531. [CrossRef] [PubMed]
96. Liang, X.; Wu, Q.; Luan, S.; Yin, Z.; He, C.; Yin, L.; Zou, Y.; Yuan, Z.; Li, L.; Song, X.; et al. A comprehensive review of topoisomerase inhibitors as anticancer agents in the past decade. *Eur. J. Med. Chem.* **2019**, *171*, 129–168. [CrossRef]
97. Calderon-Montano, J.M.; Burgos-Moron, E.; Orta, M.L.; Lopez-Lazaro, M. Effect of DNA repair deficiencies on the cytotoxicity of drugs used in cancer therapy-a review. *Curr. Med. Chem.* **2014**, *21*, 3419–3454. [CrossRef]
98. Peters, G.J. Novel developments in the use of antimetabolites. *Nucleosides Nucleotides Nucleic Acids* **2014**, *33*, 358–374. [CrossRef]
99. Chen, H.H.W.; Kuo, M.T. Improving radiotherapy in cancer treatment: Promises and challenges. *Oncotarget* **2017**, *8*, 62742–62758. [CrossRef]
100. Mansoori, B.; Mohammadi, A.; Davudian, S.; Shirjang, S.; Baradaran, B. The Different Mechanisms of Cancer Drug Resistance: A Brief Review. *Adv. Pharm. Bull.* **2017**, *7*, 339–348. [CrossRef]
101. Vitale, I.; Galluzzi, L.; Castedo, M.; Kroemer, G. Mitotic catastrophe: A mechanism for avoiding genomic instability. *Nat. Rev. Mol. Cell Biol.* **2011**, *12*, 385–392. [CrossRef] [PubMed]
102. Sia, J.; Szmyd, R.; Hau, E.; Gee, H.E. Molecular Mechanisms of Radiation-Induced Cancer Cell Death: A Primer. *Front. Cell Dev. Biol.* **2020**, *8*, 41. [CrossRef] [PubMed]
103. Huang, R.X.; Zhou, P.K. DNA damage response signaling pathways and targets for radiotherapy sensitization in cancer. *Signal. Transduct Target. Ther.* **2020**, *5*, 60. [CrossRef]
104. Balmus, G.; Pilger, D.; Coates, J.; Demir, M.; Sczaniecka-Clift, M.; Barros, A.C.; Woods, M.; Fu, B.; Yang, F.; Chen, E.; et al. ATM orchestrates the DNA-damage response to counter toxic non-homologous end-joining at broken replication forks. *Nat. Commun.* **2019**, *10*, 87. [CrossRef] [PubMed]
105. Riches, L.C.; Trinidad, A.G.; Hughes, G.; Jones, G.N.; Hughes, A.M.; Thomason, A.G.; Gavine, P.; Cui, A.; Ling, S.; Stott, J.; et al. Pharmacology of the ATM Inhibitor AZD0156: Potentiation of Irradiation and Olaparib Responses Preclinically. *Mol. Cancer Ther.* **2020**, *19*, 13–25. [CrossRef] [PubMed]

106. Hong, R.; Ma, F.; Zhang, W.; Yu, X.; Li, Q.; Luo, Y.; Zhu, C.; Jiang, W.; Xu, B. 53BP1 depletion causes PARP inhibitor resistance in ATM-deficient breast cancer cells. *BMC Cancer* **2016**, *16*, 725. [CrossRef]
107. Durant, S.T.; Zheng, L.; Wang, Y.; Chen, K.; Zhang, L.; Zhang, T.; Yang, Z.; Riches, L.; Trinidad, A.G.; Fok, J.H.L.; et al. The brain-penetrant clinical ATM inhibitor AZD1390 radiosensitizes and improves survival of preclinical brain tumor models. *Sci. Adv.* **2018**, *4*, eaat1719. [CrossRef]
108. Jin, M.H.; Oh, D.Y. ATM in DNA repair in cancer. *Pharmacol. Ther.* **2019**, *203*, 107391. [CrossRef]
109. Geng, W.; Tian, D.; Wang, Q.; Shan, S.; Zhou, J.; Xu, W.; Shan, H. DNAPKcs inhibitor increases the sensitivity of gastric cancer cells to radiotherapy. *Oncol. Rep.* **2019**, *42*, 561–570. [CrossRef]
110. Mohiuddin, I.S.; Kang, M.H. DNA-PK as an Emerging Therapeutic Target in Cancer. *Front. Oncol.* **2019**, *9*, 635. [CrossRef]
111. Wise, H.C.; Iyer, G.V.; Moore, K.; Temkin, S.M.; Gordon, S.; Aghajanian, C.; Grisham, R.N. Activity of M3814, an Oral DNA-PK Inhibitor, In Combination with Topoisomerase II Inhibitors in Ovarian Cancer Models. *Sci. Rep.* **2019**, *9*, 18882. [CrossRef]
112. Timme, C.R.; Rath, B.H.; O'Neill, J.W.; Camphausen, K.; Tofilon, P.J. The DNA-PK Inhibitor VX-984 Enhances the Radiosensitivity of Glioblastoma Cells Grown In Vitro and as Orthotopic Xenografts. *Mol. Cancer Ther.* **2018**, *17*, 1207–1216. [CrossRef] [PubMed]
113. Gavande, N.S.; VanderVere-Carozza, P.S.; Hinshaw, H.D.; Jalal, S.I.; Sears, C.R.; Pawelczak, K.S.; Turchi, J.J. DNA repair targeted therapy: The past or future of cancer treatment? *Pharmacol. Ther.* **2016**, *160*, 65–83. [CrossRef] [PubMed]
114. Saldivar, J.C.; Cortez, D.; Cimprich, K.A. The essential kinase ATR: Ensuring faithful duplication of a challenging genome. *Nat. Rev. Mol. Cell Biol.* **2017**, *18*, 622–636. [CrossRef] [PubMed]
115. Rundle, S.; Bradbury, A.; Drew, Y.; Curtin, N.J. Targeting the ATR-CHK1 Axis in Cancer Therapy. *Cancers (Basel)* **2017**, *9*, 41. [CrossRef] [PubMed]
116. Kim, H.; George, E.; Ragland, R.; Rafail, S.; Zhang, R.; Krepler, C.; Morgan, M.; Herlyn, M.; Brown, E.; Simpkins, F. Targeting the ATR/CHK1 Axis with PARP Inhibition Results in Tumor Regression in BRCA-Mutant Ovarian Cancer Models. *Clin. Cancer Res.* **2017**, *23*, 3097–3108. [CrossRef] [PubMed]
117. Vendetti, F.P.; Lau, A.; Schamus, S.; Conrads, T.P.; O'Connor, M.J.; Bakkenist, C.J. The orally active and bioavailable ATR kinase inhibitor AZD6738 potentiates the anti-tumor effects of cisplatin to resolve ATM-deficient non-small cell lung cancer in vivo. *Oncotarget* **2015**, *6*, 44289–44305. [CrossRef]
118. Kwok, M.; Davies, N.; Agathanggelou, A.; Smith, E.; Oldreive, C.; Petermann, E.; Stewart, G.; Brown, J.; Lau, A.; Pratt, G.; et al. ATR inhibition induces synthetic lethality and overcomes chemoresistance in TP53- or ATM-defective chronic lymphocytic leukemia cells. *Blood* **2016**, *127*, 582–595. [CrossRef]
119. Jin, J.; Fang, H.; Yang, F.; Ji, W.; Guan, N.; Sun, Z.; Shi, Y.; Zhou, G.; Guan, X. Combined Inhibition of ATR and WEE1 as a Novel Therapeutic Strategy in Triple-Negative Breast Cancer. *Neoplasia* **2018**, *20*, 478–488. [CrossRef]
120. Sanjiv, K.; Hagenkort, A.; Calderon-Montano, J.M.; Koolmeister, T.; Reaper, P.M.; Mortusewicz, O.; Jacques, S.A.; Kuiper, R.V.; Schultz, N.; Scobie, M.; et al. Cancer-Specific Synthetic Lethality between ATR and CHK1 Kinase Activities. *Cell Rep.* **2016**, *14*, 298–309. [CrossRef]
121. Dillon, M.T.; Boylan, Z.; Smith, D.; Guevara, J.; Mohammed, K.; Peckitt, C.; Saunders, M.; Banerji, U.; Clack, G.; Smith, S.A.; et al. PATRIOT: A phase I study to assess the tolerability, safety and biological effects of a specific ataxia telangiectasia and Rad3-related (ATR) inhibitor (AZD6738) as a single agent and in combination with palliative radiation therapy in patients with solid tumours. *Clin. Transl. Radiat. Oncol.* **2018**, *12*, 16–20. [CrossRef] [PubMed]
122. Prevo, R.; Fokas, E.; Reaper, P.M.; Charlton, P.A.; Pollard, J.R.; McKenna, W.G.; Muschel, R.J.; Brunner, T.B. The novel ATR inhibitor VE-821 increases sensitivity of pancreatic cancer cells to radiation and chemotherapy. *Cancer Biol. Ther.* **2012**, *13*, 1072–1081. [CrossRef] [PubMed]
123. Wengner, A.M.; Siemeister, G.; Lucking, U.; Lefranc, J.; Wortmann, L.; Lienau, P.; Bader, B.; Bomer, U.; Moosmayer, D.; Eberspacher, U.; et al. The Novel ATR Inhibitor BAY 1895344 Is Efficacious as Monotherapy and Combined with DNA Damage-Inducing or Repair-Compromising Therapies in Preclinical Cancer Models. *Mol. Cancer Ther.* **2020**, *19*, 26–38. [CrossRef]
124. Dai, C.H.; Wang, Y.; Chen, P.; Jiang, Q.; Lan, T.; Li, M.Y.; Su, J.Y.; Wu, Y.; Li, J. Suppression of the FA pathway combined with CHK1 inhibitor hypersensitize lung cancer cells to gemcitabine. *Sci. Rep.* **2017**, *7*, 15031. [CrossRef]

125. Trenner, A.; Sartori, A.A. Harnessing DNA Double-Strand Break Repair for Cancer Treatment. *Front. Oncol.* **2019**, *9*, 1388. [CrossRef] [PubMed]
126. Jividen, K.; Kedzierska, K.Z.; Yang, C.S.; Szlachta, K.; Ratan, A.; Paschal, B.M. Genomic analysis of DNA repair genes and androgen signaling in prostate cancer. *BMC Cancer* **2018**, *18*, 960. [CrossRef]
127. Petroni, M.; Sardina, F.; Infante, P.; Bartolazzi, A.; Locatelli, E.; Fabretti, F.; Di Giulio, S.; Capalbo, C.; Cardinali, B.; Coppa, A.; et al. MRE11 inhibition highlights a replication stress-dependent vulnerability of MYCN-driven tumors. *Cell Death Dis.* **2018**, *9*, 895. [CrossRef] [PubMed]
128. Bian, L.; Meng, Y.; Zhang, M.; Li, D. MRE11-RAD50-NBS1 complex alterations and DNA damage response: Implications for cancer treatment. *Mol. Cancer* **2019**, *18*, 169. [CrossRef]
129. Huang, F.; Mazin, A.V. A small molecule inhibitor of human RAD51 potentiates breast cancer cell killing by therapeutic agents in mouse xenografts. *PLoS ONE* **2014**, *9*, e100993. [CrossRef]
130. Alagpulinsa, D.A.; Ayyadevara, S.; Shmookler Reis, R.J. A Small-Molecule Inhibitor of RAD51 Reduces Homologous Recombination and Sensitizes Multiple Myeloma Cells to Doxorubicin. *Front. Oncol.* **2014**, *4*, 289. [CrossRef]
131. Wera, A.C.; Lobbens, A.; Stoyanov, M.; Lucas, S.; Michiels, C. Radiation-induced synthetic lethality: Combination of poly(ADP-ribose) polymerase and RAD51 inhibitors to sensitize cells to proton irradiation. *Cell Cycle* **2019**, *18*, 1770–1783. [CrossRef] [PubMed]
132. Chen, Q.; Cai, D.; Li, M.; Wu, X. The homologous recombination protein RAD51 is a promising therapeutic target for cervical carcinoma. *Oncol. Rep.* **2017**, *38*, 767–774. [CrossRef] [PubMed]
133. Zhao, Q.; Guan, J.; Zhang, Z.; Lv, J.; Wang, Y.; Liu, L.; Zhou, Q.; Mao, W. Inhibition of Rad51 sensitizes breast cancer cells with wild-type PTEN to olaparib. *Biomed. Pharmacother.* **2017**, *94*, 165–168. [CrossRef] [PubMed]
134. Zhu, J.; Chen, H.; Guo, X.E.; Qiu, X.L.; Hu, C.M.; Chamberlin, A.R.; Lee, W.H. Synthesis, molecular modeling, and biological evaluation of novel RAD51 inhibitors. *Eur. J. Med. Chem.* **2015**, *96*, 196–208. [CrossRef]
135. Roberti, M.; Schipani, F.; Bagnolini, G.; Milano, D.; Giacomini, E.; Falchi, F.; Balboni, A.; Manerba, M.; Farabegoli, F.; De Franco, F.; et al. Rad51/BRCA2 disruptors inhibit homologous recombination and synergize with olaparib in pancreatic cancer cells. *Eur. J. Med. Chem.* **2019**, *165*, 80–92. [CrossRef]
136. Chen, C.C.; Feng, W.; Lim, P.X.; Kass, E.M.; Jasin, M. Homology-Directed Repair and the Role of BRCA1, BRCA2, and Related Proteins in Genome Integrity and Cancer. *Annu. Rev. Cancer Biol.* **2018**, *2*, 313–336. [CrossRef]
137. Trenner, A.; Godau, J.; Sartori, A.A. A Short BRCA2-Derived Cell-Penetrating Peptide Targets RAD51 Function and Confers Hypersensitivity toward PARP Inhibition. *Mol. Cancer Ther.* **2018**, *17*, 1392–1404. [CrossRef]
138. Pessetto, Z.Y.; Yan, Y.; Bessho, T.; Natarajan, A. Inhibition of BRCT(BRCA1)-phosphoprotein interaction enhances the cytotoxic effect of olaparib in breast cancer cells: A proof of concept study for synthetic lethal therapeutic option. *Breast Cancer Res. Treat.* **2012**, *134*, 511–517. [CrossRef]
139. Nogueira, A.; Fernandes, M.; Catarino, R.; Medeiros, R. RAD52 Functions in Homologous Recombination and Its Importance on Genomic Integrity Maintenance and Cancer Therapy. *Cancers (Basel)* **2019**, *11*, 1622. [CrossRef]
140. Huang, F.; Goyal, N.; Sullivan, K.; Hanamshet, K.; Patel, M.; Mazina, O.M.; Wang, C.X.; An, W.F.; Spoonamore, J.; Metkar, S.; et al. Targeting BRCA1- and BRCA2-deficient cells with RAD52 small molecule inhibitors. *Nucleic Acids Res.* **2016**, *44*, 4189–4199. [CrossRef]
141. Toma, M.; Sullivan-Reed, K.; Sliwinski, T.; Skorski, T. RAD52 as a Potential Target for Synthetic Lethality-Based Anticancer Therapies. *Cancers (Basel)* **2019**, *11*, 1561. [CrossRef] [PubMed]
142. Sullivan-Reed, K.; Bolton-Gillespie, E.; Dasgupta, Y.; Langer, S.; Siciliano, M.; Nieborowska-Skorska, M.; Hanamshet, K.; Belyaeva, E.A.; Bernhardy, A.J.; Lee, J.; et al. Simultaneous Targeting of PARP1 and RAD52 Triggers Dual Synthetic Lethality in BRCA-Deficient Tumor Cells. *Cell Rep.* **2018**, *23*, 3127–3136. [CrossRef] [PubMed]
143. Weterings, E.; Gallegos, A.C.; Dominick, L.N.; Cooke, L.S.; Bartels, T.N.; Vagner, J.; Matsunaga, T.O.; Mahadevan, D. A novel small molecule inhibitor of the DNA repair protein Ku70/80. *DNA Repair (Amst)* **2016**, *43*, 98–106. [CrossRef] [PubMed]
144. Liu, Z.; Yu, M.; Fei, B.; Sun, J.; Wang, D. Identification Of Natural Compound Derivative For Inhibition Of XLF And Overcoming Chemoresistance In Colorectal Cancer Cells. *Drug Des. Devel. Ther.* **2019**, *13*, 3823–3834. [CrossRef] [PubMed]

145. Kondo, N.; Takahashi, A.; Mori, E.; Ohnishi, K.; McKinnon, P.J.; Sakaki, T.; Nakase, H.; Ohnishi, T. DNA ligase IV as a new molecular target for temozolomide. *Biochem. Biophys. Res. Commun.* **2009**, *387*, 656–660. [CrossRef]
146. Perfetti, M.T.; Baughman, B.M.; Dickson, B.M.; Mu, Y.; Cui, G.; Mader, P.; Dong, A.; Norris, J.L.; Rothbart, S.B.; Strahl, B.D.; et al. Identification of a fragment-like small molecule ligand for the methyl-lysine binding protein, 53BP1. *ACS Chem. Biol.* **2015**, *10*, 1072–1081. [CrossRef]
147. Canny, M.D.; Moatti, N.; Wan, L.C.K.; Fradet-Turcotte, A.; Krasner, D.; Mateos-Gomez, P.A.; Zimmermann, M.; Orthwein, A.; Juang, Y.C.; Zhang, W.; et al. Inhibition of 53BP1 favors homology-dependent DNA repair and increases CRISPR-Cas9 genome-editing efficiency. *Nat. Biotechnol.* **2018**, *36*, 95–102. [CrossRef]
148. Mohni, K.N.; Thompson, P.S.; Luzwick, J.W.; Glick, G.G.; Pendleton, C.S.; Lehmann, B.D.; Pietenpol, J.A.; Cortez, D. A Synthetic Lethal Screen Identifies DNA Repair Pathways that Sensitize Cancer Cells to Combined ATR Inhibition and Cisplatin Treatments. *PLoS ONE* **2015**, *10*, e0125482. [CrossRef]
149. Patterson-Fortin, J.; D'Andrea, A.D. Exploiting the Microhomology-Mediated End-Joining Pathway in Cancer Therapy. *Cancer Res.* **2020**. [CrossRef]
150. Wang, Z.; Song, Y.; Li, S.; Kurian, S.; Xiang, R.; Chiba, T.; Wu, X. DNA polymerase theta (POLQ) is important for repair of DNA double-strand breaks caused by fork collapse. *J. Biol. Chem.* **2019**, *294*, 3909–3919. [CrossRef]
151. Feng, W.; Simpson, D.A.; Carvajal-Garcia, J.; Price, B.A.; Kumar, R.J.; Mose, L.E.; Wood, R.D.; Rashid, N.; Purvis, J.E.; Parker, J.S.; et al. Genetic determinants of cellular addiction to DNA polymerase theta. *Nat. Commun.* **2019**, *10*, 4286. [CrossRef] [PubMed]
152. Ceccaldi, R.; Liu, J.C.; Amunugama, R.; Hajdu, I.; Primack, B.; Petalcorin, M.I.; O'Connor, K.W.; Konstantinopoulos, P.A.; Elledge, S.J.; Boulton, S.J.; et al. Homologous-recombination-deficient tumours are dependent on Poltheta-mediated repair. *Nature* **2015**, *518*, 258–262. [CrossRef] [PubMed]
153. Zhou, J.; Gelot, C.; Pantelidou, C.; Li, A.; Yücel, H.; Davis, R.E.; Farkkila, A.; Kochupurakkal, B.; Syed, A.; Shapiro, G.I.; et al. Polymerase Theta Inhibition Kills Homologous Recombination Deficient Tumors. *bioRxiv* **2020**. [CrossRef]
154. Tripathi, V.; Agarwal, H.; Priya, S.; Batra, H.; Modi, P.; Pandey, M.; Saha, D.; Raghavan, S.C.; Sengupta, S. MRN complex-dependent recruitment of ubiquitylated BLM helicase to DSBs negatively regulates DNA repair pathways. *Nat. Commun.* **2018**, *9*, 1016. [CrossRef] [PubMed]
155. Hengel, S.R.; Spies, M.A.; Spies, M. Small-Molecule Inhibitors Targeting DNA Repair and DNA Repair Deficiency in Research and Cancer Therapy. *Cell Chem. Biol.* **2017**, *24*, 1101–1119. [CrossRef]
156. Wanrooij, P.H.; Burgers, P.M. Yet another job for Dna2: Checkpoint activation. *DNA Repair (Amst)* **2015**, *32*, 17–23. [CrossRef]
157. Liu, W.; Zhou, M.; Li, Z.; Li, H.; Polaczek, P.; Dai, H.; Wu, Q.; Liu, C.; Karanja, K.K.; Popuri, V.; et al. A Selective Small Molecule DNA2 Inhibitor for Sensitization of Human Cancer Cells to Chemotherapy. *EBioMedicine* **2016**, *6*, 73–86. [CrossRef]
158. Orlovetskie, N.; Serruya, R.; Abboud-Jarrous, G.; Jarrous, N. Targeted inhibition of WRN helicase, replication stress and cancer. *Biochim. Biophys. Acta Rev. Cancer* **2017**, *1867*, 42–48. [CrossRef]
159. Moles, R.; Bai, X.T.; Chaib-Mezrag, H.; Nicot, C. WRN-targeted therapy using inhibitors NSC 19630 and NSC 617145 induce apoptosis in HTLV-1-transformed adult T-cell leukemia cells. *J. Hematol. Oncol.* **2016**, *9*, 121. [CrossRef]
160. Aggarwal, M.; Banerjee, T.; Sommers, J.A.; Brosh, R.M., Jr. Targeting an Achilles' heel of cancer with a WRN helicase inhibitor. *Cell Cycle* **2013**, *12*, 3329–3335. [CrossRef]
161. Sommers, J.A.; Kulikowicz, T.; Croteau, D.L.; Dexheimer, T.; Dorjsuren, D.; Jadhav, A.; Maloney, D.J.; Simeonov, A.; Bohr, V.A.; Brosh, R.M., Jr. A high-throughput screen to identify novel small molecule inhibitors of the Werner Syndrome Helicase-Nuclease (WRN). *PLoS ONE* **2019**, *14*, e0210525. [CrossRef] [PubMed]
162. Banerjee, T.; Aggarwal, M.; Sommers, J.A.; Brosh, R.M., Jr. Biochemical and cell biological assays to identify and characterize DNA helicase inhibitors. *Methods* **2016**, *108*, 130–141. [CrossRef] [PubMed]
163. Malfatti, M.C.; Gerratana, L.; Dalla, E.; Isola, M.; Damante, G.; Di Loreto, C.; Puglisi, F.; Tell, G. APE1 and NPM1 protect cancer cells from platinum compounds cytotoxicity and their expression pattern has a prognostic value in TNBC. *J. Exp. Clin. Cancer Res.* **2019**, *38*, 309. [CrossRef] [PubMed]

164. Yuan, C.L.; He, F.; Ye, J.Z.; Wu, H.N.; Zhang, J.Y.; Liu, Z.H.; Li, Y.Q.; Luo, X.L.; Lin, Y.; Liang, R. APE1 overexpression is associated with poor survival in patients with solid tumors: A meta-analysis. *Oncotarget* **2017**, *8*, 59720–59728. [CrossRef] [PubMed]
165. Poletto, M.; Malfatti, M.C.; Dorjsuren, D.; Scognamiglio, P.L.; Marasco, D.; Vascotto, C.; Jadhav, A.; Maloney, D.J.; Wilson, D.M., 3rd; Simeonov, A.; et al. Inhibitors of the apurinic/apyrimidinic endonuclease 1 (APE1)/nucleophosmin (NPM1) interaction that display anti-tumor properties. *Mol. Carcinog.* **2016**, *55*, 688–704. [CrossRef]
166. Codrich, M.; Comelli, M.; Malfatti, M.C.; Mio, C.; Ayyildiz, D.; Zhang, C.; Kelley, M.R.; Terrosu, G.; Pucillo, C.E.M.; Tell, G. Inhibition of APE1-endonuclease activity affects cell metabolism in colon cancer cells via a p53-dependent pathway. *DNA Repair (Amst)* **2019**, *82*, 102675. [CrossRef]
167. Guerreiro, P.S.; Corvacho, E.; Costa, J.G.; Saraiva, N.; Fernandes, A.S.; Castro, M.; Miranda, J.P.; Oliveira, N.G. The APE1 redox inhibitor E3330 reduces collective cell migration of human breast cancer cells and decreases chemoinvasion and colony formation when combined with docetaxel. *Chem. Biol. Drug Des.* **2017**, *90*, 561–571. [CrossRef]
168. Ma, X.; Dang, C.; Min, W.; Diao, Y.; Hui, W.; Wang, X.; Dai, Z.; Wang, X.; Kang, H. Downregulation of APE1 potentiates breast cancer cells to olaparib by inhibiting PARP-1 expression. *Breast Cancer Res. Treat.* **2019**, *176*, 109–117. [CrossRef]
169. Nickoloff, J.A.; Jones, D.; Lee, S.H.; Williamson, E.A.; Hromas, R. Drugging the Cancers Addicted to DNA Repair. *J. Natl. Cancer Inst.* **2017**, *109*, 1–13. [CrossRef]
170. Jaiswal, A.S.; Panda, H.; Law, B.K.; Sharma, J.; Jani, J.; Hromas, R.; Narayan, S. NSC666715 and Its Analogs Inhibit Strand-Displacement Activity of DNA Polymerase beta and Potentiate Temozolomide-Induced DNA Damage, Senescence and Apoptosis in Colorectal Cancer Cells. *PLoS ONE* **2015**, *10*, e0123808. [CrossRef]
171. Song, X.; Wang, S.; Hong, X.; Li, X.; Zhao, X.; Huai, C.; Chen, H.; Gao, Z.; Qian, J.; Wang, J.; et al. Single nucleotide polymorphisms of nucleotide excision repair pathway are significantly associated with outcomes of platinum-based chemotherapy in lung cancer. *Sci. Rep.* **2017**, *7*, 11785. [CrossRef] [PubMed]
172. Pajuelo-Lozano, N.; Bargiela-Iparraguirre, J.; Dominguez, G.; Quiroga, A.G.; Perona, R.; Sanchez-Perez, I. XPA, XPC, and XPD Modulate Sensitivity in Gastric Cisplatin Resistance Cancer Cells. *Front. Pharmacol.* **2018**, *9*, 1197. [CrossRef] [PubMed]
173. Hu, L.B.; Chen, Y.; Meng, X.D.; Yu, P.; He, X.; Li, J. Nucleotide Excision Repair Factor XPC Ameliorates Prognosis by Increasing the Susceptibility of Human Colorectal Cancer to Chemotherapy and Ionizing Radiation. *Front. Oncol.* **2018**, *8*, 290. [CrossRef] [PubMed]
174. Du, P.; Wang, Y.; Chen, L.; Gan, Y.; Wu, Q. High ERCC1 expression is associated with platinum-resistance, but not survival in patients with epithelial ovarian cancer. *Oncol. Lett.* **2016**, *12*, 857–862. [CrossRef]
175. Chabanon, R.M.; Muirhead, G.; Krastev, D.B.; Adam, J.; Morel, D.; Garrido, M.; Lamb, A.; Henon, C.; Dorvault, N.; Rouanne, M.; et al. PARP inhibition enhances tumor cell-intrinsic immunity in ERCC1-deficient non-small cell lung cancer. *J. Clin. Investig.* **2019**, *129*, 1211–1228. [CrossRef]
176. Cheng, H.; Zhang, Z.; Borczuk, A.; Powell, C.A.; Balajee, A.S.; Lieberman, H.B.; Halmos, B. PARP inhibition selectively increases sensitivity to cisplatin in ERCC1-low non-small cell lung cancer cells. *Carcinogenesis* **2013**, *34*, 739–749. [CrossRef]
177. Arora, S.; Heyza, J.; Zhang, H.; Kalman-Maltese, V.; Tillison, K.; Floyd, A.M.; Chalfin, E.M.; Bepler, G.; Patrick, S.M. Identification of small molecule inhibitors of ERCC1-XPF that inhibit DNA repair and potentiate cisplatin efficacy in cancer cells. *Oncotarget* **2016**, *7*, 75104–75117. [CrossRef]
178. McNeil, E.M.; Astell, K.R.; Ritchie, A.M.; Shave, S.; Houston, D.R.; Bakrania, P.; Jones, H.M.; Khurana, P.; Wallace, C.; Chapman, T.; et al. Inhibition of the ERCC1-XPF structure-specific endonuclease to overcome cancer chemoresistance. *DNA Repair (Amst)* **2015**, *31*, 19–28. [CrossRef]
179. Burkitt, K.; Ljungman, M. Phenylbutyrate interferes with the Fanconi anemia and BRCA pathway and sensitizes head and neck cancer cells to cisplatin. *Mol. Cancer* **2008**, *7*, 24. [CrossRef]
180. Chen, C.C.; Taniguchi, T.; D'Andrea, A. The Fanconi anemia (FA) pathway confers glioma resistance to DNA alkylating agents. *J. Mol. Med. (Berl)* **2007**, *85*, 497–509. [CrossRef]
181. Chirnomas, D.; Taniguchi, T.; de la Vega, M.; Vaidya, A.P.; Vasserman, M.; Hartman, A.R.; Kennedy, R.; Foster, R.; Mahoney, J.; Seiden, M.V.; et al. Chemosensitization to cisplatin by inhibitors of the Fanconi anemia/BRCA pathway. *Mol. Cancer Ther.* **2006**, *5*, 952–961. [CrossRef] [PubMed]

182. Kee, Y.; Huang, M.; Chang, S.; Moreau, L.A.; Park, E.; Smith, P.G.; D'Andrea, A.D. Inhibition of the Nedd8 system sensitizes cells to DNA interstrand cross-linking agents. *Mol. Cancer Res.* **2012**, *10*, 369–377. [CrossRef] [PubMed]
183. Xu, X.; Li, S.; Cui, X.; Han, K.; Wang, J.; Hou, X.; Cui, L.; He, S.; Xiao, J.; Yang, Y. Inhibition of Ubiquitin Specific Protease 1 Sensitizes Colorectal Cancer Cells to DNA-Damaging Chemotherapeutics. *Front. Oncol.* **2019**, *9*, 1406. [CrossRef]
184. Liang, Q.; Dexheimer, T.S.; Zhang, P.; Rosenthal, A.S.; Villamil, M.A.; You, C.; Zhang, Q.; Chen, J.; Ott, C.A.; Sun, H.; et al. A selective USP1-UAF1 inhibitor links deubiquitination to DNA damage responses. *Nat. Chem. Biol.* **2014**, *10*, 298–304. [CrossRef] [PubMed]
185. Kennedy, R.D.; Chen, C.C.; Stuckert, P.; Archila, E.M.; De la Vega, M.A.; Moreau, L.A.; Shimamura, A.; D'Andrea, A.D. Fanconi anemia pathway-deficient tumor cells are hypersensitive to inhibition of ataxia telangiectasia mutated. *J. Clin. Investig.* **2007**, *117*, 1440–1449. [CrossRef] [PubMed]
186. Kamaletdinova, T.; Fanaei-Kahrani, Z.; Wang, Z.Q. The Enigmatic Function of PARP1: From PARylation Activity to PAR Readers. *Cells* **2019**, *8*, 1625. [CrossRef]
187. Chen, K.; Li, Y.; Xu, H.; Zhang, C.; Li, Z.; Wang, W.; Wang, B. An analysis of the gene interaction networks identifying the role of PARP1 in metastasis of non-small cell lung cancer. *Oncotarget* **2017**, *8*, 87263–87275. [CrossRef]
188. Liu, Y.; Zhang, Y.; Zhao, Y.; Gao, D.; Xing, J.; Liu, H. High PARP-1 expression is associated with tumor invasion and poor prognosis in gastric cancer. *Oncol. Lett.* **2016**, *12*, 3825–3835. [CrossRef]
189. Giovannini, S.; Weller, M.C.; Repmann, S.; Moch, H.; Jiricny, J. Synthetic lethality between BRCA1 deficiency and poly(ADP-ribose) polymerase inhibition is modulated by processing of endogenous oxidative DNA damage. *Nucleic Acids Res.* **2019**, *47*, 9132–9143. [CrossRef]
190. Awada, A.; Campone, M.; Varga, A.; Aftimos, P.; Frenel, J.S.; Bahleda, R.; Gombos, A.; Bourbouloux, E.; Soria, J.C. An open-label, dose-escalation study to evaluate the safety and pharmacokinetics of CEP-9722 (a PARP-1 and PARP-2 inhibitor) in combination with gemcitabine and cisplatin in patients with advanced solid tumors. *Anticancer Drugs* **2016**, *27*, 342–348. [CrossRef]
191. Russo, A.L.; Kwon, H.C.; Burgan, W.E.; Carter, D.; Beam, K.; Weizheng, X.; Zhang, J.; Slusher, B.S.; Chakravarti, A.; Tofilon, P.J.; et al. In vitro and in vivo radiosensitization of glioblastoma cells by the poly (ADP-ribose) polymerase inhibitor E7016. *Clin. Cancer Res.* **2009**, *15*, 607–612. [CrossRef] [PubMed]
192. Jannetti, S.A.; Zeglis, B.M.; Zalutsky, M.R.; Reiner, T. Poly(ADP-Ribose)Polymerase (PARP) Inhibitors and Radiation Therapy. *Front. Pharmacol* **2020**, *11*, 170. [CrossRef] [PubMed]
193. Pirovano, G.; Jannetti, S.A.; Carter, L.M.; Sadique, A.; Kossatz, S.; Guru, N.; Demetrio De Souza Franca, P.; Maeda, M.; Zeglis, B.M.; Lewis, J.S.; et al. Targeted Brain Tumor Radiotherapy Using an Auger Emitter. *Clin. Cancer Res.* **2020**, *26*, 2871–2881. [CrossRef] [PubMed]
194. Makvandi, M.; Lee, H.; Puentes, L.N.; Reilly, S.W.; Rathi, K.S.; Weng, C.C.; Chan, H.S.; Hou, C.; Raman, P.; Martinez, D.; et al. Targeting PARP-1 with Alpha-Particles Is Potently Cytotoxic to Human Neuroblastoma in Preclinical Models. *Mol. Cancer Ther.* **2019**, *18*, 1195–1204. [CrossRef]
195. Gadducci, A.; Guarneri, V.; Peccatori, F.A.; Ronzino, G.; Scandurra, G.; Zamagni, C.; Zola, P.; Salutari, V. Current strategies for the targeted treatment of high-grade serous epithelial ovarian cancer and relevance of BRCA mutational status. *J. Ovarian Res.* **2019**, *12*, 9. [CrossRef]
196. Gadducci, A.; Guerrieri, M.E. PARP inhibitors alone and in combination with other biological agents in homologous recombination deficient epithelial ovarian cancer: From the basic research to the clinic. *Crit. Rev. Oncol. Hematol.* **2017**, *114*, 153–165. [CrossRef]
197. D'Andrea, A.D. Mechanisms of PARP inhibitor sensitivity and resistance. *DNA Repair (Amst)* **2018**, *71*, 172–176. [CrossRef]
198. La Ferla, M.; Mercatanti, A.; Rocchi, G.; Lodovichi, S.; Cervelli, T.; Pignata, L.; Caligo, M.A.; Galli, A. Expression of human poly (ADP-ribose) polymerase 1 in Saccharomyces cerevisiae: Effect on survival, homologous recombination and identification of genes involved in intracellular localization. *Mutat. Res.* **2015**, *774*, 14–24. [CrossRef]
199. Lodovichi, S.; Mercatanti, A.; Cervelli, T.; Galli, A. Computational analysis of data from a genome-wide screening identifies new PARP1 functional interactors as potential therapeutic targets. *Oncotarget* **2019**, *10*, 2722–2737. [CrossRef]

200. Antolin, A.A.; Ameratunga, M.; Banerji, U.; Clarke, P.A.; Workman, P.; Al-Lazikani, B. The kinase polypharmacology landscape of clinical PARP inhibitors. *Sci. Rep.* **2020**, *10*, 2585. [CrossRef]
201. Knezevic, C.E.; Wright, G.; Rix, L.L.R.; Kim, W.; Kuenzi, B.M.; Luo, Y.; Watters, J.M.; Koomen, J.M.; Haura, E.B.; Monteiro, A.N.; et al. Proteome-wide Profiling of Clinical PARP Inhibitors Reveals Compound-Specific Secondary Targets. *Cell Chem. Biol.* **2016**, *23*, 1490–1503. [CrossRef] [PubMed]
202. Woo, S.R.; Corrales, L.; Gajewski, T.F. Innate immune recognition of cancer. *Annu. Rev. Immunol.* **2015**, *33*, 445–474. [CrossRef] [PubMed]
203. Huang, A.; Garraway, L.A.; Ashworth, A.; Weber, B. Synthetic lethality as an engine for cancer drug target discovery. *Nat. Rev. Drug Discov.* **2020**, *19*, 23–38. [CrossRef] [PubMed]
204. Dhanjal, J.K.; Radhakrishnan, N.; Sundar, D. Identifying synthetic lethal targets using CRISPR/Cas9 system. *Methods* **2017**, *131*, 66–73. [CrossRef] [PubMed]

International Journal of *Molecular Sciences*

MDPI

Review

Inhibitors of DNA Glycosylases as Prospective Drugs

Grigory V. Mechetin [1], Anton V. Endutkin [1], Evgeniia A. Diatlova [1] and Dmitry O. Zharkov [1,2,*]

1 SB RAS Institute of Chemical Biology and Fundamental Medicine, 8 Lavrentieva Ave., 630090 Novosibirsk, Russia; mechetin@niboch.nsc.ru (G.V.M.); aend@niboch.nsc.ru (A.V.E.); e.diatlova@g.nsu.ru (E.A.D.)

2 Novosibirsk State University, 2 Pirogova St., 630090 Novosibirsk, Russia

* Correspondence: dzharkov@niboch.nsc.ru; Tel.: +7-383-363-5187

Received: 29 March 2020; Accepted: 27 April 2020; Published: 28 April 2020

Abstract: DNA glycosylases are enzymes that initiate the base excision repair pathway, a major biochemical process that protects the genomes of all living organisms from intrinsically and environmentally inflicted damage. Recently, base excision repair inhibition proved to be a viable strategy for the therapy of tumors that have lost alternative repair pathways, such as BRCA-deficient cancers sensitive to poly(ADP-ribose)polymerase inhibition. However, drugs targeting DNA glycosylases are still in development and so far have not advanced to clinical trials. In this review, we cover the attempts to validate DNA glycosylases as suitable targets for inhibition in the pharmacological treatment of cancer, neurodegenerative diseases, chronic inflammation, bacterial and viral infections. We discuss the glycosylase inhibitors described so far and survey the advances in the assays for DNA glycosylase reactions that may be used to screen pharmacological libraries for new active compounds.

Keywords: DNA repair; DNA glycosylases; drug targets

1. Introduction: Base Excision Repair

DNA in living cells is always exposed to many damaging factors of environmental and endogenous origin. These insults produce various modifications of nucleobases and lead to the formation of apurinic/apyrimidinic (AP) sites and DNA strand breaks. Accumulating damage significantly affects genomic stability and may ultimately end in mutations or cell death [1,2].

Mechanisms that correct genomic damage, commonly known as DNA repair systems, exist in all forms of life [1]. The most abundant DNA lesions—deaminated, oxidized, and alkylated bases, AP sites, single-strand breaks—are repaired by the base excision repair (BER) system [1,3]. Initially, the damaged base is recognized by a DNA glycosylase (of which, 11 are known in humans, and 8 in *Escherichia coli*), which cleaves the *N*-glycosidic bond between the nucleobase and the C1′ deoxyribose atom, forming an AP site. Afterwards, an AP endonuclease hydrolyzes the phosphodiester bond 5′ to the AP site. The repair proceeds with the incorporation of the correct dNMP by a DNA polymerase and the ligation of the remaining single-strand break (Figure 1).

DNA glycosylases belong to several major structural superfamilies and have different, sometimes overlapping, substrate specificities (Table 1). They all, however, have a common reaction mechanism: the C1′ of the damaged nucleotide is attacked by a nucleophilic moiety (either an activated water molecule in monofunctional DNA glycosylases or an enzyme's amino group in bifunctional ones), the damaged base departs, and either an AP site is generated (monofunctional glycosylases), or a Schiff base covalent intermediate forms (bifunctional glycosylases) [4–6]. The Schiff base undergoes β- or β,δ-elimination and is then hydrolyzed, leaving a nick with either an α,β-unsaturated aldehyde or a phosphate as the 3′-terminal group. To access C1′, most glycosylases flip the target nucleotide from the DNA stack into the enzyme's active site, which is equipped with a deep lesion recognition pocket, representing a convenient druggable target [7].

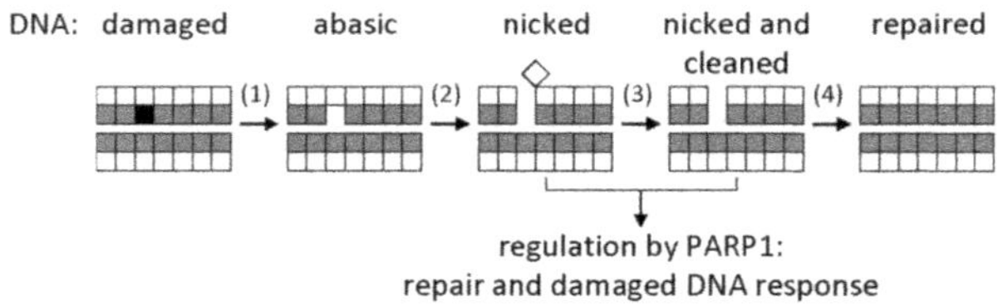

Figure 1. General base excision repair (BER) scheme. Bases are represented by shaded squares, and sugars by white squares. A damaged base (black) is excised by a DNA glycosylase (1); the resulting apurinic/apyrimidinic (AP) site is cut by an AP endonuclease (2); the deoxyribose fragment is removed by a deoxyribophosphate lyase (3); a correct dNMP is incorporated by a DNA polymerase, and the nick is sealed by a DNA ligase (4). Nicked DNA also activates signaling by poly(ADP-ribose)polymerase 1 (PARP1), which initiates poly(ADP-ribosyl)ation of many chromatin proteins to facilitate the access of DNA repair factors to the site of damage.

Table 1. Examples of DNA glycosylases found in humans, *Escherichia coli*, and other species.

Structural Superfamily	*E. coli*	Human	Other Organisms	Major Substrate Specificity
α/β fold	Ung	UNG		U in any context
	Mug	TDG		U, T, 3,N^4-ethenocytosine, oxidized/deaminated 5-methylcytosine opposite G
		SMUG1		U:G
helix–hairpin–helix		MBD4		U opposite G in CpG context
	Nth	NTHL1		oxidized pyrimidines
		OGG1		oxidized purines
	MutY	MUTYH		A opposite 8-oxoguanine
	AlkA			ring-alkylated purines, 1,N^6-ethenoadenine, hypoxanthine
			Micrococcus luteus Pdg	cyclobutane thymine dimers
helix–two-turn–helix	Nei	NEIL1		oxidized pyrimidines
		NEIL2		oxidized pyrimidines in DNA bubbles and loops
		NEIL3		oxidized pyrimidines in single-strand DNA
	Fpg			oxidized purines
	Tag			3-methyladenine
			T4 phage DenV	cyclobutane thymine dimers
HEAT repeats			*Bacillus cereus* AlkC, AlkD	ring-alkylated purines, minor groove adducts

In human cells, BER is tightly regulated at several levels. One of the best-studied players orchestrating the BER process is poly(ADP-ribose)polymerase 1 (PARP1), together with its homologs PARP2 and PARP3, which act as nick sensors and regulate the access of repair factors to the damage sites through modification of acceptor proteins and DNA ends by poly(ADP-ribose) [8,9]. PARPs attracted attention as potential targets for cancer treatment after PARP inhibitors were discovered to be highly toxic for cells with inactivated homologous recombination repair pathway [10,11]. In human tumors, recombination repair deficiency is often associated with inactivating mutations in the *BRCA1* and *BRCA2* genes, the main driver mutations in hereditary breast and ovarian cancers. BRCA1 and BRCA2 proteins regulate the DNA break response through a pathway that does not overlap with BER [12].

Blocking both these pathways is lethal for the cell, while normal cells with active recombination repair survive PARP1 inhibition. The lethal effect of PARP inhibitors is largely mediated by PARP trapping at nicks [13,14], which mainly originate from ribonucleotides misincorporated during DNA replication [15]. Several PARP inhibitors are presently approved for clinical use, and several hundred clinical trials are ongoing.

2. Inhibitors of DNA Glycosylases: General Considerations

The example of PARP inhibitors highlights the concept of synthetic lethality, which underlies most of the attempts to develop BER inhibitors into practically useful drugs. Two conditions must be satisfied for such compounds to be effective. First, the target cells must experience genotoxic stress, induced either directly by DNA-damaging factors or indirectly through some kind of metabolic stress (nucleotide pool imbalance, oxidative stress, etc.). Second, if DNA damage caused by this type of stress can be repaired bypassing BER, the bypass must be blocked by mutations or another drug used in combination with a BER inhibitor. The second requirement is often satisfied in cancers, where mutations in DNA repair genes are usually among driver mutations. Genotoxic treatment may also be tuned to produce lesions repaired predominantly by BER (such as uracil accumulated through antimetabolite treatment, uracil analogs used as drugs or prodrugs, or oxidized purines appearing through MTH1 inhibition or induced by photodynamic therapy), in which case BER impairment alone could be sufficient to effect considerable cytotoxicity.

Two considerations are crucial when assessing the cytotoxic potential of DNA glycosylase inhibitors. First, unlike the enzymes underlying common BER steps, such as break signaling by PARPs or AP site cleavage by AP endonucleases, DNA glycosylases are specific for damaged bases, and their inhibition will affect only a subset of BER reactions. This in fact may be advantageous for fine-tuning or selection of concurrently used DNA damaging agents, many of which produce specific primary lesions rather than AP sites of strand breaks [16]. Second, DNA glycosylases are often ambivalent with respect to cell-killing effects of DNA damage (as discussed below in sections about specific types of lesions and their repair): they may either counteract the damage by repairing the induced lesions or potentiate the damage by converting damaged bases to AP sites or strand breaks, which are generally more cytotoxic. Thus, the inhibition of DNA glycosylases is not always warranted for inducing synthetic lethality in cancer cells or bacteria. It is always desirable to validate a particular DNA glycosylase as a drug target by knockout or knockdown approaches in an appropriate cell line or pathogen.

The inhibitors discussed in the remaining parts of this paper are mostly small-molecule compounds. Almost all DNA glycosylases are inhibited to a certain degree by non-specific single- or double-stranded DNA, competing for binding with substrate DNA [17–21], and a number of modified nucleotides tightly bound but not cleaved when incorporated into oligonucleotides have been described [22–28]. Moreover, binding and inhibition of DNA glycosylases by polyanions such as heparin [29–32] likely stems from the ability of these enzymes to bind nucleic acids. Minor-groove ligands of various chemical nature also interfere with DNA glycosylase binding [33,34]. Despite the obvious importance of such interactions for the biological functions of DNA glycosylases, delivery and targeting problems thus far prevent the therapeutic use of oligonucleotides and other macromolecular polyanions as mass action-driven inhibitors of intracellular enzymes. However, one strategy known for a while and recently applied to DNA glycosylases is the use of prodrugs that are metabolized to nucleoside triphosphate analogs and incorporated into DNA [35,36]. For example, 1′-cyano-2′-deoxyuridine triphosphate is a good substrate for DNA polymerases and, when incorporated into DNA, inhibits *E. coli* uracil-DNA glycosylase (Ung) and human UNG, displaying nanomolar K_i values [37]. Interestingly, some lesions, such as 2-deoxyribonolactone and 5-hydroxy-5-methylhydantoin [27,38,39], demonstrate an intrinsic ability to trap bifunctional DNA glycosylases covalently on DNA, reminiscent of the PARP-trapping potential of cytotoxic PARP inhibitors. Thus, the development of nucleotides that can be incorporated into DNA and trap DNA glycosylases may be an interesting direction of the glycosylase inhibitor design.

3. Uracil in DNA: Synergism of Glycosylase Inhibitors and Antimetabolites

Antimetabolites, the class of drugs interfering with nucleotide metabolism pathways and thereby with DNA or RNA synthesis, are one of the staples of therapeutic interventions against cancer and bacterial and protozoan infections and are especially useful in combination therapy [40,41]. Many clinically used antimetabolites, such as antifolates, interfere with thymine biosynthesis and cause the accumulation of uracil (or its analogs) in genomic DNA [42,43]. The repair of drug-induced genomic uracil is a double-edge sword: while it protects cells from the effect of this non-canonical nucleobase at low levels of substitution, extensive uracil buildup and excision are toxic and may be the primary reason of cell death after exposure to antifolates [44,45]. Therefore, the inhibition of uracil repair may have different consequences depending on the level of DNA modification and, possibly, on the nature of the modification (if different from uracil).

Human cells possess four DNA glycosylases capable of excising uracil from DNA. However, for three of them (TDG, SMUG1, and MBD4), uracil either is not the main substrate or is removed only from specific contexts (for example, methylated CpG dinucleotides). The main enzyme responsible for uracil repair, UNG, is among the most important factors limiting the efficiency of antifolates and fludarabine, whose action is based on the accumulation of uracil in genomic DNA [46–48]. *UNG* knockdown in human prostate cancer cell lines increases their sensitivity to H_2O_2 and doxorubicin [49]. Non-small cell carcinoma and lung adenocarcinoma cells develop spontaneous resistance to pemetrexed, a folic acid analog inhibiting dihydrofolate reductase, thymidylate synthase, and glycinamide ribonucleotide formyltransferase, due to a significant increase in the level of UNG, and the suppression of *UNG* expression returns the sensitivity to normal [50,51]. Some uracil analogs that are accumulated in DNA (such as 5-fluorouracil) are more toxic for cells when SMUG1, rather than UNG, is downregulated [52,53]. However, due to the structural similarity between UNG, TDG, and SMUG1, low-molecular-weight inhibitors will most likely be active against all three enzymes; therefore, the nature of the glycosylase that removes uracil during treatment with antimetabolites is not of primary importance.

UNG inhibitors in combination with genotoxic stress effectively suppress the growth of *Plasmodium falciparum* [54], *Trypanosoma brucei* [55,56], and *Trypanosoma cruzi* [57,58], which makes BER a promising target for drug intervention in protozoan infections. Importantly, some inhibitors preferentially suppress the activity of UNG from infections agents but have little effect on the human enzyme (Table 2).

Low-molecular-weight UNG inhibitors are still at a preclinical stage. The literature describes three main classes of such inhibitors. All of them are competitive, mechanism-based and mimic certain features of the transition state of the UNG-catalyzed reaction [6,79]. Free uracil and its analogs inhibit UNG enzymes from various sources, presenting submillimolar to millimolar IC_{50} [59,80–85], so successful inhibitors required extensive modification of the base to ensure tight binding. Historically, the first class of compounds active against human *Plasmodium* and herpes simplex type 1 UNGs was 6-(*p*-alkylanilino)uracils, of which 6-(*p*-n-octylanilino)uracil showed the strongest affinity, with IC_{50} ~8 μM for the viral enzyme [54,59,86–88] (Table 2). Bipartite inhibitors, structurally similar to the 6-substituted derivatives, consist of a uracil base or its analog linked to a phenolic or benzoic fragment [60–62,89] (Table 2). In the structures of bipartite inhibitors bound to UNG, the uracil part occupies the uracil-binding pocket of the enzyme, while the aromatic fragment lies in the DNA-binding groove [61,62] (Table 3). Finally, triskelion inhibitors contain three functional groups at the ends of a branched linker: either one analog of uracil and two aromatic fragments, or two analogs of uracil and one aromatic fragment [63] (Table 2). Interestingly, gentamicin, a clinically used aminoglycoside antibiotic, was reported to inhibit *E. coli* Ung [64,65] (Table 2). Although the reported IC_{50} value was quite high (0.4–1.5 mM), this effect may reflect interactions between Ung and the sugar part of DNA [21,90] and suggests another possible direction for inhibitor development.

Table 2. Properties of selected DNA glycosylase inhibitors discussed in this review.

General Structure	R	Enzyme	Species	I_{50}, μM	Reference
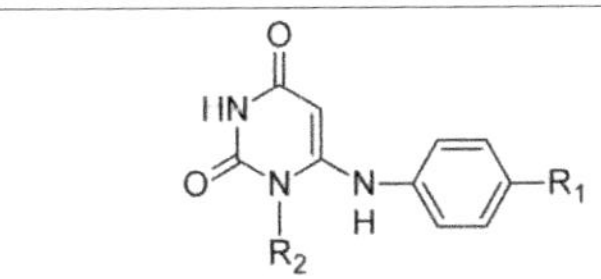	R_1 = *n*-octyl R_2 = H	UNG	HSV1	8	[59]
			human	>300	
	R_1 – *n*-octyl R_2 = 1-methoxyethyl		*Plasmodium falciparum*	17	[54]
			human	>160	
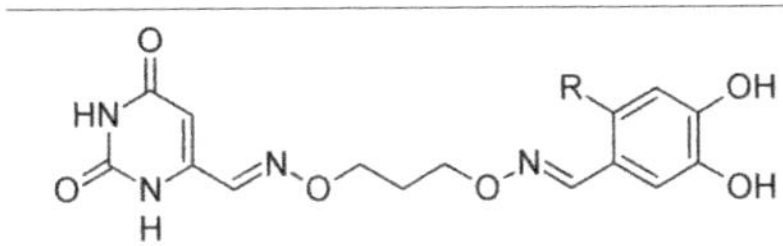 bipartite inhibitors	H	UNG	human	1.1	[60]
	OH			0.26	
	F			2.7	
	Cl			16	
	Br			40	
	NO_2			40	
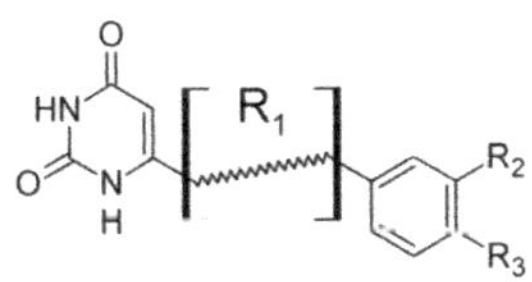 bipartite inhibitors	R_1 = 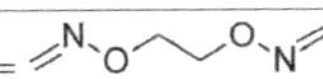 R_2 = COOH, R_3 = H	UNG	human	40	[61]
	R1 =  R_2 = COOH, R_3 = H			1.6	
	R_1 = 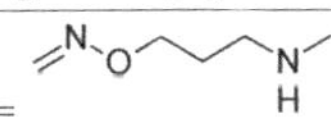 R_2 = COOH, R_3 = H			100	
	R_1 = 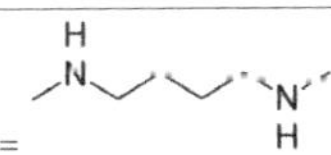 R_2 = COOH, R_3 = H			315	
	R_1 = 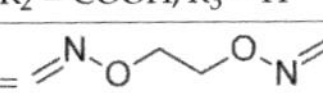 R_2 = H, R_3 = COOH			6	[62]

Table 2. *Cont.*

General Structure	R	Enzyme	Species	I_{50}, μM	Reference
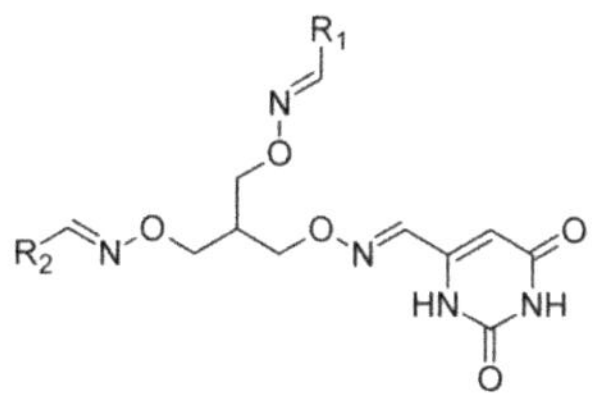 triskelion inhibitors	$R_1 = R_2 =$ 3,4-dihydroxyphenyl	UNG	human	1.6	[63]
	R_1 = 6-uracil R_2 = 3,4-dihydroxyphenyl			0.9	
	$R_1 = R_2$ = 3-carboxyphenyl			1.7	
	R_1 = 6-uracil R_2 = 3-carboxyphenyl			0.9	
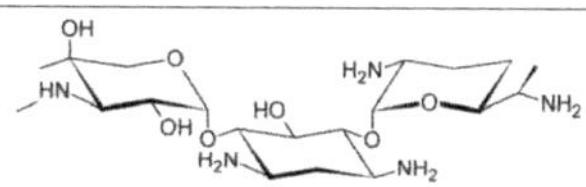 gentamicin		UNG	not specified	1500	[64]
			E. coli	420	[65]
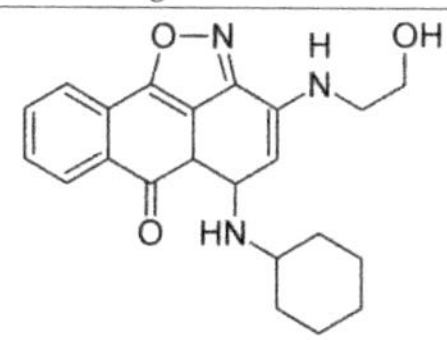		UNG	vaccinia virus	34 *	[66]
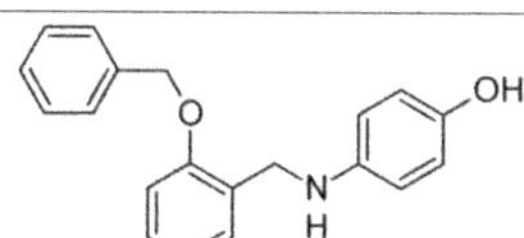				5.1 *	
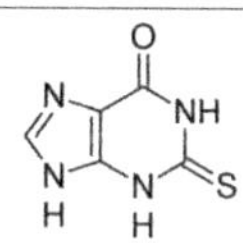 thioxanthine		Fpg	*E. coli*	17	[67]
		Fpg	*Lactococcus lactis*	100	[68]

Table 2. *Cont.*

General Structure	R	Enzyme	Species	I_{50}, μM	Reference
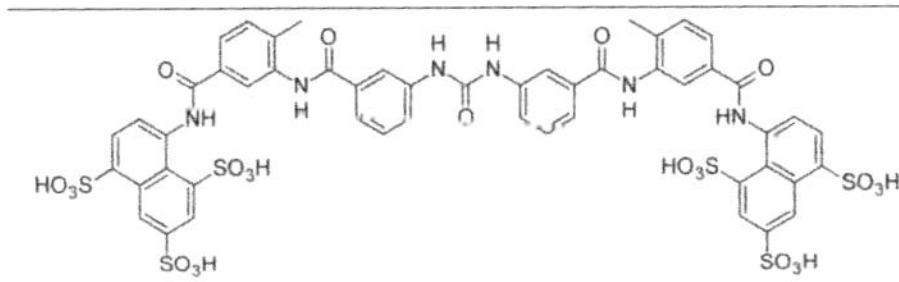 suramin		MutY	*Corynebacterium pseudotuberculosis*	16 **	[32]
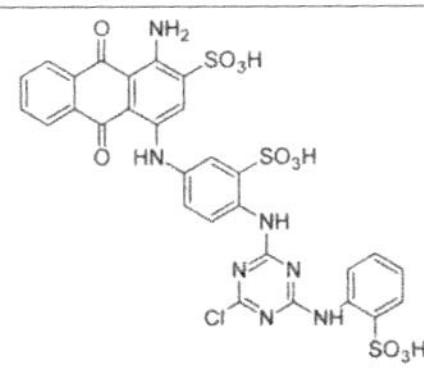 Cibacron Blue F3GA		Fpg	*E. coli*	0.005 **	[69]
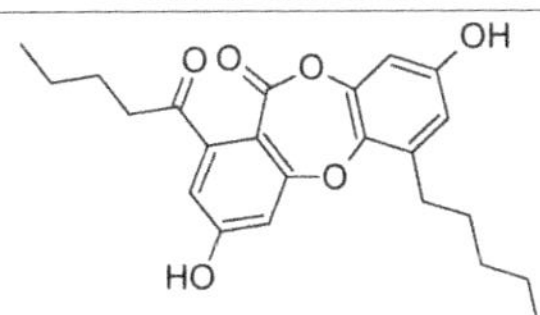 norlobaric acid		Nei2	*Mycobacterium tuberculosis*	42, 0.074 **	[70]
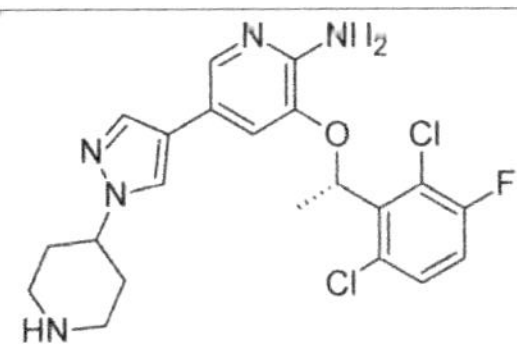 *R*-crizotinib		MTH1 ***	Human	0.33, 0.048 **	[71]

Table 2. *Cont.*

General Structure	R	Enzyme	Species	I_{50}, μM	Reference
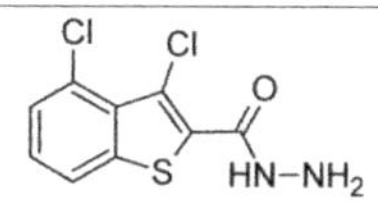		OGG1	human	0.22	[72]
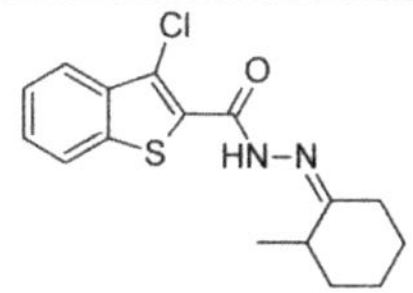				0.27	
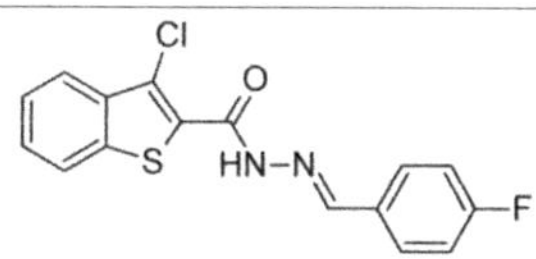				0.33	
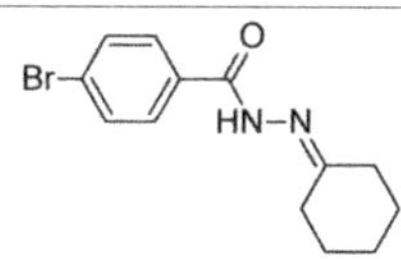				0.63	
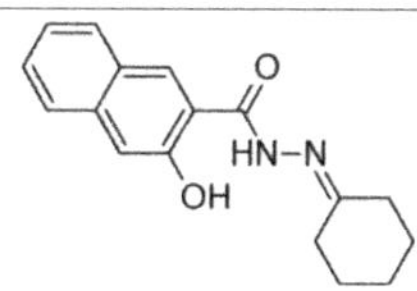				0.34	
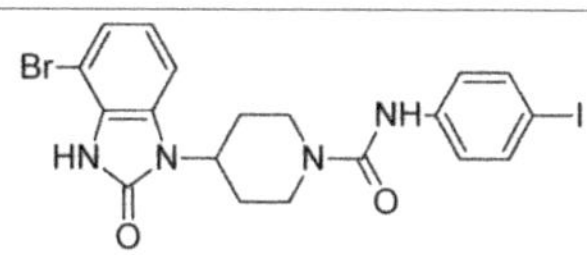 TH5487		OGG1	mouse	0.34	[73]

Table 2. *Cont.*

General Structure	R	Enzyme	Species	I_{50}, μM	Reference
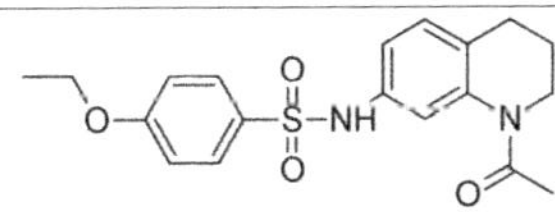		OGG1	human	2	[74]
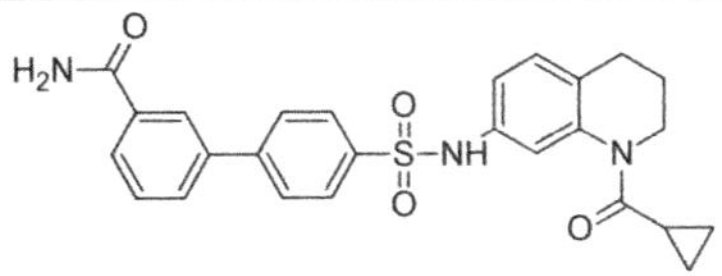				0.059	[74]
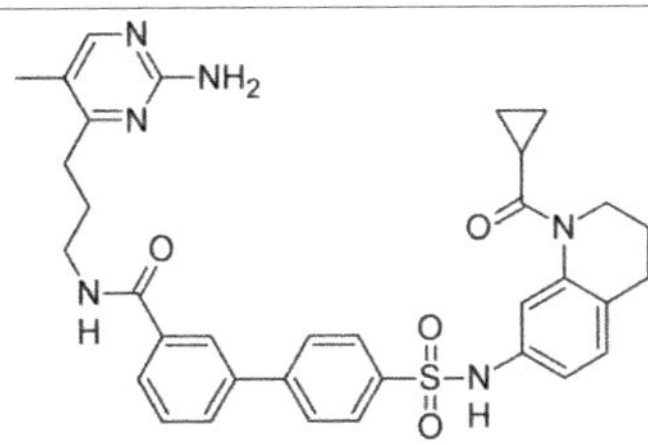 SU0383		OGG1	human	0.49	[75]
		MTH1 ***	human	0.034	

Table 2. *Cont.*

General Structure	R	Enzyme	Species	I_{50}, μM	Reference
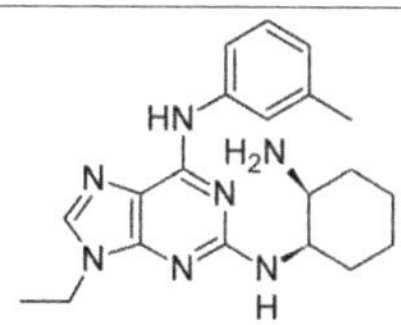		NEIL1	human	25	[76]
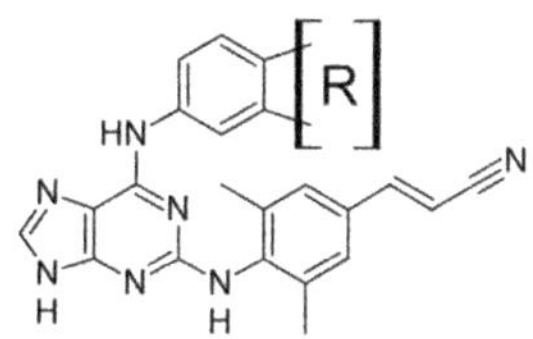	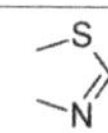			4.0	
	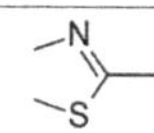			7.9	
	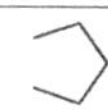			8.9	
	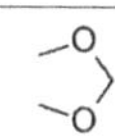			10	
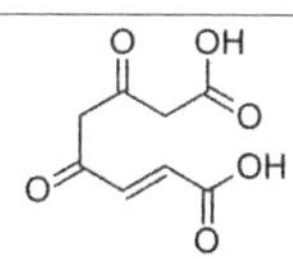 Fumarylacetoacetate		NEIL1	human	0.006	[77]
		NEIL2	human	0.032	
		OGG1	human	1.0	
		NTHL1	human	1.0	
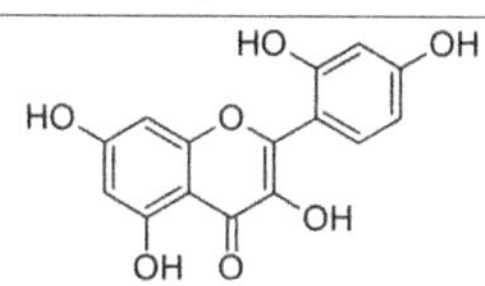 Morin		MPG	human	2.6	[78]

* IC_{50} for DNA polymerase activity in the presence of the D4/A20 complex; ** K_d or K_i directly measured; *** non-glycosylase member of the GO system (see Sections 4 and 5).

Table 3. Known structures of DNA glycosylases bound to their inhibitors.

PDB ID	Enzyme	Inhibitor	Resolution, Å	Reference
6G3Y	OGG1	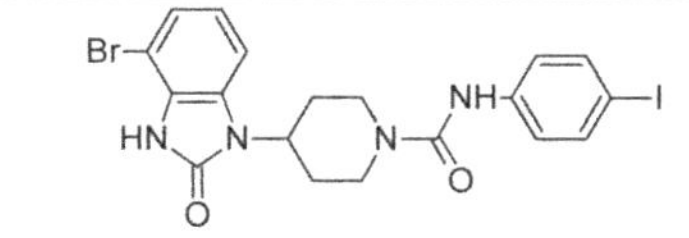	2.51	[73]
2HXM	UNG	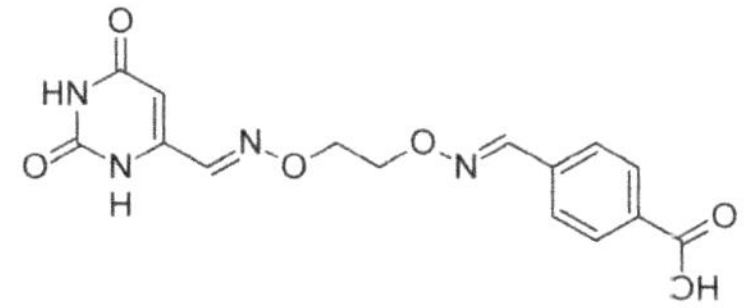	1.30	[62]
3FCF	UNG	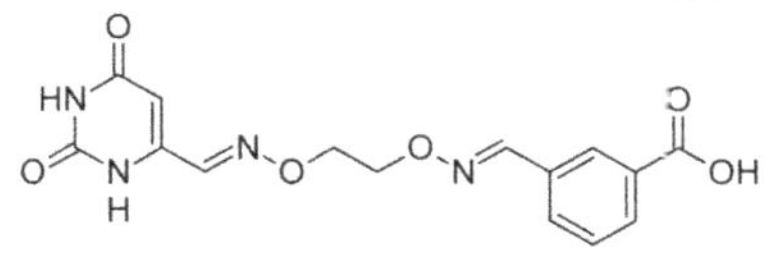	1.84	[61]
3FCI	UNG	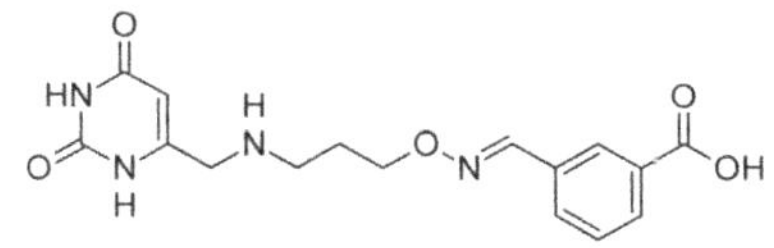	1.27	[61]
3FCK	UNG	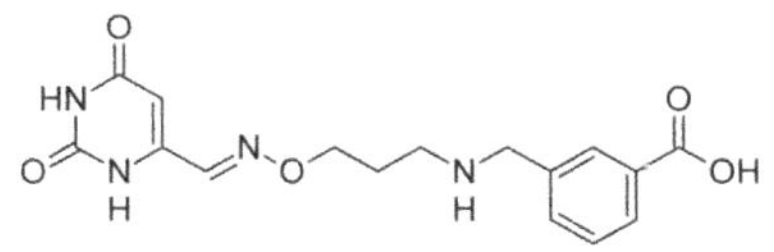	1.64	[61]
3FCL	UNG	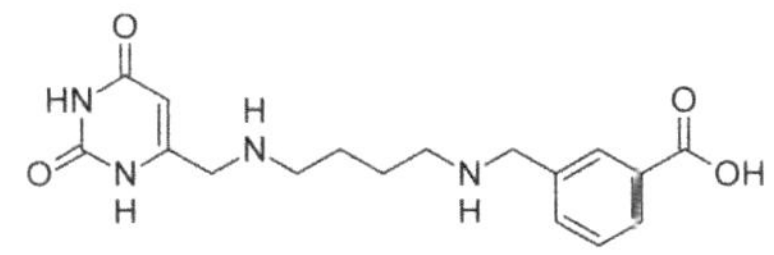	1.70	[61]

Thymine–DNA glycosylase (TDG) has recently been validated as a possible drug target in melanoma: its knockdown causes cell cycle arrest, senescence, and cell death in melanoma cell lines but not in normal cells and prevents tumor growth in a xenograft model [91]. Screening of several mid-scale compound libraries yielded about 40 inhibitors with a variety of structures and $IC_{50} > 10$ μM [91].

Uracil-DNA glycosylases from poxviruses (D4 according to vaccinia virus naming convention) provide a unique drug target. While they possess quite efficient uracil-removing activity, the main role of these enzymes is not in DNA repair, but in viral replication. D4 binds the A20 protein to form a processivity factor for poxviral DNA polymerases [92,93]. Deletion of the D4 gene causes a sharp drop in the ability of vaccinia virus to replicate in cells [94–96]. Polycyclic aromatic compounds that disrupt the D4/A20 binding interface are considered a promising class of antiviral drugs active against poxviruses [66,97,98] (Table 2).

Finally, a natural Ung inhibitor, the Ugi protein, is produced by PBS1 and PBS2 bacteriophages [99]. Although Ugi is not considered a therapeutically promising candidate, it has recently found an unexpected use in cell technologies involving CRISPR/Cas9 genome editing. A new generation of Cas9-based tools employs base editors, in which a Cas9 targeting module is fused with a cytosine deaminase to generate C→T mutations [100,101]. Repair by UNG counteracts uracil-mediated targeted mutagenesis, so co-expression of Ugi is commonly used to increase the efficiency of this gene editing procedure [100–102].

4. Oxidative Damage Repair: Key to Antibiotic Resistance?

It has been shown that BER is necessary for the survival of certain pathogenic and opportunistic bacteria (*Mycobacterium*, *Neisseria*, *Pseudomonas*, *Salmonella*) under conditions of genotoxic stress caused either by drugs or by the body's immune response [103–105]. Recently, it was discovered that oxidative stress significantly contributes to the death of bacteria exposed to antibiotics of several classes. Topoisomerase inhibitors, β-lactam antibiotics, membrane-permeabilizing agents, and aminoglycosides induce the generation of hydroxyl radicals in several divergent bacterial species through an iron-dependent Fenton reaction, increasing the lethality of these drugs [106–109], whereas reducing agents such as H_2S or NO protect bacteria from a wide range of antibiotics [110,111]. Possible reasons for the enhanced cell death include translation errors due to oxidative RNA damage [112], oxidation of the nucleotide pool followed by massive chromosome breakage at the sites of damaged nucleotide incorporation [113–115], and direct DNA damage by reactive antibiotic molecules, their metabolites, or reactive oxygen species [116,117]. Based on these observations, the systems of antioxidant defense and oxidative damage repair in bacteria are now regarded as promising targets for sensitization towards bactericidal antibiotics, which, if successful, can be a breakthrough in the current antibiotic resistance crisis.

In bacteria, several DNA glycosylases are responsible for oxidative damage repair. In *E. coli*, the best-studied enzymatic system, termed the "GO system" (for Guanine Oxidation), involves three enzymes: Fpg (MutM), MutY, and MutT, which have complementary functions in countering the mutagenic effect of 8-oxoguanine (oxoG) [118–120]. OxoG is an abundant DNA lesion that easily forms stable oxoG(*syn*):A(*anti*) Hoogsteen-type pairs, leading to characteristic G:C→T:A transversions [121,122]. Fpg is a DNA glycosylase that excises oxoG from pairs with C; such pairs appear when G is directly oxidized in DNA or when oxodGMP is incorporated opposite C during replication [123,124]. If oxoG remains in DNA and directs dAMP misincorporation, the excision of oxoG by Fpg would lead to a G→T transversion. To safeguard the cell from this mutagenic route, Fpg does not cleave oxoG:A mispairs, which are recognized by MutY, and A is excised instead of oxoG [125]. If repair DNA polymerases insert the correct dCMP opposite oxoG, the second round of repair is carried out by Fpg; otherwise, dAMP is inserted, and MutY-initiated repair is reinitiated. The third enzyme of the system, MutT, hydrolyses oxodGTP and oxoGTP to monophosphates to prevent oxoG incorporation from the oxidized nucleotide pool [126,127]. *E. coli* also possesses a homolog of Fpg, endonuclease VIII (Nei), which is not considered part of the GO system and preferentially excises oxidized pyrimidines with little opposite-base preference, although it has some activity against oxoG in vitro and prevents G:C→T:A mutations in the absence of Fpg [128–131]. Finally, endonuclease III (Nth) also removes a wide variety of oxidized pyrimidine bases [132,133].

Although the GO system has been extensively characterized in *E. coli*, little is known about its functions and the properties of its components in pathogenic bacterial species. Fpg proteins from *Salmonella enterica* [134], *Neisseria meningitidis* [135], and *Corynebacterium pseudotuberculosis* [136] have been cloned and subjected to limited biochemical characterization, which showed essentially Fpg-like properties. Several Fpg homologs from *Mycobacterium tuberculosis* and *Mycobacterium smegmatis* were characterized and found to have divergent substrate specificities resembling either *E. coli* Fpg or Nei [137–140]. For MutY, limited enzyme characterization has been done for proteins from *N. meningitidis* [141], *Helicobacter pylori* [142], and *C. pseudotuberculosis* [143,144]. The presence of a fully functional GO system with its characteristic antimutator pattern has been confirmed in vivo in their native bacterial cells for *Pseudomonas aeruginosa* [145,146], *N. meningitides* [104,135,147], *M. smegmatis* [137,148,149], and *Staphylococcus aureus* [150]. Fpg was found to be functional in vivo in *S. enterica* [134], and MutY in *H. pylori* [142].

The available information about the relevance of the GO system for the pathogenicity of bacteria shows its dual role. On one hand, it seems that this line of defense indeed assists successful primary infection. MutY deficiency has been shown to compromise mouse stomach colonization by *H. pylori* [142]. Successful macrophage infection by *Brucella abortus* requires intact *fpg* and *mutY* genes [151], and *M. tuberculosis* Fpg and Nei are required for lung colonization in a rhesus macaque model [152]. Hypervirulent *Neisseria* isolates maintain functional *fpg* and *mutY* despite having a general mutator phenotype [147]. On the other hand, hypermutability associated with GO system inactivation sometimes provides the variance for selection of highly virulent or drug-resistant strains [153–156]. This underscores the importance of a thorough characterization of the GO system for a given pathogen to assess it as a possible drug combination target.

Human homologs of Fpg and Nei (NEIL1, NEIL2, and NEIL3) are significantly different from the bacterial proteins in their sequence and structure, making realistic the development of small ligands selectively targeting the bacterial enzymes. MutY and Nth appear to be less selective targets.

Few specific inhibitors of the bacterial GO system have been reported. Fpg is weakly inhibited by free damaged base analogs such as 2,6-diamino-4-oxo-5-(*N*-methylformamido)pyrimidine, 5-nitroso-2,4,6-triaminopyridine, and 5-nitroso-4,6-diamino-2-oxopyridine [67,157]. The base analog with the strongest ability to suppress Fpg activity (IC_{50} ~10–100 μM) is 2-thioxanthine (Table 2); however, it is not a true inhibitor but rather a reagent that oxidizes cysteines in the zinc finger in Fpg, Nei, and NEIL2 [67,68]. MutY from *C. pseudotuberculosis* was reported to be sensitive to suramin, an antiprotozoan and anthelmintic drug that acts intracellularly [32] (Table 2); however, the relevant drug uptake pathways are likely absent in bacteria. Interestingly, Fpg is strongly inhibited by Cibacron Blue F3GA, a dye bearing structural resemblance to suramin [69] (Table 2). Several papers described the inactivation of Fpg, Nth, OGG1, NEIL1, and NEIL2 by NO-producing agents and suggested damage to redox-sensitive groups in the enzyme molecules [158–162]. However, the mechanisms of this reaction remain unclear, since the known redox-sensitive groups in these glycosylases are different or absent altogether. A screen of a natural product library revealed several inhibitors of *M. tuberculosis* Nei2, the best of which, norlobaric acid, has K_i = 74 nM [70] (Table 2).

5. Oxidative Damage Repair: Cancer Sensitization Strategy

OGG1 is a human DNA glycosylase that initiates the repair of oxidized purine bases, mainly oxoG and formamidopyrimidines, thus being a functional analog of Fpg. Nevertheless, in its sequence and structure, OGG1 is completely different from Fpg (Table 1). Overexpression of *OGG1* in fibroblasts, pulmonary epithelial cells, and bone marrow protects them from the toxic effects of thiotepa, carmustine, and mafosfamide, which mainly yield *N*7-alkylated purines further hydrolyzed in the cell to formamidopyrimidine derivatives [163–165]. However, it is unclear whether normal OGG1 levels can reduce the toxicity of these drugs in tumor cells. A similar effect of OGG1 has been described for cisplatin and oxaliplatin [166], although the nature of the damage removed in this case is not entirely clear. Of the antitumor agents that produce oxidative DNA damage, *OGG1* downregulation or inhibition sensitizes cells to bleomycin [167] and ionizing radiation [168].

As with uracil incorporation and repair, the activity of OGG1 may not only safeguard cells from genomic damage but also potentiate the action of DNA-damaging agents, converting damaged bases to more cytotoxic strand breaks. For instance, OGG1 downregulation protects several cancer cell lines from β-lapachone, an NAD(P)H dehydrogenase (quinone 1)-dependent redox cycling drug that produces copious amounts of intracellular H_2O_2 [169]. Hence, a strategy alternative to OGG1 inhibition may consist in saturating the BER capacity with oxidative lesions. In human cells, OGG1 together with mismatched adenine-DNA glycosylase MUTYH and nucleoside triphosphatase MTH1 (NUDT1) forms an analog of the GO system, which prevents the mutagenic effect of oxoG [170]. Recently, the knockdown or inhibition of MTH1, which hydrolyzes oxoG triphosphates and prevents their incorporation into growing DNA and RNA chains, was shown to be toxic to tumor cells, due to the accumulation of oxidized bases and DNA breaks [71,171,172]. Apparently, as in the case of PARP inhibitors, selective toxicity is due to the suppression of the last remaining pathway for the oxidative damage repair in cancer cells. Several low-molecular-weight MTH1 inhibitors were identified, including a clinically approved tyrosine kinase inhibitor, crizotinib [71,171,173] (Table 2). Interestingly, crizotinib possesses a chiral center that gives rise to *(R)*- and *(S)*- enantiomers, of which clinically used

(R)-crizotinib inhibits c-MET and ALK protein kinases, whereas the *(S)*-enantiomer preferentially binds to and inhibits MTH1 [71,174,175]. While it is still debated to which extent the cytotoxic activity of crizotinib and other MTH1 inhibitors is dependent on MTH1 and oxidative overload [172,176–178], most reports agree that oxidative damage is an important cell-killing factor, although its causes might not be limited to a direct suppression of MTH1 activity (recently reviewed in [179]). As of today, (*S*)-crizotinib is not used for patient treatment. However, another anti-cancer drug candidate, karonudib, was developed from previously found MTH1 inhibitors [172,180,181]. Presently, two Phase 1 clinical trials of karonudib are registered with the US National Library of Medicine Clinical Studies Database.

Inhibitors of OGG1 have been reported in the literature but have not yet reached clinical trials. Mechanism-based approaches had only limited success: both oxoG base and its analogs proved to be weak inhibitors [182,183], while substituted 2,6-diaminopurines performed slightly better [76]. Experimental and computational screening of small-molecule pharmacological libraries produced several hits that were expanded into inhibitors with submicromolar affinity, structurally unrelated to the OGG1 substrate [72–74] (Table 2). Combinatorial design based on the identified OGG1 and MTH1 inhibitors was used to obtain a compound with submicromolar affinity for both these enzymes [75] (Table 2).

Other DNA glycosylases that repair oxidative damage, including NEIL1, NEIL2, NEIL3, MUTYH, and NTHL1, have been targeted less successfully. Purine-analog library and general library screening produced several inhibitors for NEIL1, but their affinities were in the micromolar range, and the target selectivity was quite low when compared with the inhibition of other glycosylases [76,184] (Table 2). Fumarylacetoacetate was reported to inhibit NEIL1 and NEIL2 and to a lesser degree, OGG1 and NTHL1 (Table 2), but the structural reasons under this effect have not been established [77]. Moreover, these enzymes have low experimental support as targets for sensitization to antitumor therapy, although NEIL1 confers some resistance to ionizing radiation and antifolates [185,186].

6. Oxidative Damage Repair: Unexpected Connections

While DNA damage and its repair are well understood in the cancer paradigm, two unexpected connections of oxidative damage with other human pathologies emerged recently. OxoG and its repair by OGG1 are suspected to play a regulatory role in the inflammatory response. Several lines of evidence support this conclusion. *Ogg1*-null or -depleted mice show a significantly alleviated inflammatory response to many factors, including bacterial endotoxins, *Helicobacter* infection, foreign protein response, ragweed pollen grain extract-induced allergy, etc. [187–190]. Interestingly, however, the inflammation associated with UVB or pulmonary hyperoxia is enhanced rather than reduced in $Ogg1^{-/-}$ mice [191–193], suggesting that OGG1-dependent inflammation requires foreign antigens. After the excision of oxoG, the OGG1·oxoG complex can bind Ras family GTPases and facilitate the GDP-to-GTP exchange [194,195], which triggers the signaling pathway leading to the activation of NF-κB, the key pro-inflammatory transcription factor [196]. Moreover, OGG1 can bind oxidized G-rich promoters of pro-inflammatory genes in an enzymatically non-productive mode and facilitate their expression by attracting NF-κB [197–199]. A small-molecule OGG1 inhibitor, TH5487, was developed that competes with oxoG for binding and downregulates the inflammatory response in a mouse model [73] (Table 2). Although TH5487 has a 4-bromobenzimidazolone moiety, which is structurally similar to oxoG, the crystal structure of its complex with OGG1 [73] (Table 3) unexpectedly revealed that the oxoG-binding pocket is occupied by another moiety of TH5487, *p*-iodophenylacetamide, whereas 4-bromobenzimidazolone resides in a so-called *exo*-site, which normally binds undamaged G and serves as a transient binding site for oxoG on its way to the active site [200,201]. Thus, TH5487 functionally resembles bipartite inhibitors of UNG, simultaneously engaging two selective binding sites in the enzyme molecule. Such design may be employed to construct new potent inhibitors of OGG1 and other DNA glycosylases.

In addition, inhibition of OGG1 holds promise to prevent or delay the onset of Huntington's disease in risk groups. This hereditary condition, which belongs to the class of "trinucleotide repeat" genetic diseases, is caused by expansion of the $(CAG)_n$ repeat run in the *HTT* gene beyond the critical length of ~35 repeats [202]. Before becoming symptomatic, carriers of a pathogenic allele experience an explosive growth of the $(CAG)_n$ run up to several hundred repeats in the striatum at the early stage of the disease [203]. This expansion is triggered by the normal repair of oxoG initiated by OGG1 [204,205] and is likely caused by an imbalance of BER enzymes in this part of the brain, which leads to the accumulation of unprocessed repair intermediates [206]. In the Huntington's disease mouse model, *Ogg1* knockout suppresses the repeat number growth in the striatum and delays the onset of motor dysfunction [207]. Thus, in carriers of the pathogenic *HTT* allele, for whom the penetrance is inevitably 100%, inhibition of OGG1 may be a reasonable therapeutic strategy.

7. Alkylation Damage Repair: Dual Consequences

Alkylating antitumor agents produce many damaged bases, including O^6-alkylguanine repaired by O^6-methylguanine–DNA methyltransferase (MGMT), and ring-alkylated purines repaired predominantly by BER [16,208]. Unlike other DNA glycosylases, which impart resistance to DNA-damaging agents to cells, *N*-methylpurine-DNA glycosylase (MPG, alias alkyladenine-DNA glycosylase (AAG), and alkylpurine-DNA *N*-glycosylase, APNG) may increase the cytotoxicity of alkylating antitumor agents, removing alkylated bases from DNA to form AP sites, which are more dangerous for the cell [208–214]. A similar sensitization mechanism is also characteristic of UNG, TDG, and MBD4 DNA glycosylases when they repair C5-halogenated uracil derivatives [215–218]. On the other hand, inhibition of MPG in carcinoma cells sensitizes them to alkylating agents [219], and $Mpg^{-/-}$ murine cells are hypersensitive to 1,3-bis(2-chloroethyl)-1-nitrosourea and mitomycin C (but not to alkylating nitrogen mustards) [220]. An integrative model of temozolomide-induced DNA damage and DNA repair by MGMT and MPG in glioblastoma predicts that inhibition of both enzymes is the most successful sensitization strategy [221]. For temozolomide-resistant forms of glioblastoma, the combination of inhibition of BER enzymes and PARP-dependent signaling is effective [212,222]. In addition to the detoxification of anticancer drug adducts, MPG and OGG1 have been reported to hydrolyze a human cytomegalovirus replication inhibitor, 2-bromo-5,6-dichloro-1-(β-D-ribofuranosyl)benzimidazole, opening the possibility of antiviral action of drug combinations including DNA glycosylase inhibitors [223]. Bacterial *alkA* mutants are hypersensitive to methyl methanesulfonate [224,225]; however, alkylating agents are not among clinically used antibacterial drugs, so this vulnerability is hard to exploit.

Alkylbase-removing DNA glycosylases are the least explored group in terms of specific inhibitors. N3-substituted adenine derivatives are competitive inhibitors of bacterial Tag and mammalian MPG [226–230]. Based on this observation, a series of structural analogs has been computationally designed to inhibit TagA from *Leptospira interrogans*, the infectious agent of leptospirosis, although no experimental evidence was provided for their activity against the enzyme or the pathogen [231]. A natural flavonol, morin, inhibits MPG [78] (Table 2).

8. Assays for DNA Glycosylase Activity

Most basic research on DNA glycosylases was and is still done using radioactively labeled oligonucleotides and analyzing substrates and product by gel electrophoresis. While this assay offers the highest sensitivity, it is labor-intensive, not easily scalable, and inconvenient when screening pharmacological libraries. In the past years, a number of fluorescence-based assays to follow glycosylase activities appeared, some of them coupled with signal amplification to increase the sensitivity.

The first attempts to utilize fluorescent labels for detecting DNA glycosylase activities in a homogeneous mode were based on changes in the signal from a fluorophore incorporated next to a lesion upon the eversion or the excision of the damaged base [232,233] (Figure 2A). Although this approach has been used for inhibitor screening [75], it is not very sensitive, and fluorophores adjacent to a lesion may even inhibit the measured activity. Molecular beacon substrates developed later consist of an oligonucleotide hairpin or a duplex bearing a fluorophore and a quencher at its termini (Figure 2B). Such substrates allow measuring DNA glycosylase activities both in vitro and in living cells [76,234–237] and have been used in glycosylase inhibitor library screening [73,76]. In this case, the glycosylase must be bifunctional to nick the substrate, or else an AP endonuclease has to be present in the assay. Several types of arrays or beads with immobilized damaged oligonucleotides have been reported, in which only the damaged strand is labeled, and the cleavage produces short DNA fragments that can be washed off [238–242]. While such assays are not homogeneous, they are well suited for multiplexing and parallel screening. Fluorophores can also be incorporated into double-stranded DNA in situ through base excision, AP site cleavage, and gap filling by DNA polymerase β with a labeled dNTP [243].

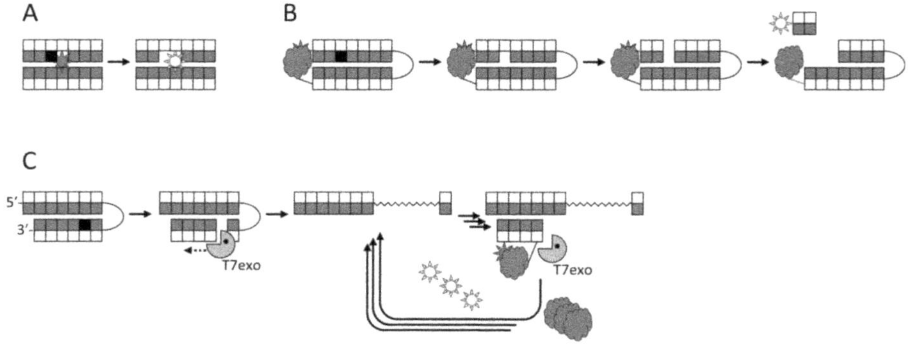

Figure 2. Schemes of several fluorescence-based approaches to DNA glycosylase detection. (**A**), fluorescent reporters adjacent to the lesion; (**B**), molecular beacons with a fluorophore–quencher pair; (**C**), T7 exonuclease-assisted signal amplification due to cyclic degradation of a probe bearing a fluorophore–quencher pair.

An interesting approach was suggested that employs a DNAzyme inactivated by a strategically placed U residue, and the excision by uracil-DNA glycosylase reactivates the DNAzyme, which cleaves a fluorescent substrate [244]. However, it cannot be applied to other glycosylases that require double-stranded substrates. In a more advanced version, base excision generates a specifically folded quadruplex, which forms a fluorescent complex with quadruplex-selective ligands [245–247].

The most sensitive fluorescent assays rely on the formation of a nick after the cleavage by a DNA glycosylase (either bifunctional or coupled with an AP endonuclease), followed by signal amplification. The amplification may be exonuclease-assisted linear isothermal, in which a beacon is annealed and degraded by a double-stranded specific exonuclease in repeated cycles [248–250] (Figure 2C). Alternatively, the signal may be enhanced by rolling-circle amplification, using any suitable assay to detect the newly synthesized DNA [251–254]. Finally, nick formation can serve as a starting point for exponential isothermal amplification [255,256] or cascade hybridization [257].

9. Conclusions

DNA glycosylases, as enzymes that initiate base excision repair, represent an attractive pharmacological target. Their structures reveal mechanism-based features, such as deep pockets for substrate base binding, indicating potential druggability, and several successful attempts of library screening produced tantalizing leads that can be explored further to develop drugs for cancer and infectious diseases. Moreover, recent findings implicate DNA glycosylases not only in genome protection but also in regulatory pathways and suggest that they can be targeted in some inflammatory and neurodegenerative processes. A number of rapid and sensitive assays for screening DNA glycosylase activities were developed in the past few years, which should facilitate the search for their inhibitors.

The most important factor that complicates the targeting of DNA glycosylases in the now well-established framework of synthetic lethality, e.g., in cancer therapy, is their dual function in cell killing. On one hand, glycosylases initiate the repair of genotoxic adducts and, in theory, should potentiate their action. On the other hand, there are many cases in which the main lethal lesions are not adducts per se but intermediates of their repair, such as AP sites or DNA breaks. Such intermediates usually accumulate if the activity of downstream BER enzymes are insufficient to process the inflicted amount of genomic lesions in full. In these situations, DNA glycosylase inhibition would protect rather than sensitize cells to genome damage. Optimally, DNA glycosylases should be targeted in some form of precision therapy, based on the general model of toxicity of various adducts and the specific knowledge of adduct spectra and downstream BER capacity in the affected cells.

Outside of the cancer field, DNA glycosylase inhibition is most likely to find its soonest clinical application in antiviral therapy, since two important groups of human pathogens, poxviruses and herpesviruses, possess their own uracil-DNA glycosylases, a validated target required for replication in host cells, and several promising drug leads are available. Inhibition of OGG1 to prevent somatic trinucleotide repeat expansion in Huntington's disease also has high priority due to the extreme morbidity and mortality of the condition and the lack of other drugs, although lead compounds capable of brain delivery have not been reported so far. The inflammation-modulating action of OGG1 inhibitors, albeit attracting considerable attention, would still require much research and mechanistic insights to produce drugs comparable with more traditional anti-inflammatory agents. Even if more basic research is required to validate DNA glycosylases as targets for antibacterial combination therapy, yet the payoff in this area may be the largest one. The prospects of bringing DNA glycosylases into the circle of drug targets ultimately depend on our understanding of their action in DNA repair and connection with other cellular pathways.

Author Contributions: Conceptualization, D.O.Z. writing—original draft preparation, G.V.M., A.V.E., E.A.D., and D.O.Z.; writing—review and editing, D.O.Z.; visualization, D.O.Z.; supervision, D.O.Z.; project administration, D.O.Z.; funding acquisition, A.V.E. and D.O.Z. All authors have read and agreed to the published version of the manuscript.

Funding: This research was funded by the Russian Science Foundation, grant 19-74-00068. Partial salary support from the Russian Ministry of Science and Higher Education (State funded budget project AAAA-A17-117020210023-1) is acknowledged. Open Access fees were paid by an intramural program of SB RAS Institute of Chemical Biology and Fundamental Medicine.

Conflicts of Interest: The authors declare no conflict of interest. The funders had no role in the design of the study; in the collection, analyses, or interpretation of data; in the writing of the manuscript, or in the decision to publish the results.

References

1. Friedberg, E.C.; Walker, G.C.; Siede, W.; Wood, R.D.; Schultz, R.A.; Ellenberger, T. *DNA Repair and Mutagenesis*; ASM Press: Washington, DC, USA, 2006; p. 1118.
2. Halliwell, B.; Gutteridge, J.M.C. *Free Radicals in Biology and Medicine*, 4th ed.; Oxford University Press: Oxford, UK, 2007; p. 704.
3. Zharkov, D.O. Base excision DNA repair. *Cell. Mol. Life Sci.* **2008**, *65*, 1544–1565. [CrossRef] [PubMed]
4. David, S.S.; Williams, S.D. Chemistry of glycosylases and endonucleases involved in base-excision repair. *Chem. Rev.* **1998**, *98*, 1221–1261. [CrossRef] [PubMed]
5. McCullough, A.K.; Dodson, M.L.; Lloyd, R.S. Initiation of base excision repair: Glycosylase mechanisms and structures. *Annu. Rev. Biochem.* **1999**, *68*, 255–285. [CrossRef] [PubMed]
6. Stivers, J.T.; Jiang, Y.L. A mechanistic perspective on the chemistry of DNA repair glycosylases. *Chem. Rev.* **2003**, *103*, 2729–2760. [CrossRef]
7. Huffman, J.L.; Sundheim, O.; Tainer, J.A. DNA base damage recognition and removal: New twists and grooves. *Mutat. Res.* **2005**, *577*, 55–76. [CrossRef]
8. Hassa, P.O.; Hottiger, M.O. The diverse biological roles of mammalian PARPs, a small but powerful family of poly-ADP-ribose polymerases. *Front. Biosci.* **2008**, *13*, 3046–3082. [CrossRef]

9. Alemasova, E.E.; Lavrik, O.I. Poly(ADP-ribosyl)ation by PARP1: Reaction mechanism and regulatory proteins. *Nucleic Acids Res.* **2019**, *47*, 3811–3827. [CrossRef]
10. Bryant, H.E.; Schultz, N.; Thomas, H.D.; Parker, K.M.; Flower, D.; Lopez, E.; Kyle, S.; Meuth, M.; Curtin, N.J.; Helleday, T. Specific killing of BRCA2-deficient tumours with inhibitors of poly(ADP-ribose) polymerase. *Nature* **2005**, *434*, 913–917. [CrossRef]
11. Farmer, H.; McCabe, N.; Lord, C.J.; Tutt, A.N.J.; Johnson, D.A.; Richardson, T.B.; Santarosa, M.; Dillon, K.J.; Hickson, I.; Knights, C.; et al. Targeting the DNA repair defect in *BRCA* mutant cells as a therapeutic strategy. *Nature* **2005**, *434*, 917–921. [CrossRef]
12. Roy, R.; Chun, J.; Powell, S.N. BRCA1 and BRCA2: Different roles in a common pathway of genome protection. *Nat. Rev. Cancer* **2011**, *12*, 68–78. [CrossRef]
13. Murai, J.; Huang, S.-Y.N.; Das, B.B.; Renaud, A.; Zhang, Y.; Doroshow, J.H.; Ji, J.; Takeda, S.; Pommier, Y. Trapping of PARP1 and PARP2 by clinical PARP inhibitors. *Cancer Res.* **2012**, *72*, 5588–5599. [CrossRef] [PubMed]
14. Hopkins, T.A.; Shi, Y.; Rodriguez, L.E.; Solomon, L.R.; Donawho, C.K.; DiGiammarino, E.L.; Panchal, S.C.; Wilsbacher, J.L.; Gao, W.; Olson, A.M.; et al. Mechanistic dissection of PARP1 trapping and the impact on in vivo tolerability and efficacy of PARP inhibitors. *Mol. Cancer Res.* **2015**, *13*, 1465–1477. [CrossRef] [PubMed]
15. Zimmermann, M.; Murina, O.; Reijns, M.A.M.; Agathanggelou, A.; Challis, R.; Tarnauskaitė, Ž.; Muir, M.; Fluteau, A.; Aregger, M.; McEwan, A.; et al. CRISPR screens identify genomic ribonucleotides as a source of PARP-trapping lesions. *Nature* **2018**, *559*, 285–289. [CrossRef] [PubMed]
16. Lawley, P.D.; Phillips, D.H. DNA adducts from chemotherapeutic agents. *Mutat. Res.* **1996**, *355*, 13–40. [CrossRef]
17. Thomas, L.; Yang, C.H.; Goldthwait, D.A. Two DNA glycosylases in *Escherichia coli* which release primarily 3-methyladenine. *Biochemistry* **1982**, *21*, 1162–1169. [CrossRef]
18. Ishchenko, A.A.; Vasilenko, N.L.; Sinitsina, O.I.; Yamkovoy, V.I.; Fedorova, O.S.; Douglas, K.T.; Nevinsky, G.A. Thermodynamic, kinetic, and structural basis for recognition and repair of 8-oxoguanine in DNA by Fpg protein from *Escherichia Coli*. *Biochemistry* **2002**, *41*, 7540–7548. [CrossRef]
19. Fedorova, O.S.; Nevinsky, G.A.; Koval, V.V.; Ishchenko, A.A.; Vasilenko, N.L.; Douglas, K.T. Stopped-flow kinetic studies of the interaction between *Escherichia coli* Fpg protein and DNA substrates. *Biochemistry* **2002**, *41*, 1520–1528. [CrossRef]
20. Zharkov, D.O.; Ishchenko, A.A.; Douglas, K.T.; Nevinsky, G.A. Recognition of damaged DNA by *Escherichia coli* Fpg protein: Insights from structural and kinetic data. *Mutat. Res.* **2003**, *531*, 141–156. [CrossRef]
21. Zharkov, D.O.; Mechetin, G.V.; Nevinsky, G.A. Uracil-DNA glycosylase: Structural, thermodynamic and kinetic aspects of lesion search and recognition. *Mutat. Res.* **2010**, *685*, 11–20. [CrossRef]
22. Schärer, O.D.; Ortholand, J.-Y.; Ganesan, A.; Ezaz-Nikpay, K.; Verdine, G.L. Specific binding of the DNA repair enzyme AlkA to a pyrrolidine-based inhibitor. *J. Am. Chem. Soc.* **1995**, *117*, 6623–6624. [CrossRef]
23. Schärer, O.D.; Verdine, G.L. A designed inhibitor of base-excision DNA repair. *J. Am. Chem. Soc.* **1995**, *117*, 10781–10782. [CrossRef]
24. Deng, L.; Schärer, O.D.; Verdine, G.L. Unusually strong binding of a designed transition-state analog to a base-excision DNA repair protein. *J. Am. Chem. Soc.* **1997**, *119*, 7865–7866. [CrossRef]
25. Schärer, O.D.; Kawate, T.; Gallinari, P.; Jiricny, J.; Verdine, G.L. Investigation of the mechanisms of DNA binding of the human G/T glycosylase using designed inhibitors. *Proc. Natl. Acad. Sci. USA* **1997**, *94*, 4878–4883. [CrossRef] [PubMed]
26. Schärer, O.D.; Nash, H.M.; Jiricny, J.; Laval, J.; Verdine, G.L. Specific binding of a designed pyrrolidine abasic site analog to multiple DNA glycosylases. *J. Biol. Chem.* **1998**, *273*, 8592–8597. [CrossRef] [PubMed]
27. Le Bihan, Y.-V.; Izquierdo, M.A.; Coste, F.; Aller, P.; Culard, F.; Gehrke, T.H.; Essalhi, K.; Carell, T.; Castaing, B. 5-Hydroxy-5-methylhydantoin DNA lesion, a molecular trap for DNA glycosylases. *Nucleic Acids Res.* **2011**, *39*, 6277–6290. [CrossRef]
28. Dai, Q.; Lu, X.; Zhang, L.; He, C. Synthesis of DNA oligos containing 2′-deoxy-2′-fluoro-D-arabinofuranosyl-5-carboxylcytosine as hTDG inhibitor. *Tetrahedron* **2012**, *68*, 5145–5151. [CrossRef]

29. Yamagata, Y.; Kato, M.; Odawara, K.; Tokuno, Y.; Nakashima, Y.; Matsushima, N.; Yasumura, K.; Tomita, K.-i.; Ihara, K.; Fujii, Y.; et al. Three-dimensional structure of a DNA repair enzyme, 3-methyladenine DNA glycosylase II, from Escherichia coli. *Cell* **1996**, *86*, 311–319. [CrossRef]
30. Zharkov, D.O.; Rosenquist, T.A.; Gerchman, S.E.; Grollman, A.P. Substrate specificity and reaction mechanism of murine 8-oxoguanine-DNA glycosylase. *J. Biol. Chem.* **2000**, *275*, 28607–28617. [CrossRef]
31. Gilboa, R.; Zharkov, D.O.; Golan, G.; Fernandes, A.S.; Gerchman, S.E.; Matz, E.; Kycia, J.H.; Grollman, A.P.; Shoham, G. Structure of formamidopyrimidine-DNA glycosylase covalently complexed to DNA. *J. Biol. Chem.* **2002**, *277*, 19811–19816. [CrossRef]
32. Eberle, R.J.; Coronado, M.A.; Peinado, R.S.; de Moraes, F.R.; Olivier, D.; Dreyer, T.; de Oliveira Lopes, D.; da Luz, B.S.R.; Azevedo, V.; Arni, R.K. The polyanions heparin and suramin impede binding of free adenine to a DNA glycosylase from *C. Pseudotuberculosis*. *Int. J. Biol. Macromol.* **2019**, *125*, 459–468. [CrossRef]
33. Li, X.; Lu, A.-L. Intact MutY and its catalytic domain differentially contact with A/8-oxoG-containing DNA. *Nucleic Acids Res.* **2000**, *28*, 4593–4603. [CrossRef] [PubMed]
34. Mechetin, G.V.; Dyatlova, E.A.; Sinyakov, A.N.; Ryabinin, V.A.; Vorobjev, P.E.; Zharkov, D.O. Correlated target search by uracil–DNA glycosylase in the presence of bulky adducts and DNA-binding ligands. *Russ. J. Bioorg. Chem.* **2017**, *43*, 23–28. [CrossRef]
35. Galmarini, C.M.; Mackey, J.R.; Dumontet, C. Nucleoside analogues and nucleobases in cancer treatment. *Lancet Oncol.* **2002**, *3*, 415–424. [CrossRef]
36. Feng, J.Y. Addressing the selectivity and toxicity of antiviral nucleosides. *Antivir. Chem. Chemother.* **2018**, *26*, 2040206618758524. [CrossRef] [PubMed]
37. Huang, H.; Stivers, J.T.; Greenberg, M.M. Competitive inhibition of uracil DNA glycosylase by a modified nucleotide whose triphosphate is a substrate for DNA polymerase. *J. Am. Chem. Soc.* **2009**, *131*, 1344–1345. [CrossRef] [PubMed]
38. Hashimoto, M.; Greenberg, M.M.; Kow, Y.W.; Hwang, J.-T.; Cunningham, R.P. The 2-deoxyribonolactone lesion produced in DNA by neocarzinostatin and other damaging agents forms cross-links with the base-excision repair enzyme endonuclease III. *J. Am. Chem. Soc.* **2001**, *123*, 3161–3162. [CrossRef] [PubMed]
39. Kroeger, K.M.; Hashimoto, M.; Kow, Y.W.; Greenberg, M.M. Cross-linking of 2-deoxyribonolactone and its β-elimination product by base excision repair enzymes. *Biochemistry* **2003**, *42*, 2449–2455. [CrossRef]
40. Peters, G.J.; van der Wilt, C.L.; van Moorsel, C.J.A.; Kroep, J.R.; Bergman, A.M.; Ackland, S.P. Basis for effective combination cancer chemotherapy with antimetabolites. *Pharmacol. Ther.* **2000**, *87*, 227–253. [CrossRef]
41. Parker, W.B. Enzymology of purine and pyrimidine antimetabolites used in the treatment of cancer. *Chem. Rev.* **2009**, *109*, 2880–2293. [CrossRef]
42. Kavli, B.; Otterlei, M.; Slupphaug, G.; Krokan, H.E. Uracil in DNA—General mutagen, but normal intermediate in acquired immunity. *DNA Repair* **2007**, *6*, 505–516. [CrossRef]
43. Sousa, M.M.L.; Krokan, H.E.; Slupphaug, G. DNA-uracil and human pathology. *Mol. Asp. Med.* **2007**, *28*, 276–306. [CrossRef] [PubMed]
44. Ingraham, H.A.; Dickey, L.; Goulian, M. DNA fragmentation and cytotoxicity from increased cellular deoxyuridylate. *Biochemistry* **1986**, *25*, 3225–3230. [CrossRef] [PubMed]
45. Van Triest, B.; Pinedo, H.M.; Giaccone, G.; Peters, G.J. Downstream molecular determinants of response to 5-fluorouracil and antifolate thymidylate synthase inhibitors. *Ann. Oncol.* **2000**, *11*, 385–391. [CrossRef] [PubMed]
46. Bulgar, A.D.; Snell, M.; Donze, J.R.; Kirkland, E.B.; Li, L.; Yang, S.; Xu, Y.; Gerson, S.L.; Liu, L. Targeting base excision repair suggests a new therapeutic strategy of fludarabine for the treatment of chronic lymphocytic leukemia. *Leukemia* **2010**, *24*, 1795–1799. [CrossRef] [PubMed]
47. Bulgar, A.D.; Weeks, L.D.; Miao, Y.; Yang, S.; Xu, Y.; Guo, C.; Markowitz, S.; Oleinick, N.; Gerson, S.L.; Liu, L. Removal of uracil by uracil DNA glycosylase limits pemetrexed cytotoxicity: Overriding the limit with methoxyamine to inhibit base excision repair. *Cell Death Dis.* **2012**, *3*, e252. [CrossRef] [PubMed]
48. Yan, Y.; Qing, Y.; Pink, J.J.; Gerson, S.L. Loss of uracil DNA glycosylase selectively resensitizes p53-mutant and -deficient cells to 5-FdU. *Mol. Cancer Res.* **2018**, *16*, 212–221. [CrossRef] [PubMed]
49. Pulukuri, S.M.K.; Knost, J.A.; Estes, N.; Rao, J.S. Small interfering RNA-directed knockdown of uracil DNA glycosylase induces apoptosis and sensitizes human prostate cancer cells to genotoxic stress. *Mol. Cancer Res.* **2009**, *7*, 1285–1293. [CrossRef]

50. Weeks, L.D.; Fu, P.; Gerson, S.L. Uracil-DNA glycosylase expression determines human lung cancer cell sensitivity to pemetrexed. *Mol. Cancer Ther.* **2013**, *12*, 2248–2260. [CrossRef]
51. Weeks, L.D.; Zentner, G.E.; Scacheri, P.C.; Gerson, S.L. Uracil DNA glycosylase (UNG) loss enhances DNA double strand break formation in human cancer cells exposed to pemetrexed. *Cell Death Dis.* **2014**, *5*, e1045. [CrossRef]
52. An, Q.; Robins, P.; Lindahl, T.; Barnes, D.E. 5-Fluorouracil incorporated into DNA is excised by the Smug1 DNA glycosylase to reduce drug cytotoxicity. *Cancer Res.* **2007**, *67*, 940–945. [CrossRef]
53. Nagaria, P.; Svilar, D.; Brown, A.R.; Wang, X.-H.; Sobol, R.W.; Wyatt, M.D. SMUG1 but not UNG DNA glycosylase contributes to the cellular response to recovery from 5-fluorouracil induced replication stress. *Mutat. Res.* **2013**, *743–744*, 26–32. [CrossRef] [PubMed]
54. Suksangpleng, T.; Leartsakulpanich, U.; Moonsom, S.; Siribal, S.; Boonyuen, U.; Wright, G.E.; Chavalitshewinkoon-Petmitr, P. Molecular characterization of *Plasmodium falciparum* uracil-DNA glycosylase and its potential as a new anti-malarial drug target. *Malar. J.* **2014**, *13*, 149. [CrossRef] [PubMed]
55. Castillo-Acosta, V.M.; Aguilar-Pereyra, F.; Vidal, A.E.; Navarro, M.; Ruiz-Pérez, L.M.; González-Pacanowska, D. Trypanosomes lacking uracil-DNA glycosylase are hypersensitive to antifolates and present a mutator phenotype. *Int. J. Biochem. Cell Biol.* **2012**, *44*, 1555–1568. [CrossRef] [PubMed]
56. Charret, K.S.; Requena, C.E.; Castillo-Acosta, V.M.; Ruiz-Pérez, L.M.; González-Pacanowska, D.; Vidal, A.E. *Trypanosoma brucei* AP endonuclease 1 has a major role in the repair of abasic sites and protection against DNA-damaging agents. *DNA Repair* **2012**, *11*, 53–64. [CrossRef] [PubMed]
57. Cabrera, G.; Barría, C.; Fernández, C.; Sepúlveda, S.; Valenzuela, L.; Kemmerling, U.; Galanti, N. DNA repair BER pathway inhibition increases cell death caused by oxidative DNA damage in *Trypanosoma cruzi*. *J. Cell. Biochem.* **2011**, *112*, 2189–2199. [CrossRef]
58. Sepúlveda, S.; Valenzuela, L.; Ponce, I.; Sierra, S.; Bahamondes, P.; Ramirez, S.; Rojas, V.; Kemmerling, U.; Galanti, N.; Cabrera, G. Expression, functionality, and localization of apurinic/apyrimidinic endonucleases in replicative and non-replicative forms of *Trypanosoma cruzi*. *J. Cell. Biochem.* **2014**, *115*, 397–409. [CrossRef]
59. Focher, F.; Verri, A.; Spadari, S.; Manservigi, R.; Gambino, J.; Wright, G.E. *Herpes simplex* virus type 1 uracil-DNA glycosylase: Isolation and selective inhibition by novel uracil derivatives. *Biochem. J.* **1993**, *292*, 883–889. [CrossRef]
60. Jiang, Y.L.; Krosky, D.J.; Seiple, L.; Stivers, J.T. Uracil-directed ligand tethering: An efficient strategy for uracil DNA glycosylase (UNG) inhibitor development. *J. Am. Chem. Soc.* **2005**, *127*, 17412–17420. [CrossRef]
61. Chung, S.; Parker, J.B.; Bianchet, M.; Amzel, L.M.; Stivers, J.T. Impact of linker strain and flexibility in the design of a fragment-based inhibitor. *Nat. Chem. Biol.* **2009**, *5*, 407–413. [CrossRef]
62. Krosky, D.J.; Bianchet, M.A.; Seiple, L.; Chung, S.; Amzel, L.M.; Stivers, J.T. Mimicking damaged DNA with a small molecule inhibitor of human UNG2. *Nucleic Acids Res.* **2006**, *34*, 5872–5879. [CrossRef]
63. Jiang, Y.L.; Chung, S.; Krosky, D.J.; Stivers, J.T. Synthesis and high-throughput evaluation of triskelion uracil libraries for inhibition of human dUTPase and UNG2. *Bioorganic Med. Chem.* **2006**, *14*, 5666–5672. [CrossRef] [PubMed]
64. Zhang, H.; Zhang, L.; Jiang, J.; Yu, R. A highly sensitive electrochemical platform for the assay of uracil-DNA glycosylase activity combined with enzymatic amplification. *Anal. Sci.* **2013**, *29*, 193–198. [CrossRef] [PubMed]
65. Tao, J.; Song, P.; Sato, Y.; Nishizawa, S.; Teramae, N.; Tong, A.; Xiang, Y. A label-free and sensitive fluorescent method for the detection of uracil-DNA glycosylase activity. *Chem. Commun.* **2015**, *51*, 929–932. [CrossRef] [PubMed]
66. Nuth, M.; Huang, L.; Saw, Y.L.; Schormann, N.; Chattopadhyay, D.; Ricciardi, R.P. Identification of inhibitors that block vaccinia virus infection by targeting the DNA synthesis processivity factor D4. *J. Med. Chem.* **2011**, *54*, 3260–3267. [CrossRef] [PubMed]
67. Speina, E.; Ciesla, J.M.; Graziewicz, M.-A.; Laval, J.; Kazimierczuk, Z.; Tudek, B. Inhibition of DNA repair glycosylases by base analogs and tryptophan pyrolysate, Trp-P-1. *Acta Biochim. Pol.* **2005**, *52*, 167–178. [CrossRef]
68. Biela, A.; Coste, F.; Culard, F.; Guerin, M.; Goffinont, S.; Gasteiger, K.; Cieśla, J.; Winczura, A.; Kazimierczuk, Z.; Gasparutto, D.; et al. Zinc finger oxidation of Fpg/Nei DNA glycosylases by 2-thioxanthine: Biochemical and X-ray structural characterization. *Nucleic Acids Res.* **2014**, *42*, 10748–10761. [CrossRef]

69. Chetsanga, C.J.; Frenette, G.P. Excision of aflatoxin B_1-imidazole ring opened guanine adducts from DNA by formamidopyrimidine-DNA glycosylase. *Carcinogenesis* **1983**, *4*, 997–1000. [CrossRef]
70. Lata, K.; Afsar, M.; Ramachandran, R. Biochemical characterization and novel inhibitor identification of *Mycobacterium tuberculosis* Endonuclease VIII 2 (Rv3297). *Biochem. Biophys. Rep.* **2017**, *12*, 20–28. [CrossRef]
71. Huber, K.V.M.; Salah, E.; Radic, B.; Gridling, M.; Elkins, J.M.; Stukalov, A.; Jemth, A.-S.; Göktürk, C.; Sanjiv, K.; Strömberg, K.; et al. Stereospecific targeting of MTH1 by (*S*)-crizotinib as an anticancer strategy. *Nature* **2014**, *508*, 222–227. [CrossRef]
72. Donley, N.; Jaruga, P.; Coskun, E.; Dizdaroglu, M.; McCullough, A.K.; Lloyd, R.S. Small molecule inhibitors of 8-oxoguanine DNA glycosylase-1 (OGG1). *ACS Chem. Biol.* **2015**, *10*, 2334–2343. [CrossRef]
73. Visnes, T.; Cázares-Körner, A.; Hao, W.; Wallner, O.; Masuyer, G.; Loseva, O.; Mortusewicz, O.; Wiita, E.; Sarno, A.; Manoilov, A.; et al. Small-molecule inhibitor of OGG1 suppresses proinflammatory gene expression and inflammation. *Science* **2018**, *362*, 834–839. [CrossRef] [PubMed]
74. Tahara, Y.-K.; Auld, D.; Ji, D.; Beharry, A.A.; Kietrys, A.M.; Wilson, D.L.; Jimenez, M.; King, D.; Nguyen, Z.; Kool, E.T. Potent and selective inhibitors of 8-oxoguanine DNA glycosylase. *J. Am. Chem. Soc.* **2018**, *140*, 2105–2114. [CrossRef] [PubMed]
75. Tahara, Y.-K.; Kietrys, A.M.; Hebenbrock, M.; Lee, Y.; Wilson, D.L.; Kool, E.T. Dual inhibitors of 8-oxoguanine surveillance by OGG1 and NUDT1. *ACS Chem. Biol.* **2019**, *14*, 2606–2615. [CrossRef] [PubMed]
76. Jacobs, A.C.; Calkins, M.J.; Jadhav, A.; Dorjsuren, D.; Maloney, D.; Simeonov, A.; Jaruga, P.; Dizdaroglu, M.; McCullough, A.K.; Lloyd, R.S. Inhibition of DNA glycosylases via small molecule purine analogs. *PLoS ONE* **2013**, *8*, e81667. [CrossRef]
77. Bliksrud, Y.T.; Ellingsen, A.; Bjørås, M. Fumarylacetoacetate inhibits the initial step of the base excision repair pathway: Implication for the pathogenesis of tyrosinemia type I. *J. Inherit. Metab. Dis.* **2013**, *36*, 773–778. [CrossRef]
78. Dixon, M.; Woodrick, J.; Gupta, S.; Karmahapatra, S.K.; Devito, S.; Vasudevan, S.; Dakshanamurthy, S.; Adhikari, S.; Yenugonda, V.M.; Roy, R. Naturally occurring polyphenol, morin hydrate, inhibits enzymatic activity of *N*-methylpurine DNA glycosylase, a DNA repair enzyme with various roles in human disease. *Bioorganic Med. Chem.* **2015**, *23*, 1102–1111. [CrossRef]
79. Stivers, J.T.; Drohat, A.C. Uracil DNA glycosylase: Insights from a master catalyst. *Arch. Biochem. Biophys.* **2001**, *396*, 1–9. [CrossRef]
80. Lindahl, T.; Ljungquist, S.; Siegert, W.; Nyberg, B.; Sperens, B. DNA *N*-glycosidases: Properties of uracil-DNA glycosidase from *Escherichia coli*. *J. Biol. Chem.* **1977**, *252*, 3286–3294.
81. Krokan, H.; Wittwer, C.U. Uracil DNA-glycosylase from HeLa cells: General properties, substrate specificity and effect of uracil analogs. *Nucleic Acids Res.* **1981**, *9*, 2599–2613. [CrossRef]
82. Leblanc, J.-P.; Martin, B.; Cadet, J.; Laval, J. Uracil-DNA glycosylase Purification and properties of uracil-DNA glycosylase from *Micrococcus luteus*. *J. Biol. Chem.* **1982**, *257*, 3477–3483.
83. Williams, M.V.; Pollack, J.D. A mollicute (mycoplasma) DNA repair enzyme: Purification and characterization of uracil-DNA glycosylase. *J. Bacteriol.* **1990**, *172*, 2979–2985. [CrossRef] [PubMed]
84. Xiao, G.; Tordova, M.; Jagadeesh, J.; Drohat, A.C.; Stivers, J.T.; Gilliland, G.L. Crystal structure of *Escherichia coli* uracil DNA glycosylase and its complexes with uracil and glycerol: Structure and glycosylase mechanism revisited. *Proteins* **1999**, *35*, 13–24. [CrossRef]
85. Duraffour, S.; Ishchenko, A.A.; Saparbaev, M.; Crance, J.-M.; Garin, D. Substrate specificity of homogeneous monkeypox virus uracil-DNA glycosylase. *Biochemistry* **2007**, *46*, 11874–11881. [CrossRef] [PubMed]
86. Pregnolato, M.; Ubiali, D.; Verri, A.; Focher, F.; Spadari, S.; Sun, H.; Zhi, C.; Wright, G.E. Synthesis and molecular modeling of novel HSV1 uracil-DNA glycosylase inhibitors. *Nucleosides Nucleotides* **1999**, *18*, 709–711. [CrossRef] [PubMed]
87. Sun, H.; Zhi, C.; Wright, G.E.; Ubiali, D.; Pregnolato, M.; Verri, A.; Focher, F.; Spadari, S. Molecular modeling and synthesis of inhibitors of herpes simplex virus type 1 uracil-DNA glycosylase. *J. Med. Chem.* **1999**, *42*, 2344–2350. [CrossRef]
88. Hendricks, U.; Crous, W.; Naidoo, K.J. Computational rationale for the selective inhibition of the herpes simplex virus type 1 uracil-DNA glycosylase enzyme. *J. Chem. Inf. Model.* **2014**, *54*, 3362–3372. [CrossRef]
89. Jiang, Y.L.; Cao, C.; Stivers, J.T.; Song, F.; Ichikawa, Y. The merits of bipartite transition-state mimics for inhibition of uracil DNA glycosylase. *Bioorgance Chem.* **2004**, *32*, 244–262. [CrossRef]

90. Jiang, Y.L.; Stivers, J.T. Reconstructing the substrate for uracil DNA glycosylase: Tracking the transmission of binding energy in catalysis. *Biochemistry* **2001**, *40*, 7710–7719. [CrossRef]
91. Mancuso, P.; Tricarico, R.; Bhattacharjee, V.; Cosentino, L.; Kadariya, Y.; Jelinek, J.; Nicolas, E.; Einarson, M.; Beeharry, N.; Devarajan, K.; et al. Thymine DNA glycosylase as a novel target for melanoma. *Oncogene* **2019**, *38*, 3710–3728. [CrossRef]
92. Stanitsa, E.S.; Arps, L.; Traktman, P. Vaccinia virus uracil DNA glycosylase interacts with the A20 protein to form a heterodimeric processivity factor for the viral DNA polymerase. *J. Biol. Chem.* **2006**, *281*, 3439–3451. [CrossRef]
93. Druck Shudofsky, A.M.; Silverman, J.E.Y.; Chattopadhyay, D.; Ricciardi, R.P. Vaccinia virus D4 mutants defective in processive DNA synthesis retain binding to A20 and DNA. *J. Virol.* **2010**, *84*, 12325–12335. [CrossRef] [PubMed]
94. Stuart, D.T.; Upton, C.; Higman, M.A.; Niles, E.G.; McFadden, G. A poxvirus-encoded uracil DNA glycosylase is essential for virus viability. *J. Virol.* **1993**, *67*, 2503–2512. [CrossRef]
95. Millns, A.K.; Carpenter, M.S.; DeLange, A.M. The vaccinia virus-encoded uracil DNA glycosylase has an essential role in viral DNA replication. *Virology* **1994**, *198*, 504–513. [CrossRef] [PubMed]
96. Ellison, K.S.; Peng, W.; McFadden, G. Mutations in active-site residues of the uracil-DNA glycosylase encoded by vaccinia virus are incompatible with virus viability. *J. Virol.* **1996**, *70*, 7965–7973. [CrossRef] [PubMed]
97. Schormann, N.; Sommers, C.I.; Prichard, M.N.; Keith, K.A.; Noah, J.W.; Nuth, M.; Ricciardi, R.P.; Chattopadhyay, D. Identification of protein-protein interaction inhibitors targeting vaccinia virus processivity factor for development of antiviral agents. *Antimicrob. Agents Chemother.* **2011**, *55*, 5054–5062. [CrossRef] [PubMed]
98. Contesto-Richefeu, C.; Tarbouriech, N.; Brazzolotto, X.; Betzi, S.; Morelli, X.; Burmeister, W.P.; Iseni, F. Crystal structure of the vaccinia virus DNA polymerase holoenzyme subunit D4 in complex with the A20 N-terminal domain. *PLoS Pathog.* **2014**, *10*, e1003978. [CrossRef]
99. Putnam, C.D.; Tainer, J.A. Protein mimicry of DNA and pathway regulation. *DNA Repair* **2005**, *4*, 1410–1420. [CrossRef]
100. Komor, A.C.; Kim, Y.B.; Packer, M.S.; Zuris, J.A.; Liu, D.R. Programmable editing of a target base in genomic DNA without double-stranded DNA cleavage. *Nature* **2016**, *533*, 420–424. [CrossRef]
101. Nishida, K.; Arazoe, T.; Yachie, N.; Banno, S.; Kakimoto, M.; Tabata, M.; Mochizuki, M.; Miyabe, A.; Araki, M.; Hara, K.Y.; et al. Targeted nucleotide editing using hybrid prokaryotic and vertebrate adaptive immune systems. *Science* **2016**, *353*, aaf8729. [CrossRef]
102. Banno, S.; Nishida, K.; Arazoe, T.; Mitsunobu, H.; Kondo, A. Deaminase-mediated multiplex genome editing in *Escherichia Coli*. *Nat. Microbiol.* **2018**, *3*, 423–429. [CrossRef]
103. Venkatesh, J.; Kumar, P.; Krishna, P.S.M.; Manjunath, R.; Varshney, U. Importance of uracil DNA glycosylase in *Pseudomonas aeruginosa* and *Mycobacterium smegmatis*, G+C-rich bacteria, in mutation prevention, tolerance to acidified nitrite, and endurance in mouse macrophages. *J. Biol. Chem.* **2003**, *278*, 24350–24358. [CrossRef] [PubMed]
104. Davidsen, T.; Tuven, H.K.; Bjørås, M.; Rødland, E.A.; Tønjum, T. Genetic interactions of DNA repair pathways in the pathogen *Neisseria meningitidis*. *J. Bacteriol.* **2007**, *189*, 5728–5737. [CrossRef] [PubMed]
105. Richardson, A.R.; Soliven, K.C.; Castor, M.E.; Barnes, P.D.; Libby, S.J.; Fang, F.C. The base excision repair system of *Salmonella enterica* serovar typhimurium counteracts DNA damage by host nitric oxide. *PLoS Pathog.* **2009**, *5*, e1000451. [CrossRef] [PubMed]
106. Dwyer, D.J.; Kohanski, M.A.; Hayete, B.; Collins, J.J. Gyrase inhibitors induce an oxidative damage cellular death pathway in *Escherichia coli*. *Mol. Syst. Biol.* **2007**, *3*, 91. [CrossRef]
107. Kohanski, M.A.; Dwyer, D.J.; Hayete, B.; Lawrence, C.A.; Collins, J.J. A common mechanism of cellular death induced by bactericidal antibiotics. *Cell* **2007**, *130*, 797–810. [CrossRef] [PubMed]
108. Nguyen, D.; Joshi-Datar, A.; Lepine, F.; Bauerle, E.; Olakanmi, O.; Beer, K.; McKay, G.; Siehnel, R.; Schafhauser, J.; Wang, Y.; et al. Active starvation responses mediate antibiotic tolerance in biofilms and nutrient-limited bacteria. *Science* **2011**, *334*, 982–986. [CrossRef] [PubMed]
109. Dwyer, D.J.; Belenky, P.A.; Yang, J.H.; MacDonald, I.C.; Martell, J.D.; Takahashi, N.; Chan, C.T.Y.; Lobritz, M.A.; Braff, D.; Schwarz, E.G.; et al. Antibiotics induce redox-related physiological alterations as part of their lethality. *Proc. Natl. Acad. Sci. USA* **2014**, *111*, E2100–E2109. [CrossRef] [PubMed]

110. Gusarov, I.; Shatalin, K.; Starodubtseva, M.; Nudler, E. Endogenous nitric oxide protects bacteria against a wide spectrum of antibiotics. *Science* **2009**, *325*, 1380–1384. [CrossRef]
111. Shatalin, K.; Shatalina, E.; Mironov, A.; Nudler, E. H_2S: A universal defense against antibiotics in bacteria. *Science* **2011**, *334*, 986–990. [CrossRef]
112. Kohanski, M.A.; Dwyer, D.J.; Wierzbowski, J.; Cottarel, G.; Collins, J.J. Mistranslation of membrane proteins and two-component system activation trigger antibiotic-mediated cell death. *Cell* **2008**, *135*, 679–690. [CrossRef]
113. Foti, J.J.; Devadoss, B.; Winkler, J.A.; Collins, J.J.; Walker, G.C. Oxidation of the guanine nucleotide pool underlies cell death by bactericidal antibiotics. *Science* **2012**, *336*, 315–319. [CrossRef] [PubMed]
114. Kottur, J.; Nair, D.T. Reactive oxygen species play an important role in the bactericidal activity of quinolone antibiotics. *Angew. Chem. Int. Ed.* **2016**, *55*, 2397–2400. [CrossRef] [PubMed]
115. Takahashi, N.; Gruber, C.C.; Yang, J.H.; Liu, X.; Braff, D.; Yashaswini, C.N.; Bhubhanil, S.; Furuta, Y.; Andreescu, S.; Collins, J.J.; et al. Lethality of MalE-LacZ hybrid protein shares mechanistic attributes with oxidative component of antibiotic lethality. *Proc. Natl. Acad. Sci. USA* **2017**, *114*, 9164–9169. [CrossRef]
116. Kang, T.M.; Yuan, J.; Nguyen, A.; Becket, E.; Yang, H.; Miller, J.H. The aminoglycoside antibiotic kanamycin damages DNA bases in *Escherichia coli*: Caffeine potentiates the DNA-damaging effects of kanamycin while suppressing cell killing by ciprofloxacin in *Escherichia coli* and *Bacillus anthracis*. *Antimicrob. Agents Chemother.* **2012**, *56*, 3216–3223. [CrossRef]
117. Ohnishi, S.; Murata, M.; Ida, N.; Oikawa, S.; Kawanishi, S. Oxidative DNA damage induced by metabolites of chloramphenicol, an antibiotic drug. *Free Radic. Res.* **2015**, *49*, 1165–1172. [CrossRef]
118. Michaels, M.L.; Tchou, J.; Grollman, A.P.; Miller, J.H. A repair system for 8-oxo-7,8-dihydrodeoxyguanine. *Biochemistry* **1992**, *31*, 10964–10968. [CrossRef]
119. Grollman, A.P.; Moriya, M. Mutagenesis by 8-oxoguanine: An enemy within. *Trends Genet.* **1993**, *9*, 246–249. [CrossRef]
120. Tajiri, T.; Maki, H.; Sekiguchi, M. Functional cooperation of MutT, MutM and MutY proteins in preventing mutations caused by spontaneous oxidation of guanine nucleotide in *Escherichia coli*. *Mutat. Res.* **1995**, *336*, 257–267. [CrossRef]
121. Moriya, M.; Ou, C.; Bodepudi, V.; Johnson, F.; Takeshita, M.; Grollman, A.P. Site-specific mutagenesis using a gapped duplex vector: A study of translesion synthesis past 8-oxodeoxyguanosine in *E. coli*. *Mutat. Res.* **1991**, *254*, 281–288. [CrossRef]
122. Wood, M.L.; Esteve, A.; Morningstar, M.L.; Kuziemko, G.M.; Essigmann, J.M. Genetic effects of oxidative DNA damage: Comparative mutagenesis of 7,8-dihydro-8-oxoguanine and 7,8-dihydro-8-oxoadenine in *Escherichia coli*. *Nucleic Acids Res.* **1992**, *20*, 6023–6032. [CrossRef]
123. Tchou, J.; Kasai, H.; Shibutani, S.; Chung, M.-H.; Laval, J.; Grollman, A.P.; Nishimura, S. 8-oxoguanine (8-hydroxyguanine) DNA glycosylase and its substrate specificity. *Proc. Natl. Acad. Sci. USA* **1991**, *88*, 4690–4694. [CrossRef] [PubMed]
124. Karakaya, A.; Jaruga, P.; Bohr, V.A.; Grollman, A.P.; Dizdaroglu, M. Kinetics of excision of purine lesions from DNA by *Escherichia coli* Fpg protein. *Nucleic Acids Res.* **1997**, *25*, 474–479. [CrossRef] [PubMed]
125. Michaels, M.L.; Cruz, C.; Grollman, A.P.; Miller, J.H. Evidence that MutY and MutM combine to prevent mutations by an oxidatively damaged form of guanine in DNA. *Proc. Natl. Acad. Sci. USA* **1992**, *89*, 7022–7025. [CrossRef] [PubMed]
126. Maki, H.; Sekiguchi, M. MutT protein specifically hydrolyses a potent mutagenic substrate for DNA synthesis. *Nature* **1992**, *355*, 273–275. [CrossRef]
127. Taddei, F.; Hayakawa, H.; Bouton, M.; Cirinesi, A.; Matic, I.; Sekiguchi, M. Radman, M. Counteraction by MutT protein of transcriptional errors caused by oxidative damage. *Science* **1997**, *278*, 128–130. [CrossRef]
128. Jiang, D.; Hatahet, Z.; Blaisdell, J.O.; Melamede, R.J.; Wallace, S.S. *Escherichia coli* endonuclease VIII: Cloning, sequencing and overexpression of the *nei* structural gene and characterization of *nei* and *nei nth* mutants. *J. Bacteriol.* **1997**, *179*, 3773–3782. [CrossRef]
129. Jiang, D.; Hatahet, Z.; Melamede, R.J.; Kow, Y.W.; Wallace, S.S. Characterization of *Escherichia coli* endonuclease VIII. *J. Biol. Chem.* **1997**, *272*, 32230–32239. [CrossRef]
130. Blaisdell, J.O.; Hatahet, Z.; Wallace, S.S. A novel role for *Escherichia coli* endonuclease VIII in prevention of spontaneous G→T transversions. *J. Bacteriol.* **1999**, *181*, 6396–6402. [CrossRef]

131. Kropachev, K.Y.; Zharkov, D.O.; Grollman, A.P. Catalytic mechanism of *Escherichia coli* endonuclease VIII: Roles of the intercalation loop and the zinc finger. *Biochemistry* **2006**, *45*, 12039–12049. [CrossRef]
132. Breimer, L.H.; Lindahl, T. DNA glycosylase activities for thymine residues damaged by ring saturation, fragmentation, or ring contraction are functions of endonuclease III in *Escherichia coli*. *J. Biol. Chem.* **1984**, *259*, 5543–5548.
133. Dizdaroglu, M.; Laval, J.; Boiteux, S. Substrate specificity of the *Escherichia coli* endonuclease III: Excision of thymine- and cytosine-derived lesions in DNA produced by radiation-generated free radicals. *Biochemistry* **1993**, *32*, 12105–12111. [CrossRef] [PubMed]
134. Suzuki, M.; Matsui, K.; Yamada, M.; Kasai, H.; Sofuni, T.; Nohmi, T. Construction of mutants of *Salmonella typhimurium* deficient in 8-hydroxyguanine DNA glycosylase and their sensitivities to oxidative mutagens and nitro compounds. *Mutat. Res.* **1997**, *393*, 233–246. [CrossRef]
135. Nagorska, K.; Silhan, J.; Li, Y.; Pelicic, V.; Freemont, P.S.; Baldwin, G.S.; Tang, C.M. A network of enzymes involved in repair of oxidative DNA damage in *Neisseria meningitidis*. *Mol. Microbiol.* **2012**, *83*, 1064–1079. [CrossRef]
136. Souza Arantes, L.; Gonçalves Vila Nova, L.; Resende, B.C.; Bitar, M.; Vale Coelho, I.E.; Miyoshi, A.; Azevedo, V.A.; dos Santos, L.L.; Machado, C.R.; de Oliveira Lopes, D. The *Corynebacterium pseudotuberculosis* genome contains two formamidopyrimidine-DNA glycosylase enzymes, only one of which recognizes and excises 8-oxoguanine lesion. *Gene* **2016**, *575*, 233–243. [CrossRef] [PubMed]
137. Jain, R.; Kumar, P.; Varshney, U. A distinct role of formamidopyrimidine DNA glycosylase (MutM) in down-regulation of accumulation of G, C mutations and protection against oxidative stress in mycobacteria. *DNA Repair* **2007**, *6*, 1774–1785. [CrossRef] [PubMed]
138. Sidorenko, V.S.; Rot, M.A.; Filipenko, M.L.; Nevinsky, G.A.; Zharkov, D.O. Novel DNA glycosylases from *Mycobacterium tuberculosis*. *Biochemistry (Mosc)* **2008**, *73*, 442–450. [CrossRef] [PubMed]
139. Olsen, I.; Balasingham, S.V.; Davidsen, T.; Debebe, E.; Rødland, E.A.; van Soolingen, D.; Kremer, K.; Alseth, I.; Tønjum, T. Characterization of the major formamidopyrimidine-DNA glycosylase homolog in Mycobacterium tuberculosis and its linkage to variable tandem repeats. *FEMS Immunol. Med. Microbiol.* **2009**, *56*, 151–161. [CrossRef]
140. Guo, Y.; Bandaru, V.; Jaruga, P.; Zhao, X.; Burrows, C.J.; Iwai, S.; Dizdaroglu, M.; Bond, J.P.; Wallace, S.S. The oxidative DNA glycosylases of *Mycobacterium tuberculosis* exhibit different substrate preferences from their *Escherichia coli* counterparts. *DNA Repair* **2010**, *9*, 177–190. [CrossRef]
141. Davidsen, T.; Bjørås, M.; Seeberg, E.C.; Tønjum, T. Antimutator role of DNA glycosylase MutY in pathogenic *Neisseria* species. *J. Bacteriol.* **2005**, *187*, 2801–2809. [CrossRef]
142. Eutsey, R.; Wang, G.; Maier, R.J. Role of a MutY DNA glycosylase in combating oxidative DNA damage in *Helicobacter pylori*. *DNA Repair* **2007**, *6*, 19–26. [CrossRef]
143. Eberle, R.J.; Coronado, M.A.; Caruso, I.P.; Lopes, D.O.; Miyoshi, A.; Azevedo, V.; Arni, R.K. Chemical and thermal influence of the $[4Fe–4S]^{2+}$ cluster of A/G-specific adenine glycosylase from *Corynebacterium pseudotuberculosis*. *Biochim. Biophys. Acta* **2015**, *1850*, 393–400. [CrossRef] [PubMed]
144. Cançado de Faria, R.; Gonçalves Vila-Nova, L.; Bitar, M.; Carvalho Resende, B.; Sousa Arantes, L.; Basso Rebelato, A.; Carvalho Azevedo, V.A.; Franco, G.R.; Machado, C.R.; dos Santos, L.L.; et al. Adenine glycosylase MutY of *Corynebacterium pseudotuberculosis* presents the antimutator phenotype and evidences of glycosylase/AP lyase activity *in vitro*. *Infect. Genet. Evol.* **2016**, *44*, 318–329. [CrossRef] [PubMed]
145. Oliver, A.; Sánchez, J.M.; Blázquez, J. Characterization of the GO system of *Pseudomonas aeruginosa*. *FEMS Microbiol. Lett.* **2002**, *217*, 31–35. [CrossRef] [PubMed]
146. Sanders, L.H.; Sudhakaran, J.; Sutton, M.D. The GO system prevents ROS-induced mutagenesis and killing in *Pseudomonas aeruginosa*. *FEMS Microbiol. Lett.* **2009**, *294*, 89–96. [CrossRef] [PubMed]
147. Davidsen, T.; Amundsen, E.K.; Rødland, E.A.; Tønjum, T. DNA repair profiles of disease-associated isolates of *Neisseria meningitidis*. *FEMS Immunol. Med. Microbiol.* **2007**, *49*, 243–251. [CrossRef] [PubMed]
148. Kurthkoti, K.; Srinath, T.; Kumar, P.; Malshetty, V.S.; Sang, P.B.; Jain, R.; Manjunath, R.; Varshney, U. A distinct physiological role of MutY in mutation prevention in mycobacteria. *Microbiology* **2010**, *156*, 88–93. [CrossRef]
149. Hassim, F.; Papadopoulos, A.O.; Kana, B.D.; Gordhan, B.G. A combinatorial role for MutY and Fpg DNA glycosylases in mutation avoidance in *Mycobacterium smegmatis*. *Mutat. Res.* **2015**, *779*, 24–32. [CrossRef]

150. Canfield, G.S.; Schwingel, J.M.; Foley, M.H.; Vore, K.L.; Boonanantanasarn, K.; Gill, A.L.; Sutton, M.D.; Gill, S.R. Evolution in fast forward: A potential role for mutators in accelerating *Staphylococcus aureus* pathoadaptation. *J. Bacteriol.* **2013**, *195*, 615–628. [CrossRef]
151. Eskra, L.; Canavessi, A.; Carey, M.; Splitter, G. *Brucella abortus* genes identified following constitutive growth and macrophage infection. *Infect. Immun.* **2001**, *69*, 7736–7742. [CrossRef]
152. Dutta, N.K.; Mehra, S.; Didier, P.J.; Roy, C.J.; Doyle, L.A.; Alvarez, X.; Ratterree, M.; Be, N.A.; Lamichhane, G.; Jain, S.K.; et al. Genetic requirements for the survival of tubercle bacilli in primates. *J. Infect. Dis.* **2010**, *201*, 1743–1752. [CrossRef]
153. Oliver, A.; Cantón, R.; Campo, P.; Baquero, F.; Blázquez, J. High frequency of hypermutable *Pseudomonas aeruginosa* in cystic fibrosis lung infection. *Science* **2000**, *288*, 1251–1253. [CrossRef] [PubMed]
154. Mandsberg, L.F.; Ciofu, O.; Kirkby, N.; Christiansen, L.E.; Poulsen, H.E.; Høiby, N. Antibiotic resistance in *Pseudomonas aeruginosa* strains with increased mutation frequency due to inactivation of the DNA oxidative repair system. *Antimicrob. Agents Chemother.* **2009**, *53*, 2483–2491. [CrossRef] [PubMed]
155. Couce, A.; Alonso-Rodriguez, N.; Costas, C.; Oliver, A.; Blázquez, J. Intrapopulation variability in mutator prevalence among urinary tract infection isolates of *Escherichia coli*. *Clin. Microbiol. Infect.* **2016**, *22*, 566. [CrossRef] [PubMed]
156. Perrin, A.; Larsonneur, E.; Nicholson, A.C.; Edwards, D.J.; Gundlach, K.M.; Whitney, A.M.; Gulvik, C.A.; Bell, M.E.; Rendueles, O.; Cury, J.; et al. Evolutionary dynamics and genomic features of the *Elizabethkingia anophelis* 2015 to 2016 Wisconsin outbreak strain. *Nat. Commun.* **2017**, *8*, 15483 [CrossRef] [PubMed]
157. Chetsanga, C.J.; Lozon, M.; Makaroff, C.; Savage, L. Purification and characterization of *Escherichia coli* formamidopyrimidine-DNA glycosylase that excises damaged 7-methylguanine from deoxyribonucleic acid. *Biochemistry* **1981**, *20*, 5201–5207. [CrossRef]
158. Wink, D.A.; Laval, J. The Fpg protein, a DNA repair enzyme, is inhibited by the biomediator nitric oxide in vitro and *in vivo*. *Carcinogenesis* **1994**, *15*, 2125–2129. [CrossRef]
159. Jaiswal, M.; LaRusso, N.F.; Nishioka, N.; Nakabeppu, Y.; Gores, G.J. Human Ogg1, a protein involved in the repair of 8-oxoguanine, is inhibited by nitric oxide. *Cancer Res.* **2001**, *61*, 6388–6393.
160. Rogers, P.A.; Eide, L.; Klungland, A.; Ding, H. Reversible inactivation of *E. coli* endonuclease III via modification of its [4Fe-4S] cluster by nitric oxide. *DNA Repair* **2003**, *2*, 809–817. [CrossRef]
161. Moritz, E.; Pauly, K.; Bravard, A.; Hall, J.; Radicella, J.P.; Epe, B. hOGG1-Cys326 variant cells are hypersensitive to DNA repair inhibition by nitric oxide. *Carcinogenesis* **2014**, *35*, 1426–1433. [CrossRef]
162. Mikhailov, A.A.; Khantakova, D.V.; Nichiporenko, V.A.; Glebov, E.M.; Grivin, V.P.; Plyusnin, V.F.; Yanshole, V.V.; Petrova, D.V.; Kostin, G.A.; Grin, I.R. Photoinduced inhibition of DNA repair enzymes and the possible mechanism of photochemical transformations of the ruthenium nitrosyl complex [RuNO(β-Pic)$_2$(NO$_2$)$_2$OH]. *Metallomics* **2019**, *11*, 1999–2009. [CrossRef]
163. Kobune, M.; Xu, Y.; Baum, C.; Kelley, M.R.; Williams, D.A. Retrovirus-mediated expression of the base excision repair proteins, formamidopyrimidine DNA glycosylase or human oxoguanine DNA glycosylase, protects hematopoietic cells from *N,N′,N″*-triethylenethiophosphoramide (thioTEPA)-induced toxicity in vitro and *in vivo*. *Cancer Res.* **2001**, *61*, 5116–5125. [PubMed]
164. Xu, Y.; Hansen, W.K.; Rosenquist, T.A.; Williams, D.A.; Limp-Foster, M.; Kelley, M.R. Protection of mammalian cells against chemotherapeutic agents thiotepa, 1,3-*N,N′*-bis(2-chloroethyl)-*N*-nitrosourea, and mafosfamide using the DNA base excision repair genes Fpg and α-hOgg1: Implications for protective gene therapy applications. *J. Pharmacol. Exp. Ther.* **2001**, *296*, 825–831. [PubMed]
165. He, Y.-H.; Xu, Y.; Kobune, M.; Wu, M.; Kelley, M.R.; Martin, W.J., II. *Escherichia coli* FPG and human OGG1 reduce DNA damage and cytotoxicity by BCNU in human lung cells. *Am. J. Physiol. Lung Cell. Mol. Physiol.* **2002**, *282*, L50–L55. [CrossRef] [PubMed]
166. Preston, T.J.; Henderson, J.T.; McCallum, G.P.; Wells, P.G. Base excision repair of reactive oxygen species-initiated 7,8-dihydro-8-oxo-2′-deoxyguanosine inhibits the cytotoxicity of platinum anticancer drugs. *Mol. Cancer Ther.* **2009**, *8*, 2015–2026. [CrossRef]
167. Wu, M.; Zhang, Z.; Che, W. Suppression of a DNA base excision repair gene, hOGG1, increases bleomycin sensitivity of human lung cancer cell line. *Toxicol. Appl. Pharmacol.* **2008**, *228*, 395–402. [CrossRef]
168. Ramdzan, Z.M.; Ginjala, V.; Pinder, J.B.; Chung, D.; Donovan, C.M.; Kaur, S.; Leduy, L.; Dellaire, G.; Ganesan, S.; Nepveu, A. The DNA repair function of *CUX1* contributes to radioresistance. *Oncotarget* **2017**, *8*, 19021–19038. [CrossRef]

169. Chakrabarti, G.; Silvers, M.A.; Ilcheva, M.; Liu, Y.; Moore, Z.R.; Luo, X.; Gao, J.; Anderson, G.; Liu, L.; Sarode, V.; et al. Tumor-selective use of DNA base excision repair inhibition in pancreatic cancer using the NQO1 bioactivatable drug, β-lapachone. *Sci. Rep.* **2015**, *5*, 17066. [CrossRef]
170. Sekiguchi, M.; Tsuzuki, T. Oxidative nucleotide damage: Consequences and prevention. *Oncogene* **2002**, *21*, 8895–8904. [CrossRef] [PubMed]
171. Gad, H.; Koolmeister, T.; Jemth, A.-S.; Eshtad, S.; Jacques, S.A.; Ström, C.E.; Svensson, L.M.; Schultz, N.; Lundbäck, T.; Einarsdottir, B.O.; et al. MTH1 inhibition eradicates cancer by preventing sanitation of the dNTP pool. *Nature* **2014**, *508*, 215–221. [CrossRef]
172. Warpman Berglund, U.; Sanjiv, K.; Gad, H.; Kalderén, C.; Koolmeister, T.; Pham, T.; Gokturk, C.; Jafari, R.; Maddalo, G.; Seashore-Ludlow, B.; et al. Validation and development of MTH1 inhibitors for treatment of cancer. *Ann. Oncol.* **2016**, *27*, 2275–2283. [CrossRef]
173. Qing, X.; Shao, Z.; Lv, X.; Pu, F.; Gao, F.; Liu, L.; Shi, D. Anticancer effect of (S)-crizotinib on osteosarcoma cells by targeting MTH1 and activating reactive oxygen species. *Anticancer. Drugs* **2018**, *29*, 341–352. [CrossRef] [PubMed]
174. Niu, Y.; Pan, D.; Shi, D.; Bai, Q.; Liu, H.; Yao, X. Influence of chirality of crizotinib on its MTH1 protein inhibitory activity: Insight from molecular dynamics simulations and binding free energy calculations. *PLoS ONE* **2015**, *10*, e0145219. [CrossRef] [PubMed]
175. Sun, H.; Chen, P.; Li, D.; Li, Y.; Hou, T. Directly binding rather than induced-fit dominated binding affinity difference in (*S*)- and (*R*)-crizotinib bound MTH1. *J. Chem. Theory Comput.* **2016**, *12*, 851–860. [CrossRef] [PubMed]
176. Dai, X.; Guo, G.; Zou, P.; Cui, R.; Chen, W.; Chen, X.; Yin, C.; He, W.; Vinothkumar, R.; Yang, F.; et al. (S)-crizotinib induces apoptosis in human non-small cell lung cancer cells by activating ROS independent of MTH1. *J. Exp. Clin. Cancer Res.* **2017**, *36*, 120. [CrossRef]
177. Ji, J.; Chen, W.; Lian, W.; Chen, R.; Yang, J.; Zhang, Q.; Weng, Q.; Hu, Z.K.J.; Chen, X.; Zou, P.; et al. (S)-crizotinib reduces gastric cancer growth through oxidative DNA damage and triggers pro-survival akt signal. *Cell Death Dis.* **2018**, *9*, 660. [CrossRef]
178. Van der Waals, L.M.; Laoukili, J.; Jongen, J.M.J.; Raats, D.A.; Borel Rinkes, I.H.M.; Kranenburg, O. Differential anti-tumour effects of MTH1 inhibitors in patient-derived 3D colorectal cancer cultures. *Sci. Rep.* **2019**, *9*, 819. [CrossRef]
179. Samaranayake, G.J.; Huynh, M.; Rai, P. MTH1 as a chemotherapeutic target: The elephant in the room. *Cancers* **2017**, *9*, 47. [CrossRef]
180. Einarsdottir, B.O.; Karlsson, J.; Söderberg, E.M.V.; Lindberg, M.F.; Funck-Brentano, E.; Jespersen, H.; Brynjolfsson, S.F.; Olofsson Bagge, R.; Carstam, L.; Scobie, M.; et al. A patient-derived xenograft pre-clinical trial reveals treatment responses and a resistance mechanism to karonudib in metastatic melanoma. *Cell Death Dis.* **2018**, *9*, 810. [CrossRef]
181. Hua, X.; Sanjiv, K.; Gad, H.; Pham, T.; Gokturk, C.; Rasti, A.; Zhao, Z.; He, K.; Feng, M.; Zang, Y.; et al. Karonudib is a promising anticancer therapy in hepatocellular carcinoma. *Ther. Adv. Med. Oncol.* **2019**, *11*, 1758835919866960. [CrossRef]
182. Morland, I.; Luna, L.; Gustad, E.; Seeberg, E.; Bjørås, M. Product inhibition and magnesium modulate the dual reaction mode of hOgg1. *DNA Repair* **2005**, *4*, 381–387. [CrossRef]
183. Mahajan, T.R.; Ytre-Arne, M.E.; Strøm-Andersen, P.; Dalhus, B.; Gundersen, L.-L. Synthetic routes to *N*-9 alkylated 8-oxoguanines; weak inhibitors of the human DNA glycosylase OGG1. *Molecules* **2015**, *20*, 15944–15965. [CrossRef] [PubMed]
184. Michel, M.; Visnes, T.; Homan, E.J.; Seashore-Ludlow, B.; Hedenström, M.; Wiita, E.; Vallin, K.; Paulin, C.B.J.; Zhang, J.; Wallner, O.; et al. Computational and experimental druggability assessment of human DNA glycosylases. *ACS Omega* **2019**, *4*, 11642–11656. [CrossRef] [PubMed]
185. Rosenquist, T.A.; Zaika, E.; Fernandes, A.S.; Zharkov, D.O.; Miller, H.; Grollman, A.P. The novel DNA glycosylase, NEIL1, protects mammalian cells from radiation-mediated cell death. *DNA Repair* **2003**, *2*, 581–591. [CrossRef]
186. Taricani, L.; Shanahan, F.; Pierce, R.H.; Guzi, T.J.; Parry, D. Phenotypic enhancement of thymidylate synthetase pathway inhibitors following ablation of Neil1 DNA glycosylase/lyase. *Cell Cycle* **2010**, *9*, 4876–4883. [CrossRef]

187. Mabley, J.G.; Pacher, P.; Deb, A.; Wallace, R.; Elder, R.H.; Szabó, C. Potential role for 8-oxoguanine DNA glycosylase in regulating inflammation. *FASEB J.* **2005**, *19*, 290–292. [CrossRef]
188. Touati, E.; Michel, V.; Thiberge, J.-M.; Avé, P.; Huerre, M.; Bourgade, F.; Klungland, A.; Labigne, A. Deficiency in OGG1 protects against inflammation and mutagenic effects associated with *H. pylori* infection in mouse. *Helicobacter* **2006**, *11*, 494–505. [CrossRef]
189. Li, G.; Yuan, K.; Yan, C.; Fox, J., III; Gaid, M.; Breitwieser, W.; Bansal, A.K.; Zeng, H.; Gao, H.; Wu, M. 8-Oxoguanine-DNA glycosylase 1 deficiency modifies allergic airway inflammation by regulating STAT6 and IL-4 in cells and in mice. *Free Radic. Biol. Med.* **2012**, *52*, 392–401. [CrossRef] [PubMed]
190. Bacsi, A.; Aguilera-Aguirre, L.; Szczesny, B.; Radak, Z.; Hazra, T.K.; Sur, S.; Ba, X.; Boldogh, I. Down-regulation of 8-oxoguanine DNA glycosylase 1 expression in the airway epithelium ameliorates allergic lung inflammation. *DNA Repair* **2013**, *12*, 18–26. [CrossRef]
191. Kunisada, M.; Yogianti, F.; Sakumi, K.; Ono, R.; Nakabeppu, Y.; Nishigori, C. Increased expression of versican in the inflammatory response to UVB- and reactive oxygen species-induced skin tumorigenesis. *Am. J. Pathol.* **2011**, *179*, 3056–3065. [CrossRef]
192. Yogianti, F.; Kunisada, M.; Nakano, E.; Ono, R.; Sakumi, K.; Oka, S.; Nakabeppu, Y.; Nishigori, C. Inhibitory effects of dietary *Spirulina platensis* on UVB-induced skin inflammatory responses and carcinogenesis. *J. Invest. Dermatol.* **2014**, *134*, 2610–2619. [CrossRef]
193. Ye, Y.; Lin, P.; Zhang, W.; Tan, S.; Zhou, X.; Li, R.; Pu, Q.; Koff, J.L.; Dhasarathy, A.; Ma, F.; et al. DNA repair interacts with autophagy to regulate inflammatory responses to pulmonary hyperoxia. *J. Immunol.* **2017**, *198*, 2844–2853. [CrossRef] [PubMed]
194. Boldogh, I.; Hajas, G.; Aguilera-Aguirre, L.; Hegde, M.L.; Radak, Z.; Bacsi, A.; Sur, S.; Hazra, T.K.; Mitra, S. Activation of Ras signaling pathway by 8-oxoguanine DNA glycosylase bound to its excision product, 8-oxoguanine. *J. Biol. Chem.* **2012**, *287*, 20769–20773. [CrossRef] [PubMed]
195. German, P.; Szaniszlo, P.; Hajas, G.; Radak, Z.; Bacsi, A.; Hazra, T.K.; Hegde, M.L.; Ba, X.; Boldogh, I. Activation of cellular signaling by 8-oxoguanine DNA glycosylase-1-initiated DNA base excision repair. *DNA Repair* **2013**, *12*, 856–863. [CrossRef] [PubMed]
196. Aguilera-Aguirre, L.; Bacsi, A.; Radak, Z.; Hazra, T.K.; Mitra, S.; Sur, S.; Brasier, A.R.; Ba, X.; Boldogh, I. Innate inflammation induced by the 8-oxoguanine DNA glycosylase-1–KRAS–NF-κB pathway. *J. Immunol.* **2014**, *193*, 4643–4653. [CrossRef]
197. Ba, X.; Bacsi, A.; Luo, J.; Aguilera-Aguirre, L.; Zeng, X.; Radak, Z.; Brasier, A.R.; Boldogh, I. 8-Oxoguanine DNA glycosylase-1 augments proinflammatory gene expression by facilitating the recruitment of site-specific transcription factors. *J. Immunol.* **2014**, *192*, 2384–2394. [CrossRef]
198. Pan, L.; Zhu, B.; Hao, W.; Zeng, X.; Vlahopoulos, S.A.; Hazra, T.K.; Hegde, M.L.; Radak, Z.; Bacsi, A.; Brasier, A.R.; et al. Oxidized guanine base lesions function in 8-oxoguanine DNA glycosylase-1-mediated epigenetic regulation of nuclear factor κB-driven gene expression. *J. Biol. Chem.* **2016**, *291*, 25553–25566. [CrossRef]
199. Pan, L.; Hao, W.; Zheng, X.; Zeng, X.; Abbasi, A.A.; Boldogh, I.; Ba, X. OGG1-DNA interactions facilitate NF-κB binding to DNA targets. *Sci. Rep.* **2017**, *7*, 43297. [CrossRef]
200. Banerjee, A.; Yang, W.; Karplus, M.; Verdine, G.L. Structure of a repair enzyme interrogating undamaged DNA elucidates recognition of damaged DNA. *Nature* **2005**, *434*, 612–618. [CrossRef]
201. Li, H.; Endutkin, A.V.; Bergonzo, C.; Fu, L.; Grollman, A.P.; Zharkov, D.O.; Simmerling, C. DNA deformation-coupled recognition of 8-oxoguanine: Conformational kinetic gating in human DNA glycosylase. *J. Am. Chem. Soc.* **2017**, *139*, 2682–2692. [CrossRef] [PubMed]
202. Orr, H.T.; Zoghbi, H.Y. Trinucleotide repeat disorders. *Annu. Rev. Neurosci.* **2007**, *30*, 575–621. [CrossRef]
203. Kennedy, L.; Evans, E.; Chen, C.-M.; Craven, L.; Detloff, P.J.; Ennis, M.; Shelbourne, P.F. Dramatic tissue-specific mutation length increases are an early molecular event in Huntington disease pathogenesis. *Hum. Mol. Genet.* **2003**, *12*, 3359–3367. [CrossRef] [PubMed]
204. Kovtun, I.V.; Liu, Y.; Bjoras, M.; Klungland, A.; Wilson, S.H.; McMurray, C.T OGG1 initiates age-dependent CAG trinucleotide expansion in somatic cells. *Nature* **2007**, *447*, 447–452. [CrossRef] [PubMed]
205. Møllersen, L.; Rowe, A.D.; Larsen, E.; Rognes, T.; Klungland, A. Continuous and periodic expansion of CAG repeats in Huntington's disease R6/1 mice. *PLoS Genet.* **2010**, *6*, e1001242. [CrossRef] [PubMed]

206. Goula, A.-V.; Berquist, B.R.; Wilson, D.M., III; Wheeler, V.C.; Trottier, Y.; Merienne, K. Stoichiometry of base excision repair proteins correlates with increased somatic CAG instability in striatum over cerebellum in Huntington's disease transgenic mice. *PLoS Genet.* **2009**, *5*, e1000749. [CrossRef]
207. Budworth, H.; Harris, F.R.; Williams, P.; Lee, D.Y.; Holt, A.; Pahnke, J.; Szczesny, B.; Acevedo-Torres, K.; Ayala-Peña, S.; McMurray, C.T. Suppression of somatic expansion delays the onset of pathophysiology in a mouse model of Huntington's disease. *PLoS Genet.* **2015**, *11*, e1005267. [CrossRef]
208. Bobola, M.S.; Kolstoe, D.D.; Blank, A.; Chamberlain, M.C.; Silber, J.R. Repair of 3-methyladenine and abasic sites by base excision repair mediates glioblastoma resistance to temozolomide. *Front. Oncol.* **2012**, *2*, 176. [CrossRef]
209. Fishel, M.L.; Seo, Y.R.; Smith, M.L.; Kelley, M.R. Imbalancing the DNA base excision repair pathway in the mitochondria; targeting and overexpressing *N*-methylpurine DNA glycosylase in mitochondria leads to enhanced cell killing. *Cancer Res.* **2003**, *63*, 608–615.
210. Rinne, M.; Caldwell, D.; Kelley, M.R. Transient adenoviral *N*-methylpurine DNA glycosylase overexpression imparts chemotherapeutic sensitivity to human breast cancer cells. *Mol. Cancer Ther.* **2004**, *3*, 955–967.
211. Fishel, M.L.; He, Y.; Smith, M.L.; Kelley, M.R. Manipulation of base excision repair to sensitize ovarian cancer cells to alkylating agent temozolomide. *Clin. Cancer Res.* **2007**, *13*, 260–267. [CrossRef]
212. Tang, J.-b.; Svilar, D.; Trivedi, R.N.; Wang, X.-H.; Goellner, E.M.; Moore, B.; Hamilton, R.L.; Banze, L.A.; Brown, A.R.; Sobol, R.W. N-methylpurine DNA glycosylase and DNA polymerase β modulate BER inhibitor potentiation of glioma cells to temozolomide. *Neuro Oncol.* **2011**, *13*, 471–486. [CrossRef]
213. Song, S.; Xing, G.; Yuan, L.; Wang, J.; Wang, S.; Yin, Y.; Tian, C.; He, F.; Zhang, L. *N*-methylpurine DNA glycosylase inhibits p53-mediated cell cycle arrest and coordinates with p53 to determine sensitivity to alkylating agents. *Cell Res.* **2012**, *22*, 1285–1303. [CrossRef]
214. Leguisamo, N.M.; Gloria, H.C.; Kalil, A.N.; Martins, T.V.; Azambuja, D.B.; Meira, L.B.; Saffi, J. Base excision repair imbalance in colorectal cancer has prognostic value and modulates response to chemotherapy. *Oncotarget* **2017**, *8*, 54199–54214. [CrossRef] [PubMed]
215. Brandon, M.L.; Mi, L.-J.; Chaung, W.; Teebor, G.; Boorstein, R.J. 5-Chloro-2'-deoxyuridine cytotoxicity results from base excision repair of uracil subsequent to thymidylate synthase inhibition. *Mutat. Res.* **2000**, *459*, 161–169. [CrossRef]
216. Turner, D.P.; Cortellino, S.; Schupp, J.E.; Caretti, E.; Loh, T.; Kinsella, T.J.; Bellacosa, A. The DNA *N*-glycosylase MED1 exhibits preference for halogenated pyrimidines and is involved in the cytotoxicity of 5-iododeoxyuridine. *Cancer Res.* **2006**, *66*, 7686–7693. [CrossRef] [PubMed]
217. Kunz, C.; Focke, F.; Saito, Y.; Schuermann, D.; Lettieri, T.; Selfridge, J.; Schär, P. Base excision by thymine DNA glycosylase mediates DNA-directed cytotoxicity of 5-fluorouracil. *PLoS Biol.* **2009**, *7*, e91. [CrossRef] [PubMed]
218. Suzuki, N.; Emura, T.; Fukushima, M. Mode of action of trifluorothymidine (TFT) against DNA replication and repair enzymes. *Int. J. Oncol.* **2011**, *39*, 263–270. [CrossRef]
219. Paik, J.; Duncan, T.; Lindahl, T.; Sedgwick, B. Sensitization of human carcinoma cells to alkylating agents by small interfering RNA suppression of 3-alkyladenine-DNA glycosylase. *Cancer Res.* **2005**, *65*, 10472–10477. [CrossRef]
220. Allan, J.M.; Engelward, B.P.; Dreslin, A.J.; Wyatt, M.D.; Tomasz, M.; Samson, L.D. Mammalian 3-methyladenine DNA glycosylase protects against the toxicity and clastogenicity of certain chemotherapeutic DNA cross-linking agents. *Cancer Res.* **1998**, *58*, 3965–3973.
221. Sorribes, I.C.; Handelman, S.K.; Jain, H.V. Mitigating temozolomide resistance in glioblastoma via DNA damage-repair inhibition. *J. R. Soc. Interface* **2020**, *17*, 20190722. [CrossRef]
222. Goellner, E.M.; Grimme, B.; Brown, A.R.; Lin, Y.-C.; Wang, X.-H.; Sugrue, K.F.; Mitchell, L.; Trivedi, R.N.; Tang, J.-B.; Sobol, R.W. Overcoming temozolomide resistance in glioblastoma via dual inhibition of NAD^+ biosynthesis and base excision repair. *Cancer Res.* **2011**, *71*, 2308–2317. [CrossRef]
223. Lorenzi, P.L.; Landowski, C.P.; Brancale, A.; Song, X.; Townsend, L.B.; Drach, J.C.; Amidon, G.L. *N*-methylpurine DNA glycosylase and 8-oxoguanine DNA glycosylase metabolize the antiviral nucleoside 2-bromo-5,6-dichloro-1-(β-D-ribofuranosyl)benzimidazole. *Drug Metab. Dispos.* **2006**, *34*, 1070–1077. [CrossRef] [PubMed]
224. Karran, P.; Hjelmgren, T.; Lindahl, T. Induction of a DNA glycosylase for *N*-methylated purines is part of the adaptive response to alkylating agents. *Nature* **1982**, *296*, 770–773. [CrossRef] [PubMed]

225. Bjelland, S.; Bjørås, M.; Seeberg, E. Excision of 3-methylguanine from alkylated DNA by 3-methyladenine DNA glycosylase I of *Escherichia coli*. *Nucleic Acids Res.* **1993**, *21*, 2045–2049. [CrossRef] [PubMed]
226. Bjelland, S.; Seeberg, E. Purification and characterization of 3-methyladenine DNA glycosylase I from *Escherichia coli*. *Nucleic Acids Res.* **1987**, *15*, 2787–2801. [CrossRef] [PubMed]
227. Riazuddin, S.; Athar, A.; Ahmed, Z.; Lali, S.M.; Sohail, A. DNA glycosylase enzymes induced during chemical adaptation of *M. luteus*. *Nucleic Acids Res.* **1987**, *15*, 6607–6624. [CrossRef]
228. Roy, R.; Brooks, C.; Mitra, S. Purification and biochemical characterization of recombinant *N*-methylpurine-DNA glycosylase of the mouse. *Biochemistry* **1994**, *33*, 15131–15140. [CrossRef]
229. Tudek, B.; VanZeeland, A.A.; Kusmierek, J.T.; Laval, J. Activity of *Escherichia coli* DNA-glycosylases on DNA damaged by methylating and ethylating agents and influence of 3-substituted adenine derivatives. *Mutat. Res.* **1998**, *407*, 169–176. [CrossRef]
230. Drohat, A.C.; Kwon, K.; Krosky, D.J.; Stivers, J.T. 3-methyladenine DNA glycosylase I is an unexpected helix-hairpin-helix superfamily member. *Nat. Struct. Biol.* **2002**, *9*, 659–664. [CrossRef]
231. Rajesh, S.S.; Sivaraman, T. Cheminformatic designing of de novo inhibitors to 3-methyl adenine DNA glycosylase I (LiTagA) from *Leptospira interrogans serovar lai strain 56601*. *Med. Chem. Res.* **2013**, *22*, 3434–3443. [CrossRef]
232. Stivers, J.T. 2-Aminopurine fluorescence studies of base stacking interactions at abasic sites in DNA: Metal-ion and base sequence effects. *Nucleic Acids Res.* **1998**, *26*, 3837–3844. [CrossRef]
233. Stivers, J.T.; Pankiewicz, K.W.; Watanabe, K.A. Kinetic mechanism of damage site recognition and uracil flipping by *Escherichia coli* uracil DNA glycosylase. *Biochemistry* **1999**, *38*, 952–963. [CrossRef]
234. Maksimenko, A.; Ishchenko, A.A.; Sanz, G.; Laval, J.; Elder, R.H.; Saparbaev, M.K. A molecular beacon assay for measuring base excision repair activities. *Biochem. Biophys. Res. Commun.* **2004**, *319*, 240–246. [CrossRef] [PubMed]
235. Mirbahai, L.; Kershaw, R.M.; Green, R.M.; Hayden, R.E.; Meldrum, R.A.; Hodges, N.J. Use of a molecular beacon to track the activity of base excision repair protein OGG1 in live cells. *DNA Repair* **2010**, *9*, 144–152. [CrossRef] [PubMed]
236. Li, C.; Long, Y.; Liu, B.; Xiang, D.; Zhu, H. Real time monitoring uracil excision using uracil-containing molecular beacons. *Anal. Chim. Acta* **2014**, *819*, 71–77. [CrossRef] [PubMed]
237. Hu, J.; Liu, M.-H.; Li, Y.; Tang, B.; Zhang, C.-Y. Simultaneous sensitive detection of multiple DNA glycosylases from lung cancer cells at the single-molecule level. *Chem. Sci.* **2018**, *9*, 712–720. [CrossRef]
238. Sauvaigo, S.; Guerniou, V.; Rapin, D.; Gasparutto, D.; Caillat, S.; Favier, A. An oligonucleotide microarray for the monitoring of repair enzyme activity toward different DNA base damage. *Anal. Biochem.* **2004**, *333*, 182–192. [CrossRef]
239. Gines, G.; Saint-Pierre, C.; Gasparutto, D. On-bead fluorescent DNA nanoprobes to analyze base excision repair activities. *Anal. Chim. Acta* **2014**, *812*, 168–175. [CrossRef]
240. Gines, G.; Saint-Pierre, C.; Gasparutto, D. A multiplex assay based on encoded microbeads conjugated to DNA NanoBeacons to monitor base excision repair activities by flow cytometry. *Biosens. Bioelectron.* **2014**, *58*, 81–84. [CrossRef]
241. Flaender, M.; Costa, G.; Nonglaton, G.; Saint-Pierre, C.; Gasparutto, D. A DNA array based on clickable lesion-containing hairpin probes for multiplexed detection of base excision repair activities. *Analyst* **2016**, *141*, 6208–6216. [CrossRef]
242. Hölz, K.; Pavlic, A.; Lietard, J.; Somoza, M.M. Specificity and efficiency of the uracil DNA glycosylase-mediated strand cleavage surveyed on large sequence libraries. *Sci. Rep.* **2019**, *9*, 17822. [CrossRef]
243. Wang, L.-J.; Ma, F.; Tang, B.; Zhang, C.-Y. Base-excision-repair-induced construction of a single quantum-dot-based sensor for sensitive detection of DNA glycosylase activity. *Anal. Chem.* **2016**, *88*, 7523–7529. [CrossRef]
244. Xiang, Y.; Lu, Y. Expanding targets of DNAzyme-based sensors through deactivation and activation of DNAzymes by single uracil removal: Sensitive fluorescent assay of uracil-DNA glycosylase. *Anal. Chem.* **2012**, *84*, 9981–9987. [CrossRef] [PubMed]
245. Wu, Y.; Wang, L.; Zhu, J.; Jiang, W. A DNA machine-based fluorescence amplification strategy for sensitive detection of uracil-DNA glycosylase activity. *Biosens. Bioelectron.* **2015**, *68*, 654–659. [CrossRef] [PubMed]

246. Lu, Y.-J.; Hu, D.-P.; Deng, Q.; Wang, Z.-Y.; Huang, B.-H.; Fang, Y.-X.; Zhang, K.; Wong, W.-L. Sensitive and selective detection of uracil-DNA glycosylase activity with a new pyridinium luminescent switch-on molecular probe. *Analyst* **2015**, *140*, 5998–6004. [CrossRef] [PubMed]
247. Ma, C.; Wu, K.; Liu, H.; Xia, K.; Wang, K.; Wang, J. Label-free fluorescence turn-on detection of uracil DNA glycosylase activity based on G-quadruplex formation. *Talanta* **2016**, *160*, 449–453. [CrossRef] [PubMed]
248. Chen, C.; Zhou, D.; Tang, H.; Liang, M.; Jiang, J. A sensitive, homogeneous fluorescence assay for detection of thymine DNA glycosylase activity based on exonuclease-mediated amplification. *Chem. Commun.* **2013**, *49*, 5874–5876. [CrossRef]
249. Wang, X.; Hou, T.; Lu, T.; Li, F. Autonomous exonuclease III-assisted isothermal cycling signal amplification: A facile and highly sensitive fluorescence DNA glycosylase activity assay. *Anal. Chem.* **2014**, *86*, 9626–9631. [CrossRef]
250. Zhao, J.; Ma, Y.; Kong, R.; Zhang, L.; Yang, W.; Zhao, S. Tungsten disulfide nanosheet and exonuclease III co-assisted amplification strategy for highly sensitive fluorescence polarization detection of DNA glycosylase activity. *Anal. Chim. Acta* **2015**, *887*, 216–223. [CrossRef]
251. Wu, Y.; Yan, P.; Xu, X.; Jiang, W. A unique dual recognition hairpin probe mediated fluorescence amplification method for sensitive detection of uracil-DNA glycosylase and endonuclease IV activities. *Analyst* **2016**, *141*, 1789–1795. [CrossRef]
252. Wang, L.-J.; Wang, Z.-Y.; Zhang, Q.; Tang, B.; Zhang, C.-Y. Cyclic enzymatic repairing-mediated dual-signal amplification for real-time monitoring of thymine DNA glycosylase. *Chem. Commun.* **2017**, *53*, 3878–3881. [CrossRef]
253. Song, J.; Yin, F.; Li, X.; Dong, N.; Zhu, Y.; Shao, Y.; Chen, B.; Jiang, W.; Li, C.-Z. Sensitive detection of formamidopyrimidine-DNA glycosylase activity based on target-induced self-primed rolling circle amplification and magnetic nanoprobes. *Analyst* **2018**, *143*, 1593–1598. [CrossRef] [PubMed]
254. Wang, J.; Wang, Y.; Liu, S.; Wang, H.; Zhang, X.; Song, X.; Huang, J. Base excision repair initiated rolling circle amplification-based fluorescent assay for screening uracil-DNA glycosylase activity using Endo IV-assisted cleavage of AP probes. *Analyst* **2018**, *143*, 3951–3958. [CrossRef] [PubMed]
255. Du, W.; Li, J.; Xiao, F.; Yu, R.; Jiang, J. A label-free and highly sensitive strategy for uracil-DNA glycosylase activity detection based on stem-loop primer-mediated exponential amplification (SPEA). *Anal. Chim. Acta* **2017**, *991*, 127–132. [CrossRef] [PubMed]
256. Wang, L.-J.; Ren, M.; Zhang, Q.; Tang, B.; Zhang, C.-Y. Excision repair-initiated enzyme-assisted bicyclic cascade signal amplification for ultrasensitive detection of uracil-DNA glycosylase. *Anal. Chem.* **2017**, *89*, 4488–4494. [CrossRef] [PubMed]
257. Wang, J.; Pan, M.; Wei, J.; Liu, X.; Wang, F. A C-HCR assembly of branched DNA nanostructures for amplified uracil-DNA glycosylase assays. *Chem. Commun.* **2017**, *53*, 12878–12881. [CrossRef] [PubMed]

MDPI
St. Alban-Anlage 66
4052 Basel
Switzerland
Tel. +41 61 683 77 34
Fax +41 61 302 89 18
www.mdpi.com

International Journal of Molecular Sciences Editorial Office
E-mail: ijms@mdpi.com
www.mdpi.com/journal/ijms

MDPI

www.ingramcontent.com/pod-product-compliance
Lightning Source LLC
Chambersburg PA
CBHW061758210926
51850CB00006B/590
9783036578811